civics+ citizenship

history

geography

economics+ business

contents

H0 Introduction to History page 1

The making of the modern world

H1 The Industrial Revolution in Britain and Australia page 15

H2 Australia: A Sacred Country and a Federation (1750–1918) page 57

H3 Asia in the Age of Imperialism: India, China and Japan (1750–1918) page 135

The modern world and Australia

H4 Australia at war: World War I page 189

History concepts + skills

H5 History How-To page 239

good humanities

Matilda Education Australia acknowledges the Aboriginal and Torres Strait Islander peoples of this nation. We acknowledge the traditional custodians on whose unceded lands we conduct the business of our company. We pay our respects to ancestors and Elders past, present and emerging. Matilda Education Australia is committed to honouring Aboriginal and Torres Strait Islander peoples' unique cultural and spiritual relationships to the land, waters and seas; and their rich contribution to society.

Introduction to History

H0

WHAT DOES A HISTORIAN DO?

page 2

historical skills
page 4
HOW DO HISTORICAL THINKING SKILLS SUPPORT RESEARCH?

learning ladder
page 6
HOW DO I USE THIS BOOK?

history overview
page 10
WHAT FACTORS SHAPED THE MODERN WORLD?

What does a historian do?

Historians study the past to understand the present. By asking questions, conducting research and sharing their discoveries, they expand humanity's knowledge of the past and how it shaped our current world and society.

The study of history

People may think history just means listing information about what happened in the past, but history is not a catalogue of facts to be memorised. History is a process and a way of thinking.

Historians study history, follow its processes and use it to inform their thinking. Historians are also researchers – they conduct research to uncover and understand historical evidence. That research can be something as simple as reading a text, or as complex as unearthing artefacts at an archaeological dig site.

Historians perform three main tasks:

1. Ask a research question, such as 'What happened here?', 'How did this event happen?' or 'Who was behind this occurence?'.
2. Examine primary historical sources to uncover relevant evidence.
3. Use the evidence to answer the research question or to tell a story about the past.

Most historians specialise in studying a particular period of history, or the history of a particular field, such as art history. This is because they need to understand a field deeply to ask appropriate research questions, and to know what kind of primary sources to examine. There is so much information to be studied that you wouldn't expect a historian who studied 18th-century India to also be an expert on medieval England.

Source 1

Archaeology is a specialised field of historical research that involves recovering and studying artefacts from the past.

A historian at work

Doctor Karen Hughes is a professional historian, and an Associate Professor of Indigenous Studies and History at Swinburne University in Melbourne.

Why did you decide to become a historian?

It happened accidently. I was making a documentary film in Arnhem Land, and while researching it I realised that what I had been taught about Aboriginal people and their history was not the truth. I felt I had grown up with a lie.

How did you become a historian?

I did a PhD, then I got my first job teaching and researching Indigenous history at Monash University in Melbourne.

What specific period or field is your focus?

I focus on the 20th century. I like to write about cross-cultural history in Australia and America, mostly between Indigenous and settler-descended peoples. I specialise in researching people's personal stories in order to understand the experiences of larger historical events and their impacts. This is called **microhistory**.

What does your work as a historian involve?

Half of my time is spent teaching Indigenous history at Swinburne University. We do a lot of interesting things, like visiting Aboriginal Elders living on Country, creating podcasts and exploring the layers of history of particular places.

The other half of my job involves researching and writing papers and books, and curating exhibitions. I often have to travel to interview people and record their oral histories, and to search for evidence in archives.

What kinds of research projects are you involved with?

I currently have a scholarship to visit the USA and write about the lives of Aboriginal women who married American servicemen in Australia during World War II. I recently consulted for the Reserve Bank of Australia, researching the Ngarrindjeri scholar, writer and inventor David Unaipon who appears on the new $50 banknote. I'm also working on a museum exhibition with the Ngarrindjeri community of South Australia, displaying rare historical photographs taken by Indigenous photographers from the 1920s to the 1970s.

Source 2

Associate Professor Karen Hughes

What do you love about being a historian?

I love the way it makes me feel connected to the past and the people who lived then. There are amazing moments when I feel like I am time-travelling, and other moments when it's like being a detective, placing all the pieces of a puzzle together through detailed forensic-style research, to reveal something that has never been fully known before.

Learning ladder H0.1

1. What is a research question?
2. Why do historians usually specialise in a particular field?
3. What is microhistory?
4. Think of a field you would choose to research if you were a professional historian. Discuss your choices as a class.

How do historical thinking skills support research?

You have learned and developed several historical thinking skills through your history studies so far. These skills are vital when conducting your own historical research and learning about the past.

Conducting historical research

You can only learn so much from a textbook – eventually, you will want to find out more and to do your own historical research. The History How-To chapter discusses the research process in depth on pages 264–65.

First, you should define your research question. This is what you are trying to learn through your research. Your research question can be broad, such as, 'How did the First Nations People of Australia trade resources?', or narrow, 'Who was the most influential VFL coach in history?'.

Research questions take some work to answer – they are rarely so straightforward that you can find the answer quickly and easily. That's why your historical thinking skills are so important! Use them to guide your research efforts, and to make the information you discover meaningful.

Source analysis

Historians analyse two types of sources to answer research questions and build narratives.

Primary sources were created at the time of a historical event, or by someone who had first-hand experience of that event. Examples include books, diaries, photographs, archives, letters, artefacts, buildings and ruins. Primary sources often show the perspectives of the people who experienced an event, and might contain unique information.

Secondary sources were created after the time or event being studied. Examples include textbooks, websites and documentaries. These sources are often part of the narratives created by historians analysing primary sources.

Primary sources aren't always reliable, as they reflect the biases of the people who created them. Secondary sources can also be biased and unreliable; websites, in particular, can be very unreliable. The way to minimise **bias** is to study many different sources.

Finding and analysing appropriate sources is one of the most important skills in historical research.

Continuity and change

History is a story of continuity and change: some things stay the same; others change.

Research can reveal information about how and why something changed, or (depending on your research question) why something was immune to change. Narratives and timelines are useful tools when researching what changed and what stayed the same.

Cause and effect

A common focus of historical research is trying to figure out *why* things happened. There are many different causes of change, such as:

- actors (individuals and groups)
- conditions (e.g. social, political and economic)
- long-term trends
- short-term triggers.

Remember: there is rarely a single cause of an effect. Most causes have even earlier causes, and many effects cause further effects in the future. Your research may lead you to one cause for an effect, but there are likely to be more – don't stop until you've identified all the relevant causes and their web of effects.

Historical significance

You can use many models to identify historical importance. This textbook uses a model developed by Australian historian Geoffrey Partington. To work out something's historical significance, ask:

Source 1

Truganini, a Palawa woman from the part of Tasmania we now call Bruny Island has been the subject of artworks, monuments, stamps and songs. Truganini is a historically significant figure as her story speaks of both the dispossession of Indigenous Tasmanians and their strategies for survival in the face of invasion.

1. How important was it to people at the time?
2. How many people were affected?
3. How deeply were people's lives affected?
4. For how long did these effects last?
5. How relevant is it to modern life?

Your research is likely to turn up information about people, places and events, any of which might be relevant. Applying Partington's model, or a similar set of questions, will help you identify which pieces of information are significant to your research.

Historical interpretations

The way we understand history is always changing. Individuals in every period of history look at the world around them, and look at their past, and **interpret** that information to make meaning.

When researching history, you will come across many different interpretations of the past, of the people who lived at that time and of the events that occurred around them. Some of these interpretations will contradict other ones, making it hard to know which (if any) are 'true'. Draw on the knowledge and information you have already gathered to decide which interpretations will be meaningful or useful in your research.

Learning ladder H0.2

1. Source analysis: Which is more useful to historical research, a primary or a secondary source?
2. Continuity and change: This year you will study the 18th, 19th and 20th centuries. List three things in modern society that have stayed the same since the 18th century.
3. Cause and effect: Do events always have one single cause, or one specific effect?
4. Historical significance: Name a method you can use to decide what is important in history.
5. Historical interpretations: Why might people from the past interpret events in different ways?

How do I use this book?

Good Humanities has been built to help you thrive as you move through the Level 9 Humanities curriculum and to enable you to demonstrate your progress in every single lesson. The history section of this book includes four chapters of Asian, Australian and world history, plus a History How-To skills section. The History How-To section is vital – you should refer to it often.

Climb the Learning ladder

Each chapter begins with a Learning ladder. The Learning ladder is your 'plan of attack' for the skills you will practice in each chapter. It lists the five historical skills you will be learning, and has five levels of progression for each of those skills.

Each skill described in the Learning ladder is of a higher difficulty than the one below it. To be able to achieve the higher-level skills, you need to be able to master the lower ones. Practising activities at all levels will help you to master more involved skills, such as evaluating. This approach is called 'developmental learning' – and it puts you in charge of your own learning progression!

Read the ladder from the bottom to the top. As you progress through the chapter, you will climb up the Learning ladder.

Source 1 The Learning ladder helps you to take charge of your own learning!

learning ladder

Step	Source analysis	Continuity and change	Cause and effect	Historical significance	Historical interpretations
step 5	I can evaluate a source	I can evaluate patterns of continuity and change	I can evaluate causes and effects	I can evaluate historical significance	I can evaluate historical interpretations
step 4	I can analyse a source	I can analyse patterns of continuity and change	I can analyse causes and effects	I can analyse historical significance	I can analyse historical interpretations
step 3	I can use the origin of a source to explain its creator's purpose	I can explain patterns of continuity and change	I can explain causes and effects	I can apply a theory of significance	I can explain historical interpretations
step 2	I can find themes in a source	I can explain why something did or did not change	I can determine causes and effects	I can explain historical significance	I can describe historical interpretations
step 1	I can list specific features of a source	I can describe continuity and change	I can recognise a cause and an effect	I can recognise historical significance	I can recognise that the past has been represented in different ways

Check your progress

Each chapter is divided into multiple sections, with each section designed to cover one lesson. Sometimes your teacher might decide to spend more or less time on a particular section. A section is either two or four pages long.

At the end of most sections, you will find a block of questions called 'Learning ladder,' which has two different types of questions or activities:

1. Show what you know

   These questions ask you to look back at the content you have read and viewed and to show your understanding of it by listing, describing and explaining.

2. Learning ladder

   These activities are linked to the Learning ladder. You can complete one of the questions or several of them. In each chapter, you will complete several activities for each level of the Learning ladder, as well as for each writing and research stage. This will sharpen your historical skills.

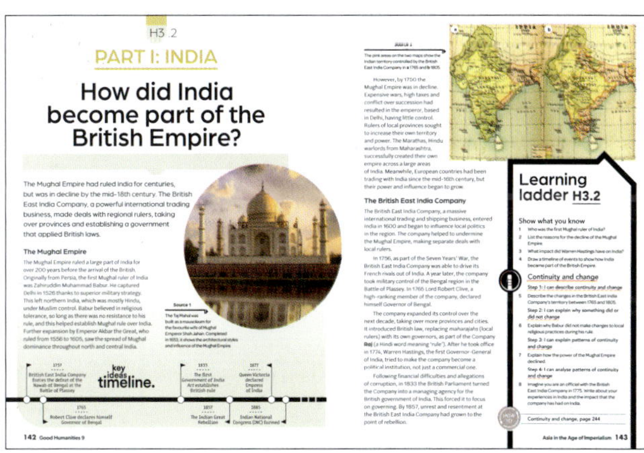

Source 2

Check your progress regularly. You can attempt one or more of the Learning ladder questions.

civics + citizenship

economics + business

The study of Humanities is more than just History and Geography – it is also the study of Civics and citizenship, and Economics and business. In every chapter of this book, you will discover either a Civics and citizenship lesson or an Economics and business lesson. School is busy and you have a lot to cover, so designing a textbook where the important Civics and citizenship and Economics and business content is placed meaningfully next to relevant History or Geography lessons makes good sense, and will help you to connect your learning.

As you work through the Civics and citizenship and Economics and business sections in this book, you will also be working your way up a Learning ladder for these subjects too.

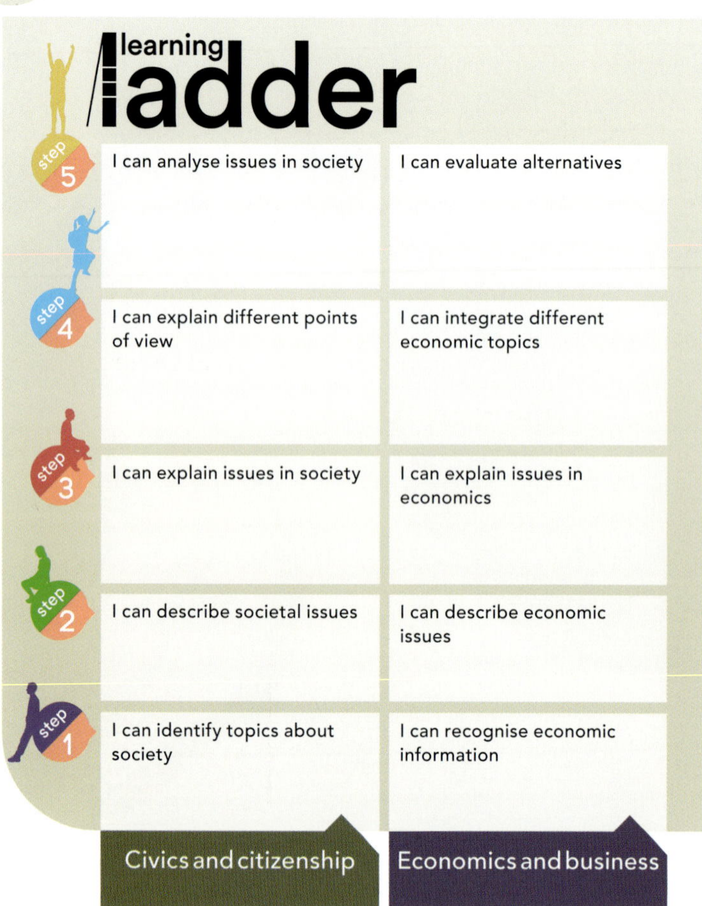

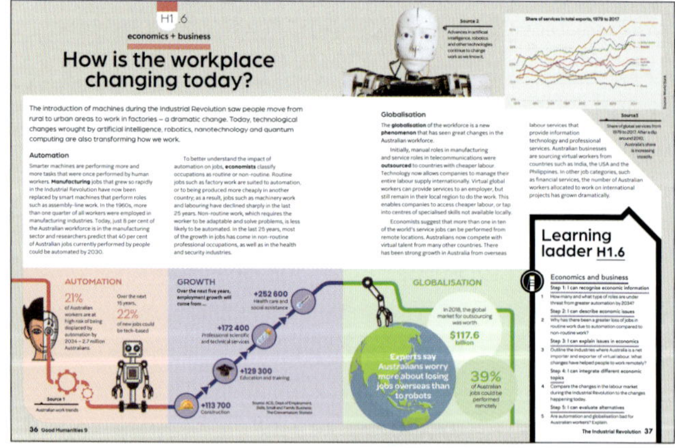

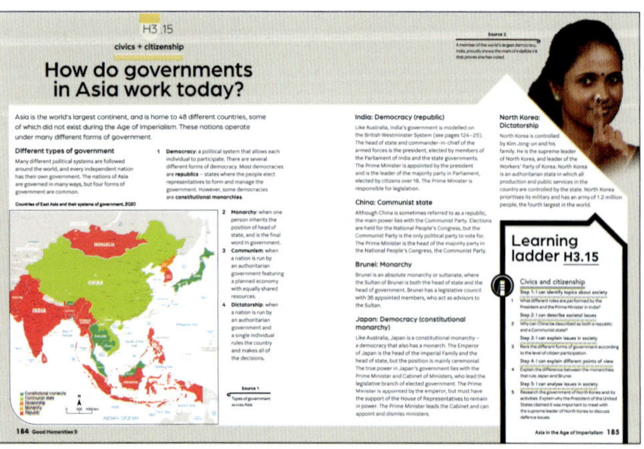

Source 3

Explore Civics and Citizenship, and Economics and Business, alongside your history course.

History How-To

In the middle of the book, you will find a skills section called 'History How-To'. This section explains how to perform each skill and the steps needed for writing and research. There are *lots* of worked examples. Refer to the How-To often, especially when answering the Learning ladder questions and completing the review activities.

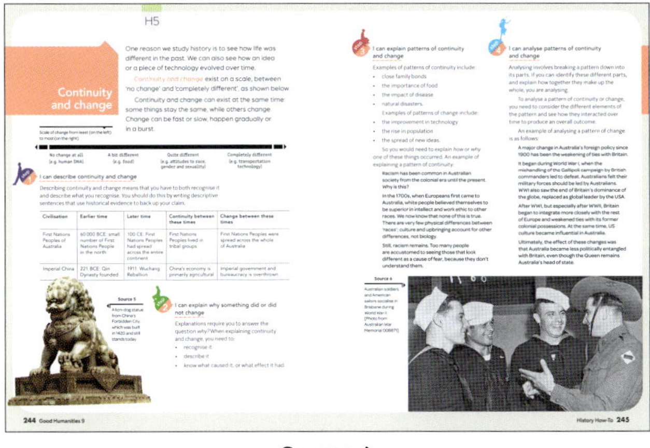

Source 4

The History How-To section is your key to success – refer to it often!

Masterclass

At the end of each chapter is a review section, called the Masterclass. The questions here are organised by the steps on the Learning ladder. You can complete all of the questions *or* your teacher might direct you to complete just some of them, depending on your progress.

Source 5

The Masterclass is your opportunity to show your progress. Take charge of your own learning and see if you can extend yourself.

Capstone

After you complete a chapter, it's time to put your new knowledge and understanding together for the capstone project to show what you *know* and what you *think*. In the world of building, a capstone is an element that finishes off an arch, or tops off a building or wall. That is what the capstone project will offer you, too: a chance to top off and bring together your learning in interesting and creative ways. It will ask you to think critically, to use key concepts and to answer 'big picture' questions. The capstone project is accessible online; scan the QR code to find it quickly.

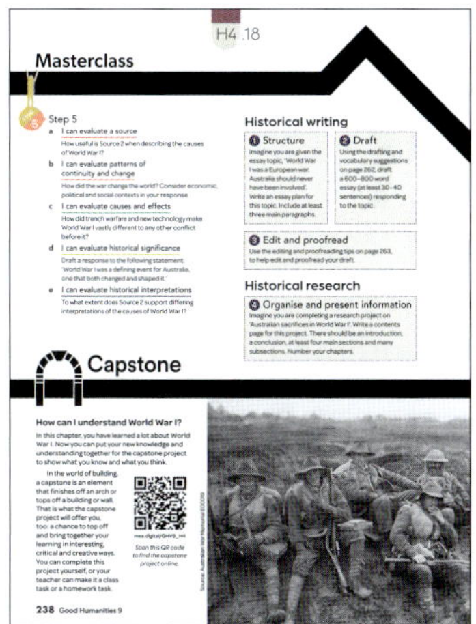

Source 6

The capstone project brings together the learning and understanding of each chapter. It provides an opportunity to engage in creative and critical thinking.

Learning ladder H0.3

1. What are the different types of questions in this textbook? Describe them in your own words.
2. How can you use the Learning Ladder to monitor your progress in Year 9 History?
3. As a class, discuss the idea of 'monitoring your own progress'. Why is this important?
4. Read through the steps of the History Learning Ladder and consider where you might already be up to for each skill, based on your prior learning.

What factors shaped the modern world?

While the world underwent many changes during the 1000 years of the Medieval Period, it changed far, far more in the following 200–300 years. The period from around the mid-18th century to the early 20th century was a time of staggering change, completely transforming almost every society and culture on Earth – whether the culture welcomed the change or not. The events and advances of this period shaped the world of the 20th century, and we continue to feel their effects today.

A period of exponential change

In maths, rates of growth and change can be described as **arithmetic** or **exponential**. Arithmetic growth changes at the same rate over time – if you draw it on a graph, it's a gentle, straight diagonal line. Exponential growth gets faster and faster over time, changing more and more – on a graph it is represented by a line that starts flat and then curves up to become extremely steep.

The period between 1750 and 1920 was one of exponential change in almost every aspect of human knowledge, science and society. Someone born in the beginning of this period, after thousands of years of gradual change, could never have imagined what the world would look like at its end.

Social, cultural, economic and political change, technological advances and conflict between powerful nations fuelled a rapid transformation of society, and shaped the world we live in today.

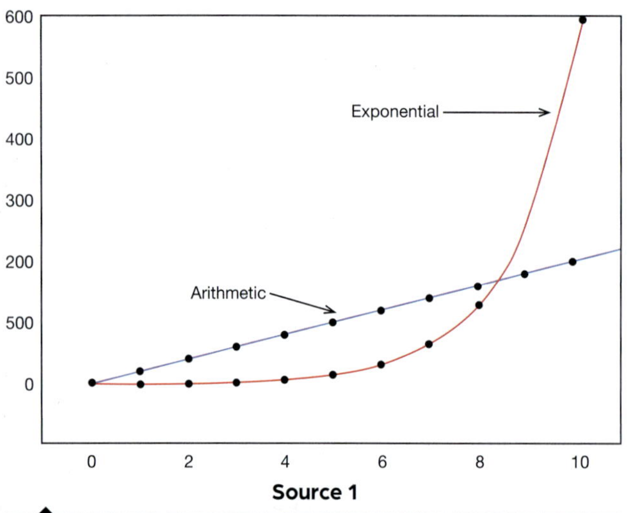

Source 1

Arithmetical change is steady and consistent, while exponential change starts slowly but speeds up so rapidly it quickly goes 'off the chart'.

Technology and industrialisation

Perhaps no change is so obvious, dramatic and wide reaching during this period than the changes in technology that followed the invention of the steam engine in 1765. For the first time, humans could generate and direct more energy than could be harnessed from natural sources like horses or the wind. This opened up incredible new possibilities, and inventors around the world created a series of new machines, each more wondrous than the last.

At the start of the 18th century, the sailing ship was the most advanced means of travel, and sending letters by ship was the best way to deliver information over long distances. By 1920, aeroplanes were powerful enough to fly across the Atlantic Ocean, and radio messages could be sent almost instantly around the world.

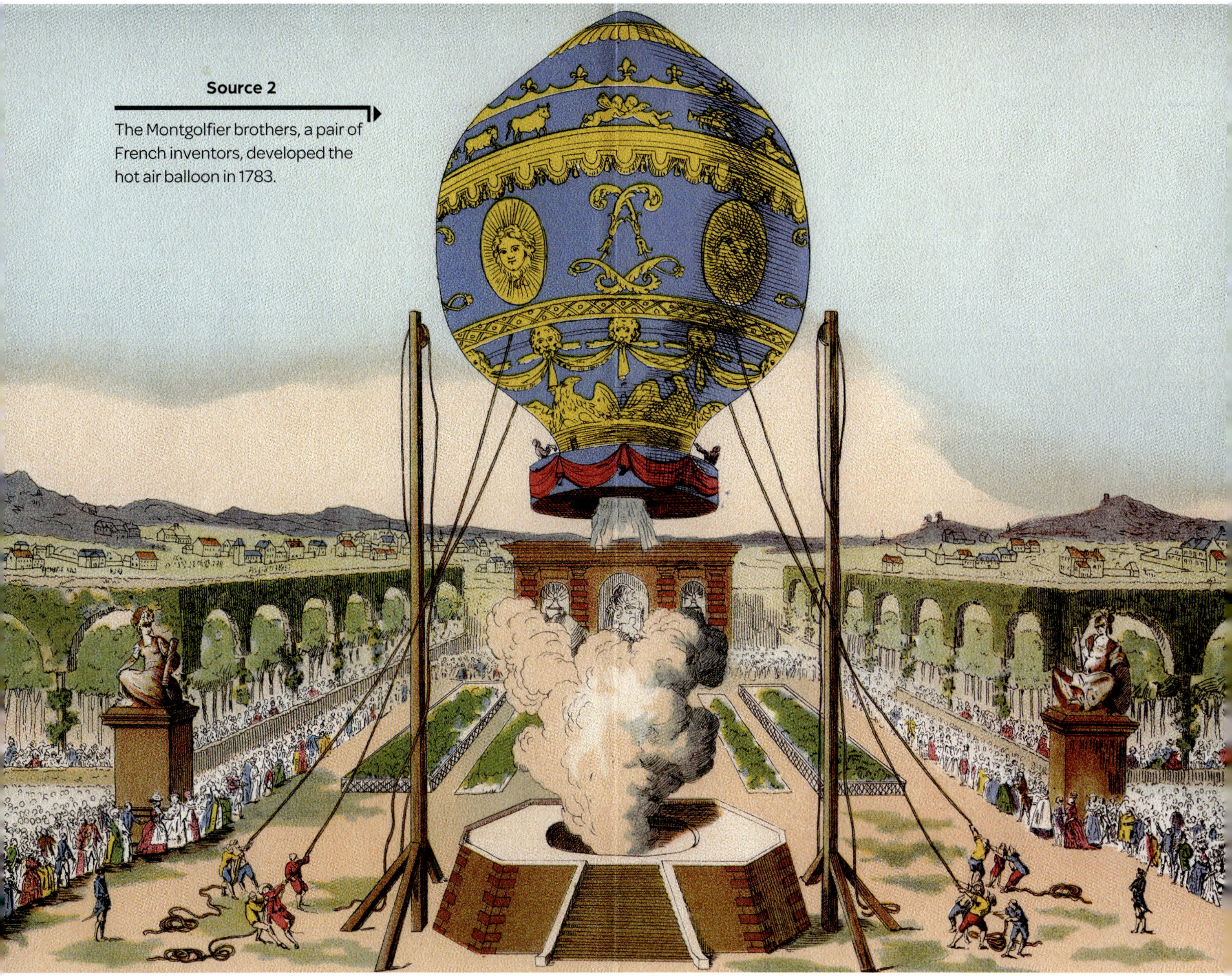

Source 2

The Montgolfier brothers, a pair of French inventors, developed the hot air balloon in 1783.

These technological advances also changed society. Wealthy nations became **industrialised**, adopting the use of advanced technology. Machines could perform tasks faster, more cheaply and in greater volume than human workers could, which led those wealthy nations to become even wealthier and more powerful. However, the wealth was not shared evenly, with many workers left without jobs. At the same time, less advanced nations became vulnerable.

Source 3

In 1903, 120 years after the first hot air balloon flight, the Wright brothers invented and flew the first aeroplane in the US.

Imperialism and nationalism

By the 18th century, the great empires of Africa and Asia had been overtaken by the rising powers of Europe. England, France, Spain, Germany and other nations used their wealth and technology to spread their power around the globe. This was an age of **imperialism**, as these European (and later American) empires used their more advanced weapons, naval power and technology to seize control of countries in Asia and Africa.

Along with the growth of empires came a growth in **nationalism**. For most of history, people had primarily identified with and been loyal to small political entities – a clan, tribe, village or city–state. Now people increasingly saw themselves as part of a country or nation, with many believing that their nation was superior to all others. This attitude underpinned the expansion of empires, but also motivated subjugated peoples to fight to reclaim their lands.

New nations and ancient peoples

The rise of the nation–state, along with the expansion of empires, meant that many new countries came into existence during this period. Some of these new nations emerged from old empires, gaining their own **sovereignty**, such as Belgium (part of the Netherlands until 1830) or Kuwait (which separated from the Khalidi Emirate in 1752).

Other nations gained their independence through revolution, such as the United States of America, which, after declaring independence in 1776, fought against Britain to ultimately win its independence in 1783. Still other nations were formed when empires invaded and **colonised** new lands – the most obvious for us is Australia, which was claimed by the British in 1770.

Source 4

Britain founded 13 colonies in North America in the early 17th century. In 1776, these colonies came together to declare their independence, and waged war against their British rulers. In 1783, the United States of America became a major new figure in international politics. This Revolutionary War painting shows the surrender of British General John Burgoyne at Saratoga in 1777. [John Trumbull, *The Surrender of General Burgoyne*, 1821]

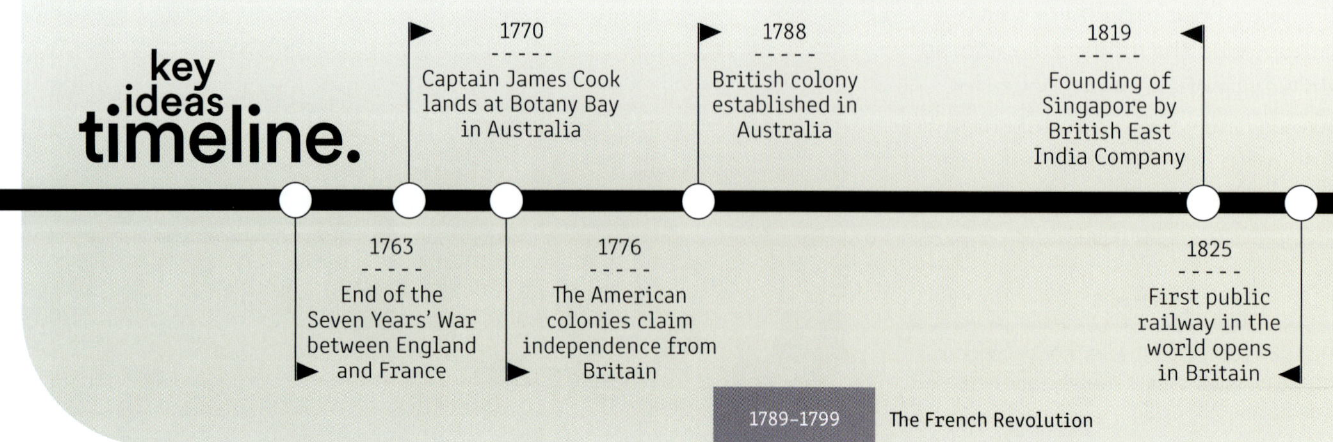

But the new lands claimed by these empires were rarely uninhabited – they were the homes of many different **indigenous** peoples with their own societies and cultures. Thanks to the racism and prejudice of the Europeans, these traditional inhabitants – such as the First Nations Peoples of North America, Australia and New Zealand – were treated as inferior, unimportant or even less than human. Their homes were taken from them, while natural resources were stripped from their lands to meet the needs of European nations. It took many, many years for most of these indigenous societies to reclaim their rights and homelands, and some peoples are still fighting for those rights today.

Global concerns, global conflicts

The expansion of empires, and the need for resources to fuel industrialisation efforts, meant that powerful nations such as Britain, France and the US came into conflict. In previous centuries, wars were a common event between powerful nations, but they occurred somewhat less often during the modern period. Leaders regarded open warfare as too great a drain on resources, and as too difficult to coordinate on the fringes of an empire. When they could, the great nations preferred to use diplomacy, spying or economic bargains to settle conflicts.

But battles and conflicts did not become a thing of the past. Empires used their soldiers and weapons to conquer less advanced nations, and wars were still fought to gain territory or increase power. In 1914, one such war erupted between European states, and expanded rapidly to involve almost every country in Europe, as well as nations all around the world. It lasted for less than five years, but by the end, the political landscape of Europe had permanently changed, and 20 million people had been killed.

Afterwards, people called it 'The War to End All Wars' – but they were wrong. In hindsight, it became known as World War I.

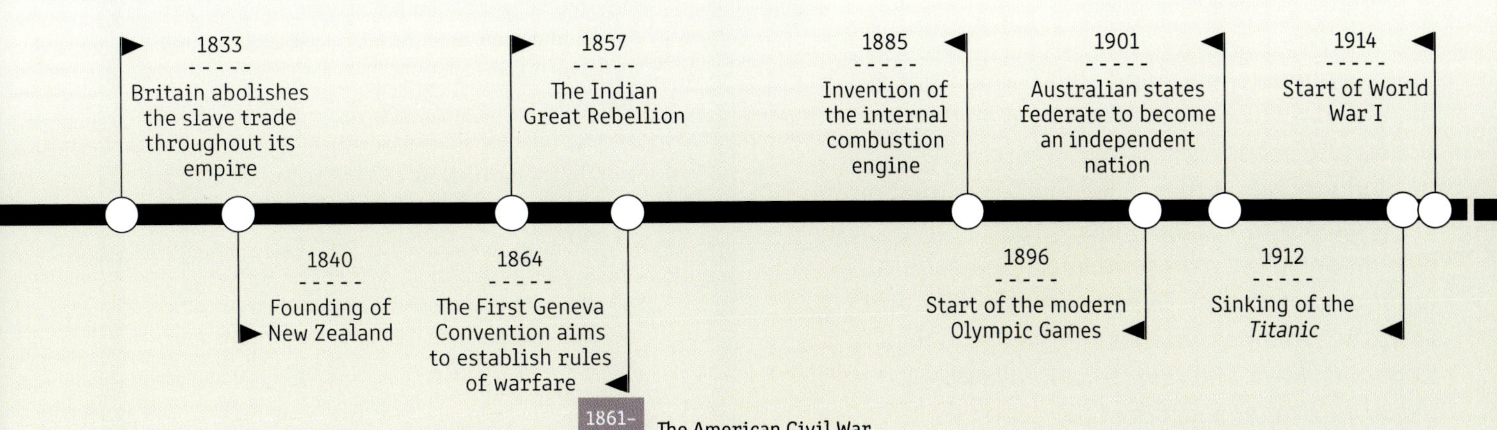

Source 5

In France, gatherings of intellectuals – and their wealthy patrons – were common from the 17th until the mid-19th century. These salons gave rise to political, economic and philosophical theories that are still discussed today. [Lemonnier, *Reading of Voltaire's tragedy of the Orphan of China in the salon of Marie Thérèse Rodet Geoffrin 1755*, painted in 1812]

Intellectual and social change

Industrialisation and warfare were not the only changes sweeping the world at this time. Following the Renaissance and the Scientific Revolution of previous centuries, the 17th and 18th centuries became known as the Age of Reason, or the Age of Enlightenment. Scientists, thinkers, philosophers and academics began to suggest new models of thought, and new ideas about how humanity could live, learn and prosper. Many of these ideas were **humanistic**, and promoted the view that human achievements and discoveries in this world were more important or helpful than religious ideas of a reward in the 'next world'.

In addition to moral and ethical philosophies, there were also practical ideas about politics, society and economics. Many of these were revolutionary ideas that called for the destruction of the old order, and the replacement of ruling families and monarchies with more democratic forms of government. Over the course of the 19th and 20th centuries, many nations let go of their traditional ways, and very structured cultures, to embrace fairer and more open societies.

Learning ladder H0.4

1. What is the difference between arithmetic and exponential change?
2. When was the hot air balloon invented?
3. How did colonisation affect indigenous people who lived in the colonised lands?
4. Why might the political philosophies of the Age of Reason have come into conflict with established religious organisations?
5. When did World War I begin? Why was it not called that at the time?
6. How might the daily life of a British worker in 1750 have been different to that of a British worker in 1910?
7. 'During this time period, women were often disadvantaged and unable to share in the advances of the time.' What evidence have you read or viewed in this section to support or rebut this statement?

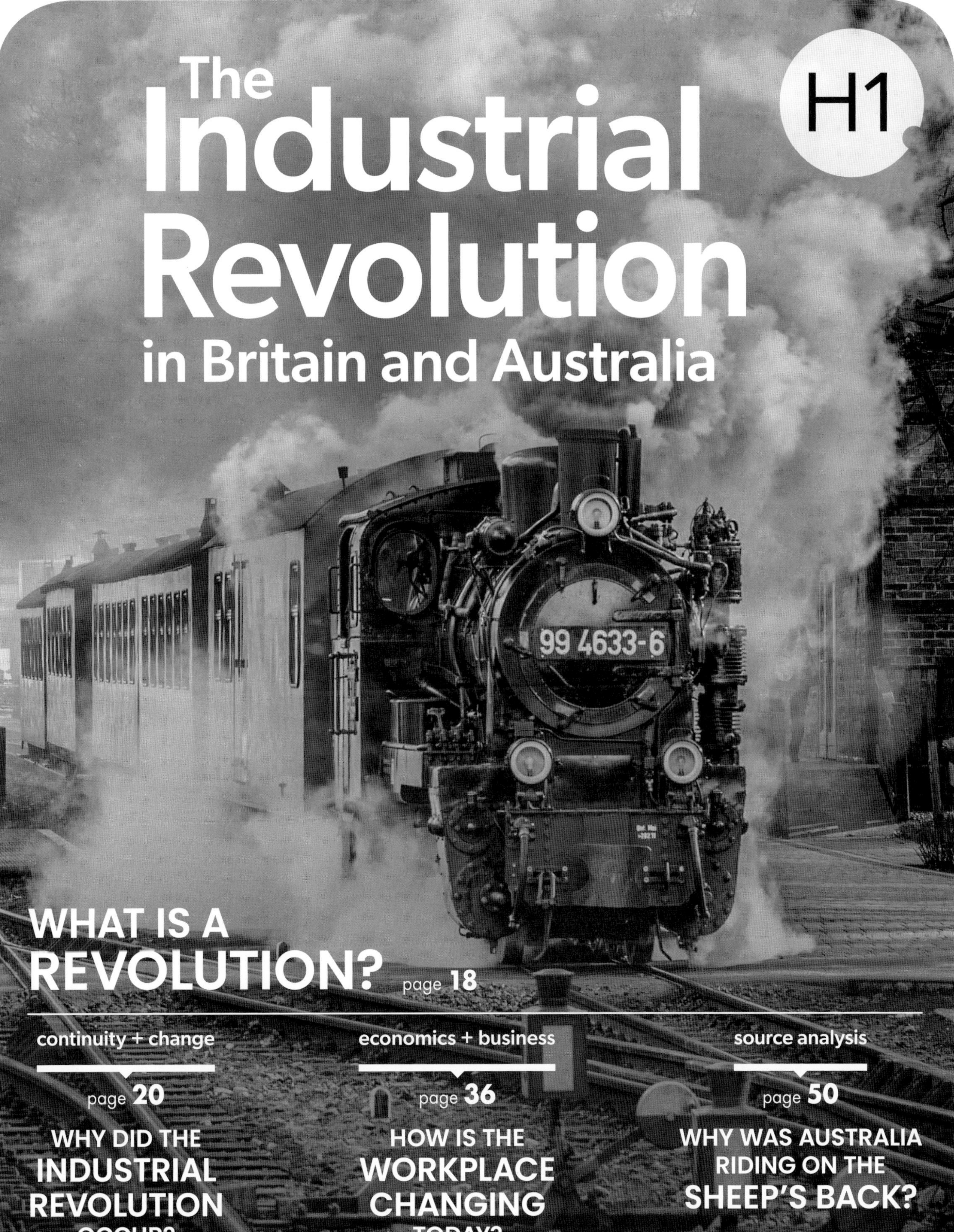

The Industrial Revolution in Britain and Australia
H1
99 4633-6
WHAT IS A REVOLUTION? page 18
continuity + change
page 20
WHY DID THE INDUSTRIAL REVOLUTION OCCUR?
economics + business
page 36
HOW IS THE WORKPLACE CHANGING TODAY?
source analysis
page 50
WHY WAS AUSTRALIA RIDING ON THE SHEEP'S BACK?

How can we understand the Industrial Revolution?

The Industrial Revolution spanned from approximately 1760 to 1820, during which new inventions and processes replaced traditional industries and systems. Despite occurring more than two centuries ago, the effects of the Industrial Revolution are still felt today. The changes that occurred during this period significantly altered politics, economics, society and technology.

learning ladder

Step	Source analysis	Continuity and change	Cause and effect
step 5	**I can evaluate a source** I can present a judgement on the usefulness of a source based on its strengths, weaknesses and limitations. I can determine whether information is missing about the event or person the source refers to.	**I can evaluate patterns of continuity and change** I answer the question 'So what?' about patterns of continuity and change. I weigh up different aspects and debate the importance of continuity or change.	**I can evaluate causes and effects** I answer the question 'So what?' about cause and effect. I weigh up different things and debate the importance of a cause or an effect.
 step 4	**I can analyse a source** I can use my own knowledge to determine the reliability of a source and can explain whether it shows a one-sided view.	**I can analyse patterns of continuity and change** I can look deeper into patterns of continuity and change and determine the factors that contribute to them.	**I can analyse causes and effects** I don't just see a cause or an effect as one thing. I can determine the factors that make up causes and effects.
 step 3	**I can use the origin of a source to explain its creator's purpose** I combine knowledge of when and where a source was created to answer the question, 'Why was it created?'.	**I can explain patterns of continuity and change** I can see beyond individual examples of continuity and change between historical periods and explain broader patterns.	**I can explain causes and effects** I can answer 'How?' or 'Why?' a cause led to an effect in the Industrial Revolution.
 step 2	**I can find themes in a source** I look a bit closer at a source and find more than just features. I find themes and patterns in a source.	**I can explain why something did or did not change** I can give a reason for why something changed or why it stayed the same.	**I can determine causes and effects** Applying what I have learnt about the Industrial Revolution, I can describe what the cause or effect of an event was.
 step 1	**I can list specific features of a source** I can look at an Industrial Revolution source and list the details I can see in it.	**I can describe continuity and change** I recognise what has stayed the same and what has changed from the Industrial Revolution until now.	**I can recognise a cause and an effect** From a supplied list, I can recognise things that were causes or effects of each other in the Industrial Revolution.

Warm up

Source 1

This rural landscape, painted by John Constable in 1816, depicts the artist's interpretation of rural life in Britain before the Industrial Revolution. [John Constable, *Wivenhoe Park*, Essex, 1816, oil on canvas (National Gallery of Art)]

Historical significance

I can evaluate historical significance
I answer the question 'So what?' about things that are supposedly important in the history of the Industrial Revolution. I weigh up factors against one another and can cast doubt on how important things are.

I can analyse historical significance
I can separate out the various factors that make something historically important in the history of the Industrial Revolution.

I can apply a theory of significance
I know a theory of significance. I use it to rank importance of changes, causes, effects and events in the history of the Industrial Revolution.

I can explain historical significance
I answer the question 'Why?' about what was important in the Industrial Revolution.

I can recognise historical significance
When shown a list of facts about the Industrial Revolution, I can work out which are important.

Historical interpretations

I can evaluate historical interpretations
I can weigh up the different historical interpretations that have been formed. I debate and challenge the interpretations that have been presented.

I can analyse historical interpretations
I can determine the factors that have led to why a historical interpretation has been formed.

I can explain historical interpretations
I can answer 'Why?' or 'How?' there are different interpretations of people and events in the past.

I can describe historical interpretations
I can provide different examples to show how people and events in the past have been interpreted.

I can recognise that the past has been represented in different ways
I can identify different views of people and events in the past.

Source analysis

1 What does Source 1 suggest about life in Britain at the start of the 19th century?

Continuity and change

2 How do you think the landscape shown in Source 1 changed after the invention of the steam locomotive in 1804?

Cause and effect

3 What effect do you think the introduction of rail transport had on British people in the 19th century?

Historical significance

4 Why was the invention of the steam-powered engine significant?

Historical interpretations

5 The spinning jenny, a machine that greatly sped up weaving production, was invented in 1764. How might the following people describe this invention and its effects on their lives?
   a Its inventor, James Hargreaves
   b A woman who works as a cottage-based spinner.

What is a revolution?

A **revolution** is a radical change in the way something is done that results in major changes and far-reaching consequences. Political revolutions may cause sudden and even violent change to the way a country is governed. Other revolutions may be slower and less obvious, but their effects can be just as dramatic. All revolutions have both intended and unintended consequences, and can create positive and negative changes.

Source 1

In 2011, Egyptian protesters overthrew their government, which had been in power for almost 30 years.

Source 2

During the scientific revolution, visionaries such as Isaac Newton made great advances in science and mathematics, changing the ways in which people viewed the world.

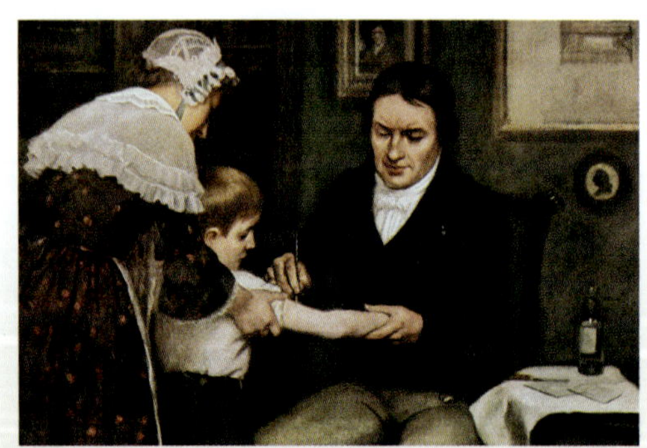

Source 3

Edward Jenner develops smallpox vaccination, 1796

Manifest

der

Kommunistischen Partei.

Veröffentlicht im Februar 1848.

London.

Source 4

The *Communist Manifesto* 1848

key ideas timeline.

- 1750 — Enclosure of land increases in Britain
- 1764 — James Hargreaves invents the spinning jenny
- 1771 — Richard Arkwright opens first factory
- 4 July 1776 — Declaration of Independence (USA)
- 1781 — James Watt invents a steam engine to power machines in factories
- 1796 — Edward Jenner develops smallpox vaccination
- 1807 — British government bans slave trade
- 1829 — George Stephenson develops the Rocket steam locomotive
- 1833 — *Slavery Abolition Act*
- 1833 — *Factory Act*
- 1838 — Brunel's SS *Great Western*, the first transatlantic steamship
- 1848 — The *Communist Manifesto* is published

Source 5

In 2015, same-sex marriage was legalised in Ireland after the world's first national vote on the issue.

Source 6

Alexander Graham Bell with his telephone

- 1856 ----- Henry Bessemer develops process to mass produce steel
- 1856 ----- Eight Hours Movement
- 1872 ----- Overland telegraph links Australia
- 1876 ----- Alexander Graham Bell patents the telephone
- 1885 ----- Karl Benz produces the first automobile
- 1903 ----- The Wright Brothers make first powered flight

Learning ladder H1.1

Show what you know

1. Categorise the events on the timeline according to political, economic, social and technological (PEST) factors.
2. In your own words, define what a revolution is.
3. Why do you think revolutions occur?
4. Predict the positive and negative consequences of revolution. Do you think people are always happy with the outcomes of a revolution? Why or why not?

Continuity and change

Step 1: I can describe continuity and change

5. Describe changes to transport technology that occurred between 1750 and 1901.

Step 2: I can explain why something did or did not change

6. Explain why the British government ended slavery.

Step 3: I can explain patterns of continuity and change

7. Explain why social changes occur over time.

Step 4: I can analyse patterns of continuity and change

8. During the Industrial Revolution, technology changed dramatically. Rank the changes outlined on the timeline in order of importance, and justify your choices.

HOW TO

Continuity and change, page 244

Why did the Industrial Revolution occur?

The causes of revolutions are complex and involve many different factors. The Industrial Revolution, which began in Britain before spreading to Europe and the rest of the world, was caused by a combination of political, economic, social and technological (PEST) factors.

Social and economic factors

In the early 19th century, Britain experienced a dramatic population increase. Immigrants from Europe and the rest of the world flocked to cities throughout Britain. By the second half of the 19th century, London's population had more than doubled. **Urbanisation** occurred, whereby an increasing number of people relocated to British cities from the countryside in search of work. This resulted in increased demands on basic resources, such as housing, infrastructure and food. The agricultural and textile industries were therefore the first areas in which **innovation** occurred to meet the growing social needs.

Source 1

How the British population changed between 1750 and 1901

	1750	1901
Population	6 million	32 million
Percentage of people living in towns	15%	85%
Life expectancy	Between 30 and 40 years for both males and females	Males: 48 years Females: 52 years

Technological factors – the agricultural revolution

Britain was also dealing with the effects of an 'agricultural revolution' that reshaped the farming industry. The land in Britain was owned by wealthy landowners or the aristocracy. They rented land to tenant farmers as well as working some of it themselves. Farm labourers worked the land for wages. Food was required to feed the growing population, so farmers had to find innovative and efficient ways of increasing food production. Most farmers in Britain still used the open field system, which had been in use since the medieval period.

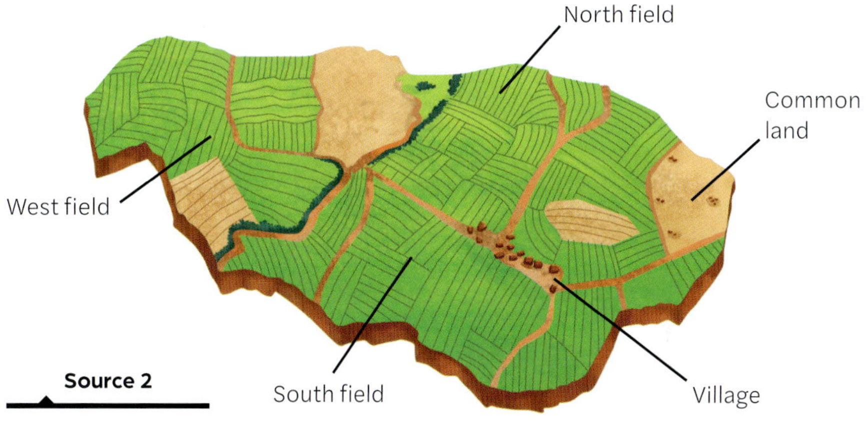

Source 2

The open field system

Under the open field system, farming land around a village was divided into three fields. Each field was divided into strips, and villagers grew their own food on their own strip of land. There was also common land where villagers could graze their animals, collect firewood, and gather fruit and herbs that were growing wild. However, the system was inefficient. Time and land was wasted, with one field always being left fallow (not in use).

From the 1740s, landowners lobbied parliament in favour of **enclosure**. Enclosure divided land into farms surrounded by fences or hedges. Villagers were given plots of land equal to their strips. By 1790, three-quarters of farming land in Britain had been enclosed.

But not everyone benefited. Tenant farmers were evicted. People who did not own their land, or could not prove ownership, lost their land and the right to access common land.

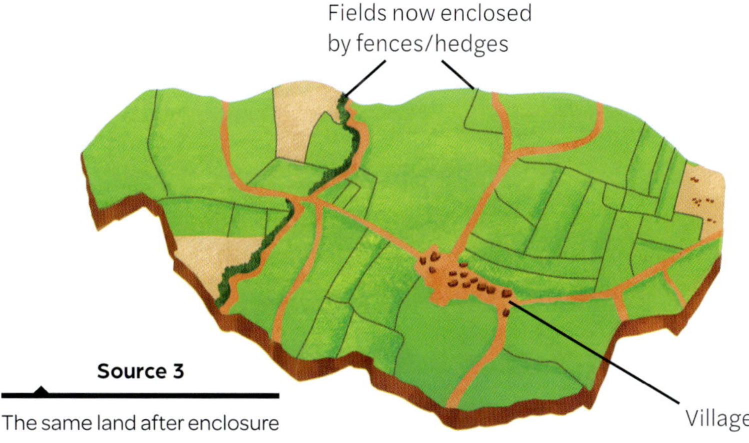

Source 3

The same land after enclosure

Despite the hardship and social unrest caused by enclosure, the ability to farm larger blocks of land resulted in innovation that increased food production. Along with innovation in animal breeding methods to increase meat and wool production, there were also improvements in crop production, such as the Norfolk crop rotation system. Introduced by Lord Charles Townshend (who earned the name Turnip Townshend), the system worked on the premise that four different crops could be grown in a field over four years, so no field was ever left fallow (Source 4). This gave farmers the ability to produce food for the growing population and, as there was winter food for cattle, for people to enjoy fresh meat and milk all year round.

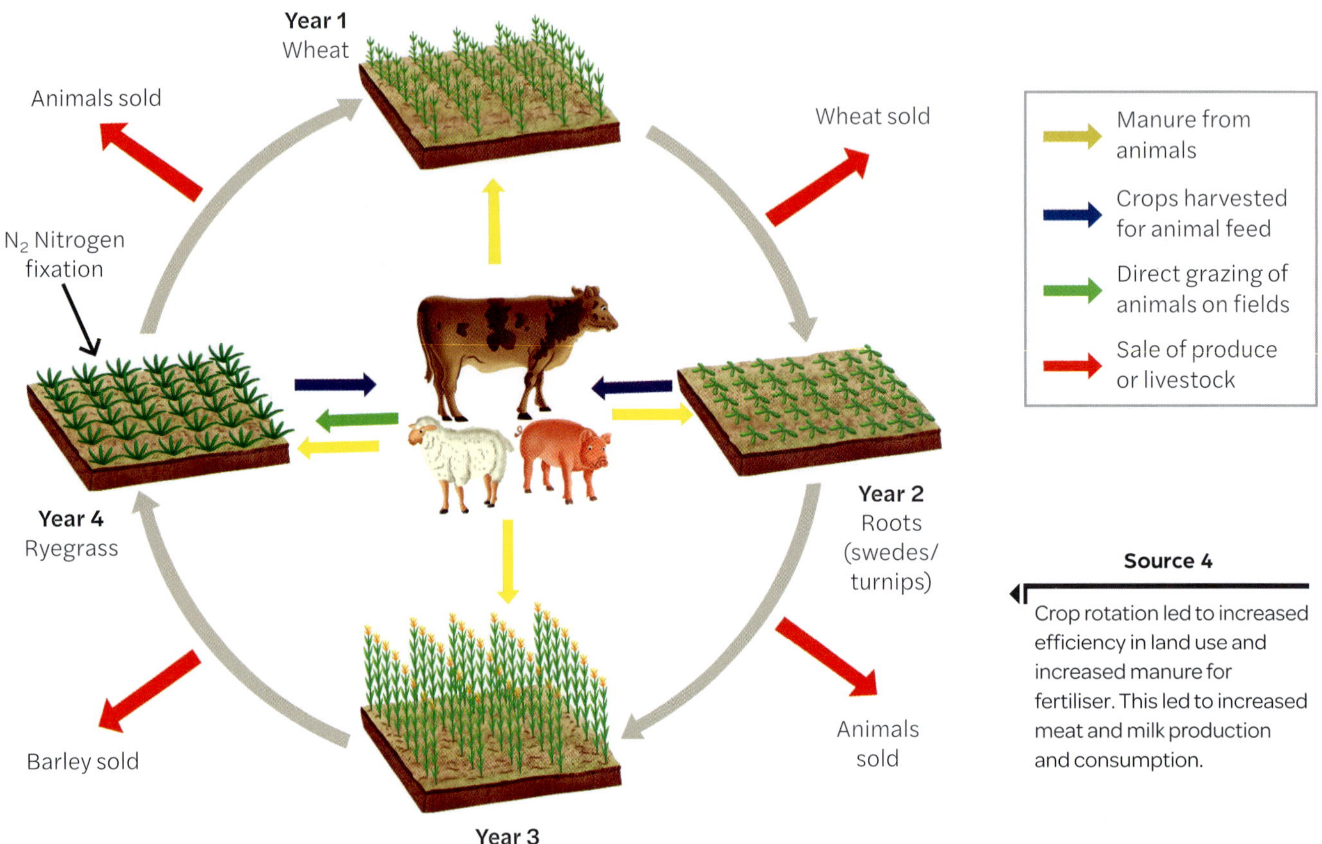

Source 4

Crop rotation led to increased efficiency in land use and increased manure for fertiliser. This led to increased meat and milk production and consumption.

Selective breeding

Experiments in selective breeding aimed to increase meat and wool production. For example, Robert Bakewell focused on selective breeding in sheep and produced a breed called the New Leicestershire, which produced good quality fleece as well as more meat.

A report to the British parliament by Sir John Sinclair in 1795 stated that the average weight of animals sold at Smithfield Market had more than doubled from 1710 to 1795, as shown in Source 5.

Livestock/ Year	1710	1795
Cattle	370 lb	800 lb
Sheep	28 lb	80 lb

Source 5

Average weight of animals sold (in pounds)

Changes in the agricultural industry meant that fewer labourers were required. Labourers in search of work moved from rural to urban areas, which led to increased urbanisation. The larger urban population gave emerging industrialists a ready-made supply of workers for their newly built factories. Increased food production meant that the growing urban population could be fed, and it also resulted in greater profits for farmers, who could invest this money into other innovations.

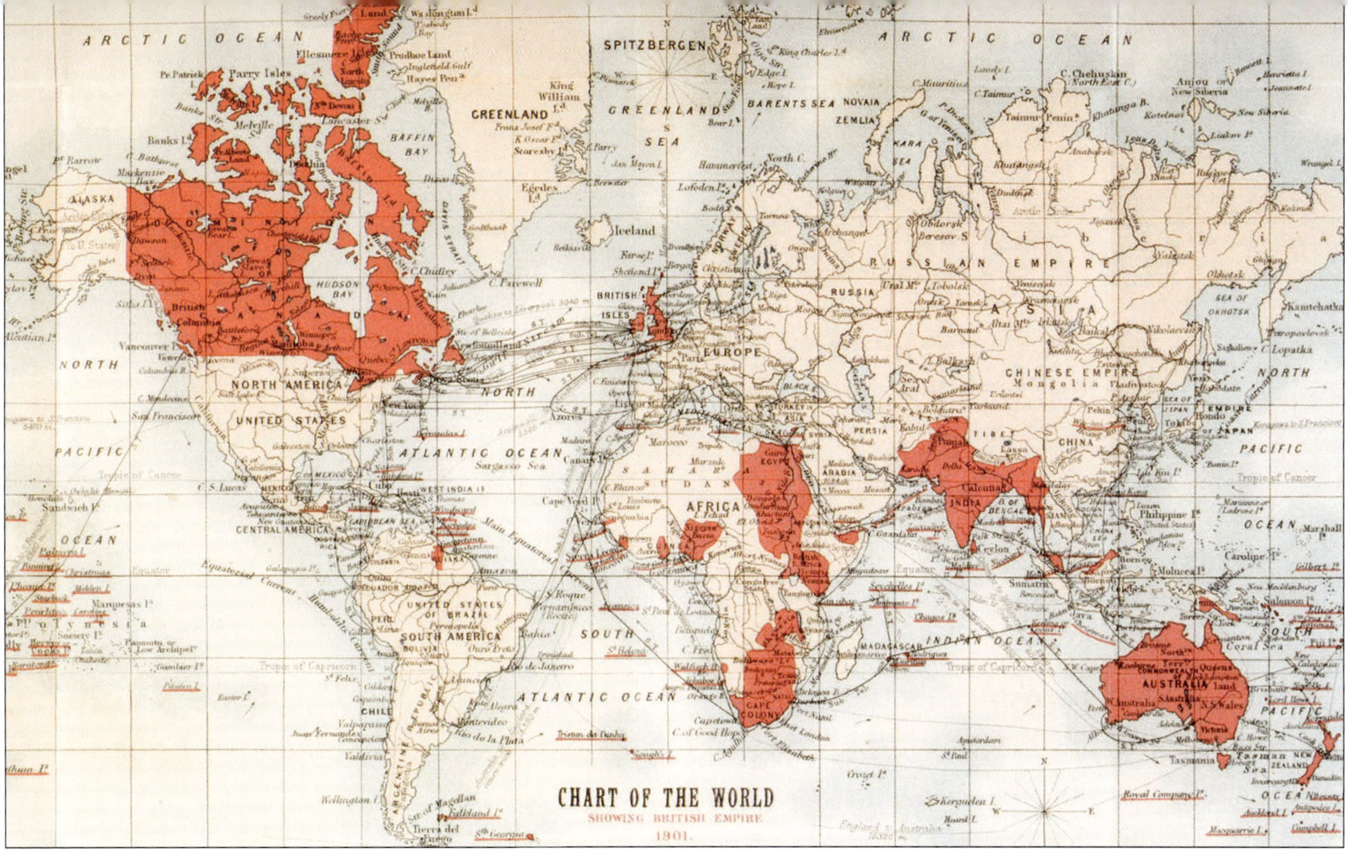

Source 6

Britain's large empire (shown here in red) gave it ready access to the resources it needed to power growing industries at home.

Political factors

Britain had a strong and stable government. This gave people the confidence to invest in businesses. Throughout the 1700s, Britain had expanded its empire, including the colonisation of Australia in 1788. Having a vast empire gave Britain access to the raw materials, such as cotton, needed in industry. The British adopted a policy of **mercantilism**, and ensured that its colonies only traded with Britain, thus increasing its wealth and power.

Imperialism is the control of countries or territories by a more powerful nation. The nation then imposes its political, economic and cultural beliefs on the peoples living in its colonies. European countries had begun expanding their empires in the 17th century and, by the late 19th century, Britain's empire covered a quarter of the world's land surface.

Learning ladder H1.2

Show what you know

1 Describe the changes to the environment brought about by enclosure.

2 How did Britain's government encourage investment?

3 Why did agricultural methods need to change?

4 Using Source 5, explain the impact of selective breeding.

5 What similarities can you see with developments in agriculture today?

Continuity and change

Step 1: I can describe continuity and change

6 Describe the changes to farming methods that occurred during the agricultural revolution.

Step 2: I can explain why something did or did not change

7 Explain why the changes in agriculture you described in question 6 occurred.

Step 3: I can explain patterns of continuity and change

8 Explain how the British population changed between 1750 and 1901.

Step 4: I can analyse patterns of continuity and change

9 What impact did innovation have for people living in Britain during the 18th century?

Continuity and change, page 244

HOW TO

What other factors contributed to the Industrial Revolution?

Changes in the way people thought about economics also contributed to the Industrial Revolution. A new approach to economic management emerged known as *laissez-faire* (a French term meaning 'let do'). This economic theory meant that individuals had more freedom to conduct their business, rather than being dictated to by government policies, such as tariffs and subsidies. Technological changes and innovations in manufacturing and transport resulted in production becoming more efficient and cheaper. This created more affordable goods, which increased demand.

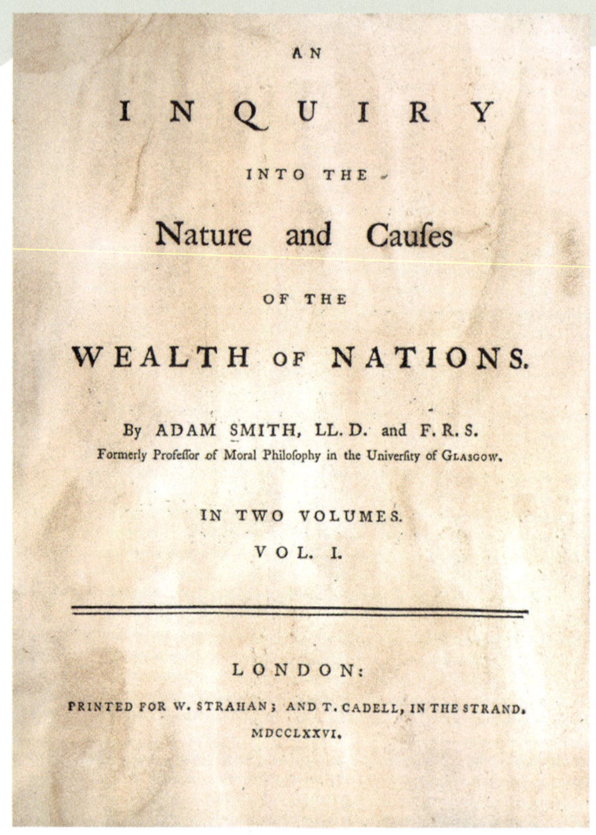
AN
INQUIRY
INTO THE
Nature and Caufes
OF THE
WEALTH OF NATIONS.
By ADAM SMITH, LL.D. and F.R.S.
Formerly Profeffor of Moral Philofophy in the Univerfity of GLASGOW.
IN TWO VOLUMES.
VOL. I.
LONDON:
PRINTED FOR W. STRAHAN; AND T. CADELL, IN THE STRAND.
MDCCLXXVI.

Source 1

In 1776, Adam Smith published his ground-breaking book about economics, commonly known as *The Wealth of Nations*.

The Enlightenment

The Enlightenment, also known as the Age of Reason, began in the mid-17th century when thinkers began to challenge the traditional teachings of religious authorities. Thinking was based on reason rather than belief, and new scientific theories and principles emerged, put forward by men such as Isaac Newton.

The Enlightenment also challenged the way that countries were governed. Ideas such as John Locke's natural rights (life, liberty and property) and his argument that the purpose of government was to protect these rights had a strong influence on the American and French Revolutions.

Advances in economic thinking

Britain had an established banking system that was much more advanced than any other in Europe. The growth of the British Empire, and the increase in trade that came with it, required financial services to support it. The Bank of England (established in 1694) ensured that there was **capital** for investment and protection for trade through insurance. Low interest rates also encouraged individuals to take risks on innovative technologies.

Source 2

This painting from 1768 shows people gathered in wonder around an experiment. [Joseph Wright, *An Experiment on a Bird in the Air Pump*, 1768, Oil on canvas.]

Merchants benefited from the fact there were no internal tariffs in Britain, and this made trading much easier. Britain's large empire meant that there was also an easily accessible international market.

The Enlightenment also gave rise to new ideas about economics, and these new ideas played an important role in the Industrial Revolution. Adam Smith wrote about a *laissez-faire* approach to the economy in his book, *The Wealth of Nations* (1776), in which he argued against the established principle of mercantilism. Smith argued that governments should not control the economy, but instead let the market forces of supply and demand dictate economic activity. This allowed individuals far greater freedom. Without the emergence of capitalism, the Industrial Revolution would not have been possible.

Capitalism

Capitalism is an economic system in which private individuals and companies own property and goods, rather than the government. The companies then compete to make a profit. Technological developments such as the steam engine enabled the factory system to develop. Private entrepreneurs and industrialists took calculated financial risks and invested in new business ventures. This meant that goods could be produced quickly and in large quantities. They were then able to generate greater revenue by exporting their goods across the British Empire, which provided a readily available market.

Britain also had the advantage of possessing the raw materials necessary for industrialisation. Coal was vital because it was the fuel used by the newly developed steam engines. Britain had large coal deposits, unlike other European nations, and this easy access reduced energy costs.

Entrepreneurs

Entrepreneurs were vitally important to the Industrial Revolution. They were talented, ambitious and from the growing middle class. They took calculated risks and invested money in innovative technologies that increased the speed of production. This meant that profits increased as production costs were reduced.

Entrepreneurs also invested in the factories and mines necessary for industrialisation. With the advent of the steam engine, the increased need for coal made mining a profitable business venture. Richard Arkwright, one such entrepreneur, established the first spinning mill in Britain. Both talented and ambitious, Arkwright was able to turn new developments into a successful business.

Source 3

Invented by James Hargreaves in 1764, the spinning jenny sped up weaving production.

Advances in technology

The increase in population meant that the cottage industry (in which goods were created within people's homes or cottages) could not keep up with demand, which stimulated the invention of new machines to speed up production.

One such innovation was the spinning jenny. Invented in 1764 by James Hargreaves, it could spin eight threads at once rather than one single thread. The spinning jenny was small enough that it could fit into people's homes. Although this sped up production, spinners were not happy and Hargreaves' home was attacked by unemployed workers.

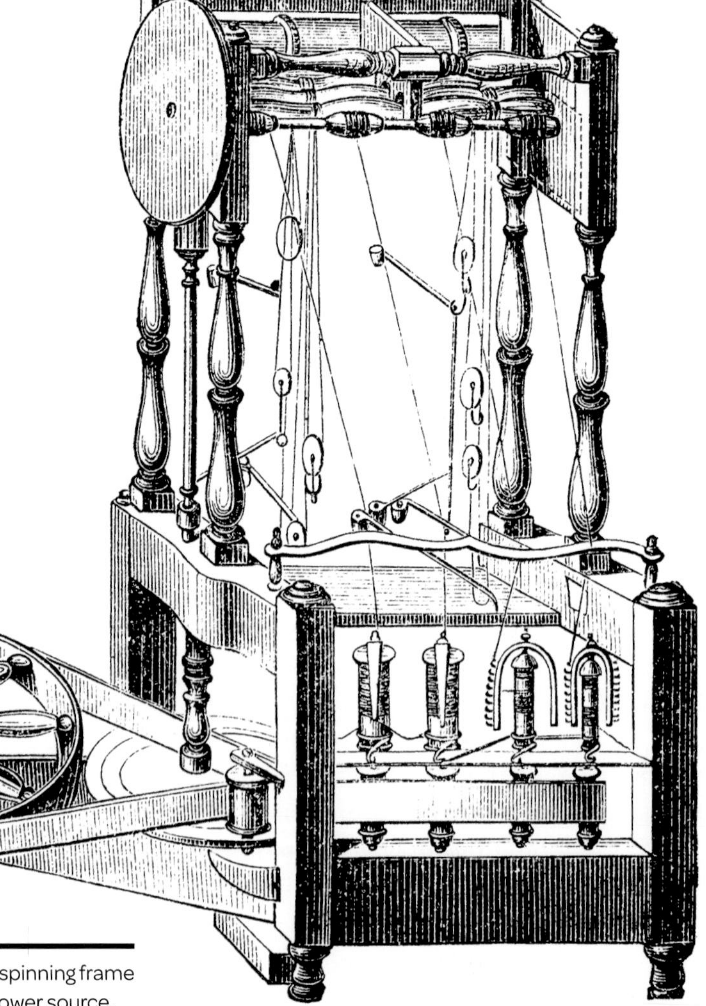

Source 4

Richard Arkwright designed the spinning frame in 1769, which used water as a power source.

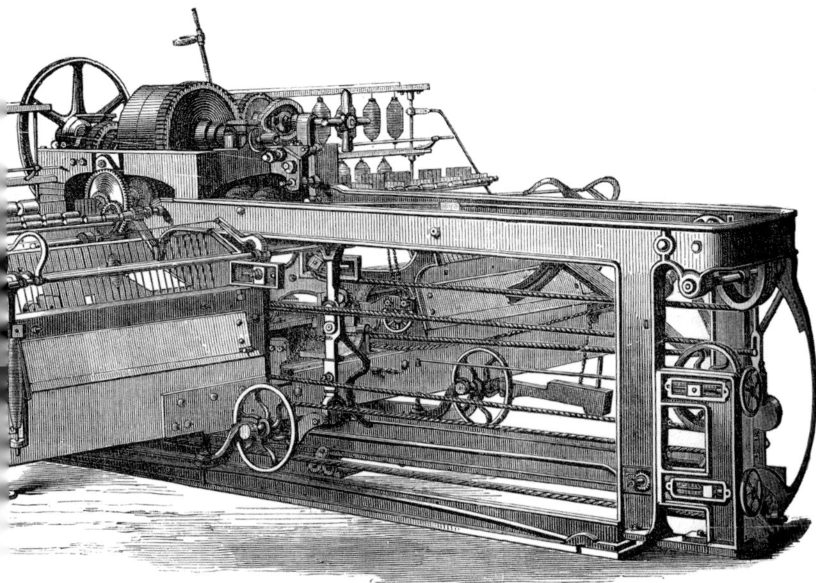

Source 5

The spinning mule, built by inventor Samuel Crompton in 1779, was used to spin cotton and other fibres in mills across Britain. It capitalised on the previous designs by Hargreaves and Arkwright to produce a better quality of thread in a higher volume.

Later spinning machines were far too big for people's homes, which created the need for factories. Richard Arkwright invented the spinning frame in 1769, which was powered by water and made much stronger thread. Samuel Crompton built on this idea in 1779 to produce thread that was both strong and fine with his spinning mule. By 1786, Edmund Cartwright had developed the power loom, which allowed even quicker cloth production.

Most of these machines could be operated by unskilled workers, with the exception of Crompton's mule. To begin with, machines were powered by water, but following the invention of the steam engine, steam was the dominant power source.

Sources of power

Before the Industrial Revolution, agriculture relied on human and horse power. Wind and water were used in the milling process. The cottage industry had run on human labour alone, so the most important development of the Industrial Revolution was the steam engine.

Early steam engines were developed by Thomas Savery (in 1698) and Thomas Newcomen (in 1712) and used to pump water from mines. In 1763, James Watt improved on their designs by adding a condensing cylinder, which made the steam engine more powerful. However, Watt's initial steam engine could not drive machinery. In 1781, Watt added a flywheel that allowed the engine to drive machinery. Known as the rotary steam engine, this revolutionised factories: they no longer needed to be built next to a strong, fast-flowing water source.

Learning ladder H1.3

Show what you know

1 List the ways that banks influenced the development of the Industrial Revolution.

2 Describe the technological changes that occurred during the Industrial Revolution.

3 Explain what *laissez-faire* means and how it applies to economics and business.

4 How did entrepreneurs contribute to the development of the Industrial Revolution?

Cause and effect

Step 1: I can recognise a cause and an effect

5 Match the cause with its effect.

Britain had no internal trading tariffs.	There was easy access to raw materials.
Bigger machines would not fit in people's homes.	They invested money, which helped industrialisation.
Britain had a large empire.	Goods could be produced more cheaply and more efficiently.
There were many entrepreneurs.	Trading was easy in Britain.
New inventions sped up production.	Factories were built.

Step 2: I can determine causes and effects

6 What industrial needs led James Watt to develop the steam engine?

Step 3: I can explain causes and effects

7 Economic factors were an important cause of the Industrial Revolution. Explain how the *laissez-faire* approach to the economy contributed to this great change.

Step 4: I can analyse causes and effects

8 List at least five of the significant causes of the Industrial Revolution in order of importance. Justify your choices.

Cause and effect, page 247

What was the effect of revolutionising transport?

One aspect of life greatly changed by the Industrial Revolution was transport. The first area of development was roads, followed by canals. However, it was the invention of railway and motor vehicles that completely revolutionised the movement of people and goods.

Roads

Travel around Britain was very slow at the start of the Industrial Revolution. Where possible, goods were transported by boat as it was considerably faster; for example, the journey from London to Edinburgh took a week by boat and 10 to 12 days by road.

Source 1

In the 1700s, some of the roads were of such poor quality, a person could travel more quickly walking between towns than taking a coach.

Source 2

The Rebecca Riots were protests against the turnpike trusts in Wales. Men dressed as women attacked gates and toll houses. It is thought the women's clothes were partly a disguse but also suggested the idea that women were entitled to act to defend their families.

To address this problem, the British government set up turnpike trusts. These groups charged local tolls and used the money for road resurfacing and repairs. This greatly increased the speed of travel, and most travellers were pleased with the improvements, especially merchants and manufacturers whose profits increased with quicker road travel. However, some protested against the new tolls. In Wales, a series of protests known as the Rebecca Riots occurred, in which men dressed as women attacked gates and toll houses (see Source 2). In response, the British parliament introduced a law stating that a person could be hanged for destroying turnpikes.

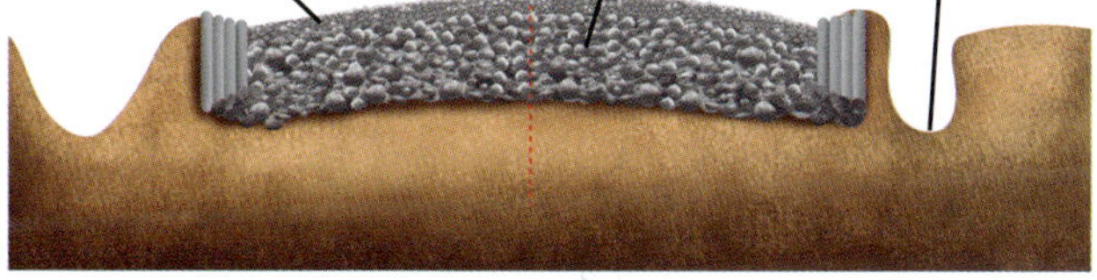

Gravel or broken stone (1-inch layer)
Centreline
Broken stone (8-inch layer)
Parallel drainage ditch

Source 3

John McAdam, a Scottish engineer, developed a system of road building in 1816 that improved the quality of roads. This diagram shows his innovative road design.

Canals

Canals were the next development. They allowed heavy goods such as coal to be transported across the country rather than via the coast. After the first canal was built, the price of coal in Manchester halved. Canals provided the best method to transport goods at a time when Britain's industry was growing. By 1840, merchants had funded the creation of over 6400 kilometres of canals.

Source 4

Canals were an effective way to move goods to and from factories.

Railways

It was the emergence of railways that completely revolutionised transport. The first railway line opened between Stockton and Darlington in 1825. Between 1830 and 1870, 11 200 kilometres of rail track were constructed. Transport became reliable and efficient.

Journey times from London (in hours)

Horse-drawn carriage		Rail
43	Edinburgh	12¼
24	Liverpool	6½
18	Exeter	4¾
11	Birmingham	3
6	Brighton	1¼

Source 5

The time difference between horse-drawn versus railway transport

The speed of rail transport meant that the British urban population was able to receive meat, fish, milk and vegetables while they were still fresh. This gave rise to the meal of 'fish and chips', which is still popular today.

Rail also had a huge impact on communication. Previously, newspapers were printed on a single sheet and transported by road, but now London newspapers could quickly reach cities around the country. The invention of the cylindrical printing press further increased the size of the newspapers, enabling double-sided printing. Thus, newspapers expanded in size and circulation.

Letters could also be sent and received quickly. This greatly improved communication, resulting in the spread of ideas, which increased people's interest in the issues of the time such as politics and social changes.

Source 6

The invention of rail transport led to goods being moved much more quickly and cheaply, which ultimately resulted in more jobs.

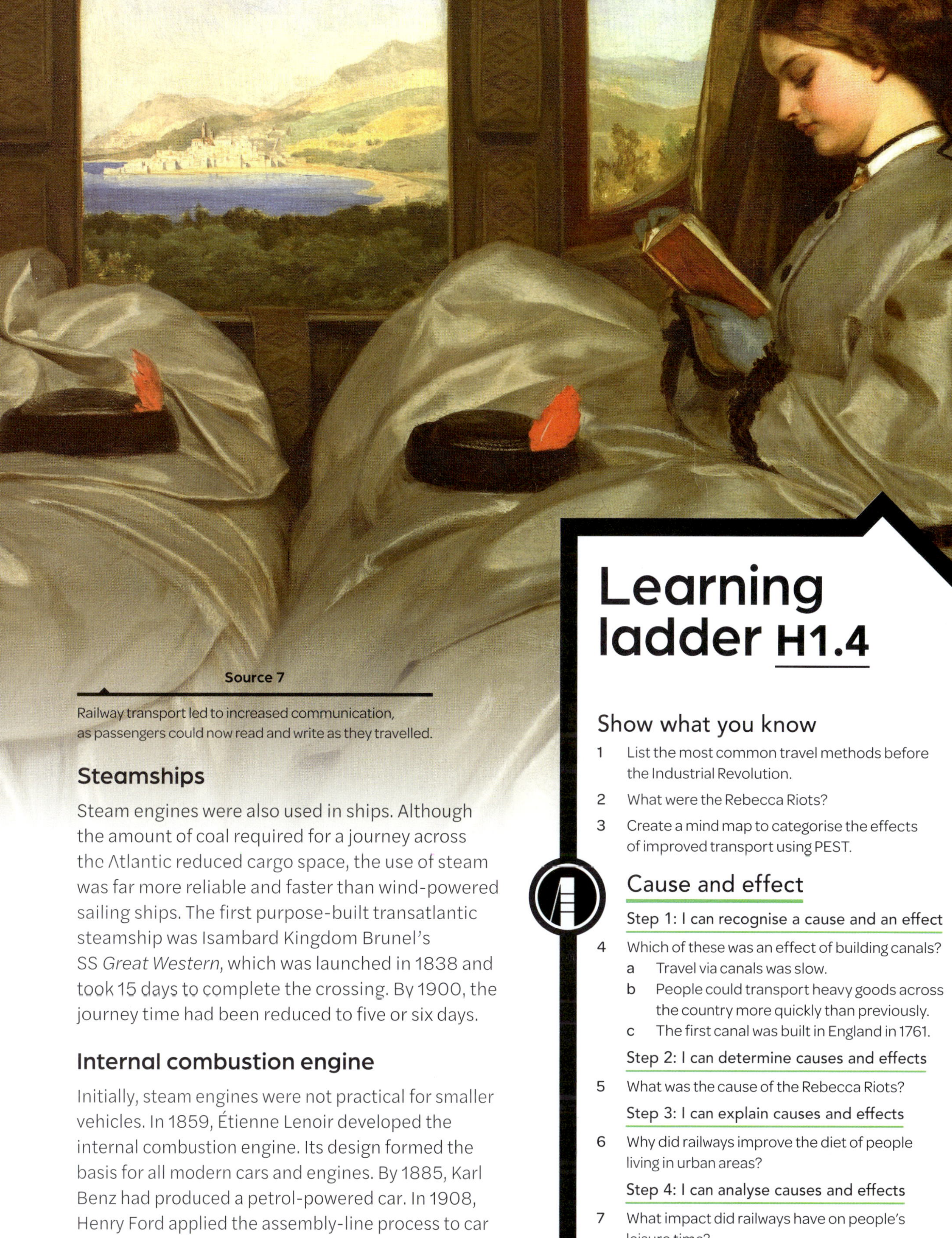

Source 7

Railway transport led to increased communication, as passengers could now read and write as they travelled.

Steamships

Steam engines were also used in ships. Although the amount of coal required for a journey across the Atlantic reduced cargo space, the use of steam was far more reliable and faster than wind-powered sailing ships. The first purpose-built transatlantic steamship was Isambard Kingdom Brunel's SS *Great Western*, which was launched in 1838 and took 15 days to complete the crossing. By 1900, the journey time had been reduced to five or six days.

Internal combustion engine

Initially, steam engines were not practical for smaller vehicles. In 1859, Étienne Lenoir developed the internal combustion engine. Its design formed the basis for all modern cars and engines. By 1885, Karl Benz had produced a petrol-powered car. In 1908, Henry Ford applied the assembly-line process to car manufacturing to produce the Model T Ford. Cars could now be manufactured cheaply and quickly, making them accessible to a wider market, not just the wealthy.

Learning ladder H1.4

Show what you know

1 List the most common travel methods before the Industrial Revolution.

2 What were the Rebecca Riots?

3 Create a mind map to categorise the effects of improved transport using PEST.

Cause and effect

Step 1: I can recognise a cause and an effect

4 Which of these was an effect of building canals?
  - a Travel via canals was slow.
  - b People could transport heavy goods across the country more quickly than previously.
  - c The first canal was built in England in 1761.

Step 2: I can determine causes and effects

5 What was the cause of the Rebecca Riots?

Step 3: I can explain causes and effects

6 Why did railways improve the diet of people living in urban areas?

Step 4: I can analyse causes and effects

7 What impact did railways have on people's leisure time?

HOW TO

Cause and effect, page 247

How did industrialisation change how people worked?

The Industrial Revolution changed people's lives radically. Before the 1740s, people in Britain lived a traditional rural lifestyle, with around 80 per cent of the population living and working in the countryside. Families worked together, and the hours of work revolved around what needed to be done on the farm. After the introduction of enclosure, many people were forced to move to urban areas to work in newly built factories.

Working conditions in factories

Work on farms involved hard physical labour from dawn until dusk. Machines, however, didn't require natural sunlight for work to be completed; nor did they require hard labour, as they were powered by water or steam. This meant that working hours increased. Workers in factories could work between 12 and 16 hours a day.

In order to make as much profit as possible, factory and mine owners paid very low wages. Workers were subject to dangerous working conditions, as machines had no safety guards and there was no government regulation. The dangerous machinery and repetitive nature of the work meant accidents were frequent and the death rate high. Women and children also worked in factories and were a source of cheap labour.

The advent of steam power also created a massive demand for coal. Although workers in mines were better paid than those working in mills, the working conditions were hard and dangerous. Deeper mines increased the risks of cave-ins, floods and explosions from using naked flames as light.

Source 1

Print from *The History of the Cotton Manufacture in Great Britain,* which was published in 1835 and written by Edward Baines. Baines was a newspaper editor for *The Leeds Mercury*. The newspaper was widely read by the mill owners in the north.

'I have a belt round my waist, and a chain passing between my legs, and I go on my hands and feet. The road is very steep, and we have to hold by a rope; and when there is no rope, by anything we can catch hold of. The pit is very wet where I work, and the water comes over our clog-tops always, and I have seen it up to my thighs; it rains in at the roof terribly. My clothes are wet through almost all day long. I have drawn till I have bathe skin off me; the belt and chain is worse when we are in the family way.'

Source 2

Testimony from a female coal-mine worker, Betty Harris, aged 37. Extract from *The First Report of Commissioners for Enquiring into the Employment and Conditions of Children in Mines and Manufactories*, 1842.

Cheap child labour

Factories

Many early mill owners employed children, as they were a very cheap source of labour as well as being small and agile. When threads broke during cloth manufacturing, children could fit underneath the machines to repair the breaks without the machine being turned off, even though this was dangerous.

Many children were 'apprentices' – orphans who were sent to work in the factories by local authorities. They had few rights and little protection.

Mines

In mines, children worked as trappers. Sitting in the dark all day, they opened and closed doors as older children pulling carts filled with coal passed through. Those carrying the coal to the surface often suffered physical deformities and long-term damage to their bodies.

'I'm a trapper in the Gawber pit. It does not tire me, but I have to trap without a light and I'm scared. I go at four and sometimes half past three in the morning, and come out at five and half past. I never go to sleep. Sometimes I sing when I've light, but not in the dark; I dare not sing then. I don't like being in the pit.'

Source 3

Testimony from Sarah Gooder, age 8, a young mine worker. Extract from *The First Report of Commissioners for Enquiring into the Employment and Conditions of Children in Mines and Manufactories*, 1842.

Source 4

Children worked in appalling conditions in the mines. The tunnels were often so low that they had to pull the carts in a bent position or on all fours, which later led to numerous health problems.

Child labour reform

Child labour had been common in England prior to the Industrial Revolution — most children had done agricultural or domestic work. However, greater numbers of children were entering the workforce at younger ages and being treated more harshly. Reformers began to demand change.

In 1833, the British government passed the *Factory Act* to improve conditions for children working in factories. The act stipulated that no children under nine years of age were allowed to work in textile factories. Children aged between nine and thirteen were only permitted to work for eight hours per day.

In 1840, Anthony Ashley Cooper, the seventh Earl of Shaftesbury, instigated the Royal Commission of Enquiry into Children's Employment, and a quote from his final report is shown in Source 5. The findings of the report were published in 1842 and included testimony from many child and female labourers (see Sources 2 and 3).

The report shocked society, and writers such as Charles Dickens published texts (such as *Oliver Twist* and *A Christmas Carol*), which protested against the use of child labour and workers' conditions. Legislation restricting the employment of children under the age of 10 in mines was passed in 1842.

'There is, however, one case of peculiar difficulty, viz., that in which all the subterranean roadways, and especially the side passages, are below a certain height, by the Evidence collected under this Commission, it is proved that there are coal mines at present in work in which these passages are so small, that even the youngest Children cannot move along them without crawling on their hands and feet, in which unnatural and constrained posture they drag the loaded carriages after them; and yet, as it is impossible, by any outlay compatible with a profitable return, to render such coal mines, happily not numerous nor of great extent, fit for human beings to work in, they never will be placed in such a condition, and consequently they never can be worked without inflicting great and irreparable injury on the health of the Children.'

Source 5

Thomas Tooke, T. Southwood Smith, Leonard Horner, Robert J. Saunders *Children's Employment Commission (Mines)* 1842, vol. XV, pp. 225–259

Learning ladder H1.5

Show what you know

1 List all the dangers people faced working in factories.
2 How many hours a day did factory employees work?
3 Why did factories change working conditions?
4 How might working in factories have changed family life?
5 Who might have been resistant to reforms to working conditions?
6 Compare differences in attitudes to health and safety practices between Industrial Revolution Britain and the present day.

Source analysis

Step 1: I can list specific features of a source

7 Using Sources 4 and 5, explain the work that children did in mines.

Step 2: I can find themes in a source

8 Consider Source 1. Why might this image present a positive perspective of working in textile factories?

Step 3: I can use the origin of a source to explain its creator's purpose

9 What was the purpose of the *Children's Employment Commission (Mines)* report of 1842?

Step 4: I can analyse a source

10 How reliable do you think the image shown in Source 4 is as a source of information about working conditions in mines? Provide evidence to support your answer.

Source analysis, page 240

H1.6

economics + business

How is the workplace changing today?

The introduction of machines during the Industrial Revolution saw people move from rural to urban areas to work in factories – a dramatic change. Today, technological changes wrought by artificial intelligence, robotics, nanotechnology and quantum computing are also transforming how we work.

Automation

Smarter machines are performing more and more tasks that were once performed by human workers. **Manufacturing** jobs that grew so rapidly in the Industrial Revolution have now been replaced by smart machines that perform roles such as assembly-line work. In the 1960s, more than one quarter of all workers were employed in manufacturing industries. Today, just 8 per cent of the Australian workforce is in the manufacturing sector and researchers predict that 40 per cent of Australian jobs currently performed by people could be automated by 2030.

To better understand the impact of automation on jobs, **economists** classify occupations as routine or non-routine. Routine jobs such as factory work are suited to automation, or to being produced more cheaply in another country; as a result, jobs such as machinery work and labouring have declined sharply in the last 25 years. Non-routine work, which requires the worker to be adaptable and solve problems, is less likely to be automated. In the last 25 years, most of the growth in jobs has come in non-routine professional occupations, as well as in the health and security industries.

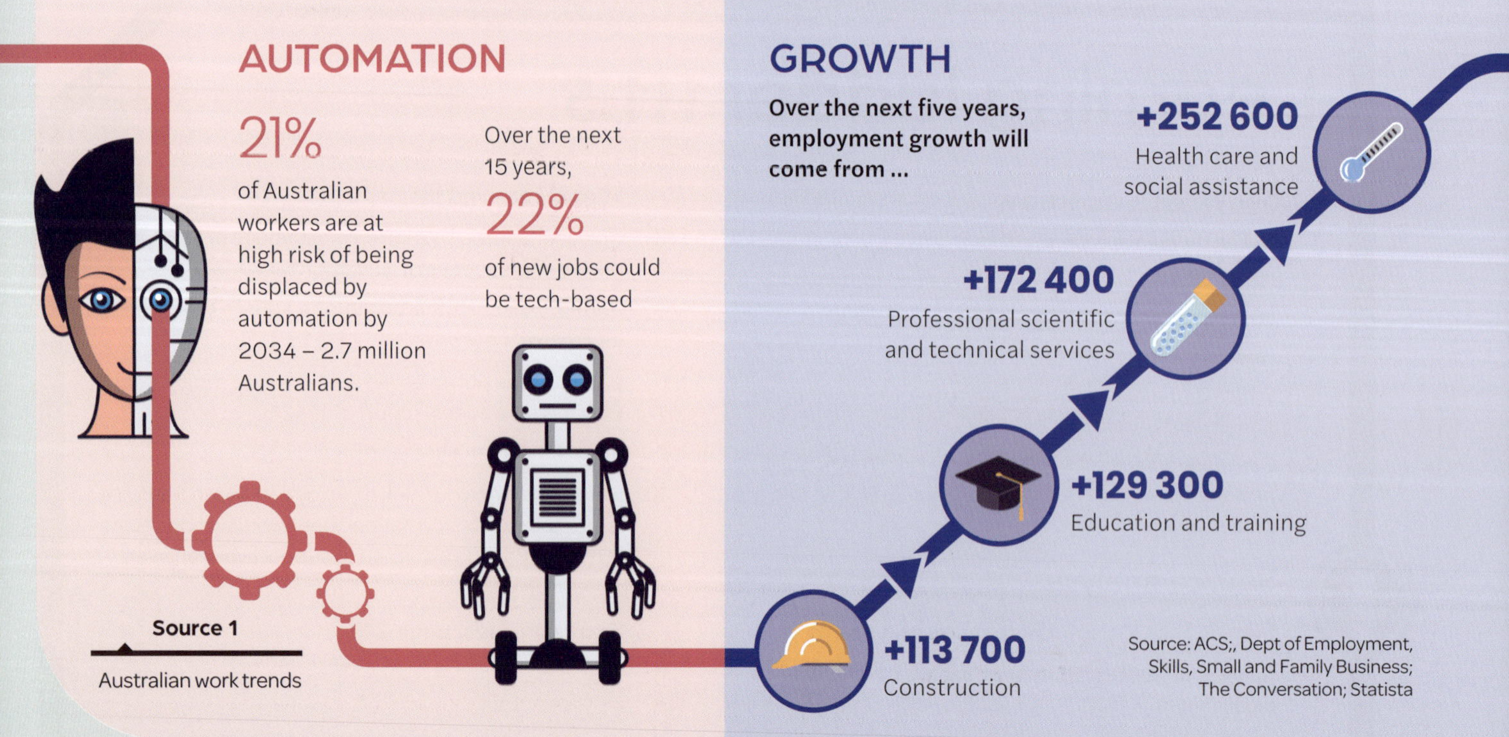

Source 1

Australian work trends

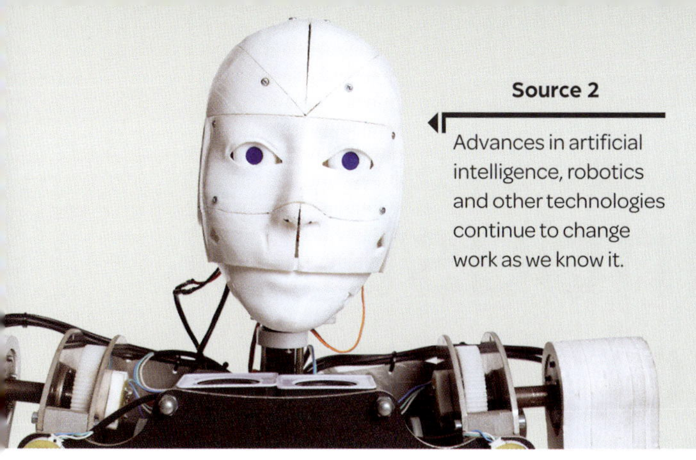

Source 2

Advances in artificial intelligence, robotics and other technologies continue to change work as we know it.

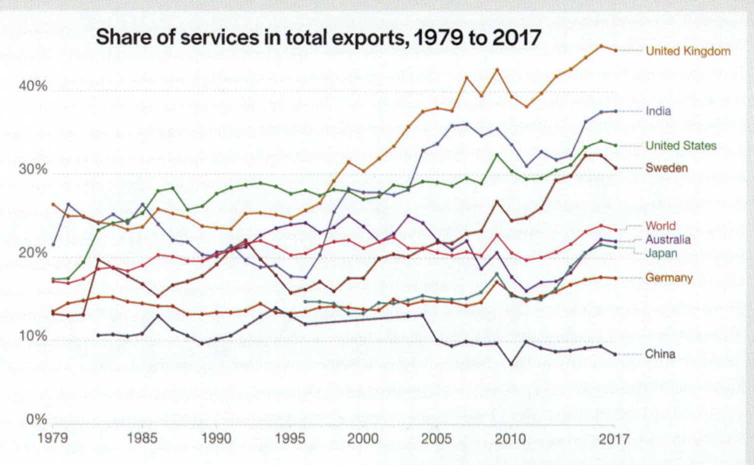

Source3

Share of global services from 1979 to 2017. After a dip around 2010, Australia's share is increasing steadily.

Globalisation

The **globalisation** of the workforce is a new **phenomenon** that has seen great changes in the Australian workforce.

Initially, manual roles in manufacturing and service roles in telecommunications were **outsourced** to countries with cheaper labour. Technology now allows companies to manage their entire labour supply internationally. Virtual global workers can provide services to an employer, but still remain in their local region to do the work. This enables companies to access cheaper labour, or tap into centres of specialised skills not available locally.

Economists suggest that more than one in ten of the world's service jobs can be performed from remote locations. Australians now compete with virtual talent from many other countries. There has been strong growth in Australia from overseas labour services that provide information technology and professional services. Australian businesses are sourcing virtual workers from countries such as India, the USA and the Philippines. In other job categories, such as financial services, the number of Australian workers allocated to work on international projects has grown dramatically.

Learning ladder H1.6

Economics and business

Step 1: I can recognise economic information

1 How many and what type of roles are under threat from greater automation by 2034?

Step 2: I can describe economic issues

2 Why has there been a greater loss of jobs in routine work due to automation compared to non-routine work?

Step 3: I can explain issues in economics

3 Outline the industries where Australia is a net importer and exporter of virtual labour. What changes have helped people to work remotely?

Step 4: I can integrate different economic topics

4 Compare the changes in the labour market during the Industrial Revolution to the changes happening today.

Step 5: I can evaluate alternatives

5 Are automation and globalisation bad for Australian workers? Explain.

How did the Industrial Revolution change living conditions?

Industrialisation meant that most opportunities for work lay in the cities compared to the country. By 1801, the urban population of England and Wales was 31 per cent of the total population; by 1851, 50 per cent of the population lived in cities, and by 1881 the figure had risen to 68 per cent. Houses were built quickly, cheaply and as close to factories as possible, as most people started work at 6 am and had to walk to work.

In order to make a profit, building companies constructed small, narrow, terraced houses, squeezing in as many dwellings as they could without much thought for their occupants.

The rapid development of towns and cities occurred without adequate government supervision, and no provision was made for sanitation. Houses did not have running water or indoor toilets. **Privies** were shared by many, and the waste was collected in cesspits. The cesspits frequently overflowed and contaminated drinking water supplies. These unsanitary conditions led to epidemics of cholera and typhus.

Source 1

The rapid development of towns without government regulation resulted in crowded living conditions and poor sanitation, which meant that diseases could spread quickly.

Source 2

A contemporary cartoon highlights factory workers' deplorable living conditions, which enabled diseases such as cholera to thrive and spread.

It was not only working conditions that concerned reformers, but also the conditions in which people lived. Flora Tristan, French socialist and women's rights advocate, wrote in her journal in 1842:

'Most of the workers lack clothing, a bed, furniture, a fire, wholesome food, and often even potatoes! They are shut up twelve to fourteen hours a day in mean rooms where they breathe in, along with foul air, cotton, wool, and linen fibers, particles of copper, lead iron, etc., and frequently go from insufficient nourishment to excessive drinking. These unfortunates are also pale, rickety, and sickly; they have thin, feeble bodies with weak arms, wan complexions, and dull eyes; one cannot help thinking that all them must have lung disease. ... In English factories there isn't any singing, chatting, or laughter, such as there is in ours. The master does not want his workers distracted for a minute by any reminders of life...'

Source 3

Source: From *Flora Tristan, Utopian Feminist: Her Travel Diaries and Personal Crusade*, Doris Beik (transl), Indiana University Press, 1993. Originally published *Promenades dans Londres*, Flora Tristan, 1840.

Reforms did not come about for another six years, when the British government, in response to a severe cholera epidemic, introduced the *Public Health Act* of 1848. This act was a milestone in public health history, as it forced local governments to take responsibility for drainage, water supplies and paving to improve the health of the population.

Learning ladder H1.7

Show what you know

1 What can you learn from Flora Tristan about the health of workers?
2 Why did building companies build houses near factories? What did this lead to?
3 Using Sources 1 and 2 , explain what living conditions were like.
4 Why did towns and cities develop so rapidly during this period?

Continuity and change

Step 1: I can describe continuity and change

5 Describe how industrialisation changed where people lived.

Step 2: I can explain why something did or did not change

6 Explain why living conditions were not subject to reform until the mid-19th century.

Step 3: I can explain patterns of continuity and change

7 Explain why living conditions changed during the 19th century. Provide evidence.

Step 4: I can analyse patterns of continuity and change

8 Why did many people move to new locations during the 19th century? Did everyone move? Why or why not?

Continuity and change, page 244

Who challenged the status quo?

Just as the Enlightenment brought about a new way of thinking based on reason and science, the Industrial Revolution gave rise to a number of free thinkers. Robert Owen, Friedrich Engels and Mary Wollstonecraft all challenged the social **status quo**, while Charles Darwin's scientific theories, seemingly radical at the time, are today widespread and generally accepted.

Robert Owen

Robert Owen became owner of the New Lanark mills in Scotland in 1799. As an advocate of **socialism**, he believed looking after the welfare of his workers was as important as making a profit for his company.

This contrast in attitude to many factory owners at the time was met with suspicion. Owen was highly critical of child labour and, following his purchase of the New Lanark mills, stopped the employment of children under 10. Instead, he set up a school for them to attend. Physical punishments were banned and the working hours of children over 10 were limited so they could continue their education.

Owen also provided his workers with housing and ensured they had access to doctors. Despite the fears of his partners that this would reduce profits, the New Lanark mills were a commercial success.

However, Owen alienated other factory owners when he began to call for mill profits to be shared with workers. He was also credited with devising the slogan '8 hours work, 8 hours recreation and 8 hours rest' that was adopted by the 8 Hours Movement in Victoria in 1856.

Source 2

In 2013, this street art image of Mary Wollstonecraft appeared on the wall of New Unity Unitarian Church where she worshipped in the 18th century.

Source 1

The New Lanark cotton mills, now a world heritage site

Friedrich Engels

German philosopher Friedrich Engels was a **socialist** and co-authored *The Communist Manifesto* with Karl Marx in 1848. Engels wrote:

> 'Nobody troubles about the poor as they struggle helplessly in the whirlpool of modern industrial life. The working man may be lucky enough to find employment, if by his labour he can enrich some member of the middle classes. But his wages are so low that they hardly keep body and soul together. If he cannot find work, he can steal, unless he is afraid of the police; or he can go hungry and then the police will see to it that he will die of hunger in such a way as not to disturb the equanimity of the middle classes.'
>
> **Source 3**
>
> Friedrich Engels, *The Condition of the Working Class in England*, 1844

Mary Wollstonecraft

Mary Wollstonecraft lacked a formal education, so she studied on her own and taught herself instead. She became a major British writer and philosopher, who advocated for equality and education for women. Her publisher, Joseph Johnson, held weekly dinners where she met radical thinkers such as Thomas Paine and William Godwin, whom she later married.

In her book, *A Vindication of the Rights of Woman* (1792), Wollstonecraft argued that men and women should be treated equally, and that if women were afforded the same opportunities and education as men, they could equally contribute to society. This led to the establishment of many women's rights groups across Europe and North America.

Charles Darwin

In 1859, Charles Darwin published *On the Origin of Species*. The English clergy saw the book and Darwin's theory of evolution as a threat to everything that mattered most to British society.

Darwin had just left university in 1831 when he was presented with an opportunity to sail around the world on a scientific expedition. Darwin set sail with the HMS *Beagle* in December 1831. His observations on the Galapagos Islands helped him formulate his ideas regarding evolution. Darwin theorised that different species of plants and animals had evolved to suit their environment to ensure their ongoing survival. He called this 'natural selection'. However, his theory shocked the Church, as it challenged the Christian doctrine that God had created all plants and animals.

Learning ladder H1.8

Show what you know

1. Identify the ways that Robert Owen looked after his workers.
2. Describe Mary Wollstonecraft's beliefs.
3. Summarise Darwin's beliefs on evolution.
4. Did Engels view industrialisation as positive or negative? Provide evidence from this spread.

Historical significance

Step 1: I can recognise historical significance

5. Which was more significant – Engels' *The Condition of the Working Class in England* or Darwin's *On the Origin of Species*?

Step 2: I can explain historical significance

6. Explain why Mary Wollstonecraft was a significant individual.

Step 3: I can apply a theory of significance

7. Apply Partington's Theory to explain the significance of Robert Owen and his treatment of workers.

Step 4: I can analyse historical significance

8. Explain why Darwinism was a historically significant idea, giving at least three reasons. Write a paragraph explaining how these reasons make Darwinism important.

Historical significance, page 251

What groups fought for social and political change?

The 19th century was a period of social inequality. While wealthy men had power, working class men were prohibited from voting and discouraged from forming trade unions, and women of all classes had very limited rights. These latter groups sought to make themselves heard during the Industrial Revolution, calling for social change.

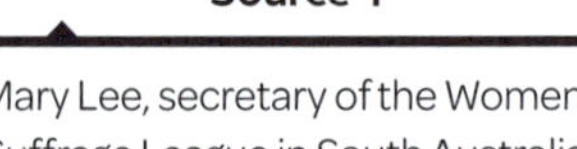

Source 1

Mary Lee, secretary of the Women's Suffrage League in South Australia

The women's suffrage movement

During the 19th century in Britain and Australia, women had only limited rights. Society dictated that a woman's place was in the home, raising her children and supporting her husband. Once she married, her legal rights were transferred to her husband. Many men believed that women did not have the intellectual capacity to understand politics and that they were too emotional to make important decisions. However, social attitudes began to change slowly.

In many nations, women began campaigning for **suffrage**, or the right to vote. In Britain, when their demands continued to be ignored by the British parliament, some campaigners became more militant and used violent tactics. These women became known as **suffragettes**.

In comparison, Australian **suffragists** used non-violent methods to attract support for their campaigns and were closely linked to the **temperance movement**. The first women's suffrage society was formed in Victoria in 1884, and in 1894 South Australia became the first place in the world to pass legislation allowing women to both vote *and* stand for parliament. British women would have to wait until 1918 for the vote.

Source 2

Inspired by success in countries such as Australia and New Zealand, women in the USA campaigned for suffrage also.

Source 3

The Tolpuddle martyrs

Trade unions

The changes brought about by the agricultural revolution and industrialisation had greatly affected people's lives and caused a great deal of social unrest. In 1799, the British government, fearing the influence of the French Revolution coupled with pressure from employers, banned **trade unions**.

Despite this repression, workers continued to meet illegally to fight for better working conditions, and the laws banning trade unions were repealed in 1824. However, workers still faced a long struggle for better working conditions, and the trade union movement grew throughout the 19th century.

Tolpuddle Martyrs

The fall in farm labourers' wages prompted George Loveless, a ploughman and Wesleyan preacher, to form a 'friendly society' to support the farm workers of Tolpuddle, Dorset. But, fearing that workers who had joined would be sacked, they decided to swear an oath of secrecy. It was this oath that allowed the government to arrest this particular trade union, having passed a law in 1797 prohibiting secret oaths.

In February 1834, six men were sentenced to seven years' transportation to Australia as an 'example to the working-class men'. However, there was such a public outcry against their sentence that they were pardoned in 1836 and given free passage home from Van Diemen's Land (Tasmania) in 1837.

Chartists

Chartism was a movement in Britain that demanded electoral reform in order to secure the vote for all working-class men. The Chartists named themselves after the *People's Charter*, an 1838 document that called for an end of the domination of the British political system by wealthy landowners, who were the only men eligible to become members of parliament.

Although the Chartists' petition of 1842 was rejected, they won several small victories: the property qualifications were lowered in 1867, and by 1884 two-thirds of all men could vote. However, a universal suffrage bill for all men was not passed until 1918.

Several key Chartist leaders, such as John Basson Humffray, were transported to Australia. They later became involved in Australian protests against unfair working conditions and demands for political equality, such as the Eureka Rebellion.

'... in the long run Chartism by no means failed ... the principles of the Charter have gradually become parts of the British constitution ... its restricted platform of political reform, though denounced as revolutionary at the time, was afterwards substantially adopted by the British State ... before all the Chartist leaders had passed away, most of the famous Six Points became the law of the land ... the Chartists have substantially won their case. England has become a democracy, as the Chartists wished, and the domination of the middle class ... is at least as much a matter of ancient history as the power of the landed aristocracy.'

Source 4

One interpretation of the impact of the Chartist movement on British politics came from Mark Hovell, from his book, *The Chartist Movement*, 1918.

'The movement's failures lay in the direction of securing legislation, or national approbation for its leaders. Judged by its crop of statutes and statues, Chartism was a failure. Judged by its essential and generally overlooked purpose, Chartism was a success. It achieved, not the Six Points, but a state of mind. This last achievement made possible the renascent trade union movement of the 'fifties, the gradually improving organization of the working classes, the Labour Party, the co-operative movement, and whatever greater triumphs labour will enjoy in the future.'

Source 5

A differing view on the Chartists came from Julius West in *A History of the Chartist Movement*, 1920, Constable & Company Ltd, pp. 294–95.

Source 6

Luddites protested against the use of machinery in favour of skilled workers, and from 1811–1817 gangs targeted and destroyed machines.

Source 7

The Chartists argued for six main points: votes for every man, a secret ballot for voting, payment of members or parliament, equal sizes of electoral districts, an annual election for parliament and the abolition of property qualifications for members of parliament.

Luddites

The Luddites were protesters who objected to the use of machinery in the textile industry, as they reduced the need for skilled spinners and weavers. From 1811 until 1817, gangs of Luddites set about smashing machines. In Yorkshire, they targeted shearing machines.

Severe punishments were introduced, and the British government made it a capital crime to wreck machines. This resulted in fewer riots after 14 Luddites were hanged in 1813 for attacking William Cartwright's mill.

Learning ladder H1.9

Show what you know

1 Identify the demands of the women's suffrage movement.

2 What did the Chartists want?

3 Why was the British government so swift to suppress the Luddites?

4 Research a trade union. How have its aims changed or stayed the same since the Industrial Revolution?

Historical interpretations

Step 1: I can recognise that the past has been represented in different ways

5 Which political reform group developed in the 1830s?

Step 2: I can describe historical interpretations

6 Consider Source 4. What is Hovell's interpretation of the success of the Chartist movement?

Step 3: I can explain historical interpretations

7 Consider Source 5. Explain why West's interpretation of the Chartist movement differs from Hovell's.

Step 4: I can analyse historical interpretations

8 Was the Chartist movement a failure? Provide evidence to support your response.

HOW TO

Historical interpretations, page 255

'Am I not a man and a brother?'

Slavery existed around the world long before the Industrial Revolution. But with the expansion of European empires, particularly in the Americas where cotton, tobacco and sugar were grown, those in power demanded more and more slaves to work the land.

The slave trade

By 1750, a third of the British merchant navy was involved in the slave trade. Goods that were desirable in Africa, such as manufactured goods, guns and gunpowder, were traded to African kings for slaves. Several kingdoms in western Africa grew rich from the slave trade, with British guns giving them an advantage in warfare over rival tribes. Populations fell dramatically as young, healthy children were removed from their families.

The journey between Africa and America was known as the 'middle passage' and took around five weeks. Slaves were crammed onto the ships, chained together and brought up on deck twice a day for food and exercise. Outbreaks of disease such as dysentery and smallpox could wipe out a quarter of the slaves before they reached their destination. Malnutrition and scurvy were common because of the poor food they received.

Slaves were considered property and nothing more. In 1782, the captain of the slave ship *Zong* ordered 130 sick slaves to be thrown overboard to prevent the spread of disease. He did this so that the owners would be able to claim insurance on the dead slaves.

Source 1

A painting of a sugar plantation worked by slaves in the British colony of Antigua, 1823

On arrival, the slaves would be inspected and then sold at auction. The best slaves would fetch £60 or more each. Others were sold for merely £1. Any slaves who were too weak or sick to sell were left to die on the waterfront. Once sold, slaves faced a lifetime of misery working from dawn until dusk on the sugar, cotton and tobacco plantations.

Source 3

Medallion produced by Josiah Wedgwood in support of the abolition movement. Benjamin Franklin declared that the medallion's effectiveness was 'equal to that of the best written Pamphlet, in procuring favour to those oppressed People'.

The abolition movement

The horrors of the slave trade resulted in a growing abolition movement. In May 1787, the Society for Effecting the Abolition of the Slave Trade was formed. Thomas Clarkson, a leading member of the anti-slavery campaign, investigated slavery conditions. He was supported by William Wilberforce, a British MP.

In 1807, the British government banned the sale of slaves throughout its empire; however, slavery itself was not banned until 36 years later.

'I am sure that the immediate abolition of the slave trade is the first, the principal, the most indispensable act of policy, of duty, and of justice that this country has to take. There are, however, arguments set up to [defend the slave trade]. The slave system, it is supposed, has taken such deep root in Africa that it is absurd to think of it being eradicated. "We are friends," they say, "to humanity. We are second to none of you in our zeal for the good of Africa – but the French will not abolish – the Dutch will not abolish. We wait, therefore, till they join us or set us an example." How, sire, is this enormous evil ever to be eradicated, if every nation waits?'

Source 2

Extract from a speech made by the British Prime Minister, William Pitt, 2 April 1792

Learning ladder H1.10

Show what you know

1. Identify the goods that were traded for slaves.
2. Why did Britain's leaders claim that slavery was necessary to sustain the Empire?
3. Suggest a possible impact of the British Empire banning the sale of slaves in 1807, but not banning slavery itself for another 36 years.

Source analysis

Step 1: I can list specific features of a source

4. List the reasons Pitt argued to abolish slavery.

Step 2: I can find themes in a source

5. Consider Sources 2 and 3. What ideas are similar in both?

Step 3: I can use the origin of a source to explain its creator's purpose

6. Why did Pitt make his speech in April 1792?

Step 4: I can analyse a source

7. How reliable is Pitt's speech in Source 2 regarding views on the slave trade?

Source analysis, page 240

How did the Industrial Revolution affect the environment?

The rapid development that occurred during the Industrial Revolution came at a great cost. Industrialisation had, and continues to have, a devastating environmental impact: it put pressure on natural resources as land was cleared, rivers were dammed and as pollution increased from the burning of fossil fuels (such as coal, oil and natural gas).

Source 1

A 'pea souper': a very thick, yellow, green or black fog that occurred during the 19th and 20th centuries because of air pollution from coal fires. This image depicts the thick fog at Ludgate Circus, London, in the late 19th century.

Deforestation

At the beginning of the Industrial Revolution, forests were cleared for timber and to increase the land available for farming. But as industrialisation gained momentum, land was cleared for factories and to make way for urban development.

Deforestation still occurs, as economic factors further drive the need for agricultural land. This has extended the impact on the environment because land degradation can lead to problems such as erosion and loss of habitat. These further endanger vulnerable species and increase the risk of natural disasters, such as flooding and landslides.

Air pollution

The invention of steam-driven machines created a demand for coal. The negative impact of this was that the smoke from coal-driven factories created significant air pollution. When the smoke mixed with fog, it created a deadly smog. During the 1800s this led to numerous problems in London, such as traffic accidents and the deaths of animals at markets. It continued to be a problem into the mid-20th century; in 1952, five days of thick smog led to the deaths of thousands of Londoners. Today, countries such as China and India continue to suffer from poor air quality as a result of reliance on fossil fuels for manufacturing.

In the past 60 years, it has become apparent that the Earth's climate is changing. The increased levels of carbon dioxide, produced when fossil fuels are burnt, have resulted in **global warming**. This has led to climatic changes such as warmer temperatures, rising sea levels and melting glaciers, all of which are having a significant impact. However, our increased awareness of the damage being done to the planet has resulted in increased efforts to control emissions.

Source 3

The infamous 1952 smog almost brought London to a standstill and led to the death of thousands.

Water pollution

Not only has the air suffered significant pollution, but water supplies are often affected. During the Industrial Revolution, lack of sanitation led to water supplies becoming polluted, and to dangerous epidemics of diseases such as cholera, typhus and typhoid.

Waste from industries also contaminated water, which has led to further environmental problems. An example of this is the King River in Tasmania, where toxic waste from copper mining (begun in 1880) led to deforestation.

'And what cities! ... smoke hung over them and filth impregnated them, the elementary public services – water supply, sanitation, street-cleaning, open spaces, and so on – could not keep pace with the mass migration of men into the cities, thus producing, especially after 1830, epidemics of cholera, typhoid and an appalling constant toll of the two great groups of 19th century urban killers – air pollution and water pollution or respiratory and intestinal disease.'

Source 2

Extract from E.J. Hobsbawm, *The Pelican Economic History of Britain, Volume 3, Industry and Empire*, Harmondsworth, 1969, p. 86

Learning ladder H1.11

Show what you know

1 Describe the ways that the Industrial Revolution had an impact on the environment.

2 What does Source 2 tell you about epidemics?

3 Think about the different industries you have studied. Create a concept map to show the long-term environmental impacts they have had.

4 How have attitudes towards the environment changed since the Industrial Revolution?

Cause and effect

Step 1: I can recognise a cause and an effect

5 What causes smog?

Step 2: I can determine causes and effects

6 What has been the effect of deforestation on the environment?

Step 3: I can explain causes and effects

7 Explain the long-term effects that the Industrial Revolution have had on the environment.

Step 4: I can analyse causes and effects

8 Which environmental impact has been the most significant? Provide evidence to support your answer.

Cause and effect, page 247

Why was Australia riding on the sheep's back?

Australia was founded as a penal colony to solve the problem of overcrowded prisons in Britain. Rapid industrialisation and urbanisation had resulted in rising crime rates among poverty-stricken workers, many of whom were new to urban life. The British brought the technology of the Industrial Revolution with them to Australia; within a hundred years, manufacturing had developed to the point of self-sufficiency in New South Wales and Victoria.

Industry in the Australian colony was initially dependent on human and wind power. However, that soon changed as Australia benefited from the rapid technological advances being made in Britain. The first steam engine to be used in Australia powered a flour mill in Sydney in 1813. Reliant at first on British technology, steam engines began to be manufactured in Australia from the 1830s.

Source 1

Ex-convict George Howell and his son George Jr. built Howells' Mill in Parramatta in 1828. The combined wind and water mill was the largest in the area at the time.

Source 2

Slums were common in inner-city suburbs such as Fitzroy and Collingwood.

A number of factors encouraged the development of Australian manufacturing: the length of time it took for manufactured goods to reach Australia, machines needing to be adapted to Australian conditions, and a lack of interest from the British government in responding to the colonies' needs forced colonists to develop their own industries to meet local demands.

In contrast to Britain, there was no shift from rural to urban areas. Factories using steam power could be established anywhere, so manufacturing industries were established in capital cities, where communications and infrastructure were already beginning to develop.

Although Melbourne had been laid out in a wide street grid by Robert Hoddle, its rapid growth meant that slums developed next to the factories of inner Melbourne. These houses were squalid and ramshackle, with no running water, toilets or sewers. Waste was thrown onto the streets and flowed into street channels. This, along with the waste from factories and manure from horses (the main mode of transport at the time), was all eventually washed into the Yarra River. By the 1840s, the Yarra was so polluted that it could not be used for domestic purposes.

Source 3

Despite Melbourne's streets being well laid out, open drains, human sewage and manure from horses polluted the city and created an unhealthy environment.

Gold and industry

The discovery of gold brought a flood of prospectors in the mid-19th century. As surface gold dwindled, miners used steam engines to reach gold deeper underground. Those who had profited from the gold rush also began to invest. This aided the development of local industry and manufacturing which, in turn, sustained cities such as Ballarat and Bendigo, which no longer needed to rely on gold to continue to thrive.

The wool industry

Pastoralist John Macarthur introduced merino sheep to Australia. These sheep, used to the warm climates of Spain, flourished in Australia. By 1815, Australia was exporting 30 tonnes of wool per year. By 1849, this had increased to around 14 800 tonnes. The prosperity brought by wool exports gave rise to the idiom that Australia was 'riding on the sheep's back'.

The gold rush forced sheep farmers to enclose their land, as workers sought their fortunes on the gold fields. Developments in mechanical sheep-shearing meant that, by 1887, it took just a few days to learn how to shear. These mechanical shears were powered first by steam and later by electricity.

Railways in Australia

Victoria was the first colony in Australia to open a railway line. It opened on 12 September 1854, and was a short track linking Flinders Street with Port Melbourne. The development of railways in Australia brought the same benefits to Australia that it had to Britain. Where railways went, towns prospered, and social and economic links developed between rural and urban Australia.

A national rail network was planned from 1855, but this was hampered by the fact that each colony used different gauge systems. This resulted in passengers and freight having to transfer from one train to another at state borders. It was not until 1955 that Australia's mainland interstate track was standardised. Victoria had the most developed rail network and, by 1891, 4670 kilometres of track had been constructed.

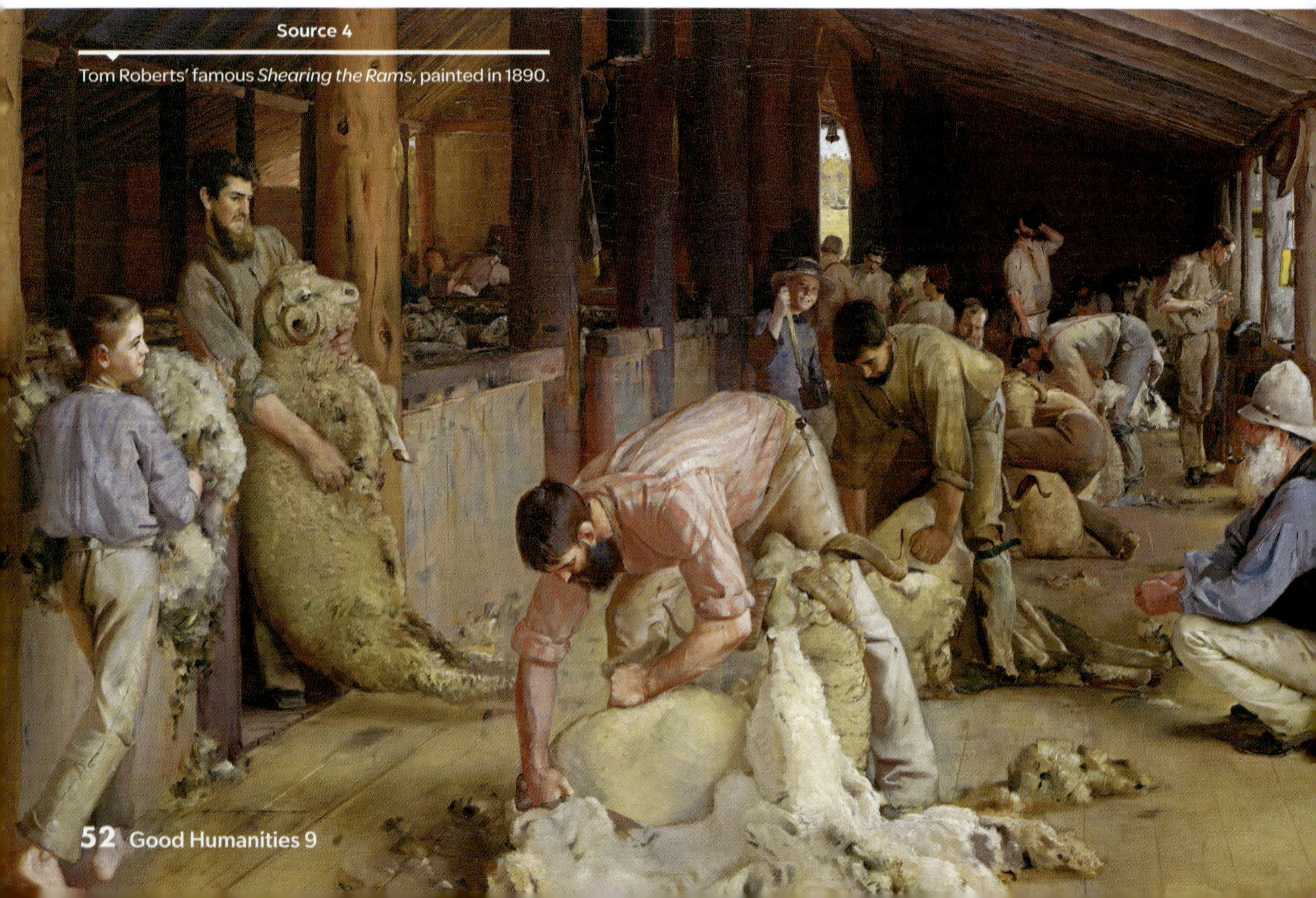

Source 4

Tom Roberts' famous *Shearing the Rams*, painted in 1890.

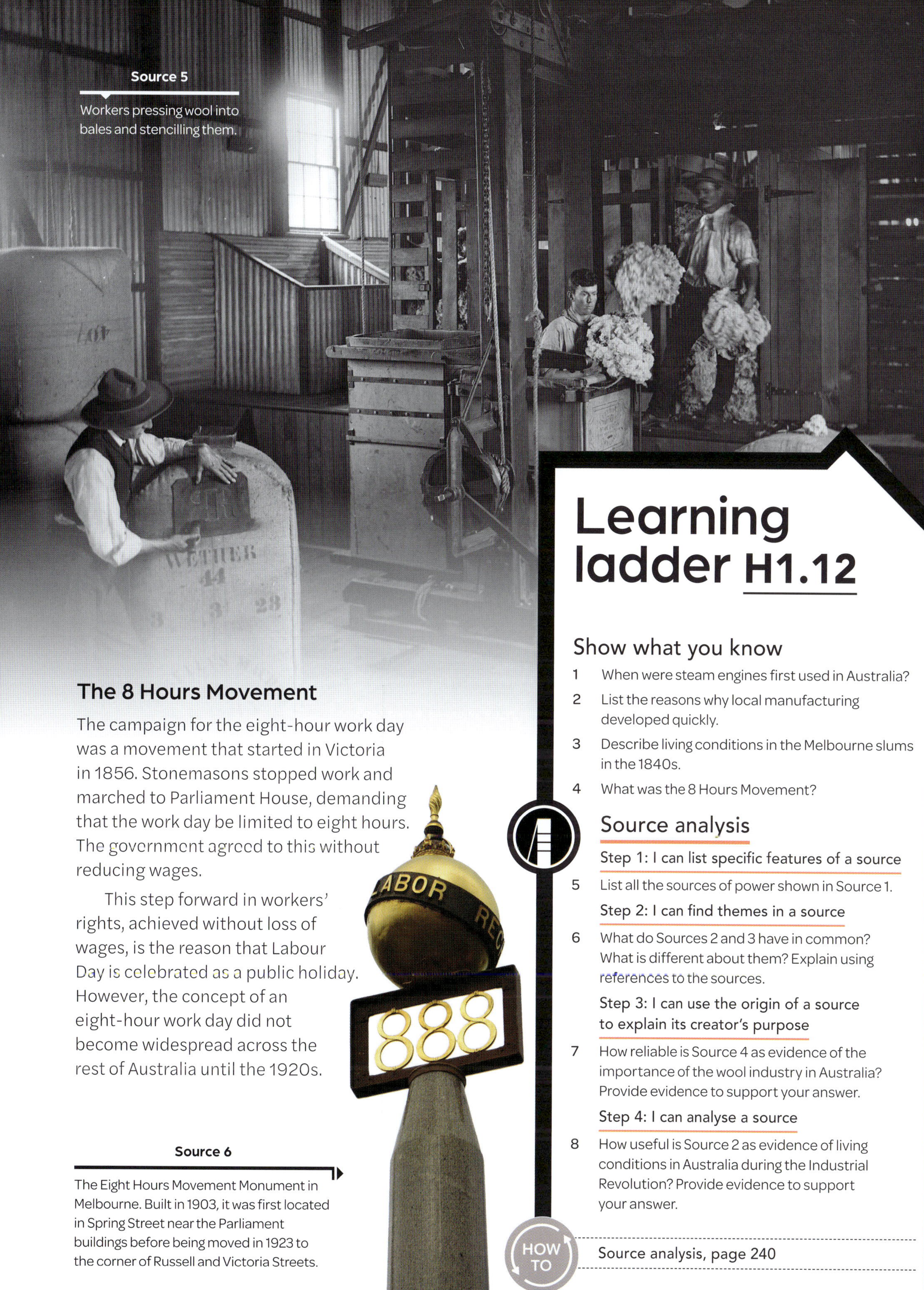

Source 5

Workers pressing wool into bales and stencilling them.

The 8 Hours Movement

The campaign for the eight-hour work day was a movement that started in Victoria in 1856. Stonemasons stopped work and marched to Parliament House, demanding that the work day be limited to eight hours. The government agreed to this without reducing wages.

This step forward in workers' rights, achieved without loss of wages, is the reason that Labour Day is celebrated as a public holiday. However, the concept of an eight-hour work day did not become widespread across the rest of Australia until the 1920s.

Source 6

The Eight Hours Movement Monument in Melbourne. Built in 1903, it was first located in Spring Street near the Parliament buildings before being moved in 1923 to the corner of Russell and Victoria Streets.

Learning ladder H1.12

Show what you know

1 When were steam engines first used in Australia?

2 List the reasons why local manufacturing developed quickly.

3 Describe living conditions in the Melbourne slums in the 1840s.

4 What was the 8 Hours Movement?

Source analysis

Step 1: I can list specific features of a source

5 List all the sources of power shown in Source 1.

Step 2: I can find themes in a source

6 What do Sources 2 and 3 have in common? What is different about them? Explain using references to the sources.

Step 3: I can use the origin of a source to explain its creator's purpose

7 How reliable is Source 4 as evidence of the importance of the wool industry in Australia? Provide evidence to support your answer.

Step 4: I can analyse a source

8 How useful is Source 2 as evidence of living conditions in Australia during the Industrial Revolution? Provide evidence to support your answer.

HOW TO

Source analysis, page 240

Masterclass

Learning ladder

Work at the level that is right for you or level-up for a learning challenge!

Source 1

George Pinwell, 'Death's Dispensary', *Fun Magazine III*, August 18,1866

'Richard Arkwright, whose spinning machines revolutionised the manufacture of cotton, was, perhaps even more importantly, a business genius of the first order. The founder of the modern factory system, he was the creator of a new industrial society that transformed England from a nearly self-sufficient country, her economy based on agriculture and domestic manufactures, into the workshop of the world.'

Source 2

Extract from R.S. Fitton, *The Arkwrights: Spinners of Fortune*, 1989

Is this a holy thing to see,
In a rich and fruitful land,
Babes reduced to misery,
Fed with cold and usurous hand?
Is that trembling cry a song?
Can it be a song of joy?
And so many children poor?
It is a land of poverty!
And their sun does never shine.
And their fields are bleak & bare.
And their ways are fill'd with thorns.
It is eternal winter there.

Source 3

An extract from the poem 'Holy Thursday', by William Blake, *Songs of Innocence*, 1789

Step 1

a **I can list specific features of a source**

Identify all the details you can see in the cartoon in Source 1.

b **I can describe continuity and change**

How did Britain's agricultural industry change between 1750 and 1850? What aspects of the industry, if any, stayed largely the same?

c **I can recognise a cause and an effect**

Which of the following effects was mostly likely to have been caused by Britain's population increasing?

- Transportation became faster.
- Production became more efficient.
- Agricultural methods changed to grow/harvest more food.

d **I can recognise historical significance**

Place the following developments in order from most important to least important.

- Isambard Kingdom Brunel built railways lines, bridges and the first iron ships.
- James Watt's steam engine could power machines in 1781.
- Railways brought fresh foods into towns.
- Samuel Morse developed the telegraph in 1837.
- Karl Benz produced the first automobile.

e **I can recognise that the past has been represented in different ways**

Read Source 2. To what does the author attribute the Industrial Revolution?

Step 2

a **I can find themes in a source**

Read Source 3, 'Holy Thursday' by William Blake. What do you think the themes of the poem are?

b **I can explain why something did or did not change**

Why did pastoralists in Australia enclose their land?

c **I can determine causes and effects**

What conditions in factories caused the *Factory Act* to be passed in 1833?

Source 4

A cartoon depicting the slums of Melbourne's underbelly

BEHIND THE SCENES

d **I can explain historical significance**

Why was the invention of the telegraph significant for Australia?

e **I can describe historical interpretations**

Consider Source 2. Describe what Fitton concludes was Richard Arkwright's contribution to the Industrial Revolution.

Step 3

a **I can use the origin of a source to explain its creator's purpose**

Look at the cartoon in Source 4. Why do you think this source was produced?

b **I can explain patterns of continuity and change**

Explain how the environment has changed over time because of the Industrial Revolution.

c **I can explain causes and effects**

Explain why the Chartist movement developed.

d **I can apply a theory of significance**

Consider James Hargreaves' spinning jenny.

i How important was the jenny to weavers at the time?

ii How deeply were the lives of weavers affected by this innovation?

iii Which different people would have been affected by this invention?

iv How long do you think the the effects of the spinning jenny lasted?

v Do people still use the spinning jenny today?

e **I can explain historical interpretations**

Consider Source 2. Why does the author think that Richard Arkwright was the creator of a new society?

Masterclass

Step 4

a I can analyse a source

How reliable is William Blake's depiction of British life in Source 3?

b I can analyse patterns of continuity and change

Why did industrialisation initially result in shorter life expectancies? How has this changed over time?

c I can analyse causes and effects

Why did Britain industrialise before other countries? Use historical evidence to support your explanation.

d I can analyse historical significance

What impact did the trade union movement in Britain have upon workers' rights in Australia?

e I can analyse historical interpretations

Source 2: Why does the author believe Arkwright was important to England's development as an industrial nation?

Step 5

a I can evaluate a source

How useful is Source 3 for studying living and working conditions in Industrial England?

b I can evaluate patterns of continuity and change

What aspects of life do you think were lost by the coming of the railway age?

c I can evaluate causes and effects

How did new ideas and technological developments cause changes during the 18th and 19th centuries?

d I can evaluate historical significance

Which invention was more significant: the steam engine or the internal combustion engine? Provide evidence to support your answer.

e I can evaluate historical interpretations

In Source 2, the author suggests that Richard Arkwright was the creator of a new industrial society. To what extent do you think this is an accurate interpretation?

Historical writing

1 Structure

Imagine you are given the essay topic, 'What were the social effects of industrialisation in the 18th and 19th centuries?'. Write an essay plan for this topic. Include at least three main paragraphs.

2 Draft

Using the drafting and vocabulary suggestions on page 262, draft a 600–800 word essay (at least 30–40 sentences) responding to the topic.

3 Edit and proofread

Use the editing and proofreading tips on page 263 to help edit and proofread your draft.

Historical research

4 Organise and present information

Imagine you are completing a research project on 'Living conditions in London during the Industrial Revolution'. Write a contents page for this project. There should be an introduction, a conclusion, at least four main sections and many subsections. Number your chapters.

Capstone

How can I understand the Industrial Revolution?

In this chapter, you have learnt a lot about the Industrial Revolution. Now you can put your new knowledge and understanding together for the capstone project to show what you know and what you think.

In the world of building, a capstone is an element that finishes off an arch or tops off a building or wall. That is what the capstone project will offer you, too: a chance to top off and bring together your learning in interesting, critical and creative ways. You can complete this project yourself, or your teacher can make it a class task or a homework task.

mea.digital/GHV9_H1

Scan this QR code to find the capstone project online.

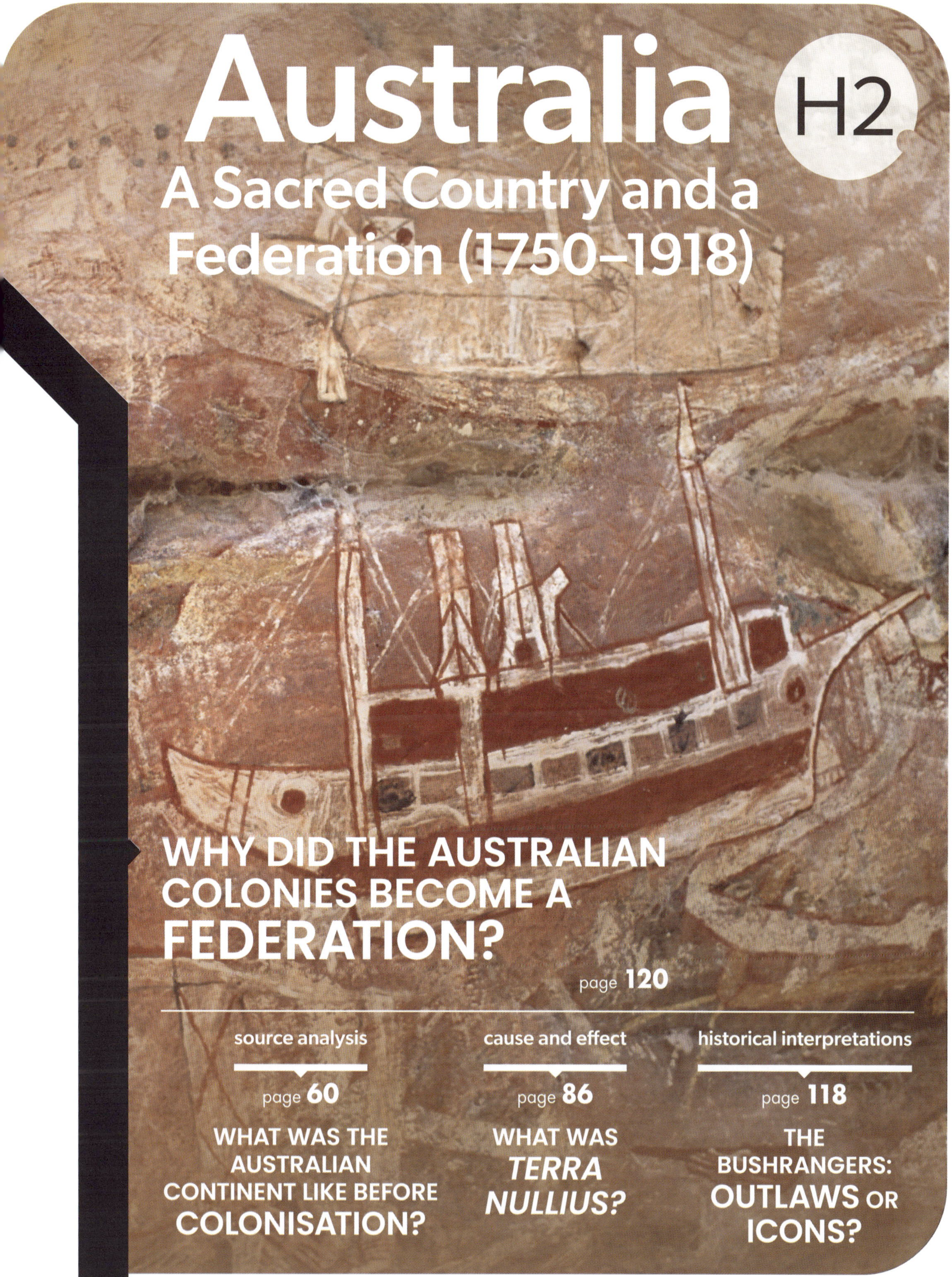

Australia H2

A Sacred Country and a Federation (1750–1918)

WHY DID THE AUSTRALIAN COLONIES BECOME A FEDERATION?

page 120

source analysis

page 60

WHAT WAS THE AUSTRALIAN CONTINENT LIKE BEFORE COLONISATION?

cause and effect

page 86

WHAT WAS *TERRA NULLIUS?*

historical interpretations

page 118

THE BUSHRANGERS: OUTLAWS OR ICONS?

How can we understand Australia?

In 1750 Australia was, as it still is, a sacred place to First Nations Peoples, who thrived in one of the most harsh and dry continents on Earth. Colonisation brought them into contact and conflict with an industrial empire undergoing massive political, economic and technological changes. The society that developed eventually became a federation. However, Australia is still dealing with the ongoing effects of colonisation today.

learning ladder

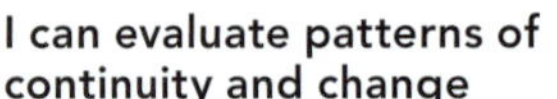

	Source analysis	Continuity and change	Cause and effect
step 5	**I can evaluate a source** I can present a judgement on the usefulness of a source based on its strengths, weaknesses and limitations. I can determine whether information is missing about the event or person the source refers to.	**I can evaluate patterns of continuity and change** I answer the question 'So what?' about patterns of continuity and change. I weigh up different aspects and debate the importance of continuity or change.	**I can evaluate causes and effects** I answer the question 'So what?' about cause and effect. I weigh up different things and debate the importance of a cause or an effect.
step 4	**I can analyse a source** I can use my own knowledge to determine the reliability of a source and can explain whether it shows a one-sided view.	**I can analyse patterns of continuity and change** I can look deeper into patterns of continuity and change and determine the factors that contribute to them.	**I can analyse causes and effects** I don't just see a cause or an effect as one thing. I can determine the factors that make up causes and effects.
step 3	**I can use the origin of a source to explain its creator's purpose** I combine knowledge of when and where a source was created to answer the question, 'Why was it created?'.	**I can explain patterns of continuity and change** I can see beyond individual examples of continuity and change between historical periods and explain broader patterns.	**I can explain causes and effects** I can answer 'How?' or 'Why?' a cause led to an effect in Australia.
step 2	**I can find themes in a source** I look a bit closer at a source and find more than just features. I find themes and patterns in a source.	**I can explain why something did or did not change** I can give a reason for why something changed or why it stayed the same.	**I can determine causes and effects** Applying what I have learned about Australia, I can describe what the cause or effect of an event was.
step 1	**I can list specific features of a source** I can look at an Australian source and list the details I can see in it.	**I can describe continuity and change** I recognise what has stayed the same and what has changed from the colonial period until now.	**I can recognise a cause and an effect** From a supplied list, I can recognise factors that were causes or effects of each other in Australian history.

Source 1

Warlugulong, 1976, by Clifford Possum Tjapaltjarri and Tim Leura Tjapaltjarri. Sacred creation stories have many layers that connect people to sacred sites including history, law, lore, spirituality and culture.

This painting depicts a number of the sacred stories of the Anmatyerr people, including stories about the Creation Spirit Lungkata, the blue-tongue lizard man, creating the first big fire. Another story shown here is that of Lungkata's sons, who broke customary law by not sharing their hunted kangaroo with him and were burned.

Warm up

Source analysis

1 Source 1: What different perspectives does *Warlugulong* present as a historical source?

Continuity and change

2 Source 1: How does the painting, its story and its current context reflect aspects of continuity and change?

Cause and effect

3 Source 1: What was the result of Lungkata's sons' behaviour?

Historical significance

4 Consider a historically significant event in Australia's history. Is this event changing in significance as society's values shift? If so, how and for whom?

Historical interpretations

5 Source 1: What does the event described show about Lungkata? What does it show about his sons' views? What could be the message or teaching within the source and the lore associated with it?

Historical significance	Historical interpretations
I can evaluate historical significance I answer the question 'So what?' about things that are supposedly important in the history of Australia. I weigh up factors against one another and can cast doubt on how important things are.	**I can evaluate historical interpretations** I can weigh up the different historical interpretations that have been formed. I debate and challenge the interpretations that have presented.
I can analyse historical significance I can separate the various factors that make something historically important in the history of Australia.	**I can analyse historical interpretations** I can determine the factors that have led to why a historical interpretation has been formed.
I can apply a theory of significance I know a theory of significance. I use it to rank importance of changes, causes, effects and events in the history of Australia.	**I can explain historical interpretations** I can answer 'Why?' or 'How?' there are different interpretations of people and events in the past.
I can explain historical significance I answer the question 'Why?' about what was important in Australian history.	**I can describe historical interpretations** I can provide different examples to show how people and events in the past have been interpreted.
I can recognise historical significance When shown a list of facts about Australian history, I can work out which are important.	**I can recognise that the past has been represented in different ways** I can identify different views of people and events in the past.

PART I: A SACRED COUNTRY (1750–1770)

What was the Australian continent like before colonisation?

The Australian landscape in 1750 was rugged yet beautiful. Much of the land was arid and unyielding. The country was sacred to the First Nations Peoples of Australia, and they managed it sustainably.

The First Nations Peoples of Australia, the Aboriginal and Torres Strait Islander Peoples, were highly skilled in dealing with environmental challenges. These included significant droughts, fires and floods.

In recent years, new evidence has shown that First Nations Peoples had semi-permanent residences, economic trade routes, technology, and land-management and food-production methods. Many historians now recognise that 18th-century First Nations Peoples were advanced cultures with considerable populations.

Source 1

First Nations Peoples' land management

Source 2

First Nations Peoples' food production included using eel traps to harvest eels.

The search for the great southern land

In 150 CE, Ptolemy of Alexandria proposed that there was a hypothetical *terra australis nondum cognita*, a 'great unknown southern land'. The European quest for a mythical 'great southern continent' began in earnest when Captain Luis Váez de Torres sailed the Torres Strait in 1606.

Many European nations searched the Pacific Ocean for the southern continent. Some reached Australia, and attempted to claim or **annex** portions of the land on behalf of their nations. Ultimately, the British colonised the land and displaced its original inhabitants.

key ideas timeline.

For more than 65 000 years, First Nations Peoples inhabit the land.

Evidence of First Nations Peoples' history includes:
- semi-permanent residences
- land management
- food-production methods.

1606

Luis Váez de Torres (Spanish) lands on the Islands and Strait that still bear his name

Willem Janszoon (Dutch) lands on the Gulf of Carpentaria

1616

Dirk Hartog (Dutch) lands on the west coast of Australia and leaves an engraved pewter plate

Source 3

The beautiful Australian outback is also subject to extreme weather, making it tough for those who do not know Country.

Language matters

In the past, historians used words such as *settled*, *settlement* and *settler* to describe the British occupation of Australia. These words gloss over the way in which the land was taken from the First Nations Peoples. Modern historians prefer words such as *colonist* and *invader*, which acknowledge violence and **dispossession**.

Historians also use the word **genocide** to describe the experience of First Nations Peoples of Australia. Genocide is defined by the United Nations as any of these five acts directed towards a national, ethnic, racial or religious group:

- killing members of the group
- causing serious bodily or mental harm to members of the group
- deliberately inflicting on the group conditions of life calculated to bring about its physical destruction
- imposing measures intended to prevent births
- forcibly transferring children.

Source 4

Van Diemen's Land, present-day Tasmania. Created by Sidney Hall in 1828.

Source: Longman & Co.

1642–1644
Abel Tasman (Dutch) names Van Diemen's Land (Tasmania), claiming it as part of Nieuw Holland (New Holland)

1688
William Dampier (British) camps at Karrakatta Bay near Broome

1699
William Dampier (British) lands where Hartog did, renaming it Shark Bay

1770–1771
James Cook (British) lands at Botany Bay. Cook incorrectly declares Australia as *terra nullius*.

Learning ladder H2.1

Show what you know

1 How did the name 'Australia' originate?

2 What words are now used instead of *settled*, *settlement* or *settler*, and why?

3 Hypothesise why many European nations searched for the 'great southern land'.

Source analysis

Step 1: I can list specific features of a source

4 Source 3: List the natural features on this image that point to it being both 'rugged and unyielding' and able to sustain life.

Step 2: I can find themes in a source

5 Source 1: How would an intimate knowledge of the land enable survival?

Step 3: I can use the origin of a source to explain its creator's purpose

6 Consider the nationalities of explorers on the timeline. Which countries were involved in trying to name, claim and colonise Australia?

Step 4: I can analyse a source

7 Consider the United Nations' definition of 'genocide'. Why is the forcible transfer of children a form of genocide?

HOW TO
Source analysis, page 240

What were the cultures of First Nations Peoples?

Millennia before Europeans set foot on the continent of Australia, the First Nations Peoples had clearly defined political, social and legal structures. Their unique traditions of time, history and sacred stories are the oldest on Earth.

Source 1

This engraved drawing, *New Hollanders*, is the earliest known European image of Australia's First Nations Peoples, the Bardi, in a 1698 edition of the explorer William Dampier's journal.

Political and legal structures

European explorers in the 1770s found the cultures of the First Nations Peoples of Australia to have complex political, social and legal structures. However, the enormous diversity of First Nations People – more than 500 at the time of European contact – means that generalisations can oversimplify the complexity of these structures.

In addition, some concepts of time, law and social conduct are considered by First Nations Peoples to be sacred. They can only be shared with the initiated, and cannot be described in a textbook.

Australia's pre-colonial population

In the 1930s, English anthropologist A.R. Radcliffe-Browne estimated Australia's First Nations Peoples' population before 1770 to be 300 000. In the 1980s, historian Noel Butlin revised this estimate to 1 000 000 people in total. Both scholars indicate that, after colonisation, the First Nations population dropped by as much as 96 per cent by the 1920s.

This drastic reduction in population followed similar patterns to when the Spanish and Portuguese conquistadors came into contact with the Indigenous peoples of the American continents, bringing with them diseases, massacres and coerced assimilation.

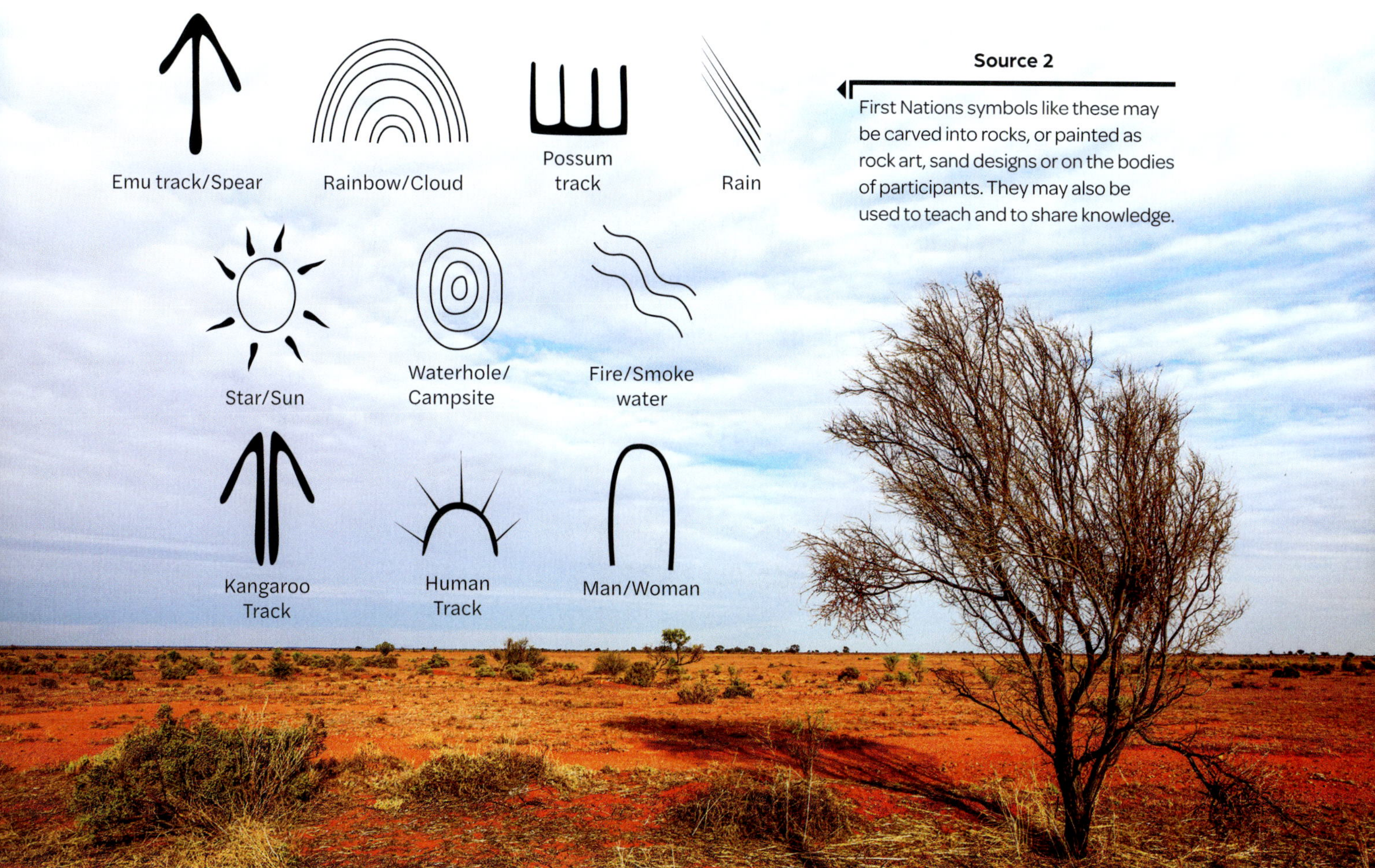

Source 2

First Nations symbols like these may be carved into rocks, or painted as rock art, sand designs or on the bodies of participants. They may also be used to teach and to share knowledge.

Social structures

The nation and clan social structures of the First Nations Peoples of Australia are generally based on a three-part system of kinship, involving **moiety**, **totem** and **skin name**.

- Moiety is a system of dividing the community into groups that reflect environmental, regional, matrilineal (through the mother), patrilineal (through the father) and generational factors. Such generations hark back through time to ancestors and Creation Spirits.
- Totem includes national, clan, family and personal factors. Apart from personal factors, totems are inherited and members are accountable to them.
- Skin name relates to a person's genetic matrilineal or patrilineal lines. Each generation in a community is given a number, which increases over time until reaching a maximum, after which it cycles back to the beginning again.

This elaborate kinship system had laws preventing people with certain totems and skin names from marrying one another, as they were considered siblings. This ensured the continuing genetic health of the community.

Men's and Women's Business

A Council of Elders from respective genders initiate, teach and regulate spiritual and legal matters, called **Men's** and **Women's Business**. Elders still perform this vital community role today, and are called 'Uncles' and 'Aunties'. The Council would also discuss general community business. Heads of Families have significant leadership roles in keeping order within clans and families, negotiating with Elders and other heads in disputes.

Legal structure

Spiritual and legal matters were (and sometimes still are) handled by people of authority, known as Elders or Heads of Family. The system of Aboriginal Customary Law (ACL) is still practised by some of Australia's First Nations Peoples today, and the Koori Courts are a recognised part of the Victorian legal system.

Concepts of time

According to First Nations Peoples, time is circular. History is not fixed but is in flux, and is something that moves across the past, present and future. Furthermore, in First Nations cultures, a re-enactment of an event might be perceived to be the same as the original event. This means that the past can be experienced in the present moment. This circularity can be seen in the narrative structure of sacred lore, and can affect whether events are seen as significant.

Oral history traditions rely on consistency. An initiated person telling a sacred story must tell it the same way every time, or they will be corrected by a group of Elders, who were told the story by their Elders in turn.

Source 3

Dancers performing a traditional story-based dance at the Laura Dance Festival.

For example, a ritual dance of a brolga at a Men's Business corroboree:

- may be learned, after initiation, to perfection
- may link the dancer back to ancestors who performed the same chants and dances
- may link the performer and even the viewer to the ancestral Creation Spirit, who may take the form of a brolga in some sacred stories.

The Mabo decision

In 1992, the High Court of Australia acknowledged that Eddie Mabo (an Ailan Pasin/Torres Strait Islander man) and his people could exercise their traditional land claims under Ailan Pasin laws. The court reviewed evidence from the First Nations People of Mer that they had lived in the islands for thousands of years in established villages and had a defined social and leadership structure.

The key aspect of the Mabo decision was the articulation of the Torres Strait laws of inheritance on the islands. This established a clearly defined understanding of sovereign ownership of the land, acknowledging that First Nations Peoples were and continued to be the owners of their lands.

Learning ladder H2.2

Show what you know

1 In First Nations societies, who is in charge and what are they in charge of?

2 Which kinship systems exist in your culture? What information do they reveal?

3 Consider Sources 2 and 3. How do First Nations Peoples record and perform historical events?

Historical significance

Step 1: I can recognise historical significance

4 What aspects of the Mabo decision made it such a landmark case?

Step 2: I can explain historical significance

5 What are the different ways that oral histories can be explained from both First Nations' and European perspectives?

Step 3: I can apply a theory of significance

6 How does First Nations Peoples' understanding of time differ from the European understanding of time?

Step 4: I can analyse historical significance

7 What can the size of the First Nations Peoples' population before and after Europeans came to Australia tell us about colonisation?

Historical significance, page 251

What can language reveal about the First Nations Peoples of Australia?

Language is closely linked to identity, both for individuals and for a nation as a whole. The First Nations Peoples of Australia use language groups to define their clans and nations.

Tribal boundaries in Aboriginal Australia, Norman B. Tindale, 1974

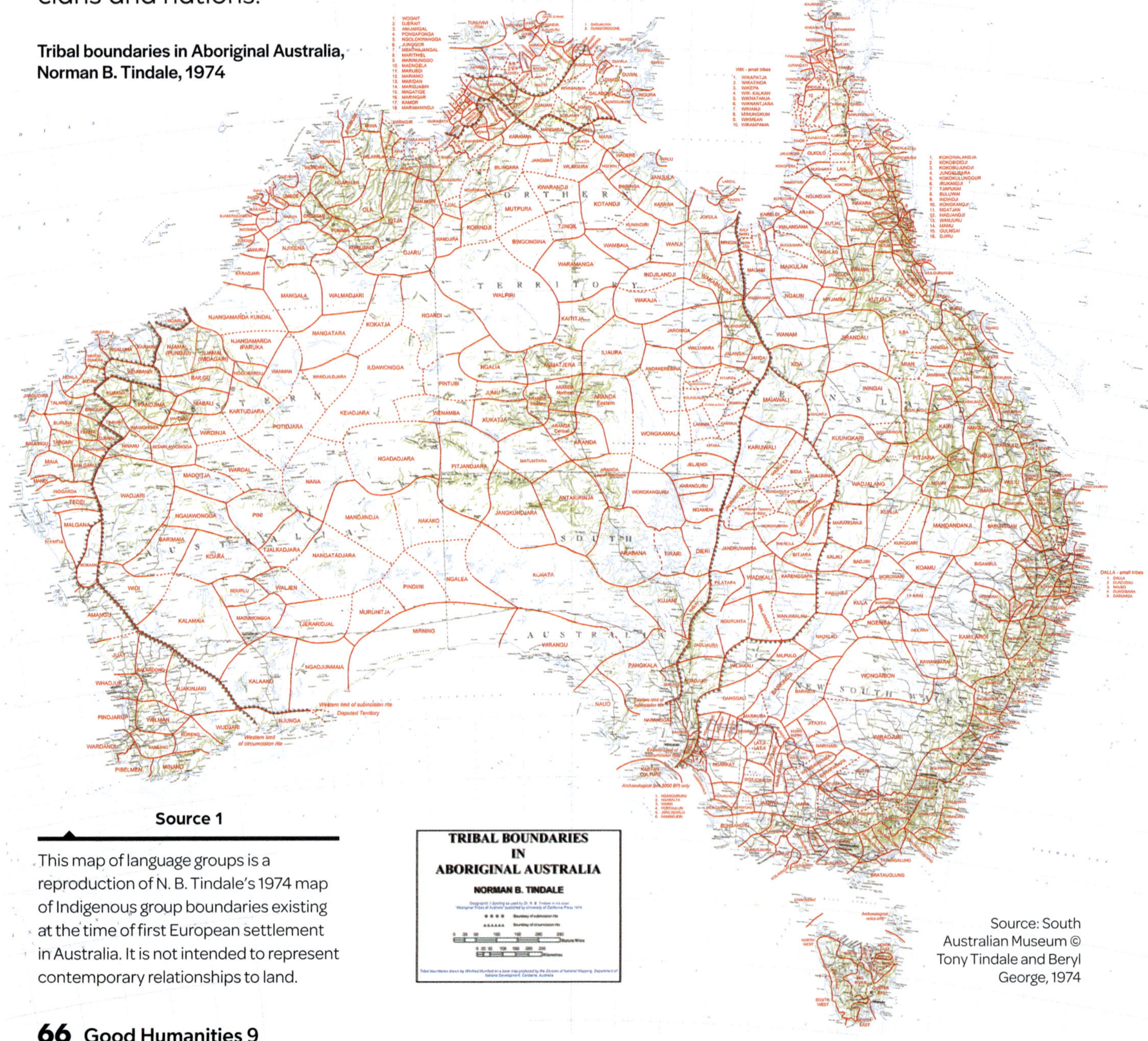

Source 1

This map of language groups is a reproduction of N. B. Tindale's 1974 map of Indigenous group boundaries existing at the time of first European settlement in Australia. It is not intended to represent contemporary relationships to land.

Source: South Australian Museum © Tony Tindale and Beryl George, 1974

Lieutenant William Dawes, an officer in the original British colony of New South Wales, was one of the most learned British colonists regarding the local Aboriginal languages. He learned the Dharuk (sometimes spelled Dharug or Darug) language primarily from Patyegarang, a 15-year-old girl of the Eora nation.

Dawes' *Notebooks on the Aboriginal Language of Sydney (1790–1791)* are some of the first written records of a First Nations language, and represent a genuine attempt by a *berewalgal* (Eora word for the English/European colonists) to understand the 29 clans of the Eora nation of Warrane/Warrang (Port Jackson/Sydney Cove).

Language is one of the living artefacts of the places colonised by the British and other empires. A number of First Nations Peoples' words became part of the new colonial identity and are still used today in Australian English.

A great deal of research has been done on the languages, trade borders and connections between First Nations Peoples. A study from 1988–1994 shows that there is no universal agreement on the borders of the language groups and nations. First Nations Peoples often knew several languages for diplomatic and economic purposes. Knowledge of different tribal languages was important for trade, sacred purposes and the application of Aboriginal Customary Law.

Source 2

Some of William Dawes' translations of Dharuk terms, from his *Notebooks on the Aboriginal Language of Sydney (1790–1791)*

Recorded Dharuk	Dawes' English translations
Be-re-wal-gal	The name given to us by the natives (berewal = a great distance off)
Booroodel, Maugoran	Booroong says these people are unfriendly to us.
Booroowunne	The name of a male stranger
Carreweer	The name of a female stranger
Dje-ra-bar, Je-rab-ber	The Natives frequently called us by the name they give the musket.

Source 3

Several everyday terms originally stemmed from First Nations Peoples' languages.

English	Dharuk today	Meaning
Corroboree	Garriberri	Dancing event
Dingo	Dingu	Dog
Cooee	Guwawi	Call of location
Waratah	Warada	Type of flower

Learning ladder H2.3

Show what you know

1 When did William Dawes create his notebooks about the First Nations languages around Sydney?

2 Source 2: What do the Dharuk words given to the British and European colonists indicate about an Indigenous understanding of the colonists or invaders?

3 Dawes was an engineer of Port Jackson's military defences, and an astronomer. How might knowledge of First Nations Peoples have helped him in these roles?

Source analysis

Step 1: I can list specific features of a source

4 What do you notice about Source 1? What information does it convey?

Step 2: I can find themes in a source

5 Source 1: Why do the sizes of the territories vary and what might this indicate about the sustainability of the land?

Step 3: I can use the origin of a source to explain its creator's purpose

6 Given its features and limitations, what might be the purpose of the map in Source 1? (You may need to do further research into its creation.)

Step 4: I can analyse a source

7 Given the number of nations shown on Source 1, why is it difficult to make generalisations about Australia's First Nations Peoples?

Source analysis, page 240

H2.4

What were the economic structures of the First Nations of Australia?

Some of the First Nations Peoples of Australia operated on the basis that spirituality, lore, laws, Country and people are interconnected. For these communities, it is almost meaningless to apply terms like 'economy'. However, some travelled via trade routes to exchange goods and enact law at gatherings such as corroborees.

Economics involves the production, consumption and transfer of wealth, such as in the exchange of goods and services. The Anangu people of the Northern Territory instead have the concept of 'Tjukurpa', where the land, soil, vegetation, waterways, animals and people are interrelated. Tjukurpa means that no one can own part of the system, making the concept of economics irrelevant.

However, we can apply the concept of economics to Australia's pre-colonial societies if we consider it more broadly. 'Wealth' can be defined as being well fed and fit, sustainably managing food resources and enhancing or increasing food sources. We can also look at how goods, food and materials were exchanged.

Source 1

Bogong moths were one of the seasonal food sources of First Nations Peoples.

Gatherings

Many sacred sites are close to water and have a good supply of food. As such, these places were carefully maintained and used to maximise agricultural production. The seasonal production and migrations of food sources, such as eels and bogong moths, enabled large gatherings of people to occur. Such gatherings often coincided with ceremonies, the trade of goods, marriages and applications of customary law.

Trade and exchange

The exchange of goods is a basic function of economics, and there is evidence that pre-colonial First Nations Peoples traded rare items such as granite chips (to make small tools), ochre (for body and rock art) and preserved foods (such as smoked eel).

A number of trade routes existed across Australia, often crossing tribal borders known as 'songlines'. Trade might occur between individuals, within a clan or at large gatherings.

Willem Janszoon (Dutch, 1604), Jan Carstenzoon (Dutch, 1623) and James Cook (British, 1770–1771) all attempted to trade with local people, but were often met with resistance or evasiveness. When First Nations Peoples at Cape York tried to trade for muskets and other items, the Europeans refused and in some cases attempted to kidnap Indigenous men. Such actions earned Europeans various names meaning 'devils' in different languages and dialects of the Cape.

Health and wellbeing

Colonists observed that the First Nations Peoples of Australia were well fed, fitter and probably healthier than their European counterparts. After just 3 or 4 hours of effort, First Nations People could secure as much food as a European worker over 10–12 hours.

> 'From what I have said of the Natives of New Holland they may appear to be to some to be the most wretched people upon Earth, but in reality they are far more happier than we Europeans; being wholly unacquainted not only with the superfluous but the necessary Conveniences so much sought after in Europe, they are happy in not knowing the use of them. They live in a Tranquillity which is not disturbed by Inequality of Condition: The Earth and sea of their own accord furnishes them with all things necessary for life, they covet not Magnificent Houses, Household-stuff &c, they live in a warm and fine Climate and enjoy a very wholesome Air, so that they have little need of Clothing and this they seem to be fully sensible of, for many to whom we gave Cloth &c to, left it carelessly upon the Sea beach and in the woods as a thing they had no manner of use for. In short they seem'd to set no Value upon any thing we gave them, nor would they ever part with anything of their own for any one article we could offer them; this in my opinion argues that they think themselves provided with all the necessarys of Life and that they have no superfluities.'

Source 2

Extract from James Cook's Journal of HMS *Endeavour*, 1768–1771. [Text is verbatim and includes grammatical errors]

When Cook tried to trade with the First Nations Peoples, he was unsuccessful. Language barriers prevented him from trading during initial interactions. Later attempts were foiled, as on each occasion the First Nations evaded Cook and his party.

Source 3

Eels being wrapped in leaves and paper bark, ready to be smoked and preserved. Smoked eels were one of the goods traded by First Nations Peoples at gatherings such as corroborees.

Learning ladder H2.4

Show what you know

1 What does 'economics' mean?

2 Why does Cook say the First Nations Peoples of Australia may appear to some to be 'wretched' in Source 2? Do you think this was an accurate description? Why or why not?

3 Hypothesise what the 'Conveniences' might be that Cook refers to in his journals.

4 Why didn't the First Nations Peoples accept the Europeans' gifts or trade with them?

Historical interpretations

Step 1: I can recognise that the past has been represented in different ways

5 How does a European understanding of economics apply to the First Nations societies of Australia?

Step 2: I can describe historical interpretations

6 Why does Cook describe his own society using negative terms such as 'Inequality of Condition'?

Step 3: I can explain historical interpretations

7 What do Cook's descriptions of the First Nations societies of Australia reveal about his own attitudes and biases?

Step 4: I can analyse historical interpretations

8 Why did Cook and other European explorers view the food consumption and health of First Nations Peoples as indicators of wealth, and a determinant of whether land was worthy of colonising?

Historical interpretations, page 255

Were the First Nations Peoples the first Australian farmers?

European explorers described the First Nations Peoples of Australia as nomads and hunter-gatherers. This inaccurate description was used to justify the claim that Australia was *terra nullius* – an uninhabited land. In fact, recent analysis of explorers' journals and colonial artworks show that First Nations Peoples had semi-permanent residences, used complex techniques to increase food production and built structures similar to silos to store food.

Seasonal villages and food production

The Wadawurrung people of Victoria lived in groups of hundreds of people – as large as villages in Industrial Revolution England – on a seasonal or semi-permanent basis. These large groups engineered weirs and dams, built grain storage facilities, and planted grain and *murnong* (yam daisy) fields.

The seasonal abundance of certain foods allowed societies to trade food and hold ceremonies, as there was enough food to feed a large number of people. The timing of large gatherings, including cultural and spiritual events, hinged on the abundance of foods such as yams, Bogong moths and eels, which were trapped in weirs and then smoked to preserve them.

Some explorers also noticed structures that resembled silos, which were used to store grain or other food. The archaeological evidence of pulverised grains on grindstones, animal remains and ochre from 32 000 years ago are some of the oldest evidence of baking technology in the world.

Source 1

The fish traps at Brewarrina.

Source 2

Joseph Lycett, *Aboriginal People Cooking and Eating Beached Whales* (1817). This painting shows the food-gathering practices of the Awabakal and Worimi people of Muloobinba (later Newcastle).

Dating Australia's First Nations sites

The oldest First Nations site to be dated is in the Northern Territory at Madjedbebe, which shows a range of 50 000 to 60 000 years of occupation.

Measurements such as this are often done using radiocarbon dating, which measures radioactive decay of isotopes within an organic item. However, it has a limit of around 50 000 years. To investigate an object's age beyond this limit, a method called optically stimulated luminescence can date the soil around an object to 100 000 years.

These methods can be used together to validate or check each other.

Bush tucker

'Bush tucker' is a modern phrase used to describe Australia's native edible plants and (in some cases) animals. One of the rules of collecting bush tucker is not to take all the plants, but to leave some behind so you know exactly where to look for such plants in future seasons. A traditional story may tell the lore of the location and how to collect it. These stories may be painted, told, danced and chanted.

Source 3

Plants such as quandong, desert fig, pencil yam, black wattle and ruby saltbush are still enjoyed as bush tucker.

In the Northern Territory, some rock art serves not just spiritual but also practical purposes, such as showing the location of the nearest water source, or providing a 'menu' of animals to hunt and plants to gather in the area. Some of these rock art menus have been used by archaeologists and environmental scientists to judge how and when habitats changed. The depiction of certain animals, which lived in habitats that no longer exist, show what the place was like in the past.

Dingo or cattle dog?

Debates about the modern dingo often link it to wolves and domesticated dogs. However, *Canis dingo* has recently been identified as a separate species of animal by zoologists analysing historical samples.

Source 4

The thylacine or Tasmanian tiger (below) is believed to have been made extinct on the mainland by the dingo (right), and on Tasmania by humans who hunted it for its distinctive fur.

Dingoes can be trained, and there is evidence from the stories of First Nations Peoples that they assisted in hunting by 'rounding up' prey.

Most likely introduced to Australia around 4000 years ago, dingoes are thought to have preyed upon both thylacines (Tasmanian tigers) and Tasmanian devils, driving them to extinction on the mainland. As dingoes were not introduced to Tasmania, the thylacine survived there until the middle of the 20th century.

Source 5

This 1817 painting by Joseph Lycett, a former bank forger, shows First Nations People using firesticks to flush out animals for hunting.

Firestick farming

Archaeological evidence shows that First Nations Peoples deliberately enhanced the productivity of their land through 'firestick farming', or 'cold burning', which involved burning the undergrowth of forests. This practice is still used by Indigenous rangers to manage Country.

These practices reduced the amount of fuel available for bushfires in the hottest months, preventing catastrophic firestorms. They also created more hunting grounds, flushed out animals to hunt and increased the productivity of the land. Some native plants actually require fire to release their seeds or to sprout, so fires were lit in order to increase their abundance.

Australia is a dry country, but early colonial artists often depicted the landscape as lush and green. For many years, these landscapes were interpreted as representing the artist's romantic longing for home while living in an unfamiliar land. However, historian Bill Gammage challenged these interpretations in 2012. He argued that the painters were accurately depicting the abundant pastures created by successful Indigenous land management, including the use of firestick farming.

Learning ladder H2.5

Show what you know

1 What technologies enhanced the regularity and collection of food for First Nations Peoples?

2 How does firestick farming work, and what dangers did firestick farming prevent?

3 What unique features of the dingo assisted Australia's First Nations Peoples?

Source analysis

Step 1: I can list specific features of a source

4 What does Source 2 depict the Awabakal and Worimi people doing? What does it show about the environment around Muloobinba (later Newcastle)?

Step 2: I can find themes in a source

5 Source 1:. How do the fish and eel traps help the people? Is this hunting, gathering or farming?

Step 3: I can use the origin of a source to explain its creator's purpose

6 Source 2: A beached whale could feed large numbers of people. What other cultural purposes could a gathering such as this serve?

Step 4: I can analyse a source

7 What does Australia look like in the early pastoral image shown in Source 5? What land-management techniques and hunting technologies are being used?

Source analysis, page 240

What technologies did First Nations Peoples develop in Australia?

As well as developing sophisticated farming and food-production methods, First Nations Peoples created stone tools, weapons, navigation techniques and bush medicines.

Source 1

Grindstone used by ancient Australian First Nations Peoples

Stone tools

For tens of thousands of years, First Nations Peoples used countless large and small tool-making traditions. At the time Europeans arrived on the continent, First Nations Peoples were making many different tools from the flakes of hard stones.

Small stones were quarried and traded over long distances, and were 'hafted' (attached as handles) to spears and axe handles using resins (such as heated spinifex resin). This was a careful use of a precious resource; rather than hammering chips from one big stone to make just a few tools, this technique allowed people to make multiple tools from a single stone.

First Nations Peoples also developed an extension for the throwing ends of their spears. The woomera is a device that makes throwing easier when attached to a spear. It catapults spears beyond the distance of mere hand throwing, and greatly increases speed and accuracy.

Source 2

A woomera is a device that extends the arm for throwing a spear. The nodule at the end fitted into the butt of the spear like a 'ball and socket' joint.

Source 3

How to use a woomera to throw a spear. The extension of the thrower's arm by the woomera significantly extends the range of the weapon when it is released.

Boomerangs

There are many types of boomerangs. Two of the best-known examples are the returning boomerang and the hunting boomerang.

The returning boomerang shows the ingenuity of Australia's First Nations Peoples. Used to hunt waterfowl, the returning aspect meant that hunters would not have to search far to retrieve the boomerang in the water. It was not designed to strike, but to simulate the flight of a bird of prey and cause the waterfowl to take flight, providing targets for spears and stones for other hunters on the ground.

The hunting boomerang was designed to be thrown into a flock of birds, or at other animals living in trees, and strike them down. Some versions could take down more than one bird in a single throw. Heavier versions were used to hunt kangaroos.

Source 4

The hunting boomerang, which does not return, was designed to be thrown into a flock of birds. The longer end of the boomerang could hit multiple birds, stunning or killing them.

Navigation techniques

For millenia, people have used the positions of the stars to find where they are. Travellers also used the stars to work out the direction they needed to travel, particularly in areas devoid of landmarks, such as deserts and seas.

The First Nations Peoples of Australia used star maps to safely cross the 10 major deserts of Australia when travelling along their trade routes, as well as to pilot boats through the Torres Strait and across the Indonesian archipelago. Songs about constellations safely led people across their sacred country, and each nation, language group and clan may have named certain constellations according to their sacred lore.

Source 5

European astronomers identified the stars Castor and Pollux as part of the Gemini constellation. To the Wergaia people of Victoria, those stars are the brothers Yuree, the fan-tailed cuckoo, and Wanjel, the long-necked tortoise.

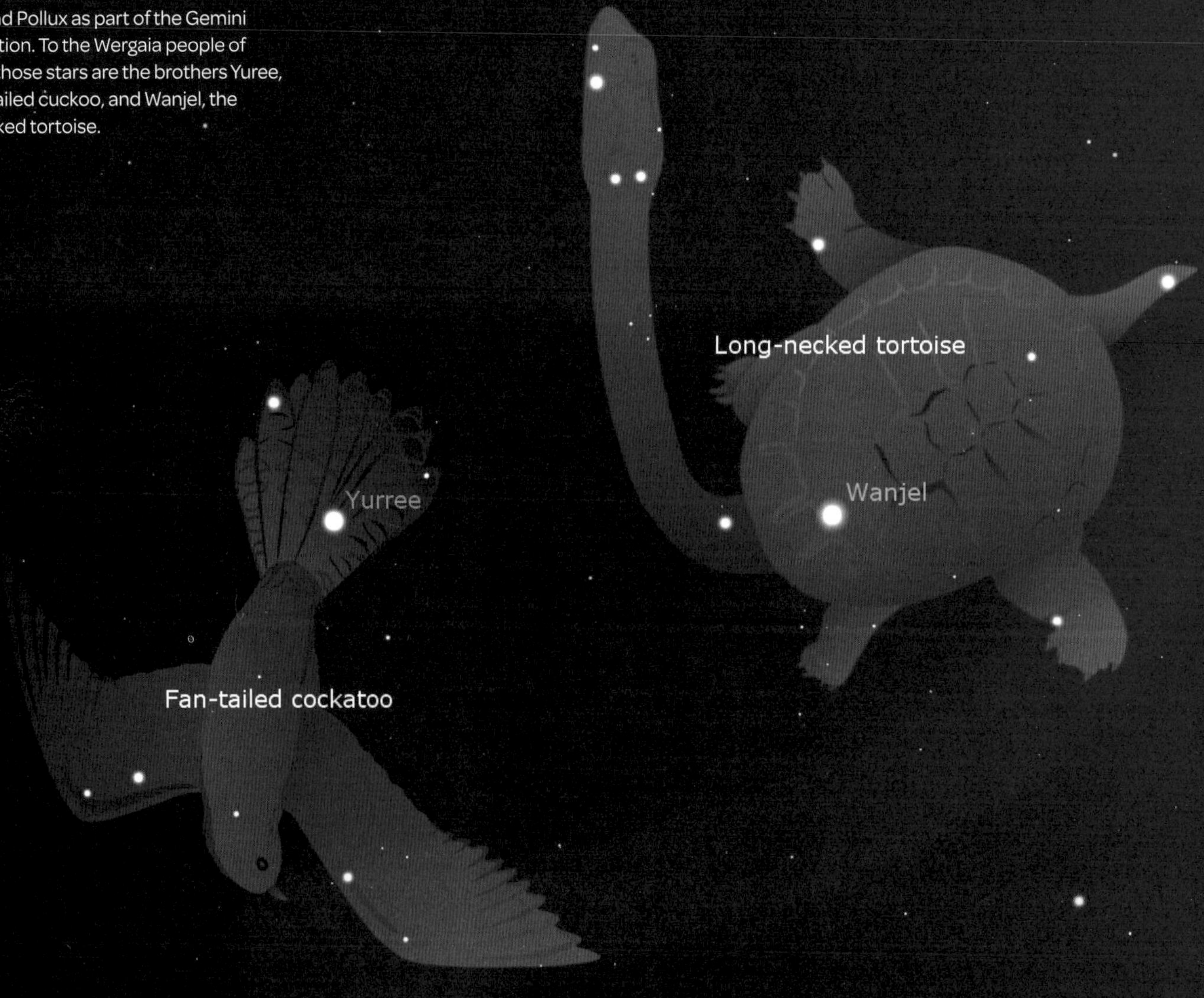

Toxic plant harvesting and bush medicine

Macrozamia cycad is a plant with a highly toxic seed and kernel. First Nations Peoples safely removed the toxins from this valuable food source, using methods such as washing the plant over a week in a woven basket or letting mould process the poison. The food yield in carbohydrates, protein and calories was significant and worth the effort taken to harvest it.

First Nations Peoples knew about the toxicity and the medicinal uses of a variety of plants, insects and venoms. For example:

- ant jaws could be used to close skin wounds, acting like stitches or sutures
- tea-tree billabong bathing assisted with skin conditions like psoriasis
- certain plants were made into tinctures, ointments, poultices and drinks.

Modern pharmaceutical companies are now researching many First Nations medicines. First Nations Peoples also knew how to set broken bones, preventing gangrene and potential limb amputation.

Source 6

First Nations Peoples' methods allowed them to safely harvest and eat *macrozamia cycad* while avoiding its poisons.

Learning ladder H2.6

Show what you know

1 What does Source 1 reveal about First Nations Peoples' societies and their technology?

2 What technological innovation was added to the spear? What advantages did this give to the thrower?

3 Describe two types of boomerangs and what they were used for.

4 How would a navigator use Source 5? What would you need to know to do so?

Continuity and change

Step 1: I can describe continuity and change

5 Describe some medical knowledge that the First Nations Peoples of Australia have shared.

Step 2: I can explain why something did or did not change

6 How and why have European uses and applications of boomerangs changed since colonisation? Research one current use as an example.

Step 3: I can explain patterns of continuity and change

7 What do stone tool cultures indicate about technology and social changes in response to environmental changes?

Step 4: I can analyse patterns of continuity and change

8 Which First Nations medicines are available at your local pharmacy? What does this suggest about the variety and uses of bush medicine?

Continuity and change, page 244

H2 .7

What interactions did First Nations Peoples have with other cultures?

The arrival of the British had a massive and lasting impact on the First Nations Peoples of Australia, but they were far from the first foreigners to visit Australia. Ailan Pasin ('island custom') or Torres Strait Islander peoples, Indonesian traders and other European explorers interacted with the First Nations long before the British did.

Torres Strait Islander peoples

The Torres Strait is the region of ocean between Australia and the large island of Papua New Guinea, around 150 kilometres north of Cape York. There are more than 250 small islands in the Strait, many of which have been inhabited for more than 2000 years.

Far further back in history, when sea levels were lower, an extensive land bridge existed between Cape York, Papua New Guinea and Indonesia, which made it much easier for people and wildlife to move between Australia and the islands. The land bridge submerged over 8000 years ago, because of rising sea levels.

Thanks to the land bridge and the ease of travel between the islands, the Torres Strait Islander peoples have interacted with the First Nations Peoples of northern Australia for thousands of years. These neighbouring cultures learned and borrowed from each other while retaining their own identity, customs and social structures.

Source 1

Many First Nations Peoples and Torres Strait Islander peoples consider themselves to be connected through ties of blood, kinship and history.

Source 2

Rock art at Malarrak, in the Wellington Range, Western Arnhem Land. This rare depiction of a building could be a Makassan smokehouse, used to cure *trepang*.

Indonesians

Fishermen from the Makassar region of Indonesia began visiting the northwestern regions of Australia centuries ago – perhaps in 1750, perhaps earlier. These fishermen were *trepangers,* meaning that they harvested *trepangs*, the aquatic animals also known as sea cucumbers.

The Makassans interacted with the Yolngu people of the region. Evidence suggests that they traded materials such as cloth, metal, rice and canoes with the Yolngu in exchange for the right to gather sea cucumbers from the local waters. These interactions were recorded in Yolngu songs, stories and dances, and some Makassarese words made their way into the Yolngu language. However, recent historical research suggests that these interactions were not always peaceful, and that violent conflicts occurred when Yolngu groups felt they were being exploited.

Contact between the Makassans and Australia's First Nations Peoples continued for hundreds of years, and did not end until the Australian government passed restrictive policies in 1906.

European explorers

Ever since the ancient mathematician Ptolemy of Alexandria proposed a 'great unknown southern land' in 150 CE, Europeans believed that a southern continent lay waiting to be explored. The European search for the great southern continent began in earnest when Captain Luis Váez de Torres sailed the northern Torres Strait in mid-1606.

A few months before Torres, on 26 February 1606, Willem Janszoon accidentally landed on the northern coastline, becoming the first European to visit mainland Australia. The Dutch seafarer and his crew thought they were on a previously uncharted southern part of New Guinea.

Captain Dyrck Hartoochz (Dirk Hartog), another Dutchman, was the second European to land on Australia on 25 October 1616, on an island in what is now called Shark Bay in Western Australia. He left behind a pewter plate, which he engraved with a message about his landing. The plate, which is Australia's oldest evidence of European contact, is currently in the Rijksmuseum in Amsterdam.

It was not until 1644 that the name 'New Holland' was first applied to the northwest coast of Australia by the Dutch explorer Abel Tasman, who claimed it for Holland by having a crewman swim ashore and plant his country's flag. Tasman named the small island he had discovered after his patron – calling it 'Van Diemen's Land'. Later the island was renamed Tasmania after Tasman himself.

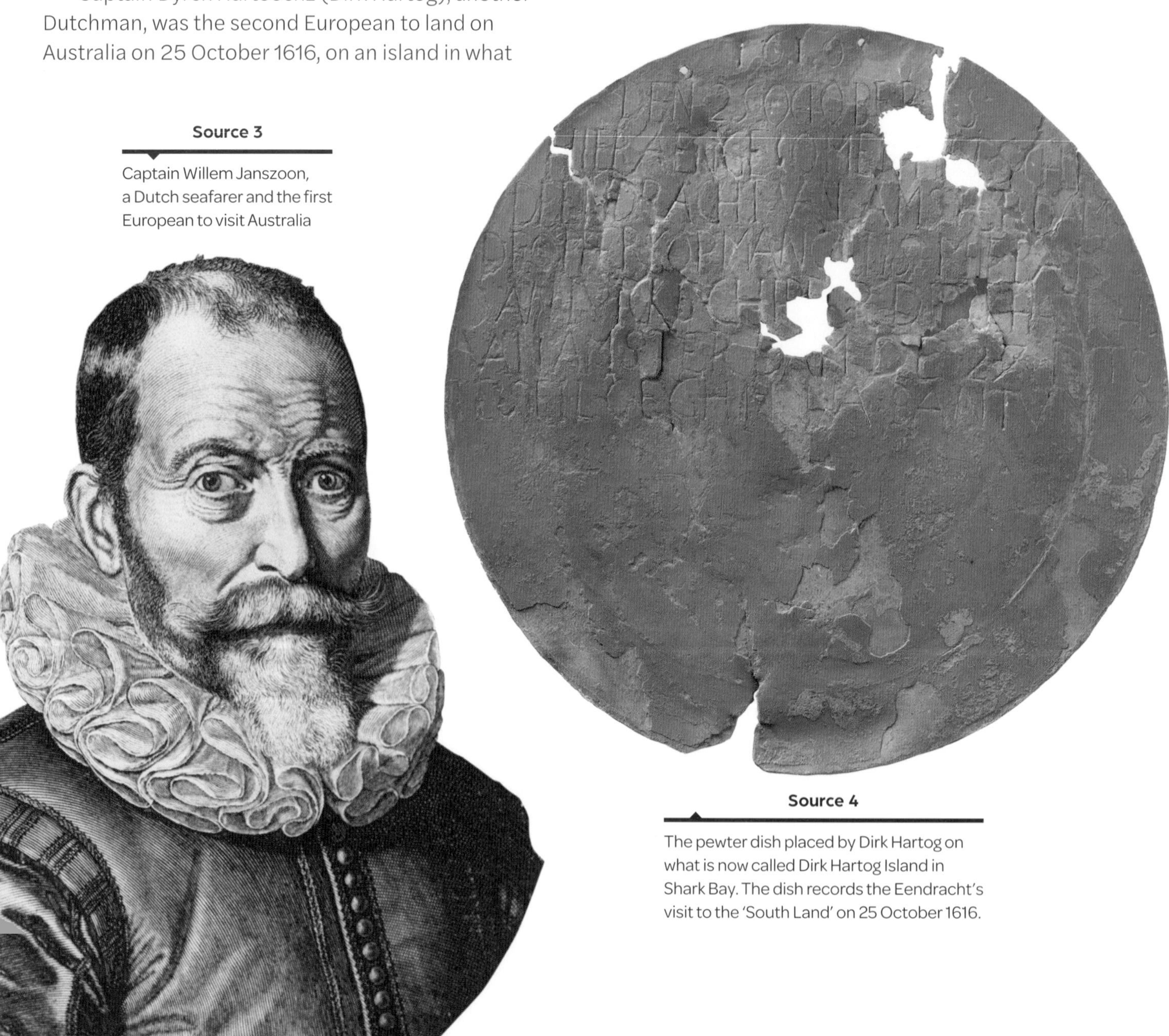

Source 3

Captain Willem Janszoon, a Dutch seafarer and the first European to visit Australia

Source 4

The pewter dish placed by Dirk Hartog on what is now called Dirk Hartog Island in Shark Bay. The dish records the Eendracht's visit to the 'South Land' on 25 October 1616.

Early British contact

William Dampier, an explorer and former pirate, became the first Englishman to set foot on Australian soil, camping in the lands of the Bardi nation in north-west Australia in 1688. In his journal, Dampier wrote of the Bardi people: 'They are people of good stature but are very thin and lean, I judge for want of food ... They build their weirs of stone across the bay; they search those weirs for what the sea has left behind'.

However, when his journal was published, these observations were changed by his publisher to, 'The inhabitants of this country are the miserabilist people in the world... brutes', and other derogatory remarks.

Dampier's remarks about the First Nations Peoples of Australia may have influenced later explorers, and set a negative tone for later interactions. They may have also influenced the decision to declare New Holland as *terra nullius* – 'a vacant land'.

How did Australia obtain its name?

Europeans called the land they believed to exist in the southern hemisphere *terra australis nondum cognita*, which is Latin for the 'unknown southern land'. But when Europeans finally reached Australia, they disagreed over what to call it.

Dutch explorers labelled it Nieuw Holland, Nova Hollandicus or Hollandia Nova (New Holland) on the incomplete maps of the time. It was Matthew Flinders who, in 1801–1803, circumnavigated and mapped what he discovered to be an island. In 1804 he gave it the name 'Australia', slightly altering the Latin *australis*, meaning 'southern'.

But Flinders was not alone – he was accompanied by Bungaree, a Kuringgai man who acted as a diplomat and intermediary when the expedition met with coastal communities. Bungaree became the first member of the First Nations People to circumnavigate the continent. Not only that, he was the first person in the world to ever be referred to as 'an Australian' in print.

Source 5

Lieutenant Matthew Flinders called the land 'Australia' after sailing around it.

Learning ladder H2.7

Show what you know

1 What happened to the identity of the land on which the European explorers landed?

2 How did different European nations, kingdoms and empires 'claim' lands that already had people living on them for thousands of years?

3 What is the difference between a 'claim', a 'colony' and a 'settlement'?

Historical significance

Step 1: I can recognise historical significance

4 Research Ptolemy's theory of a southern continent. What was his hypothesis for its existence?

Step 2: I can explain historical significance

5 What does the name 'Australia' mean, and why do you think the name was changed from New Holland to 'Australia'?

Step 3: I can apply a theory of significance

6 What is the significance of Bungaree, a member of the First Nations, being the first person to be called an Australian?

Step 4: I can analyse historical significance

7 Considering that Janszoon, Tasman, Hartog and others explored the region, why did the Dutch not attempt to establish a colony in Australia? Use the information from this chapter and other research to form a hypothesis.

Historical significance, page 251

PART II: AUSTRALIA'S COLONIAL PERIOD (1770–1901)

How did colonisation change the continent of Australia?

The continent of Australia was forever changed when the first British penal colony disrupted the sacredness of Country. As the British expanded their colonies, they fought with First Nations Peoples to seize the bushlands and goldfields. The early colonial period featured a great deal of conflict, between colonists and First Nations Peoples, as well as between the colonists themselves.

Source 1

William Bradley, *First Interview with the Native Women at Port Jackson New South Wales* (c. 1802)

key ideas timeline.

1788 — Captain Arthur Phillip arrives with First Fleet to establish British Penal Colony at Sydney Cove

1790 — Hawkesbury and Nepean Wars in NSW

1801–03 — Captain Matthew Flinders (British) and Bungaree (Kuringgai) circumnavigate Australia

1810 — First Nations Peoples begin to be forced into missions

1824 — The Black War of Tasmania begins

1824 — The British government formally names the great southern land 'Australia'

1835 — Melbourne is established in what is now Port Phillip Bay

1838 — The Myall Creek Massacre occurs in New South Wales

1851 — Port Phillip colony separates from NSW, and is named Victoria. The Australian gold rush begins

1856 — All men, including First Nations Peoples, given the right to vote in South Australia

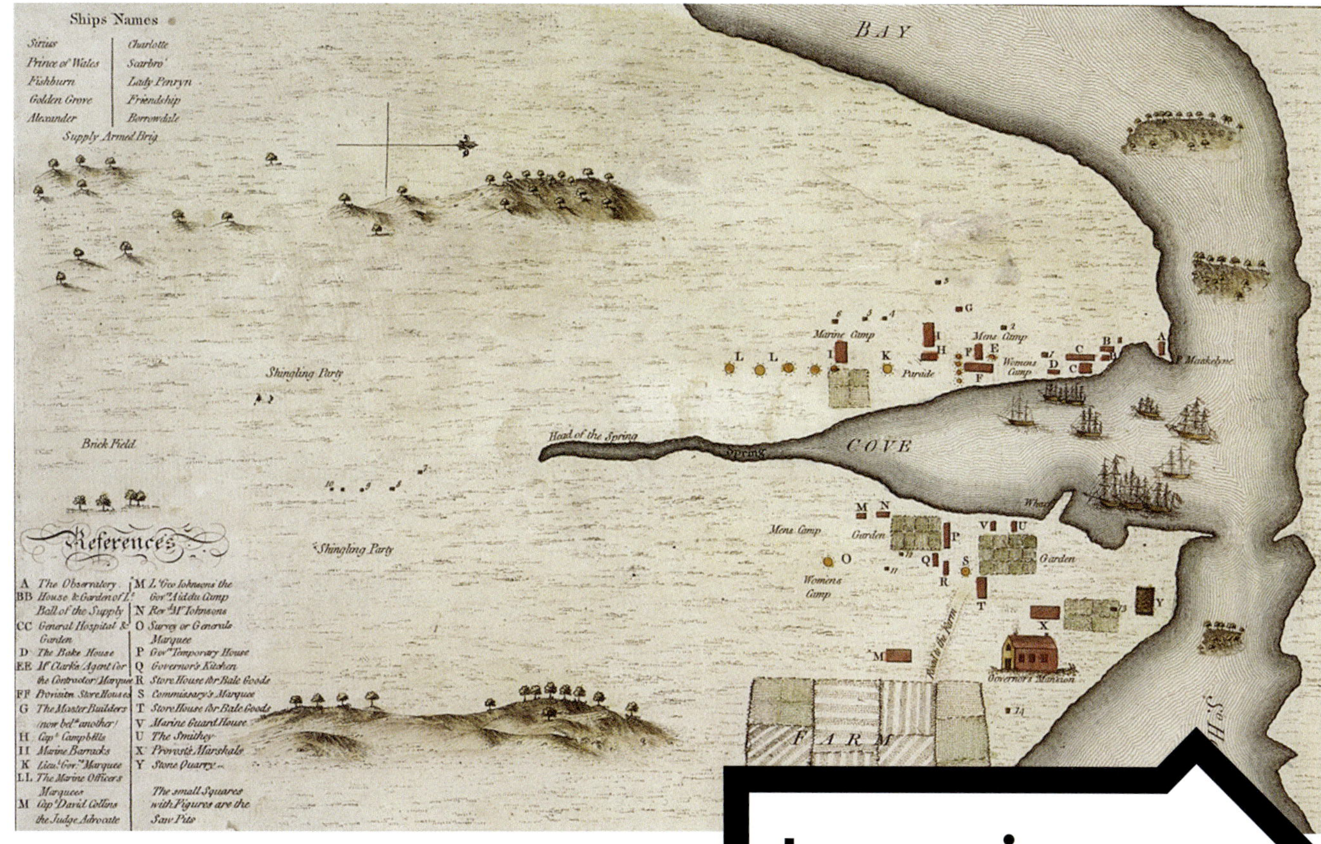

Source 2

Engraving of the settlement at Port Jackson, Sydney Harbour

1869
The *Aboriginal Protection Act* is passed in Victoria

1863
'Blackbirded' Pacific Islander men and women arrive in Queensland

1872
The *Education Act* makes public education free, secular and compulsory in Victoria

1895
South Australian women, including First Nations Peoples, given right to vote

Learning ladder H2.8

Show what you know

1 When did Victoria become a separate colony from New South Wales?

2 Which colony first allowed the First Nations Peoples of Australia the right to vote?

Source analysis

Step 1: I can list specific features of a source

3 Source 2: What infrastructure of the Port Jackson settlement can be seen in this engraving?

Step 2: I can find themes in a source

4 Source 2: Describe Port Jackson in 1788. What do you notice about the pace of its establishment during the first year?

Step 3: I can use the origin of a source to explain its creator's purpose

5 How could Source 2 have been used militarily by either the colony or a competing nation?

Step 4: I can analyse a source

6 Source 1: Why would an interview require women to be forcefully dragged and held on a boat under threat of muskets? Does the interaction appear peaceful, coercive or violent?

HOW TO

Source analysis, page 240

How did James Cook reach Australia?

Captain James Cook, a British sailor and cartographer, set sail for Tahiti on the HMS *Endeavour* in 1769 with orders to observe the transit of Venus from Tahiti. He also carried with him secret orders to expand the British Empire by discovering and seizing the southern continent – even if he had to do so illegally.

Britain's secret agenda

The British Empire's finances had been severely drained by the Seven Years' War with France (1754–1763) and the American Revolution (1775–1783). Their need for resources and cheap labour made the British keen to claim newly discovered lands, such as Canada, and to find the mythical great southern continent.

James Cook was officially commissioned by the British Royal Society to sail to Tahiti, in order to set up an observatory and observe an astronomical event called the transit of Venus. But before he left, on the 30th of July 1768 the British Admiralty also issued Cook secret instructions to sail to 40 degrees latitude and search for 'New Holland'. He was then to claim both the land and its people.

Source 1

Captain Cook's Landing at Botany, A.D. 1770.
[Created 1872, artist unknown]

Cook's Pacific travels

Cook arrived in Tahiti, where a Polynesian navigator called Tupaia (from the Ra'iatea or Society Islands) joined him for the next part of the voyage, largely through uncharted territory. Tupaia drew a map of the known islands in the Pacific Ocean (see Source 2). He shared his traditional methods of navigation with Cook.

Using Tupaia's map and his own considerable navigation skills, Cook piloted the *Endeavour* to 40 degrees latitude and then onwards, until he reached Aotearoa (New Zealand) on 6 October 1769. After circumnavigating Aotearoa and realising that it was an island, and not New Holland, Cook left to continue his mission.

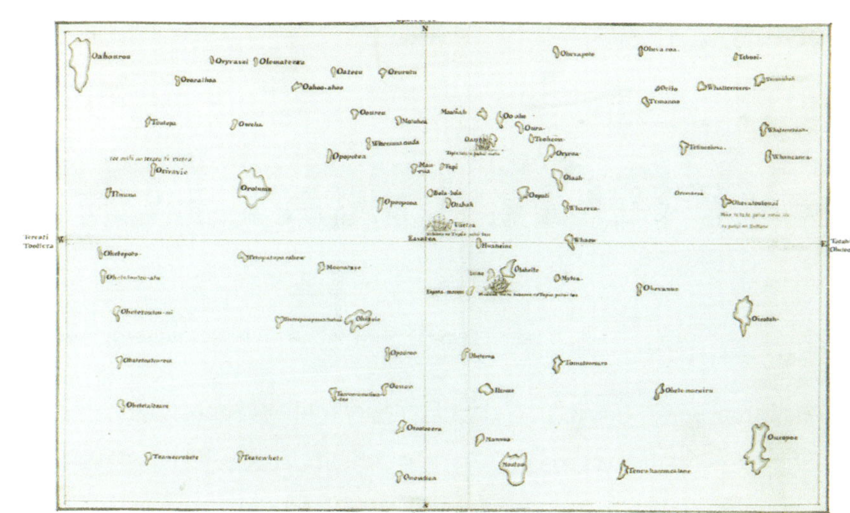

Source 2

This map of the Pacific Ocean islands was drawn in 1769 by Tupaia, a noble and high priest of the war god Oro and expert navigator, using traditional Polynesian methods. [Chart of the Society Isles discovered by Captn. Cook, 1769, from Thomas Conder, Alexander Hogg, & George William Anderson (1784)]

HMS *Endeavour* reaches New Holland

Cook made landfall on 29 April 1770 at Botany Bay, New Holland. He received a hostile reception, much like the one he received in Aotearoa – two Gweagal men aimed their spears at the boat and were shot at. Cook raised the British flag, called the land 'New South Wales' and seized it without the consent of the Indigenous inhabitants.

Cook sailed further up the east coast of Australia, but all his attempts to engage the native people were unsuccessful. Even after the *Endeavour* ran aground on the Barrier Reef and had to make a long landing for repairs, Australia's First Nations Peoples eluded Cooks' attempts to make contact.

Source 3

According to Gweagal oral history, this shield was taken from a First Nations man named Cooman during the 1770 encounter at Botany Bay. It is held by the British Museum, and Cooman's descendants are campaigning for its return.

Learning ladder H2.9

Show what you know

1. What was the official reason for Cook's voyage?
2. What were Cook's secret instructions?
3. What benefits did Tupaia provide to Cook's voyage?

Cause and effect

Step 1: I can recognise a cause and an effect

4. What made Cook realise that Aotearoa (New Zealand) was not the great southern continent?

Step 2: I can determine causes and effects

5. Source 1: Why was Tupaia included in the official landing party?

Step 3: I can explain causes and effects

6. Why do you think Cook claimed Australia 'without the consent of the natives'?

Step 4: I can analyse causes and effects

7. What major developments or events led the British government, admiralty and monarch to approve Cook's voyage, during which he 'discovered' Australia? (You may need to conduct further research to answer this.)

Cause and effect, page 247

What was *terra nullius*?

In the 18th century, the British failed to recognise that the First Nations Peoples of Australia were advanced cultures who actively managed the land. The British declared the land *terra nullius*. This declaration was made to remove the legal requirement to negotiate with the inhabitants of the land, and instead allowed the land to be seized with minimal cost to the British.

Two sets of laws

Aboriginal Customary Law (ACL) connects every part of the land: its lore, fauna and flora, Creation Spirits, ancestors, Elders and people, both past and present. Aboriginal people own the land, as it owns them. Councils of male and female Elders governed and regulated spiritual and social matters through Men's and Women's Business, as well as combined community or family business.

This was vastly different from the British system of government, where the **monarch** granted authority to ministers and governors. This same system extended to the colonies; however, during England's colonial expansion, further delineation of the laws was required. British agents were bound to follow the law; in practice, this often meant they looked for ways to stretch or rewrite laws to suit themselves.

Terra nullius

A Latin phrase meaning 'nobody's land', *terra nullius* is the legal term applied to land that is deemed uninhabited, unoccupied or not subject to the sovereignty of any state. The British application of *terra nullius* to Australia in the 18th century was a highly debated topic for two centuries. The Mabo decision finally struck out this application in 1992 and recognised Native Title (see page 65).

The laws surrounding colonisation were analysed in 1765 by Sir William Blackstone, an English judge. He defined the laws for unclaimed land described in Source 2.

Source 1

Joseph Lycett, *View of the Heads at the Entrance to Port Jackson New South Wales* (1824)

> '(1) that if an uninhabited country be discovered and planted by English subjects, all the English laws then in being ... are immediately there in force ...
>
> 'But in conquered or ceded countries, that have already laws of their own, the King may indeed alter and change those laws; but, till he does actually change them, the ancient laws of the country remain ... Our American plantations are principally of this latter sort, being obtained in the last century either by right of conquest and driving out the natives ... or by treaties ...'

Source 2

Sir William Blackstone's legal analysis

The British alleged that the lack of land cultivation and structures meant that the continent of Australia was uninhabited. However, this is contradicted by the existence of First Nations Peoples' songlines, which defined the borders of language groups, nations and clan lands, as well as complex laws around land inheritance. Furthermore, European explorers' records describe how First Nations Peoples managed the land for food production, and had silos and semi-permanent villages.

Nevertheless, the British push to claim new land was too strong. Before Cook discovered New Holland, he was secretly instructed:

> 'You are also with the Consent of the Natives to take Possession of Convenient Situations in the Country in the Name of the King of Great Britain: Or: if you find the Country uninhabited take Possession for his Majesty by setting up Proper Marks and Inscriptions, as first discoverers and possessors.'

Source 3

Secret Instructions to James Cook, contained in the letterbook carried on HMS *Endeavour* (1768)

Consent was never given, but Cook used the principle of *terra nullius* to claim the land anyway.

Conquest or settlement?

The concept of *terra nullius* is first mentioned in colonial sources in an 1819 debate between Supreme Court Judge Barron Field and Governor Lachlan Macquarie, who wanted to tax the colonists.

Field argued that Australia was not 'conquered' but 'freely settled' under *terra nullius*. Although the British Parliament agreed with Field, violent conflicts between colonists and First Nations Peoples contradict any notion of Australia being peacefully 'settled'.

Today, we recognise that the First Nations Peoples of Australia had established cultures and ownership of the land before the first British explorers arrived.

Learning ladder H2.10

Show what you know

1 How does ACL view the connection between the people and the land?

2 How does First Nations Peoples' view of the law and land differ to European land ownership laws?

3 What does *terra nullius* mean?

Cause and effect

Step 1: I can recognise a cause and an effect

4 How did declaring the land *terra nullius* lead to the colonisation of Australia?

Step 2: I can determine causes and effects

5 How did Sir William Blackstone provide the legal justification for the colonisation of other people's land?

Step 3: I can explain causes and effects

6 How did the inability of European colonisers to recognise the governing structures of the First Nations Peoples of Australia affect their view of the land?

Step 4: I can analyse causes and effects

7 How did Judge Field use *terra nullius* to defeat Governor Macquarie?

Cause and effect, page 247

Why was the first Australian colony established?

From Cook's 'discovery' of Australia to the landing of the First Fleet in 1788, 18 years passed. During that time, the British Empire suffered devastating financial losses due to war with America. They needed new sources of income and a new place for their prisoners, as their gaols and prison ships were overflowing. The British looked to Australia to expand their empire.

The initial proposal

James Mario Matra, an American who was loyal to England, proposed that the British government establish a colony in New South Wales (NSW) to house loyalists from the USA, to exploit the land and resources and as a place to send convicts. Both Matra and Joseph Banks had accompanied Cook on his 1770 voyage, and Banks supported Matra's proposal. Prime Minister William Pitt saw Matra's plan as the only viable option.

Source 1

The Founding of Australia by Captain Arthur Phillip R.N. Sydney Cove, January 26th 1788. [Original oil sketch by Algernon Talmage, 1937]

Economic incentives

The Seven Years War left Britain with £133 million of debt, and around £10 million per year in interest. The American Revolution had cost £250 million and, when it lost, Britain had forfeited substantial income, through taxation, exports and access to natural resources. They also lost a location to which to send criminals for the punishment of transportation.

Sending convicts to NSW thus solved a number of the empire's problems. The convicts would provide cheap labour to build the colony and it was hoped the new colony would provide a good return on investment.

Social incentives

Britain's Industrial Revolution led to the **mechanisation** of some agricultural jobs (see Chapter H1). Former farm labourers sought work in the cities, but there were not enough jobs for everyone.

Low wages, high unemployment and widespread poverty led to higher rates of crime and, therefore, more prisoners. Because the prisons in England were overflowing, the government began using prison hulks – ships converted into floating prisons. However, these hulks were often overcrowded and diseases were rife.

Meanwhile, English judges were still sentencing criminals to transportation, but there was nowhere to send the convicts. The NSW colony promised to take the pressure off the penal system.

The First Fleet

The First Fleet consisted of nine transport ships (six carrying convicts and three carrying supplies) and two warships. These carried Captain Arthur Phillip, his guards and officers, and approximately 850 convicts to NSW. They arrived at the place the Eora called Warrane, later known as Sydney Cove, on 26 January 1788, where they established the first colony in Australia. Phillip's journal noted that Sydney Harbour was an ideal harbour from military and commercial perspectives, as it was able to host a thousand ships.

Source 2

One of many prison hulks anchored in Portsmouth Harbour, England, in the early 19th century.

Learning ladder H2.11

Show what you know

1 Name three problems it was hoped the new colony would solve.

2 Source 1: How would the scene of British soldiers alighting from ships and holding a flag-raising ceremony have appeared to the local First Nations Peoples?

Historical significance

Step 1: I can recognise historical significance

3 List three or more different causes that contributed to the British establishing a colony in New South Wales.

Step 2: I can explain historical significance

4 What were Matra, Banks and Pitt's roles in making the case for establishing a colony?

Step 3: I can apply a theory of significance

5 What was the rationale for using a convict labour force to establish the colony?

Step 4: I can analyse historical significance

6 Explain why the American Revolution was also a significant event in the colonisation of Australia.

HOW TO

Historical significance, page 251

Who were the convicts?

The convict population consisted of British citizens from across the empire, including Africa, Asia, the Americas and India. The diversity of ethnicities, social classes and varied professions created a labour force suited to establishing a colony.

Convict demographics

Three-quarters of the convict population were English, with the remainder Irish, Scottish or from the rest of the British Empire. Transportees were mostly men, but there were also some women and children. Approximately 163 500 convicts were sent to Australia from 1788 to 1868.

Crime and punishment

Most convicts were sentenced to crimes related to poverty, such as stealing food and goods, selling stolen goods, highway robbery and 'indecent public displays of affection'. A few convicts were soldiers who had committed military offences.

Skilled criminals, who could write important documents or had learned a trade, were put to work. Some were given a 'Ticket of Leave', which let them move relatively freely around the colony, or were pardoned when they finished their sentences.

Some convicts who completed their sentences became freed men and women known as **emancipists**.

Female convicts

British women in the colonial period were treated as second-class citizens, and married women had no legal status as separate entities to their husbands. Women were expected to obey their husbands and to bear children. In contrast, the women of the First Nations of Australia held equal status and power in their communities from the distant past until today.

From 1788–1853, 29 960 women were transported; some were convicts while others were convicts' wives. They could not vote or hold positions of authority, and were not encouraged to get an education.

Source 1

Captain Watkin Tench's 'The Landing of the Convicts at Botany Bay', from his book *A Narrative of the Expedition to Botany Bay*, published in 1789

As legal divorce was only an option for the wealthy, poorer people conducted wife-selling auctions to dissolve their marriages. Many women insisted on this illegitimate means of divorce, as it was placed on public record. Governor Macquarie of New South Wales ended the practice of wife-selling in 1811.

Currency lads and lasses

The 'currency lads and lasses' were the new generation of children born in the colony, known collectively as the 'currency'. This negative label was used to distinguish British-born 'sterling' children from the colony-born children.

'The currency' were apparently fitter, taller and healthier than most children from England. They and their parents embraced the freedom that the colony offered and wanted to leave their convict pasts behind. They even developed their own unique Australian accent within a generation, a fusion of southeast English and Irish accents.

Source 2

Augustus Earle, *Female penitentiary or factory, Parramatta*, NSW

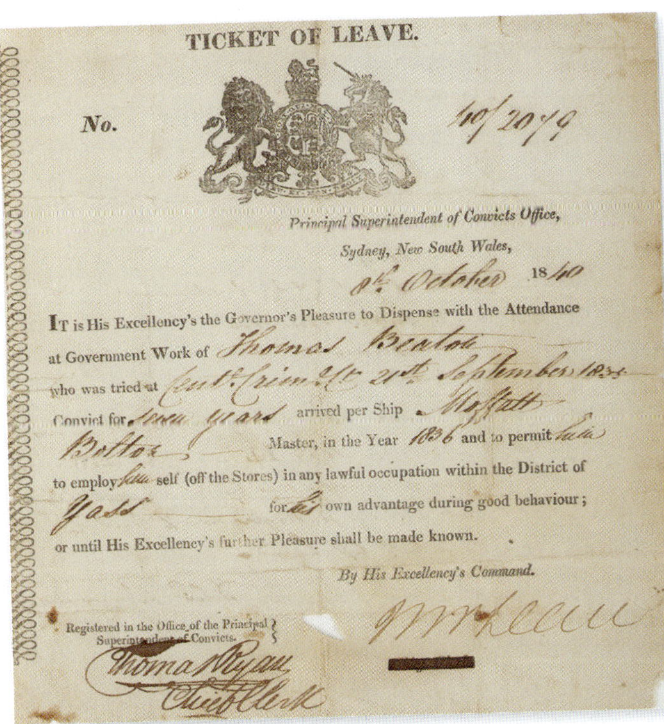

TICKET OF LEAVE.

No. 40/2079

Principal Superintendent of Convicts Office,
Sydney, New South Wales,
8th October 1840

IT is His Excellency's the Governor's Pleasure to Dispense with the Attendance at Government Work of Thomas Beaton who was tried at Cent. Crim. Ct. 21st September 1835 Convict for seven years arrived per Ship Moffatt Bolton Master, in the Year 1836 and to permit him to employ him self (off the Stores) in any lawful occupation within the District of Yass for his own advantage during good behaviour; or until His Excellency's further Pleasure shall be made known.

By His Excellency's Command.

Registered in the Office of the Principal Superintendent of Convicts.
Thomas Ryan
Chief Clerk

Source 3

With a Ticket of Leave, you were free to work, marry and move around the colony, while still serving your sentence.

Learning ladder H2.12

Show what you know

1 Create a dot-point summary of the crimes for which people were sent to New South Wales.
2 Why was a Ticket of Leave desirable to convicts?
3 Roughly how old is the unique Australian accent?

Historical interpretations

Step 1: I can recognise that the past has been represented in different ways

4 Were all convicts treated the same or did some receive special status and privileges?

Step 2: I can describe historical interpretations

5 What was the public view of convict women?

Step 3: I can explain historical interpretations

6 Why were people keen to renounce their convict pasts? Is this true of people with convict heritage today?

Step 4: I can analyse historical interpretations

7 Why might a poor woman insist on being sold at a wife-selling-auction?

Historical interpretations, page 255

Who were the free settlers?

British colonists who chose to start a new life in New South Wales were known as free settlers. Believing themselves to be superior to convicts and emancipists, they viewed themselves as near the top of the developing colonial social order.

Squatters and exclusives

First arriving in 1793, the free settlers were granted the largest and best landholdings, were elected to government, had links to the military and had connections to people of influence in England. Also known as 'exclusives' and 'squatters', the free settlers tried to create a class system in which they enjoyed a higher status to former convicts.

In addition to receiving large tracts of land seized from the First Nations Peoples of Australia, free settlers could 'squat' (rent and run livestock) on other Crown Land in addition to their own, were supplied with food until they could be self-sufficient, were given seeds and farming implements, and were provided with convicts as free labour.

Source 1

Frederick McCubbin's *The Pioneer* (1904) is one of the most famous depictions of early Australian colonial life. In the background, the development of the colony of NSW is shown through the progression from wagon to cottage to city.

John Macarthur: hero of the fleece

Squatter John Macarthur was a prominent soldier, entrepreneur and politician. He is best known as the pioneer of the Australian wool industry. His high-quality wool won numerous awards and earned him the title, 'the hero of the fleece'. Macarthur wanted to establish the free settlers as the 'landed gentry' of the colony, proposing a government structure like that of Britain's House of Lords.

Macarthur significantly developed the wool industry by introducing more Merino sheep to Australia. High-quality wool exports, along with whaling, sealing and coal industries, established New South Wales' role as a primary resource for nations undergoing industrialisation. The success of agriculturalists such as Macarthur led the colony closer to food and economic security. His house, Elizabeth Farm, still exists to this day as a national trust. This indicates his importance to colonial history.

Source 2

Caroline Chisholm (1808–1877) was a free settler and philanthropist who helped colonial women and immigrants, as well as attempting to reform unjust land ownership laws.

Caroline Chisholm: free settler and philanthropist

Caroline Chisholm and her husband travelled to NSW in 1838 and found the colony's single women in a deplorable state. Many young girls were homeless, surviving via begging or prostitution. In 1841, Chisholm successfully petitioned Governor George Gipps to provide an old barracks, where she housed, fed and taught women. She also petitioned Gipps to write letters recommending immigrants for farm labour, and even escorted them into regional parts of the colony.

Chisholm's attempts to obtain land for immigrants disturbed other free settlers, who harnessed anti-Catholic sentiment against her. She spent 1846–1854 in England, where she presented to the House of Lords on the colony and Irish matters.

In 1854, Chisholm went to Victoria, where she assisted immigrants travelling to the goldfields and advocated for reforms to unlock land for smaller farmers. She was able to challenge the existing 'squattocracy' in her push for a fairer vision of Australian society.

Learning ladder H2.13

Show what you know

1 Why did the free settlers try to deny the emancipists certain rights?

2 What rights did the free settlers have that were denied to the emancipists?

3 Why was colonial Australia described as 'riding on the sheep's back'?

Historical significance

Step 1: I can recognise historical significance

4 Which important industries were the first to be established in the colony of New South Wales?

Step 2: I can explain historical significance

5 Use political, economic, social and technological (PEST) factors to explain the factors that might encourage a person to relocate their family from England to the Australian colonies.

Step 3: I can apply a theory of significance

6 Why is Macarthur's contribution to agriculture recognised above coal mining, whaling and sealing?

Step 4: I can analyse historical significance

7 Why was Caroline Chisholm's contribution to colonial society considered to be so important in two phases of colonial expansion?

Historical significance, page 251

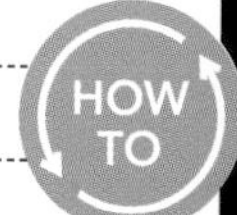

How was the colony governed?

New South Wales was governed as a prison as well as a society. Governors were often military men, who were used to enforcing harsh discipline. Soldiers were charged with enforcing the code of the penal colony, the magistrates' harsh judgements and the governors' orders.

The power of the military

Until 1850, British troops came from England to defend the colonies and enforce the law. In theory, one of the troops' duties was to guard New South Wales against external attack, but that never occurred. Their main job was to maintain civil order, often by putting down convict uprisings and protests, and to suppress First Nations Peoples' resistance efforts.

After finishing their tour of duty, some soldiers returned to England, while others stayed in the colony as free settlers. Many of the officers were well educated, and several of the colony's intellectuals, judges and future governors emerged from the New South Wales Corps.

Separated socially from the convicts, and backed up by their weapons and training, the military forces were often corrupt. The soldiers formed an **oligarchy**, seizing control and monopolising the colony. They had first pick of the supplies on any ships, and shamelessly increased the price of goods. They took food from government storage, dismissed magistrates and assumed these positions for themselves. Even the grain that soldiers sold back into government stores fetched a higher price than that of the free settlers and emancipated convicts.

Lieutenant John Macarthur was a soldier before he came a wool pioneer, and he was just as powerful as the governor, in many ways. He had the soldiers and judiciary on his side, owned the choicest land in the colony, and played major roles in colonial politics, farming and industry.

Source 1

Soldiers of the British Empire in colonial Australia

Source 2

This lithograph of the arrest of Governor Bligh was displayed outside Government House during the Rum Rebellion. A propaganda piece, it shows Bligh being dragged from under his bed; in reality, the soldiers searched for hours to find him.

The Rum Rebellion

From the day the First Fleet landed, a system of barter developed in the colony. An alternative 'black market' economy flourished because there was not enough currency and food was strictly rationed. Rum and liquor from overseas was traded as a new currency – one controlled by the military, which came to be known as the 'Rum Corps'.

In 1807, Governor William Bligh became unpopular with soldiers when he prohibited the trade and barter of rum. He also blocked John Macarthur's attempts to import stills (to make his own rum), and opposed Macarthur's land claims. When Bligh later challenged Macarthur over one of Macarthur's ships, it was the last straw. Macarthur and his allies in the military decided to take control of the colony.

In 1808, on 26 January (the colony's foundation day), the New South Wales Corps stormed Government House and placed Bligh under house arrest for two years. Major George Johnston led the Corps, and appointed himself 'Lieutenant-Governor', while Macarthur adopted the role of Colonial Secretary.

When word of the rebellion reached Britain, Johnston was recalled to England and court-martialled, while Bligh was exonerated. As Macarthur would likely have faced charges himself, he joined Johnston to support his defence. The entire New South Wales Corps was recalled, and Macarthur stayed in England until 1817, when he received permission to return to New South Wales on the condition that he stay out of public affairs.

Source 3

Arthur Phillip was the first governor of New South Wales. His reformist approach towards convicts and desire to establish a relationship with the First Nations Peoples of Australia distinguished him as progressive for his time. [Picture by H. Macbeth-Raeburn, 1936]

The colonial governors

In the British Empire, governors were public officials appointed by the monarch (or their cabinet) to oversee colonies and manage their administration. It was a powerful role; governors were the supreme authority in each colony, with the power to implement Britain's laws and oversee all functions of government.

New South Wales was the starting point of colonisation, so its governors were (at first) the governors of the entire continent of Australia. As the new colonies were established, they were assigned their own governors, each as powerful as each other, but all were ultimately answerable to the British government.

By the end of Governor Macquarie's term (1810–1821), New South Wales was the largest and most prosperous colony in Australia. Its exports of coal, wool and other resources had established it as a resource economy. From an original population of around 1000 Europeans in 1788, the colony had grown to 75 000 colonists by 1831. Additionally, many First Nations People lived in the colony at the time.

Early governors of the New South Wales colony

Name	Term of office	Major events
Arthur Phillip	1787–1792	Phillip founded the colony on 26 January 1788. He initially tried to establish good relations with the First Nations Peoples. Later he changed policy, ordering the hunting of Pemulwuy (see page 100).
Francis Grose William Paterson	1792–1794	Lieutenant-Governors Grose and Paterson temporarily took shared control of the colony.
John Hunter	1795–1799	Hunter was an artist, and made more than 100 paintings of plants, animals, fish and scenes around Sydney. He was unpopular with the military, and was recalled to England in 1799.
Philip Gidley King	1800–1806	King oversaw the establishment of new colonies in Van Diemen's Land and sent expeditions to explore the coastline of what is now called Victoria.
William Bligh	1806–1809	Bligh disbanded the oligarchy of soldiers, made liquor illegal as currency and established a 'futures market' promoting small farmers. He was an unpopular figure, and was arrested during the Rum Rebellion of 1808.
Lachlan Macquarie	1810–1821	Macquarie reformed the soldiery and the police, outlawed wife-selling and funded public works. He also escalated conflicts with the First Nations Peoples of Australia.

Representative government

As the numbers of free settlers increased, the lack of a civil (non-military) government became a political issue within New South Wales. The British Parliament responded in 1823 by passing legislation that established a Legislative Council – a civilian body that would advise the governor. The legislation also regulated the courts and judiciary, and any bill put forward by the governor had to be approved by the chief justice to become law.

In 1842, the Council passed the *Constitution Act*, which introduced a level of representative government to the colony. The Legislative Council now consisted of 24 elected members, all wealthy landowners, along with 12 members appointed by the governor. In 1856, New South Wales finally implemented its own bicameral (two-house) parliament. Men gained the right to vote in 1858, while women won the right to vote in 1902.

Source 4

Woollarawarre Bennelong, was a Wangal man abducted by Governor Arthur Phillip. Bennelong was taught how to speak English, and ultimately became an ambassador between the colonists and the local First Nations Peoples. [Print c. 1798, artist unknown]

Source 5

The NSW Parliament House in Sydney was originally built in 1816 as a hospital. It was expanded in 1843 for its new purpose as the meeting place for the colony's Legislative Council.

Learning ladder H2.14

Show what you know

1 How did the New South Wales Corps soldiers amass so much power?

2 What evidence can you find of the soldiers' corruption and privilege?

3 Which major economic industries were established in the early years of the colony?

Historical interpretations

Step 1: I can recognise that the past has been represented in different ways

4 What was the artist attempting to show in Source 2?

Step 2: I can describe historical interpretations

5 What would the display of Source 2 outside Government House have done to both Bligh and the position of governor?

Step 3: I can explain historical interpretations

6 Were the governors at the top of colonial society? What evidence challenges this interpretation?

Step 4: I can analyse historical interpretations

7 What was the Rum Rebellion really about? What does calling it 'the Rum Rebellion' 50 years later indicate about the causes of the rebellion?

Historical interpretations, page 255

What were the Frontier Wars?

As the colonies grew, tensions escalated between the colonists and the First Nations Peoples of Australia, resulting in many violent conflicts. The resistance of First Nations Peoples took many forms, such as taking livestock, burning crops, attacking colonists and continuing their society, laws and languages as a surviving resistance. The ongoing conflict between the colonists and the First Nations Peoples is now referred to as the Frontier Wars.

Source 1

This image, titled 'Australian Aborigines – War', appeared in the *Melbourne Post* on 27 May 1867.

Early conflicts

Governor Arthur Phillip initially tried to establish good relations with the First Nations Peoples of Australia and ordered that they be well-treated, punishing colonists who did not respect his orders. However, this did not prevent conflict, as the colony continued to take resources and steal traditional lands. First Nations Peoples responded to this in various ways, sometimes with violence, sometimes by trying to take back stolen animals and resources.

Conflict escalated in the 1790s and the early 1800s, as soldiers supported the colonists with force. Convicts and colonists raided First Nations Peoples' lands, killing warriors and abducting and sexually assaulting women. First Nations warriors retaliated, killing the perpetrators as 'payback'. They also resisted land claims by spearing colonists and their livestock.

This became a vicious cycle of violence. The colonists retaliated in various ways: shooting people on the spot, and poisoning, torturing and even massacring First Nations Peoples. The colonists did not always separate the innocent from the guilty – some of their victims were those who were easiest to find, rather than the actual perpetrators.

Source 2

Governor Arthur Phillip was speared as 'payback' for kidnapping Bennelong and forcing him to live as a prisoner. Phillip prevented his soldiers from retaliating. [Detail from Drawing 23 of the Watling Collection, by the Port Jackson Painter, c. 1790.]

Resistance and violence

Targeted retaliation or 'payback' is a traditional punishment under Aboriginal Customary Law, often carried out by a *Carradhy* or 'clever man'. Much of the resistance from the First Nations Peoples of Australia came from *Carradhy* spearing a specific colonist as payback for heavy crimes. The guns of the time were not as accurate as spears thrown by skilled warriors, so a spearing was often more lethal than a bullet wound.

Other warriors resisted the loss of their land and traditional food supplies, which the colonists were taking for their livestock. When the numbers of native species were reduced, some First Nations Peoples speared the colonists' livestock for food, which resulted in retaliation from the colonists.

The skirmishes between the two groups lasted anywhere from a few months to a decade, and continued from 1788 until the 1930s. This long period of ongoing violence is now referred to by many historians as the Frontier Wars, recognising that the conflict went far beyond a few isolated instances.

Source 3

Pemulwuy was responsible for a number of payback attacks on colonists. He escaped death and incarceration on several occasions, and claimed to have spiritual protection from guns.

Pemulwuy

Pemulwuy was a First Nations *Carradhy* who became famous for his guerrilla warfare against the colonists. Claiming immunity to bullets, he was caught but escaped captivity. He set fire to crops, stole valuable tools and organised resistance forces of Eora men to attack the colonists in multiple raids for over a decade.

In 1790, Pemulwuy used a 'death spear' to kill John McIntyre, Governor Phillip's groundsman, as payback for McIntyre's murder of several First Nations People. In response, Governor Phillip ordered colonial troops to either find Pemulwuy or enact collective punishment on his people.

Pemulwuy continued to lead resistance efforts for more than a decade. In 1795, he was severely wounded by colonial forces, but evaded capture. In 1801, Governor King issued an order to bring Pemulwuy in dead or alive, and he was killed by a British sailor in 1802.

Source 4

'Death spears' had a special blade made from chips of stone set in resin. The bladed head inflicts a long gash and leaves chips inside the wound.

Truganini – Queen, diplomat and bushranger

The Black War (1824–1831), one of the most violent frontier conflicts, was an attempt by colonists in Tasmania to wipe out the Palawa people. Brutal attacks by colonists and whalers, as well as a program of forced resettlement, almost annihilated the Palawa, who had numbered approximately 4000 before colonisation.

Truganini was a Palawa chieftain's daughter from Bruny Island, south of Hobart. By 1830, when she was 18, most of her people had been murdered or abducted. She was then approached by George Augustus Robinson, a preacher who acted as a **conciliator** between the colonists and the Palawa.

Robinson persuaded Truganini and her people to relocate to a mission on Flinders Island, where he insisted they would be safe. Believing Robinson to be sincere, Truganini travelled with him to meet other Palawa and helped convince them to relocate. By 1835, almost all the surviving Palawa – around 200 people – had moved to Flinders Island. But the mission was not the sanctuary Robinson promised – conditions were harsh and the colonists tried to convert the Palawa to Christianity.

Robinson sailed to Melbourne in 1838 to manage the Port Phillip Protectorate (see page 108). Truganini went with him to help negotiate with the Kulin nations, but she no longer trusted Robinson and did little to aid the Protectorate. In fact, she ran away in 1840 to join a gang of bushrangers, which robbed colonists and killed two whalers. They were captured in 1841 and Truginini was sent back to Flinders Island.

Source 5

Truganini aided the resettlement of the Palawa people, ultimately preventing their complete extermination.

She was later relocated to Oyster Cove, near Hobart, with the rest of the surviving Palawa – approximately 50 people. She died there in 1876, outliving all of her husbands.

Once believed an extinct cultural group, the Palawa people have recently enjoyed a revival of their community and even their language. This would never have been possible if Truganini had not protected her people from the Europeans' attempted genocide.

Frontier massacres

Not every Frontier War conflict was a battle; most were massacres in which unarmed people on both sides were killed. If the British military or colonists could not find the specific warriors responsible for an act of resistance, they often punished any group of First Nations Peoples they could find, even if those people had nothing to do with the original attack. These massacres were acts of genocide, as defined by the United Nations – the killing of a group because of their national, ethnic, racial or religious status.

In 1838, the Waterloo, Slaughterhouse and Myall Creek massacres occurred in northern NSW, near Moree. The Myall Creek massacre was particularly grotesque, as the Wererai people were raped, burned alive and decapitated. What was unusual about the massacres of 1838 is that they resulted in prosecution; seven of the 10 accused stockmen were hanged for the murder of an innocent child. Massacres continued after this but the perpetrators used more discreet means of killing, such as poisoned flour, and they were more careful to destroy the evidence (usually by burning the bodies).

Source 6

Commemorative stone at the site of the Myall Creek massacre. Myall Creek was added to the National Heritage List in 2008.

Casualties

The casualty rate from the Frontier Wars is hard to determine because of under-reporting. 'Unpleasant' events were often omitted in official soldier, colonist and police reports of the time. Through remaining documents, such as journals and letters, historians have pieced together some of the violence that occurred, but much is probably still missing.

Conservative estimates put the colonist deaths at between 2500–5000, and the deaths of First Nations Peoples from 20 000–60 000. Some historians estimate that the combined casualties of the Frontier Wars came to nearly 100 000 people.

Learning ladder H2.15

Show what you know

1 Describe the battle occurring in Source 1.

2 Write a short profile on either Pemulwuy or Truganini. How did they protect their people?

3 What is 'collective punishment'? How did the colony retaliate against attacks?

Cause and effect

Step 1: I can recognise a cause and an effect

4 What caused the Frontier Wars?

Step 2: I can determine causes and effects

5 Why do you think Phillips' policy towards First Nations Peoples changed? How did Pemulwuy play a part in his new attitude?

Step 3: I can explain causes and effects

6 Why is there limited evidence about First Nations Peoples' history of resistance and massacres in Australia?

Step 4: I can analyse causes and effects

7 How did the Myall Creek Massacre change the way in which massacres were subsequently conducted, investigated and prosecuted?

Cause and effect, page 247

How was the colony of Victoria established?

For over 40 000 years, the First Nations Peoples of Victoria lived in large, semi-permanent groups of over 500 people. The Victorian colonies, first at Sullivan Bay and then at Port Phillip Bay, disrupted these traditional homelands amid the signing of a contentious treaty.

From Naarm to Port Phillip Bay

The region we now call south central Victoria is the traditional home of the Kulin nations, an alliance of the Wurundjeri, Boonwurrung, Wathaurong, Taungurong and Dja Dja Wurrung Peoples. Those living around Melbourne spoke the Woi wurrung language, and what we know as Port Phillip Bay, they called Naarm. Boon Wurrung was spoken along the Mornington Peninsular and Wathaurong was spoken around Geelong.

Source 1

Sullivan Bay was the site of Victoria's first penal colony from 1803–4, which was located in Port Phillip Bay, near what is known today as Sorrento.

Source 2

Rock art from Bunjil's sacred cave in the Grampians

According to the Boonwurrung people, the sea became angry because the people broke the laws of the land, and rose to flood the plains. The people asked the powerful Creation spirit Bunjil (the Sea Eagle) to help, but he only intervened once they promised to return to their traditional way of life. Bunjil then walked into the sea and reversed the flooding, creating the bay called Naarm in the process.

Thousands of years later, in 1802, British explorer John Murray became the first European to sail into the bay. He named it Port King, after Governor Philip Gidley King. Three years later, King renamed it Port Phillip Bay in honour of Captain Arthur Phillip, the First Fleet Captain and colonial governor.

First colonisation attempts

Commercial and political reasons were behind Governor King's push to set up a colony in the newly discovered port, including the desire to:

- establish a seal trade
- use the region's plentiful natural resources
- claim and secure Bass Strait for the British, as King was worried the French could claim it.

The British government approved the establishment of a new colony in 1803, and an expedition of about 400 soldiers and convicts, led by Lieutenant-Governor David Collins, set sail from England. However, Collins was not happy with the site, named Sullivan Bay, because it lacked timber and enough water. He obtained the King's approval to transfer the colony to Van Diemen's Land (Tasmania), and eventually settled in a place he named Hobart Town.

By 1836, Van Diemen's Land was independent from NSW, but the new colony was struggling. The farm land was suffering from overuse and there had been food shortages. The colonists shot kangaroos to supplement their diet, which was one of the causes of the 'Black War' with the Palawa. Governor Richard Bourke obtained approval to establish a new colony, and two men – John Batman and John Pascoe Fawkner – led efforts to colonise Port Phillip Bay.

John Batman

John Batman was a farmer and bounty hunter who had attacked the Palawa during Tasmania's Black War. Despite this, Batman was considered progressive for his era, and was willing to employ and negotiate with First Nations Peoples. The need for farmable land spurred Batman to travel to Port Phillip Bay to stake a claim. In 1835, he had lawyers draft a treaty, which was very unusual for the period, and he set sail for the mainland.

Batman met with a group of *Ngurungaeta* (headmen or leaders) from several tribes of the Kulin nations, including the Duttigallar tribe, whose lands he hoped to claim. He apparently negotiated terms with them, and the treaty was 'signed' with a ritual pouring of sand through hands somewhere near the junction of the Merri Creek and the Yarra River.

The treaty was one of the largest claims in all the colonies, for 600 000 acres from the Yarra River to Djilong (Geelong) and beyond. In exchange, Batman gave the *Ngurungaeta* '20 pairs of blankets, 12 tomahawks, 30 knives, 12 pairs of scissors, 10 looking glasses, 50 handkerchiefs, 12 shirts, 4 flannel jackets, 4 suits of clothes and 50 pounds of flour' as stated in the treaty.

Debate continues over the signing of the treaty between Batman and the *Ngurungaeta*. Some historians argue that the Kulin saw the ceremony as a gift, and did not realise the Europeans wanted to claim the land. However, First Nations oral history suggests that the *Ngurungaeta* were well aware of what they were doing, and signed the treaty to limit the spread of the colonists onto their lands. This position argues that the Kulin nation entered a *tanderrum* with the colonists – a ceremonial agreement that allows safe passage and a temporary sharing of land.

Source 3

An artist's impression of the signing of Batman's treaty by *Ngurungaeta* of the Duttigallar people of the Kulin nations

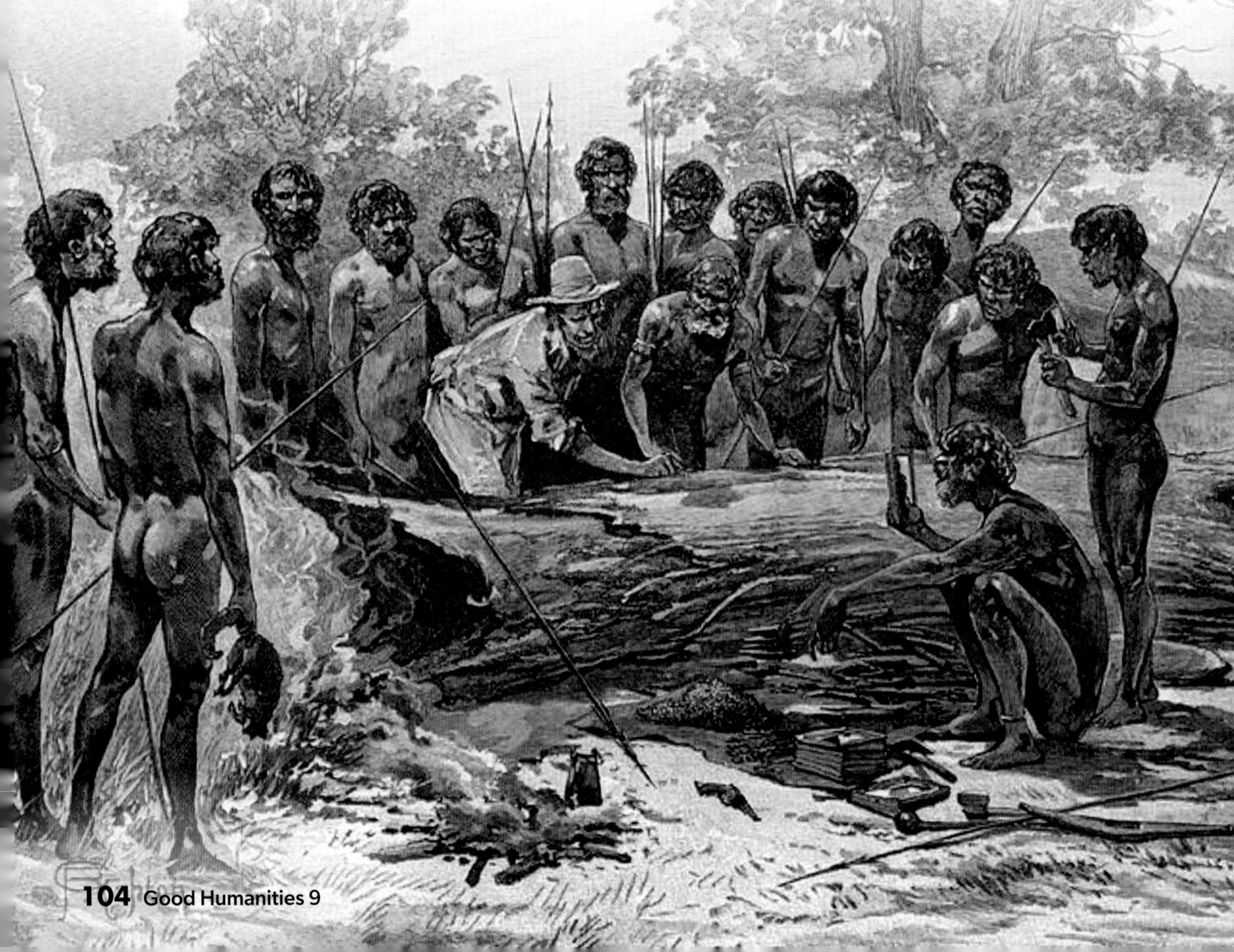

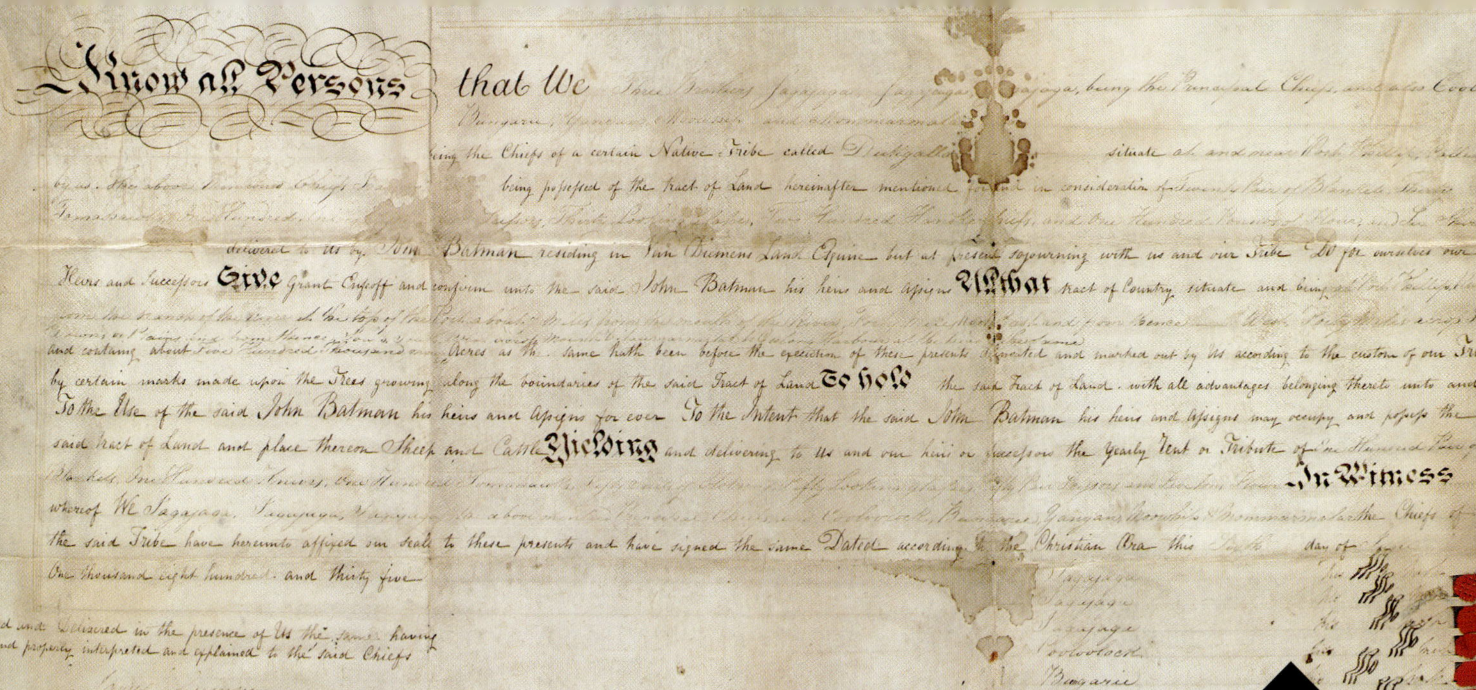
Know all Persons that We [illegible]

[illegible] being the Chiefs of a certain Native Tribe called [illegible] situate at and near Port Phillip [illegible]

[illegible] being possessed of the tract of Land hereinafter mentioned for and in consideration of [illegible]

[illegible] delivered to Us by John Batman residing in Van Diemens Land Esquire but at present sojourning with us and our Tribe Do for ourselves our Heirs and Successors Give Grant Enfeoff and confirm unto the said John Batman his heirs and assigns All that tract of Country situate and being at [illegible]

[illegible] and containing about Five Hundred Thousand [illegible] Acres as the same hath been before the execution of these presents delineated and marked out by Us according to the custom of our Tribe by certain marks made upon the Trees growing along the boundaries of the said Tract of Land To Hold the said Tract of Land with all advantages belonging thereto unto and To the Use of the said John Batman his heirs and assigns for ever To the Intent that the said John Batman his heirs and assigns may occupy and possess the said tract of Land and place thereon Sheep and Cattle Yielding and delivering to Us and our heirs or successors the Yearly Rent or Tribute of [illegible] In Witness whereof We Jagajaga [illegible] the Chiefs of the said Tribe have hereunto affixed our seals to these presents and have signed the same Dated according to the Christian Era this [illegible] day of [illegible] One thousand eight hundred and thirty five

[illegible] Sealed and Delivered in the presence of Us the same having [illegible] and properly interpreted and explained to the said Chiefs

Source 4

Reproduction of a transcript of John Batman's treaty with the Kulin Nation *Ngurungaeta*.

Problems with Batman's treaty

The Batman treaty was legally problematic for a number of reasons.

- It did not include representatives of peoples from Djilong, which meant a second treaty with the Wathaurong was needed.
- It did not have the approval of the British Crown, parliament or the governor.
- It did not align with the British legal definition of Australia as a 'settlement', and the declaration of the land as *terra nullius*. A treaty was only required if the land was conquered in battle.

The treaty was therefore declared void, and an act of 'trespass' on vacant Crown land, by NSW Governor Richard Bourke on 26 August 1835.

Batman's treaty acknowledged the prior rights of the First Nations Peoples to Australian land. It even offered to pay them rent. Today, there are requests by First Nations Peoples for their own treaties with the Australian government.

Learning ladder H2.16

Show what you know

1 By whom, how and why was Naarm created?

2 Why did Collins request permission to move the colony from Sullivan Bay?

3 Why was the Batman treaty declared void by the government?

Historical interpretations

Step 1: I can recognise that the past has been represented in different ways

4 How is the signing of the Batman treaty portrayed in art such as Source 3?

Step 2: I can describe historical interpretations

5 Identify at least two reasons why First Nations Peoples might have accepted a treaty with colonists.

Step 3: I can explain historical interpretations

6 John Batman believed he had the authority to make a treaty with First Nations Peoples. Was he correct? Explain your answer.

Step 4: I can analyse historical interpretations

7 Research the full text of Batman's 1835 treaty. Suggest at least two ways in which the treaty and its signing were beneficial for First Nations–colonist relations, and at least two ways in which they were harmful.

Historical interpretation, page 255

How was Melbourne founded?

Governor Bourke nullified Batman's treaty with the Duttigallar people in 1836, and the new colony came under the control of NSW. Melbourne was set up as the colony's southern outpost by surveying land, dividing it into plots and selling it off.

John Pascoe Fawkner

John Pascoe Fawkner was the child of a British convict. He set up in Van Diemen's Land and became an entrepreneur: publishing, baking, transporting, building and selling alcohol without a licence.

In the race to colonise Port Philip Bay, Fawkner formed a syndicate in Launceston to purchase the *Enterprize*, a large coal schooner capable of carrying 55 tons of cargo. On 29 August 1835, they sailed the *Enterprize* up the Yarra before landing and erecting what was to become Melbourne's first building.

Tensions arose when Batman and his followers returned from Van Diemen's Land to discover that Fawkner had already started building. The two men eventually agreed to parcel out land in order to avoid conflict. Fawkner went on to become a major Melbourne businessman and one of the founding members of Victoria's first parliament in 1851.

Source 1

An artist's impression of John Pascoe Fawkner's landing in Port Phillip Bay in 1835, near what is now Melbourne's Docklands.

Source 2

The First Nations Peoples of Victoria are shown overlooking Collins Street, Melbourne, in this 1839 picture.

Establishing Melbourne

Melbourne's streets were laid out in a grid pattern by surveyor Robert Hoddle, who wanted the new city to have an organised structure. However, the city began as a sprawl of tents and huts on the banks of the Yarra River, and it took several years for the 'Hoddle grid' to be completed.

Governor Bourke visited the outpost in March 1837, as Hoddle was surveying, and sanctioned the new town. He also officially named it after the Second Viscount Melbourne – before this the town had several unofficial names, including Bearbrass, Bareport and Batmania. Melbourne was legally incorporated as a 'town' in 1842, and as a city in 1847.

Conflict with First Nations Peoples

As the colonists spread out and began to clear land to graze livestock, the First Nations People lost their hunting grounds, murnong pastures, traditional lands and sacred sites. They speared livestock for food, as well as in retaliation. This not only threatened the squatters' incomes, but meant they ran low on food. This created a culture of fear, as had happened in Van Diemen's Land.

Once again, this led to violence. Many First Nations people were killed in the colonists' extra-judicial reprisals and collective punishments. The conflict was so bad that the British Parliament established the Port Phillip Protectorate in 1838, in response to the violence committed against Victoria's First Nations Peoples.

Learning ladder H2.17

Show what you know

1 Who founded the city of Melbourne – John Batman, John Pascoe Fawkner or Governor Richard Bourke?

2 How was the new colony received by the First Nations Peoples of Victoria?

3 Why did the First Nations Peoples fight the squatters?

Continuity and change

Step 1: I can describe continuity and change

4 What was different about the founding of Melbourne, as opposed to Sydney?

Step 2: I can explain why something did or did not change

5 How is the Hoddle grid still present in the urban planning of modern-day Melbourne? Provide evidence to support your answer.

Step 3: I can explain patterns of continuity and change

6 How did Batman's treaty, and the action it involved, represent the desires of First Nations Peoples at the time, and today?

Step 4: I can analyse patterns of continuity and change

7 Analyse a map of Melbourne's Docklands. Identify five features of the area that have changed since Fawkner's landing and suggest why they changed.

Continuity and change, page 244

How did law and order change in Victoria?

Early colonial law focused on controlling convicts and severely punishing acts of rebellion. Slowly, it took on an organised structure with police officers, gaols and even rehabilitation. However, law enforcement also targeted and suppressed First Nations Peoples.

The Port Phillip Protectorate

In 1835, preacher George Augustus Robinson negotiated the surrender of 150–200 Palawa people in Van Diemen's Land. Hoping to achieve a similar surrender in Victoria, Governor George Gipps established reservations, missions and schools, called the Port Phillip Protectorate, and put them under Robinson's control.

The Protectorate's goal was to 'civilise' First Nations Peoples. They were given new names, taught to read and write English and encouraged to elect their own Native Police. In this way, Robinson hoped to replace their traditional cultures with British culture, Christian religion and European notions of work – notions that were exploitative and very different to those of First Nations societies.

Robinson and his Protectorate failed to achieve any significant progress. After a decade of mediocre management and limited success, the Protectorate was closed down.

The Native Police

Alexander Maconochie, a Scottish penal reformer, introduced the idea of a Native Police force in 1837. As with the Protectorate, the idea was that the Native Police would be 'civilised' and assimilated into British society. In Victoria, the Native Police were initially stationed in Narre Warren, in what was known as the Police Paddock. Narre Warren was originally called Narre Narre Warren, which means 'special place'; it was also the site of an Aboriginal reserve.

Source 1

A Native Police force (1842). Established in Port Phillip Bay and Queensland, these forces recruited young First Nations men.

Source 2

An illustration showing the attempt to prevent the landing of the convicts, published in *The Argus* on 6 March 1849

Maconochie's attempts to assimilate the First Nations members of the Native Police ultimately failed. The men retained many of their traditional customs and would often abandon their posts in bushlands. Nevertheless, they proved to be a powerful means of defeating the First Nations' guerrilla resistance, as they enabled colonists to avoid becoming recipients of 'payback'.

The Anti-Transportation League

Among the colonists, there was increasing disgust at having to receive convicts from England. Not because they felt transportation was morally wrong, but because they felt convicts took jobs and brought crime into the colony.

In 1849, 10 000 Victorian colonists marched in the streets to stop a shipment of 500 convicts from landing. In 1850–51, the eastern colonies formed the Anti-Transportation League to put an end to transportation. The League's advocacy was successful, and transportation was discontinued in 1851. The last convicts from England arrived in 1853.

Law in a new colony

In the early years of Melbourne, the settlers paid for their own police force, which worked alongside the military to maintain law and order.

When the Port Phillip District officially became the colony of Victoria in 1851, Superintendent Charles Joseph Latrobe became Lieutenant-Governor of Victoria. The Victorian Police was established in January 1853, with a force of 875 men. This police force was given jurisdiction over the entire colony, under the control of the Chief Commissioner of Police.

The new police force also managed the colony's gaols and prisons, which incorporated more of Alexander Maconochie's concepts of reform. Prisoners in Melbourne Gaol could earn privileges, and even their freedom, in exchange for good behaviour. This marked the beginning of a value shift from incarceration to rehabilitation, which is still part of Australia's criminal justice system.

Learning ladder H2.18

Show what you know

1 What was the Port Phillip Protectorate? What were its purposes?

2 Source 1: How are the Native Police being promoted in this image? Analyse the fashions evident in the photograph and who is seated and who is standing.

3 Why did the Anti-Transportation League want to stop convicts being sent to Australia?

4 Why would First Nations People have joined the Native Police?

Continuity and change

Step 1: I can describe continuity and change

5 Which of Maconochie's reforms have been maintained in the present day?

Step 2: I can explain why something did or did not change

6 Why was the structure of Victoria's Native Police copied by other Australian colonies?

Step 3: I can explain patterns of continuity and change

7 The old Melbourne Gaol stood on a hill overlooking the city. Why was its presence on the horizon both a warning and a deterrent?

Step 4: I can analyse patterns of continuity and change

8 How have Australian legal systems continued to interact with First Nations Peoples and their customary laws and culture? (You may need to conduct further research.)

Continuity and change, page 244

What was life like during the Victorian gold rush?

The discovery of gold in Australia in 1851 caused a huge influx of people to the goldfields. Thousands of miners worked claims, enduring tough conditions in the hopes of striking it rich. Population shifts created a cultural melting pot, Victoria's economy was supercharged, and the lands and food sources of Victoria's First Nations Peoples were devastated.

The gold rush in Victoria

Following finds in NSW in 1851, gold was also discovered in Victoria around Warrandyte, Clunes and Ballarat. Miner James Esmond sent news to the *Geelong Advertiser* that Clunes was rich in gold. The story was published on 22 July and, with that, the Victorian gold rush had begun.

Soon there was a flood of people arriving from Europe, Asia and America, all hoping to cash in. Every month, 10 000–20 000 people arrived in Victoria with no idea of how to mine or how to live in the bush. By 1852, the population of the colony soared to 168 231.

Once again, the colonists' activities profoundly affected the lives of the local First Nations Peoples, the Wathaurong and the Dja Dja Wurrung. Mining desecrated sacred sites, hunting grounds were lost and traditional food sources moved to other areas. Poorly marked tunnels and shafts entombed people during the night. Increased contact with Europeans brought violence, alcoholism and disease. However, mining also brought business opportunities for First Nations Peoples, as their knowledge of the land was of great assistance to people seeking claims.

Source 1

Prospectors in Dargo, in the Gippsland region of Victoria. Prospectors would build huts over their claim, making them easier to watch and protect.

Source 2

Victorian police officers of the 1860s

Traps and diggers

Governors FitzRoy and LaTrobe were dismayed at the exodus of people from the cities and farms to the goldfields, which left Australia's pastoral industries short of labour. To help retain labour, and bring in more revenue, they set a gold-mining licence fee of 30 shillings a month. The governors also recruited an order of police known as 'traps', who wore a distinctive blue uniform, to check that miners carried valid licences and documentation.

Many of the miners, known as 'diggers', hid their valuable licences to prevent them from becoming damaged by their wet, messy work. However, this meant that when traps went on licence raids, many diggers were fined for not being able to produce their licence on the spot. Traps were allowed to keep half the fines they collected, so they frequently went on 'digger hunts'. The traps were violent during these raids, and miners might be flogged, have their tents burned or be chained to a log for a day or a night without access to food, water or a toilet.

The traps performed other police duties on the goldfields, such as breaking up fights or protecting gold shipments. Some of them were ex-convicts, with a reputation for brutality, which made them ideal for fending off bushrangers while gold was transported from the diggings to the ports.

The hard mining life

Miners endured tough living conditions. By day, they worked in hard, wet and frequently dark conditions, while by night they lived in canvas tents or other very basic accommodation. Mining work offered no guarantee of turning up enough gold to pay the monthly licence fee; so many miners took their chances to mine without one. Miners would dig a claim until any gold ran out, before starting again elsewhere. Unethical prospectors would mine at an angle into another person's claim, or find out about a pending claim and try to 'jump' it first.

To maximise efficiency, industrialised mining companies used steam-powered drilling, crushing and smelting machines. Once these machines were developed, many single or small-operation miners were put out of work. A number of First Nations People also worked on the goldfields. Some became wealthy from guarding transports, from guiding miners through their lands or from their own gold discoveries.

Source 3

Ballarat during the Victorian gold rush

Victoria's economic boom

Gold from the mines earned Victoria a lot of money. In 1852, the gross domestic product of Victoria was £16.1 million, making it the most profitable Australian colony at the time. The government revenues in 1853 were £3.2 million, which enabled the colony to build infrastructure, such as trains, and to become industrialised.

During the decade of the gold rush, mining was the second largest income source for the colonial economy, although this also included coal and other minerals. When mining dried up after a decade, the economy transitioned back to wool and other commodities.

The population boom on the goldfields led to a dramatic rise in food production to feed the miners. Subsidiary industries emerged to serve the miners, such as general stores, carpenters, doctors, dentists, lawyers and wagon makers. Merchants and professionals of all kinds flocked to the goldfields. The large population of Chinese miners brought with them food stalls, professional laundries, 'joss houses' (temples) and opium dens.

In the cities, the wealthy middle classes enjoyed a new innovation from England – department stores that offered shopping as a leisure activity. Australia's first department store was Appleton & Jones, established in Sydney in 1835, which became David Jones in 1838. Melbourne's first department store was the Coles Book Arcade, established in 1873; it included a monkey exhibit, a band, a library and a portrait studio among many other 'departments'.

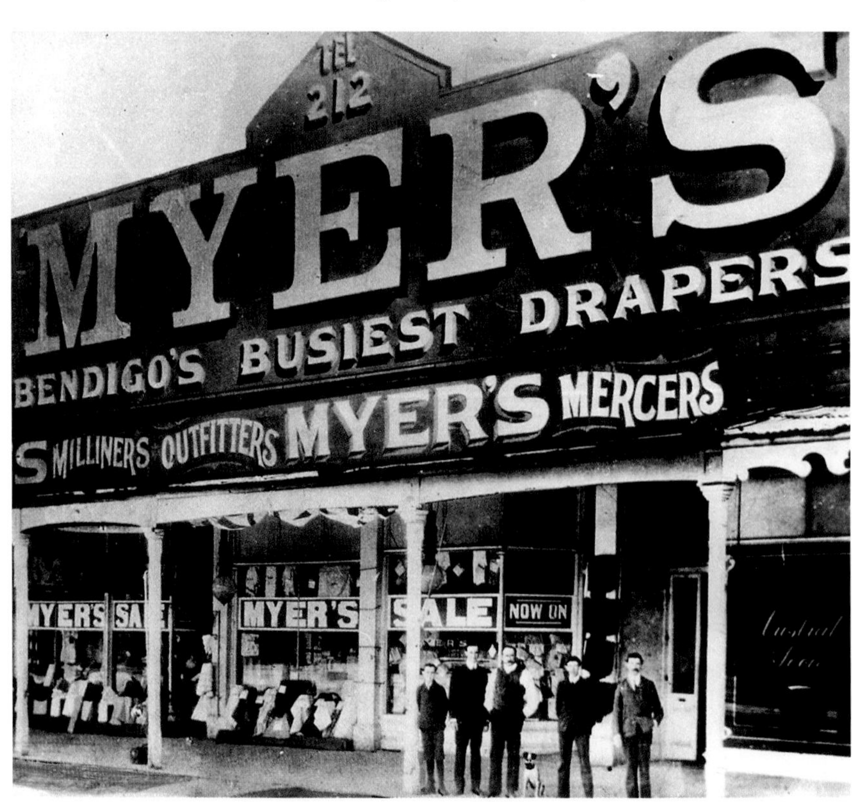

Source 4

In 1899, the Baevski brothers set up the first Myer's store in Bendigo (they changed their surname to Myer soon after arriving in Australia from Russia). After success in regional Victoria, they bought land in Melbourne and opened an eight-storey store in Bourke Street, called Myer Emporium, in 1914.

Cultural change and ethnic tension

At its foundation, Victoria was chiefly populated by people of British descent, in addition to the resident First Nations Peoples. The population exploded during the gold rush – from roughly 4000 people in 1839 to 12 000 in 1841 and 328 000 by 1861. Victoria's ethnic diversity expanded too; people from Africa, Asia and Europe came to the goldfields and made Victoria their new home.

A large number of Chinese immigrants followed the discovery of gold to Australia, which they named *Xin Jin Shan,* or New Gold Mountain. Some were men looking to provide for their families back in China, and left as soon as they had earned sufficient funds; others stayed in Australia and established families. Arriving in the tens of thousands, Chinese-born immigrants constituted 3.3 per cent of the total colonial Australian population (note that First Nations Peoples were not included in census population counting).

Source 5

A family of Chinese immigrants in Australia, circa 1891

The Chinese immigrants tended to keep to themselves and recreated a Southern Chinese way of life. Their foreign religions, customs, strong work ethic and mining successes attracted hostility from other miners and white settlers. In 1855, the *Chinese Immigration Act* limited the number of Chinese immigrants allowed to enter Victoria; undeterred, Chinese travellers landed in New South Wales and walked to the Victorian goldfields.

Chinese miners were harassed by both officials and other miners. They were required to pay a residency licence, in addition to the standard gold licence, and their claims might be 'jumped' by other miners if their residence papers weren't produced upon request. Acts of violence were common. In 1857, a full-scale riot against the Chinese erupted at Buckland River, with evictions, beatings, mine dispossessions and deaths.

Racist sentiments continued against other people as the gold rush waned. In fact, racism expanded from 1860 as Hindu and Muslim 'Ghan cameleers' connected towns in the Australian interior with camel trains and hawker carts. The men labelled as 'Ghans' were actually from many parts of Hindustan, British India and Afghanistan. They were segregated in fenced areas called 'Ghantowns', which housed some of the earliest mosques in Australia. Racial mixing was a social taboo at the time but a number of Ghan cameleers married both Indigenous and non-Indigenous women.

Tensions between white settlers and other ethnic groups shaped the first Act made by the Australian government – the *Immigration Restriction Act 1901*, which became known as 'the White Australia Policy'.

Learning ladder H2.19

Show what you know

1 Near which three Victorian towns was gold first discovered?

2 How did the traps' policing of licences create hostility?

3 How did the Chinese miners fare on the goldfields?

Cause and effect

Step 1: I can recognise a cause and an effect

4 Identify two causes and two effects of the Victorian gold rush.

Step 2: I can determine causes and effects

5 How did the gold rush affect the infrastructure of Victoria?

Step 3: I can explain causes and effects

6 How did the influx of immigrants during the gold rush create a new set of problems for the colonies?

Step 4: I can analyse causes and effects

7 How did the Australian gold rush economically disadvantage pastoral and agricultural industries?

Cause and effect, page 247

What was the Eureka Rebellion?

Miners on the Australian goldfields grew increasingly restless under oppressive licensing fees and what they viewed as a corrupt police and legal system. Tensions finally reached breaking point in Ballarat at the Eureka Stockade in 1854, in a battle that left 22 people dead.

Source 1

A miner's gold licence from 1853

Goldfield politics

There was considerable political unrest among the diggers on the Australian goldfields. Prospectors were upset at having to pay burdensome licence fees, while Chinese miners were further taxed by residency licences. The police conducted regular 'licence hunts', and were frequently accused of accepting bribes, extorting money and imprisoning people without proper trials. Inspired by English trade unionists and political movements (such as the Chartists, see page 44), Australian miners banded together to form unions.

The miners of Ballarat formed the Ballarat Reform League in response to a number of political issues. John Humffray, the secretary of the League, believed that the miners were not fairly represented in the political system. He and other leaders were largely unsuccessful in their efforts to address key concerns such as the miners' right to vote, the removal of restrictions on land purchases, and the reform of the entire system administering the gold fields. The diggers' growing resistance to the oppressive licensing led to uprisings at Sofala in 1852, at Bendigo in 1853 and – most dramatically – at Ballarat in 1854.

Source 2

J.B. Henderson, 1854, *Battle of the Eureka Stockade*

Murder and arson

James Scobie, a Scottish miner, was killed under suspicious circumstances in Ballarat on 6 October 1854. James Francis Bentley, owner of the Eureka Hotel, was accused of his murder but was quickly acquitted. The miners were outraged, and some accused the magistrate who heard the case of taking bribes. A group of 5000 men and women met to talk over the case, and afterwards a small number of them set the Eureka Hotel on fire before being arrested.

Lieutenant-Governor Charles Hotham dismissed petitions for the arsonists to be released. Sensing the growing unrest, he deployed 150 British Army soldiers to Ballarat to help maintain public order.

The Eureka Stockade

On 31 November 1854, after another police licence hunt, 1500 men and women marched to Bakery Hill in Ballarat and built a **stockade**. They burned mining licences, demanded political change, sang the French Revolutionary Anthem and raised a flag showing the Southern Cross, which was later called the Eureka flag. They remained there guarding the stockade for the next two days, but their numbers dramatically declined to between 120 and 200 rebels.

At 3 am on 3 December, 300 soldiers and police attacked the stockade. The miners withstood ferocious bayoneting, musket and pistol fire for 25 minutes; even those injured continued to be stabbed. The diggers were overwhelmingly defeated, with 22 casualties recorded. Of the 120 survivors, 12 were injured and all were arrested and charged with treason.

Aftermath

Thirteen of the Eureka rebels were tried for treason in February 1855. The colony's newspaper, *The Argus*, covered the trial, reporting that the jury's sympathies were with the miners. All of them were acquitted, and charges against the others were dropped.

Later that same year, a Royal Commission recommended the removal of gold licence fees. They were replaced with an export tax on gold, along with a £1 annual fee for miners to receive some legal rights and limited political/voting rights. Peter Lalor, a leader of the stockade, became an elected representative and has since had an electorate named after him.

Source 3

The remnants of the original Eureka flag, which was cut into pieces and hidden by participants. Years later, it was reassembled and is now on display at the Eureka Centre in Ballarat.

Learning ladder H2.20

Show what you know

1 What does Source 2 reveal about the Eureka Stockade conflict?

2 Why did the soldiers stab people who were injured and no longer fighting?

3 Hypothesise why the Eureka flag (Source 3) was cut up and then hidden.

4 Why were the miners exonerated by the jury?

Historical significance

Step 1: I can recognise historical significance

5 How long did the Eureka Stockade conflict actually last?

Step 2: I can explain historical significance

6 How did the miners turn the rebellion and court case into a lasting victory?

Step 3: I can apply a theory of significance

7 Which government reforms were implemented as a result of the rebellion?

Step 4: I can analyse historical significance

8 Why do some unions still fly the Eureka flag today? (You may need to research which particular unions fly the flag.)

Historical significance, page 251

How is Coranderrk a symbol of success and resistance?

One of the outcomes of the colonisation of Victoria, and the genocide committed against its First Nations Peoples, was a growing discontent among First Nations communities. They wanted to resist government control over their lives, but were willing to work with the colonists to prevent further loss of life and culture.

Coranderrk

Coranderrk was a reserve (settlement), intended as a 'new home' for displaced First Nations People. In 1863, 931 hectares of land were allocated at the junction of the Yarra River and Badger Creek, roughly 65 km northeast of Melbourne. William Barak and Simon Wonga, two Wurundjeri leaders, led their people across the Dandenong Ranges to the site, and built Coranderrk Aboriginal Station.

Coranderrk was initially home to 40 Kulin people, but within 12 years its population had tripled. Its First Nations community combined their traditional knowledge of Country with the new plants, animals and farming techniques introduced by the Europeans. For a time, Coranderrk was the most productive agricultural land in Victoria; it even won an agricultural prize at the Melbourne International Exhibition in 1881. Coranderrk became a popular tourist destination for Victorian colonists, who purchased goods such as woven baskets, boomerangs and possum skin rugs.

The success of Coranderrk was supported by John Green, the Protector of Aboriginal Peoples in Victoria. Progressive for the time, Green encouraged self-governance and autonomy for Coranderrk's community.

Simon Wonga

Simon Wonga was a Wurundjeri man of the Woi wurrung Clan, born around 1824; as a boy, he witnessed his father Billibellary signing the Batman Treaty. When he was 19, he badly injured his foot, and was taken in by the Assistant Protector of Aborigines for treatment. During his recovery, he tried to learn all he could about the colonists and their ways.

Billibellary died in 1846, making Simon the *Ngurungeata* of his clan. Instead of taking up this role, he continued working for the colonists to learn about their culture. Once Wonga had learned as much as he could, he accepted his role as *Ngurungaeta* and began campaigning for a home for his displaced people. He successfully argued for the establishment of Coranderrk, with the help of some progressive colonists.

Wonga was a central figure within the Coranderrk community. He continued to campaign for his people's rights, often walking to Melbourne to petition for more support. With the help of John Green, he was able to protect the hard-won gains of his people; for a time, the residents at Coranderrk were able to live in peace. Wonga passed away in 1874 and was succeeded as *Ngurungaeta* by William Barak.

William Barak

William Barak was born in 1824 and, like his cousin Simon Wonga, he was a Wurundjeri man of the Woi wurrung Clan. He became an accomplished negotiator and leader, and during his time at Coranderrk he also recorded First Nations culture through story and art.

Source 1

Simon Wonga

When Barak became the clan leader in 1874, he tried to address the deteriorating living conditions at the reserve. He wrote many letters and petitions and, like his cousin, he walked to Melbourne and back to meet with politicians, including the premier.

Closure and impact

By the 1870s, colonists wanted to take over the land at Coranderrk. They undermined John Green, who resigned in 1874. With Green gone, living conditions at Coranderrk deteriorated; the government cut off food supplies, refused to maintain housing and no longer supplied medicines.

In 1886, the Victorian Government passed the *Aboriginal Protection Law Amendment Act*. This forced all First Nations People aged 15–35 with any European ancestors to vacate government reserves. The Act tore apart Coranderrk's community, and further mismanagement reduced the population to only 31 by 1893. The residents could not overcome the growing pressure from colonial farmers and developers, and Coranderrk officially closed in 1924.

The story of Coranderrk is not one of failure, but of resistance. Its success, and the political achievements of its leaders, laid the foundations for the Victorian First Nations political activism of the 20th century. Barak, the last recognised *Ngurungaeta* of the Wurundjeri people, left a powerful legacy that continues to inspire First Nations leaders today.

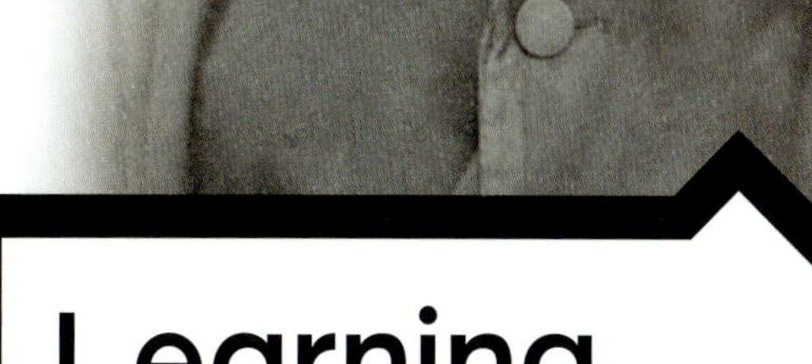

Source 2

William Barak

Source 3

William Barak's grave at Coranderrk Cemetery. The cemetery was returned to the Wurundjeri people in 1991.

Learning Ladder H2.21

Show what you know

1 What source of income, other than farming, helped to support Coranderrk?

2 Why did colonial farmers want First Nations Peoples to leave Coranderrk?

Historical significance

Step 1: I can recognise historical significance

3 Why was it necessary for governments to establish Aboriginal Reserves?

Step 2: I can explain historical significance

4 Why was it significant that the Coranderrk community applied their traditional knowledge to European crops and livestock?

Step 3: I can apply a theory of significance

5 Use Partington's theory (see page 251) to explain why William Barak was a significant historical figure.

Step 4: I can analyse historical significance

6 What was the significance of the *Aboriginal Protection Law Amendment Act 1886*?

HOW TO

Historical significance, page 251

The bushrangers: Outlaws or icons?

Frequent gold transports proved tempting to criminals, as did naïve travellers using quiet bush tracks. Bushrangers such as Ned Kelly were notorious outlaws in colonial Australia. Some viewed them as brutal criminals, others as brave rebels.

Source 1

Ned Kelly, the most notorious bushranger of the 19th century, wore a suit of metal armour during his final gun battle with police.

Highway robbery

Bushrangers were the highwaymen of Australia, holding up travellers on bush roads and robbing them of their possessions. Some operated in gangs and robbed police transports, banks and even an entire ship on one occasion. However, not all bushrangers fitted this description. Almost any criminal who hid out in bushland was referred to as a 'bushranger'. The term was applied to petty criminals, escaped convicts, First Nations rebels and others.

The Kelly gang

Ned Kelly came from a poor Irish-immigrant family who lived in the regional towns of Wallan, Beveridge Avenel and Greta. He had several run-ins with the police as a teenager over the theft of livestock, and was an accomplice to bushranger Harry Power.

In 1878, after a fight with police, Ned Kelly went on the run and formed the Kelly gang with his brother and two friends. After killing three policemen in an ambush at Stringybark Creek, the gang turned to robbing banks and terrorising towns along the Murray River. In his 1879 manifesto, known as the Jerilderie Letter, Kelly claimed that persecution by the police had forced him into a life of crime, and that small farmers were oppressed by a government controlled by banks and large landholders.

On 28 June 1880, after murdering a suspected informant, the Kelly gang took over the town of Glenrowan. They wanted to derail a train filled with police coming to arrest them, and built suits of metal armour to protect themselves. The gang held the people of Glenrowan hostage while waiting for the train. Because the train was sent from Melbourne rather than Benalla, the expected 12-hour siege turned into a 30-hour ordeal, so the gang released a few people over time. One of those released hostages was Thomas Curnow, who informed the coming train of the ambush. The police attacked the gang at the inn; Kelly and his men were defeated.

Ned Kelly was tried on multiple charges of murder, robbery and other crimes, in a court case that was heavily sensationalised by the media of the time. He was found guilty and hanged on 11 November 1880. His iconic armour and political manifesto made him famous to this day. However, his violent crimes and murders are often forgotten or overlooked.

Jessie Hickman

The heyday for bushrangers was the gold rush era. After the death of Ned Kelly, fewer people turned to robbery in the bush. As well as being one of Australia's very few female bushrangers, Jessie Hickman was one of its last – she robbed farms, stole cattle and evaded police until the late 1920s.

Born in 1890 in central NSW, Hickman joined a travelling circus at the age of eight and became a roughrider (a person who roped cattle and performed stunts on horseback). When the circus closed in 1910, she turned to gambling and theft to support herself. She served two terms in gaol, then began stealing and selling cattle. There are stories of her stealing cattle from the police, escaping from custody while on a moving train and even killing her husband (in self-defence).

In the 1920s, Hickman relocated to the Wollemi region of the Blue Mountains, where she led a gang of cattle thieves. She was charged with cattle rustling in 1928, but the charges were dropped when the evidence (the cattle) went missing. She eventually left her life of crime to live on a farm, and died of a brain tumour in 1938.

Jimmy Governor

Jimmy Governor was a First Nations man, most likely of the Wunumara people, who lived in the Talbragar region of NSW in the 1890s. A literate man who worked with the police as a tracker, Governor and his family were nevertheless bullied and treated badly by local white settlers, especially once he married a European woman.

After being abused by the wife of a wealthy landowner, Governor and his friend Jacky Underwood killed the landowner and his family on 20 July 1900, then went on the run. Styling themselves as bushrangers, Governor and Underwood (along with some of Governor's brothers), committed more crimes in the area over the next three months, including murders, robberies, burglaries and assaults.

A reward of £1000 was offered for Governor's capture, and 2000 volunteers and police hunted for them. He was captured on 27 October 1900, and was hanged on 18 January 1901. His story has been retold in modern times by creators who sought to understand the abuse and oppression that led to his terrible crimes.

Source 2

Prison photo of Jesse Hickman, taken 15 August 1913

Learning ladder H2.22

Show what you know

1. What were some of the crimes that bushrangers committed?
2. Why were many different criminals, not just highwaymen, referred to as 'bushrangers'?
3. Why has Ned Kelly received so much attention from scholars, filmmakers and tourists?

Historical interpretations

Step 1: I can recognise that the past has been represented in different ways

4. What people within the colony (if any) might have seen Ned Kelly as a hero?

Step 2: I can describe historical interpretations

5. How were Ned Kelly's actions and his trial portrayed by the media of his day?

Step 3: I can explain historical interpretations

6. Kelly's 'Jerilderie Letter' is an important document in Australian history. Find a copy and read the key sections in pairs or groups. How does Kelly justify his actions? How does he criticise society? Do you believe him?

Step 4: I can analyse historical interpretations

7. Many books, films and TV shows have told the stories of bushrangers, such as *The Chant of Jimmy Blacksmith, Mad Dog Morgan* and *The True History of the Kelly Gang*. To what degree do they depict the truth, or a romanticised version of the past?

Historical interpretations, page 255

PART III: FEDERATION (1901–1914 AND BEYOND)

Why did the colonies become a federation?

As an Australian identity began to emerge, issues such as immigration, tariffs, the need for a national army and even different sized rail gauges drove political movements to create a federal government. Legislators such as Henry Parkes argued for a federated Australia, which was realised in 1901 to great celebration.

Source 1

This was possibly the first cartoon supporting an Australian federation. Featured in *Melbourne Punch*, 1860, it shows all of the Australasian colonies (including New Zealand) banding together.

Independent colonies

By 1900 – the dawn of the 20th century – six British colonies existed in Australia: New South Wales, Victoria, Queensland, Tasmania, South Australia and Western Australia. All were independent, self-governing bodies with their own militias. They were no longer answerable to the Governor-General of New South Wales, but instead reported to the British Parliament. The notion of being 'Australian' – rather than a British colonist – started to be celebrated in popular songs and poetry.

key ideas timeline.

- 1880–1900 – Calls for federation begin; conventions are held to debate a new constitution
- 1900 – *The Commonwealth of Australia Constitution Act* is presented to the British Parliament, passed on 5 July
- 1901 – The government passes its first Act, the *The Immigration Restriction Act*
- 1901 – The Constitution of Australia begins on 1 January
- 1908 – Agreement reached to establish a new capital at Canberra
- 1913 – The foundation stone of the Australian Parliament House is laid

1869–1970s the Stolen Generations suffer forced removals from their families

Calls to bring the colonies together as a federation – a single country with a national government – began in 1880, when NSW Premier Henry Parkes proposed a federal council to manage matters that were relevant to all the colonies. The Federal Council of Australasia was formed in 1886, with representatives of all of the mainland colonies except for New South Wales; it also included representatives from Fiji and New Zealand. Henry Parkes and other early legislators, such as Victorian politician Alfred Deakin, continued to promote the idea of federation and a national government.

Source 2

This 1888 cartoon urges women, symbolising the different colonies, to use federation as a means to deport the Chinese. While acceptable at the time, it is racist and offensive by modern standards.

Support for a new federation

The colonies held conventions throughout the 1890s, which debated the form a national **constitution** might take and how it would be governed. A series of legislators drew from both the British and American government systems to devise a middle path between Westminster and Washington, known as the 'Washminster' system.

Many people who supported federation saw economic benefits from removing inter-colonial trade tariffs, and hoped that measurement and transport would be standardised. For example, every colony used a different track gauge for its railways, so trains could not travel from one colony to another; passengers had to get out and change trains at the border. Military considerations included a perceived need for national armed forces. Federation would help solve these problems.

There were other, less positive arguments for federation. In the 1880s, following the end of the gold rush, there was a rise in anti-Chinese and anti-immigrant sentiment. One of the most persuasive arguments for federation was that a united immigration policy would strengthen all borders against non-white foreigners. First Nations Peoples were not included in the federation debate at all.

The Australian flag

The competition to design the new Australian flag was held by the journal *Review of Reviews*, which offered a £200 prize to the winners. The 32 823 entries gathered over eight months were displayed in the Royal Exhibition Building, and five winners were selected by Prime Minister Sir Edmund Barton and Lady Hopetoun, the governor-general's wife.

The Australian flag bears the Union Jack, the Commonwealth seven-pointed star (which replaced the original six-pointed star) and the Southern Cross on a blue background. The Union Jack represents Australia's British colonial heritage, while the star points represent its seven states and territories. The symbol of the Southern Cross is also sacred to many of Australia's First Nations Peoples. Today the blue flag is one of three internationally recognised Australian flags, along with the Aboriginal and Torres Strait Islander flags.

Source 3

The Australian flag owes its design to a public competition.

Opposition to federation

Not everyone supported federation. Politicians in NSW and Victoria, who had profited from an earlier start and economic prosperity from the gold rushes, feared losing prestige and were reluctant to share their wealth with their poorer counterparts. Western Australia stalled on offering a referendum, until gold miners in the region threatened a rebellion similar to the Eureka Stockade.

Queensland's government was also reluctant; as a young colony, it feared domination by the more established and wealthier colonies. There was also concern that a national government might prohibit the importation and exploitation of workers from the Pacific Islands, who were central to Queensland's sugar cane industry.

Some figures in the powerful labour movement supported federation, but others were against it. There was concern within the movement that a national government would reinforce the power of banks and wealthy landowners, rather than give power to workers. There was also religious opposition; many pro-federation leaders were Protestants, which meant some Catholic leaders were automatically opposed to federation.

Who could vote?

From 1891 to 1893, it was decided by popular vote that **referenda** would decide the form of the national constitution, and would also be the means by which it could be later changed. The referenda did not include all Australians. Voting was optional, and there were restrictions based on ownership of property; Queensland and Western Australia even restricted Indigenous people who owned property from voting. Women in South Australia and Western Australia had the right to vote, but women in the other colonies did not. Others who were excluded were people on welfare assistance.

Some groups, such as the First Nations Peoples of Australia and the South Sea Islanders, had the right to vote as British subjects, but were denied this right by local governments. (Queensland did not allow First Nations Peoples to vote until 1965.)

Federation

The Commonwealth of Australia Constitution Act was presented to the British Parliament in 1900, and was passed on 5 July. Royal assent was given by Queen Victoria four days later.

On 1 January 1901, the Constitution of Australia began, with the new territories and states combining to become the Commonwealth of Australia. Australia was a constitutional monarchy with a bicameral system of an upper house and a lower house, both elected by secret ballot. The Governor-General represented the British monarch, and this position came to be seen as symbolic.

Source 4

The Royal Exhibition Building was the place where, on 9 May 1901, the first Australian Parliament was sworn in, with 12 000 dignitaries and their families present.

For 27 years, NSW and Victoria debated whether the nation's capital city should be Melbourne or Sydney. In 1908 it was decided that Australia's capital would be Canberra (meaning 'meeting place' in the local Ngunnawal language), and that it would be placed between Sydney and Melbourne. American architect Walter Burley-Griffin won a competition to design the city in 1912, and the first foundation stone of Australia's Parliament House was laid in 1913.

Learning ladder H2.23

Show what you know

1 Consider Source 1. Why is New Zealand included in the cartoon?

2 What city is Australia's national capital?

3 Name some of the significant figures involved in the federation of Australia.

Continuity and change

Step 1: I can describe continuity and change

4 Which major concerns led to the federation of Australia?

Step 2: I can explain why something did or did not change

5 Which parliamentary systems influenced Australia's leading legislators at the time?

Step 3: I can explain patterns of continuity and change

6 How did the British monarchy retain a voice and power in the independent nation of Australia?

Step 4: I can analyse patterns of continuity and change

7 What symbol of the British monarchy can be seen in Source 4? How else was Queen Victoria honoured (i.e. what was named after her and her family)?

HOW TO

Continuity and change, page 244

How does Australia's government work?

On 1 January 1901, the six colonies became states of the Commonwealth of Australia, retaining their separate parliaments and handing over matters concerning the entire nation to a new federal government, which would operate under the guidelines set out in the Constitution.

Westminster system

The Westminster system is England's parliamentary system of government, named after the Palace of Westminster, the seat of the British Parliament. This system was the framework used to establish parliaments in the six Australian colonies before federation, as well as the Australian parliamentary system in 1901.

The Westminster system is a series of procedures for law making. It includes many specific roles, processes and concepts, including:

- a sovereign, or head of state such as a governor or governor-general, who holds the power to sign off on laws passed by parliament
- a head of government, either prime minister (for Australia), premier (for each state) or chief minister (for each territory)
- a parliamentary opposition with an official leader
- an executive branch with the power to administer or implement the law
- a legislative branch with the power to make laws. Legislatures are usually bicameral, involving two houses of parliament. Queensland has Australia's only unicameral government, with an elected Legislative Assembly and governor
- a judicial branch to establish common laws that address gaps or confusion with the statutory laws made by parliament
- parliamentary privilege, which gives Members legal immunity for any statements made in parliament
- an independent civil service that advises government ministers and implements decisions.

Source 1

Australia's parliament

Australia's Constitution

The Commonwealth of Australia Constitution Act 1900 was an Act of the Parliament of the United Kingdom. The Act enabled Australia's Constitution – a set of principles that state how power is shared between a federal parliament and six state parliaments.

The Constitution separates the government's powers into different branches. This prevents one group from having power over both the law-making and law-judging systems. The federal government has three branches:

- executive: the Prime Minister, along with senior government ministers and the Governor-General, have the power to administer or implement the law
- legislature: parliament has the power to make the law
- judiciary: the courts have the power to interpret and apply the law.

Source 2

The original Parliament House in Canberra was opened in 1927 and hosted federal parliament until 1988, when it was replaced by the new Parliament House.

Each of Australia's six states and two territories also has a parliament that makes laws on state matters, such as health, education and transport. The powers of state governments are also separated into branches.

Changes to the Constitution can only be made by a referendum, where a majority of Australian voters and a majority of states vote to approve the changes proposed by parliament.

Federal parliament

Australia's federal parliament consists of two houses: the **House of Representatives** and the **Senate**.

The House of Representatives has 151 members, each representing one local area of Australia known as an **electorate**. They are elected for a term of three years. The government is formed by the **political party** with the most members in the House of Representatives, which then holds executive power.

The Senate has 76 members, elected for a term of six years. Half of the **senators** face election every three years. Each of the six states has 12 senators, and the Australian Capital Territory and Northern Territory both have two senators.

Representing electorates

Members of Parliament (MPs) provide a direct link between the people in their electorate and the Parliament. On average, each federal electorate has 100 000 voters.

Each MP has an office in their electorate where they can hear the concerns of the voters. Some groups of voters may try to influence (or lobby) the local member to present their interests about a special issue. In parliament, MPs are expected to represent their voters on matters of interest to their electorate, such as a major road construction or the closure of a local industry.

MPs must represent their electorate well; at the end of their term, they face an election where voters decide whether to vote for them or not.

Learning ladder H2.24

Civics and citizenship

Step 1: I can identify topics about society

1 List three ways in which Australia's government is influenced by the Westminster system.

Step 2: I can describe societal issues

2 Describe how the interests of citizens are represented in government, and how their concerns can be heard.

Step 3: I can explain issues in society

3 Why do you think there is a separation of powers under the Australian Constitution?

Step 4: I can explain different points of view

4 Research and explain the role of the Opposition in Australian government.

Step 5: I can analyse issues in society

5 The USA, Canadian and New Zealand constitutions all recognise Indigenous people, but the Australian Constitution does not. Research what actions are being debated as a way to formally recognise Indigenous Australians.

What was the White Australia Policy?

At the time of federation, British subjects and allies, including citizens of India, the Pacific Islands and Japan, were able to travel anywhere within the empire. However, many Australians regarded non-white immigrants as dangerous interlopers who would steal Australian jobs. The new government acted to restrict immigration along racial lines.

Immigration restrictions

The first Act of Australia's federal parliament demonstrates public sentiment at the time of federation. The country's leaders saw immigration as an issue of paramount importance, and gave long speeches on the benefits of making Australia into a '[white] working man's paradise'.

The Immigration Restriction Act 1901 was part of a package of reforms that became known as the 'White Australia Policy'. This policy mandated a dictation test for any non-British prospective immigrant – a test that was biased and unfair. An immigration officer could select *any* European language at random and ask the applicant to translate it into English. Furthermore, a pass in one language could mean that the applicant just had to re-sit the test in another language, until they ultimately failed. In the first eight years, only 52 people passed this test; after 1909, immigration officers ensured that *nobody* passed.

The Act was very popular, and was in place for nearly 60 years. However, the worst parts of the Act were relaxed after World War II, and the dictation test was abolished in 1958.

Blackbirding

Another aspect of the White Australia Policy was its impact on the Pacific Islanders working in Queensland. As Queensland's sugar and cotton industry developed during the 19th century, plantation owners went to islands such as Vanuatu and Melanesia to 'recruit' Islanders for plantation work. Plantation owners duped, kidnapped or coerced the Islanders into boarding their ships – a process known as **blackbirding**. Each Islander was charged an Indenture Bond (a type of contract fee) from their already very low wages.

The 'blackbirded' Islanders endured horrendous conditions, and were treated as slaves. A third of them died from disease, malnutrition and mistreatment; many were buried in unmarked graves. They worked in sweltering heat, were underfed, would often be physically punished and were segregated from the community.

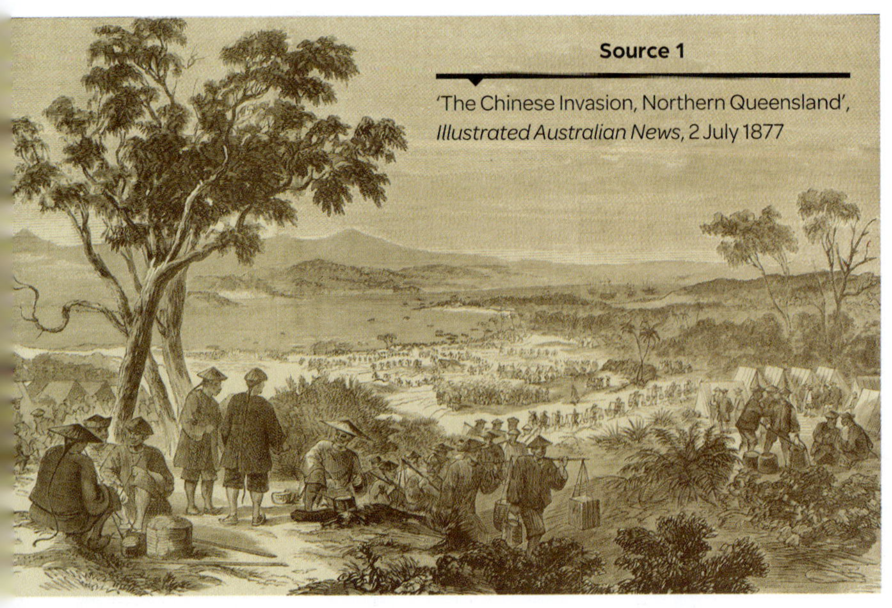

Source 1

'The Chinese Invasion, Northern Queensland', *Illustrated Australian News*, 2 July 1877

Source 2

South Sea Islander labourers on a Queensland pineapple plantation in the 1890s

The Pacific Islander Labourer Act 1901 was part of the White Australia Policy reforms package. It ordered the mass deportation of around 10 000 Islanders. The money earned by deceased Islanders should have gone to their next of kin, but the government only paid 15 per cent of their earnings. The Pacific Islanders Association unsuccessfully petitioned the Australian government multiple times over their wage dispute from 1902–04.

After the deportations, only 2500 Islanders remained in Queensland; their descendants make up the Australian South Sea Islander community today. On 7 September 2000, Queensland Premier Peter Beattie emotionally read the *Queensland Government Recognition Statement*, which acknowledged the Australian South Sea Islanders as a distinct cultural group and recognised their history of unjust treatment and discrimination.

Source 3

Badge distributed by the Australian Natives' Association to promote a 'White Australia'. This group used 'native' to mean 'Australian-born'; it had nothing to do with the First Nations Peoples of Australia.

Learning ladder H2.25

Show what you know

1 What is the *Immigration Restriction Act 1901* better known as? How long was it in place?

2 What was unfair about the dictation test administered to prospective immigrants?

3 What is blackbirding?

Source analysis

Step 1: I can list specific features of a source

4 Consider Source 1. What does the illustration show about the concerns of 'White Australia'? Describe the details of the image.

Step 2: I can find themes in a source

5 How are Chinese miners depicted in Source 1? What features highlight the illustrator's prejudiced attitudes?

Step 3: I can use the origin of a source to explain its creator's purpose

6 Source 3: How did this badge's creators define the term 'natives'? Why do you think they decided on that definition?

Step 4: I can analyse a source

7 What information does Source 2 provide about the lives of the South Sea Islanders?

HOW TO

Source analysis, page 240

Who were the Stolen Generations?

From federation up until the 1970s, Australian government agencies and church missions forcibly removed many First Nations children in an attempt at assimilating them into 'white society'. These children later became known as the Stolen Generations. The United Nations defines the forcible transfer of children as genocide.

Source 1

A large number of Victorian Stolen Generations children were sent to the Ballarat Orphanage. Operating for over 100 years, more than 4000 children lived at the orphanage. Researchers believe 10 to 15 per cent of them were from the Stolen Generations.

Aboriginal 'protection'

In a continuation of one of Australia's most atrocious periods, the *Victorian Aboriginal Protection Act 1869* (as well as a later Act in 1886) began a nationwide policy of child removal, carried out under various state and territory acts.

The Act focused on removing mixed-race and fair-skinned Aboriginal and Torres Strait Islander children away from their families, so that they could be brought up in missions and other institutions, fostered out or adopted by white families. Named 'half-castes', 'quarter-castes' or 'octaroons' (terms now understood to be racist and insulting), it was thought that these children would fit into white society easily due to their lighter skin.

Homes Are Sought For These Children

A GROUP OF TINY HALF-CASTE AND QUADROON CHILDREN at the Darwin half-caste home. The Minister for the Interior (Mr Perkins) recently appealed to charitable organisations in Melbourne and Sydney to find homes for the children and rescue them from becoming outcasts.

I like the little girl in centre of group, but if taken by anyone else, any of the others would do, as long as they are strong

Source 2

A note from a potential adoptive parent on an advertisement for foster homes for the children of First Nations Peoples. The advertisement sought homes for these children, in an attempt to 'rescue them from becoming outcasts'.

Underpinning this policy of **assimilation** was the racist belief that First Nations Peoples were incapable of surviving on their own in modern, British Australia. Another motive was to instil a strong work ethic through education and the Christian faith. This was meant to prepare the children for lives of menial labour, as First Nations boys were considered suitable only for manual labour, and girls only for domestic labour.

Broken families

The policy of child removal spread throughout Australia by the early 20th century, and was actively upheld for over six decades. Each state passed similar acts, and appointed Protection Boards and commissioners to oversee the separation of First Nations families. These officials assumed the right to determine a person's Indigeneity.

Children from a very young age were taken from many different family structures. Police raided happy and loving 'mixed' families just as often as they raided encampments on the fringes of white society. During these raids, children endured the trauma of being forcibly removed from their parents, or hiding from the police while their parents lied about their whereabouts. Parents had to witness their children being arrested and taken away by the police.

Decades later, these experiences were retold by First Nations artists, such as in Archie Roach's song 'Took the children away' and Doris Pilkington Garimara's book *Follow the Rabbit Proof Fence*, which was later turned into an award-winning film.

Transgenerational trauma

Once children had been forcibly removed from their families, they were cut off from their culture. They were forbidden from using their traditional names and language; parents who tried to contact their children were charged exorbitant fees for their children's removal and accommodation.

Source 3

Stolen Generations children attending a school at Mornington Island, Queensland, 1950

This process of cultural dislocation had profound ongoing physical, mental and social effects upon entire communities. This is known as **transgenerational trauma** – pain that affects multiple generations. A national report, conducted in 1997, included confidential testimony from a number of Stolen Generations children. John, who was removed from his family as an infant in the 1940s and sent to the Bomaderry Children's Home at Nowra, said:

> 'I was definitely not told that I was Aboriginal. What the Sisters told us was that we had to be white. It was drummed into our heads that we were white. It didn't matter what shade you were. We thought we were white. They said you can't talk to any of them coloured people because you're white.'

Source 4

Extract from *Bringing them Home: The Report of the National Inquiry into the Separation of Aboriginal and Torres Strait Islander Children from Their Families*, p. 144.
[Confidential evidence 436, New South Wales, Cultural Heritage of the Stolen Generation]

THREE GENERATIONS
(Reading from Right to Left)

1. Half-blood—(Irish-Australian father; full-blood Aboriginal mother).
2. Quadroon Daughter—(Father Australian born of Scottish parents; Mother No. 1).
3. Octaroon Grandson—(Father Australian of Irish descent; Mother No. 2).

Source 5

Stolen children were classed according to anthropological principles that are now understood to be racist and unscientific. Calipers were used to measure children's skulls and noses. They were also classed according to their skin colour, from lightest- to darkest-skinned.

Source 6

Kevin Rudd's apology to the Stolen Generations in 2008 (left), during which two First Nations women express their sadness (right).

An ongoing legacy

From a modern perspective, it is difficult to appreciate the extent to which 'protection' policies damaged the lives and societies of the First Nations Peoples of Australia. Protection Boards had almost total control over the lives of these people, including their movement, fostering, access to healthcare and education, and later their marriages, wages and travel. For over six decades, children were not only removed from the love and care of their families, but also were wholly denied access to their culture and heritage, as well as often being subjected to abuse and terrible conditions.

This cultural dislocation was catastrophic and caused immeasurable harm to the estimated 50 000–100 000 infants and children who are now known as the Stolen Generations. Many of the social problems facing the First Nations Peoples of Australia, which affect mental health, alcohol and other addictions and high rates of suicide, are directly related to the policy of child removal.

Former Australian Prime Minister Kevin Rudd officially apologised to the Stolen Generations on 13 February 2008. On June 11 that same year, Canadian Prime Minister Stephen Harper apologised for a similar policy enacted from 1870, in which the children of the First Nations Peoples of Canada were forced into Residential Schools.

Learning ladder H2.26

Show what you know

1 What justifications were given for taking First Nations children away from their parents?

2 What did this policy do to families with children who had both First Nations and European parents?

3 Research and compare historical rates of First Nations child removal with rates today. How have they changed? Why are First Nations children still being removed from their families?

Historical interpretations

Step 1: I can recognise that the past has been represented in different ways

4 Source 5: What do the terms 'half-blood', 'quadroon' and 'octaroon' refer to?

Step 2: I can describe historical interpretations

5 Source 2: What does the 'X' drawn on the child and the accompanying note indicate about which children were most desirable to foster parents?

Step 3: I can explain historical interpretations

6 What does it mean to read Source 5 from right to left? How does it change if read from left to right? What is the implied progression?

Step 4: I can analyse historical interpretations

7 Read a transcript of Prime Minister Kevin Rudd's 2008 Apology and view media of Rudd's delivery. How do these sentiments expressed contrast with the understanding and intention of the original laws that permitted this to happen?

Historical interpretations, page 255

Masterclass

Learning ladder

Work at the level that is right for you or level-up for a learning challenge!

Source 1

An 1883 photograph of two First Nations warriors from the Victorian region, along with their weapons and war implements

'... they had seen several of the native chiefs, with whom, as they said, they had exchanged all sorts of things for land; but that I knew could not have been, because unlike other savage communities, or people, they have no chiefs claiming or possessing any superior right over the soil: theirs only being as the heads of families. [...] I therefore looked upon the land dealing spoken of as another hoax of the white man, to possess the inheritance of the uncivilised natives.'

Source 2

An 1835 quote by escaped convict William Buckley, who lived with the Wathaurung for more than 30 years, regarding John Batman's treaty negotiations. [From John Morgan, 1852, *The life and adventures of William Buckley*.]

Source 3

Ned Kelly's suit of armour

Step 1

a I can list specific features of a source

List the different weapons and artefacts shown in Source 1.

b I can describe continuity and change

Who were the First Nations Peoples of Naarm, or Port Phillip Bay?

c I can recognise a cause and an effect

What did the 'discovery' of Australia by Europeans mean for its First Nations Peoples?

d I can recognise historical significance

What is the significance of the date 26 January? How is this date recognised by different groups of Australians?

e I can recognise that the past has been represented in different ways

i How did the First Nations Peoples of Australia react to the European landings?

ii How did the Europeans view their landings in Australia?

Step 2

a I can find themes in a source

Source 2: How did William Buckley describe the First Nations People of the lands John Batman hoped to claim?

b I can explain why something did or did not change

Why did some British citizens sell all their goods and travel to the colonies as free settlers?

c I can determine causes and effects

What were the causes for the establishment of a colony in New South Wales (Australia) and the Port Phillip Protectorate?

d I can explain historical significance

What was the significance of the Batman Deed–Duttigallar Treaty in 1835? What is its significance today?

e I can describe historical interpretations

Source 2: How does William Buckley describe both the First Nations Peoples and the Europeans?

Step 3

a I can use the origin of a source to explain its creator's purpose

What do you think the purpose of the Source 1 photograph was at the time it was taken?

b I can explain patterns of continuity and change

Consider the map on page 66. How did colonisation change the internal borders of Australia?

c I can explain causes and effects

What caused the massacres and collective punishments of the Frontier Wars?

d I can apply a theory of significance

Why did the Port Phillip Protectorate seek to encourage First Nations Peoples' participation in colonial society? Were these goals shared by others in the colony?

e I can explain historical interpretations

Source 2: What do Buckley's comments add to the explanations of John Batman and the Wurundjeri *Ngurungaeta* about signing Batman's treaty?

Step 4

a I can analyse a source

Source 3: What was the purpose of the armour worn by the Kelly Gang, given that it made it harder for them to aim and shoot their guns?

b I can analyse patterns of continuity and change

How did features of the *Immigration Restriction Act 1901* demographically engineer Australian society? What were the intended outcomes of the Act?

c I can analyse causes and effects

What caused the gold rushes? What effects did the gold rushes have on Australian society?

Masterclass

d **I can analyse historical significance**

Why do Australia's infamous bushrangers still attract the interests of both historians and the public?

e **I can analyse historical interpretations**

Why might the stories in modern Australian family histories differ from the contemporary accounts of convict and other colonial women?

Step 5

a **I can evaluate a source**

Source 3: What necessitated the creation of this armour? Why was a reward of £8000 pounds offered in 1879 for the capture of the Kelly gang?

b **I can evaluate patterns of continuity and change**

Given that Australia was established partly as a penal colony, how did ideas about imprisonment change during the 19th century? Did they change the prison system for worse or for better?

c **I can evaluate causes and effects**

What were the impacts of the Eureka Stockade on Australia's political system?

d **I can evaluate historical significance**

What were the contributions of Chinese miners and other non-European groups to Australian society?

e **I can evaluate historical interpretations**

What did the anti-Chinese riots on the goldfields in NSW and Victoria show about Australian society at the time?

Historical writing

1 Structure

Imagine you are given the essay topic, 'Every argument in favour of federation could be met with equally valid counterarguments'. Write an essay plan for this topic. Include at least three main paragraphs.

2 Draft

Using the drafting and vocabulary suggestions on page 262, draft a 600–800 word essay (at least 30–40 sentences) responding to the topic.

3 Edit and proofread

Use the editing and proofreading tips on page 263, to help edit and proofread your draft.

Historical research

4 Organise and present information

Imagine you are completing a research project on the practice of naming Australian streets, parks, buildings and other places after major colonial figures. Write a contents page for this project. There should be an introduction, a conclusion, at least four main sections and many subsections. Number your chapters.

Capstone

How can I understand pre- and post-colonial Australia?

In this chapter, you have learned a lot about the history of Australia. Now you can put your new knowledge and understanding together for the capstone project to show what you know and what you think.

In the world of building, a capstone is an element that finishes off an arch or tops off a building or wall. That is what the capstone project will offer you, too: a chance to top off and bring together your learning in interesting, critical and creative ways. You can complete this project yourself, or your teacher can make it a class task or a homework task.

mea.digital/GHV9_H2

Scan this QR code to find the capstone project online.

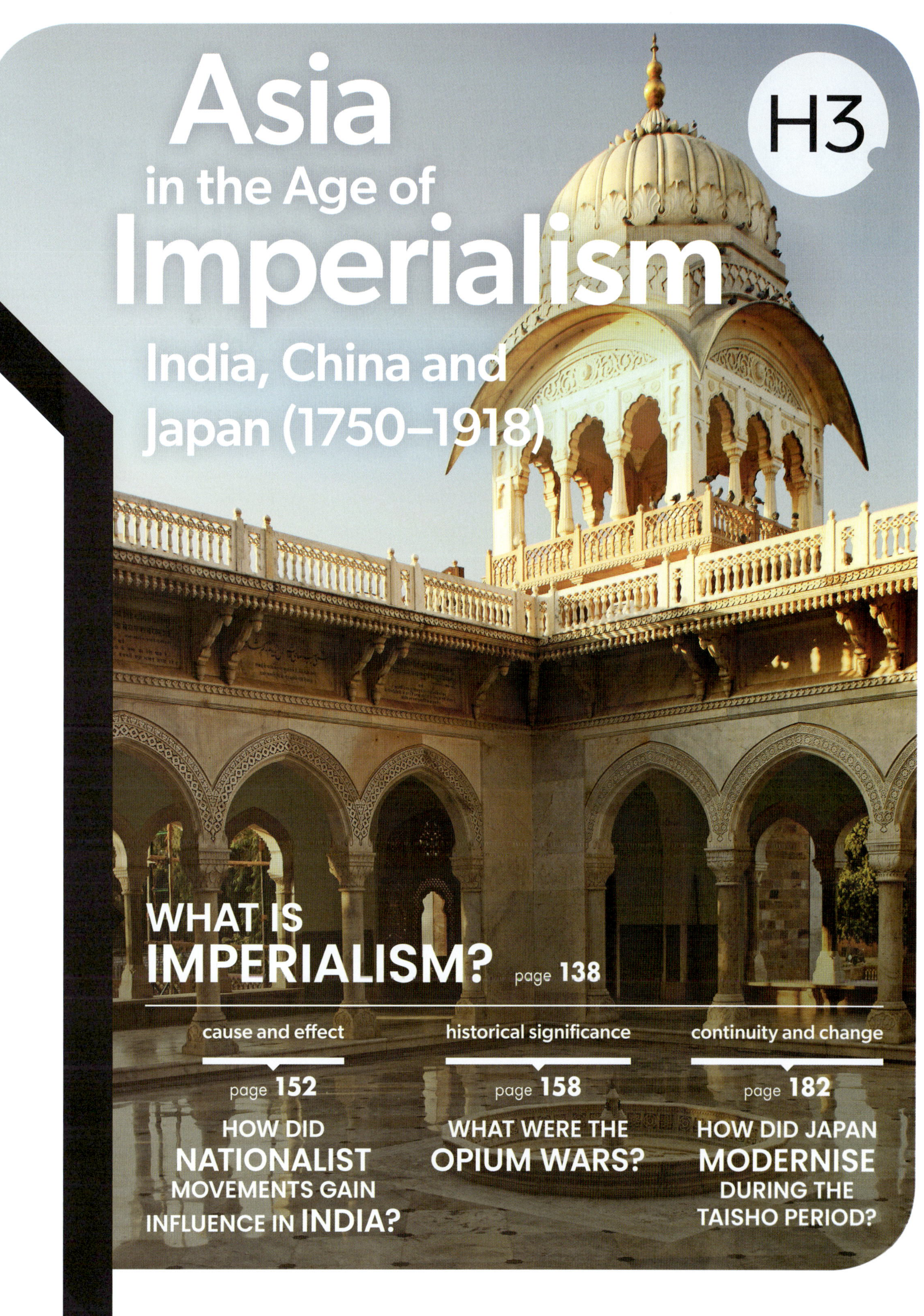
H3
Asia
in the Age of
Imperialism
India, China and
Japan (1750–1918)
WHAT IS
IMPERIALISM? page 138
cause and effect
page 152
HOW DID NATIONALIST MOVEMENTS GAIN INFLUENCE IN INDIA?
historical significance
page 158
WHAT WERE THE OPIUM WARS?
continuity and change
page 182
HOW DID JAPAN MODERNISE DURING THE TAISHO PERIOD?

How can we understand the Age of Imperialism?

The 18th and 19th centuries have been called the 'Age of Imperialism'. It was a period when powerful Western nations made great efforts to influence, colonise or take control of other countries, especially in Africa and Asia. British and American imperialism caused massive changes to the societies of India, China and Japan, and those changes still have effects today.

learning ladder

Step	Source analysis	Continuity and change	Cause and effect
 step 5	**I can evaluate a source** I can present a judgement on the usefulness of a source based on its strengths, weaknesses and limitations. I can determine whether information is missing about the event or person the source refers to.	**I can evaluate patterns of continuity and change** I answer the question 'So what?' about patterns of continuity and change. I weigh up different aspects and debate the importance of continuity or change.	**I can evaluate causes and effects** I answer the question 'So what?' about cause and effect. I weigh up different things and debate the importance of a cause or an effect.
 step 4	**I can analyse a source** I can use my own knowledge to determine the reliability of a source and can explain whether it shows a one-sided view.	**I can analyse patterns of continuity and change** I can look deeper into patterns of continuity and change and determine the factors that contribute to them.	**I can analyse causes and effects** I don't just see a cause or an effect as one thing. I can determine the factors that make up causes and effects.
 step 3	**I can use the origin of a source to explain its creator's purpose** I combine knowledge of when and where a source was created to answer the question, 'Why was it created?'.	**I can explain patterns of continuity and change** I can see beyond individual examples of continuity and change between historical periods and explain broader patterns.	**I can explain causes and effects** I can answer 'How?' or 'Why?' a cause led to an effect in Asia in the Age of Imperialism.
 step 2	**I can find themes in a source** I look a bit closer at a source and find more than just features. I find themes and patterns in a source.	**I can explain why something did or did not change** I can give a reason for why something changed or why it stayed the same.	**I can determine causes and effects** Applying what I have learnt about Asia in the Age of Imperialism, I can describe what the cause or effect of an event was.
 step 1	**I can list specific features of a source** I can look at an Age of Imperialism source and list the details I can see in it.	**I can describe continuity and change** I recognise what has stayed the same and what has changed from the Age of Imperialism until now.	**I can recognise a cause and an effect** From a supplied list, I can recognise things that were causes or effects of each other in Asia in the Age of Imperialism.

Source 1

An editorial illustration from 1902. Propaganda images such as these framed Western imperialism as almost a holy duty to spread civilisation around the world.

Warm up

Source analysis

1 Refer to Source 1. How did England view its colonial mission? Who are the 'barbarians' and how are they depicted?

Continuity and change

2 How can the effects of European colonisation in Asia be seen today? Has your family been affected, or have you travelled in parts of Asia that were previously colonised by Europeans?

Cause and effect

3 To what extent was conflict in the Asian region influenced by the Industrial Revolution in Europe?

Historical significance

4 Modern Britain has attempted to address some of its actions from the era of colonial expansion, such as handing back Hong Kong to China. To whom might these acts be significant?

Historical interpretations

5 How was European colonialism viewed in the 19th century? How is it viewed today?

Historical significance	Historical interpretations
I can evaluate historical significance I answer the question 'So what?' about things that are supposedly important in the history of Asia in the Age of Imperialism. I weigh up factors against one another and can cast doubt on how important things are.	**I can evaluate historical interpretations** I can weigh up the different historical interpretations that have been formed. I debate and challenge the interpretations that have been presented.
I can analyse historical significance I can separate the various factors that make something historically important in the history of Asia in the Age of Imperialism.	**I can analyse historical interpretations** I can determine the factors that have led to why a historical interpretation has been formed.
I can apply a theory of significance I know a theory of significance. I use it to rank importance of changes, causes, effects and events in the history of Asia in the Age of Imperialism.	**I can explain historical interpretations** I can answer 'Why?' or 'How?' there are different interpretations of people and events in the past.
I can explain historical significance I answer the question 'Why?' about what was important in Asia in the Age of Imperialism.	**I can describe historical interpretations** I can provide different examples to show how people and events in the past have been interpreted.
I can recognise historical significance When shown a list of facts about Asia in the Age of Imperialism, I can work out which are important.	**I can recognise that the past has been represented in different ways** I can identify different views of people and events in the past.

What is imperialism?

Imperialism is a policy of one nation extending its rule over other nations to form an empire. The imperialist nation does this by force or by taking political and economic control.

Throughout most of human history, powerful nations have sought to expand their borders by taking control of other territories, and to gain wealth by exploiting these territories' resources. Many great empires have existed in the past, such as the Roman Empire, the Khmer Empire and the Ottoman Empire. These empires **annexed** other nearby nations or kingdoms, holding great power for a time before falling.

As sailing technology opened up more of the world, European nations such as Spain, Britain and the Netherlands began establishing trade networks with distant countries, particularly in the Asia–Pacific region. The 18th and 19th centuries saw conquering nations turn these trade networks into international empires, so that they could gain more profit and power.

Because of this, this period is sometimes called the 'Age of Imperialism'. It was a period not only of international trade and diplomacy, but also of brutal conflicts. European nations saw the non-European nations in their empires as resources to be exploited, rather than societies in their own right. They used their military and technological advantages to suppress other nations and force their citizens into submission.

Source 1

This map, produced by a Scottish biscuit manufacturer in 1900, shows the extent of the British Empire (shown in red) at the end of the 19th century.

The great empires

The most dominant empire during the Age of Imperialism was the British Empire, which became the largest and most powerful empire in human history. Britain began establishing colonies around the Caribbean and Americas in the 16th century, while building trade networks into Asia. Britain developed an **expansionist** agenda after winning the Seven Years' War, an international conflict that went from 1756 to 1763. From then on, it began using its economic and technological advantages to take over many other countries, especially around Asia and Africa.

The USA also emerged as a major imperialist power during the latter part of the 19th century. After winning its freedom in 1783, the former British colony was **isolationist** and opposed to empire-building. This changed with the Monroe Doctrine, an 1850 policy opposing further European colonisation. In practice, this policy meant that the USA began establishing its own empire, taking over territories such as Hawaii, Cuba and the Philippines. It also meant that the USA tried to undermine European trade in Asia and Africa.

Other nations such as France, Belgium and Russia also established international empires. But after the Age of Imperialism ended in the mid-20th century, the British and American empires proved to be the ones that had the longest and most far-reaching influence over the rest of the world.

Source 2

An 1892 caricature of businessman Cecil Rhodes, a major proponent of British imperialism. Rhodes led British efforts to exploit Africa and help found the colonies that eventually become the nations of Zimbabwe and South Africa.

Imperialism in Asia

Asia is the largest continent in the world, and it became a focus of the European powers during the Age of Imperialism. Countries such as Portugal, France and Britain fought to exploit the region's natural resources and claim the wealth of Asian societies for themselves.

Over the course of the period, several Asian nations became the main targets of British and American imperialism and were permanently reshaped by the actions of these Western powers.

Source 3

Richard Paton, *The Battle of Quiberon Bay, 20 November 1759*, artwork undated

Source 4

Michael Angelo Hayes, (artist) and James Henry Lynch, (lithographer), *The 18th (Royal Irish) Regiment of Foot at the storming of the fort of Amoy, 26 August 1841* (1840s)

key ideas timeline.

- **1763** The Seven Years' War concludes, leaving Britain the major power in Europe
- **1783** American colonies successfully revolt and leave British rule
- **1819** Singapore founded as an outpost of the British Empire
- **1839** The Opium Wars occur between China and Britain
- **1842** Treaty of Nanking cedes Hong Kong to the British
- **1849** British East India Company annexes Punjab
- **1854** United States forces Japan to open its borders to trade
- **1863** France establishes a protectorate over Cambodia
- **1867** Canada established as self-governing nation ruled by Britain

Source 5

Proclamation of Canadian Federation (1867)

BY THE QUEEN!

A PROCLAMATION

For Uniting the Provinces of Canada, Nova Scotia, and New Brunswick, into one Dominion, under the name of CANADA.

VICTORIA R.

WHEREAS by an Act of Parliament, passed on the Twenty-ninth day of March, One Thousand Eight Hundred and Sixty-seven, in the Thirtieth year of Our reign, intituled, "An Act for the Union of Canada, Nova Scotia, and New Brunswick, and the Government thereof, and for purposes connected therewith," after divers recitals it is enacted that "it shall "be lawful for the Queen, by and with the advice of Her Majesty's "Most Honorable Privy Council, to declare, by Proclamation, that "on and after a day therein appointed, not being more than six "months after the passing of this Act, the Provinces of Canada, "Nova Scotia, and New Brunswick, shall form and be One Domi-"nion under the name of Canada, and on and after that day those "Three Provinces shall form and be One Dominion under that "Name accordingly;" and it is thereby further enacted, that "Such Persons shall be first summoned to the Senate as the Queen "by Warrant, under Her Majesty's Royal Sign Manual, thinks fit "to approve, and their Names shall be inserted in the Queen's "Proclamation of Union:"

We, therefore, by and with the advice of Our Privy Council, have thought fit to issue this Our Royal Proclamation, and We do ordain, declare, and command that on and after the First day of July, One Thousand Eight Hundred and Sixty-seven, the Provinces of Canada, Nova Scotia, and New Brunswick, shall form and be One Dominion, under the name of CANADA.

And we do further ordain and declare that the persons whose names are herein inserted and set forth are the persons of whom we have by Warrant under Our Royal Sign Manual thought fit to approve as the persons who shall be first summoned to the Senate of Canada.

Given at our Court, at Windsor Castle, this Twenty-second day of May, in the year of our Lord One Thousand Eight Hundred and Sixty-seven, and in the Thirtieth year of our reign.

GOD SAVE THE QUEEN.

1884 — France makes Vietnam a colony

1885 — King Leopold of Belgium establishes the Congo Free State

1898 — United States annexes Hawaii, Philippines, Guam and Puerto Rico

1898 — Cecil Rhodes establishes Rhodesia in Africa

India, China and Japan

In this chapter we will focus on how imperialism affected three nations:

- India was called 'the brightest jewel in the imperial crown'. Britain took direct political control over India after a series of wars; its government, called 'the Raj', ruled over the Indian people.
- China was not taken over by foreign nations in the same way as India, but it was forced to sign agreements that caused it to lose significant economic and political **autonomy**. Britain took over Hong Kong where it wielded great social and political influence over Chinese citizens.
- Japan suffered a similar fate after signing an agreement with the USA, which forced it to open its borders to international trade. However, it resisted foreign influence and instead grew its own empire.

Learning ladder H3.1

Show what you know

1. Define the terms 'imperialism' and 'colonialism'.
2. Who were the European colonial powers in the period from 1750 to 1918?
3. Why were European nations so interested in establishing a presence in Asia?

Cause and effect

Step 1: I can recognise a cause and an effect

4. What did Asian and African countries have that the European and American empires wanted?

Step 2: I can determine causes and effects

5. What was the impact of colonisation upon colonised cultures?

Step 3: I can explain causes and effects

6. Why did the US policy of opposing European colonisation lead it to establish its own empire?

Step 4: I can analyse causes and effects

7. Analyse the impact that reliable, long-range sailing technology had upon the ability of European nations to exert power on Asian nations.

HOW TO

Cause and effect, page 247

PART I: INDIA

How did India become part of the British Empire?

The Mughal Empire had ruled India for centuries, but was in decline by the mid-18th century. The British East India Company, a powerful international trading business, made deals with regional rulers, taking over provinces and establishing a government that applied British laws.

The Mughal Empire

The Mughal Empire ruled a large part of India for over 200 years before the arrival of the British. Originally from Persia, the first Mughal ruler of India was Zahiruddin Muhammad Babur. He captured Delhi in 1526 thanks to superior military strategy. This left northern India, which was mostly Hindu, under Muslim control. Babur believed in religious tolerance, so long as there was no resistance to his rule, and this helped establish Mughal rule over India. Further expansion by Emperor Akbar the Great, who ruled from 1556 to 1605, saw the spread of Mughal dominance throughout north and central India.

Source 1

The Taj Mahal was built as a mausoleum for the favourite wife of Mughal Emperor Shah Jahan. Completed in 1653, it shows the architectural styles and influence of the Mughal Empire.

key ideas timeline.

- 1757 — British East India Company forces the defeat of the Nawab of Bengal at the Battle of Plassey
- 1765 — Robert Clive declares himself Governor of Bengal
- 1833 — The first *Government of India Act* establishes British rule
- 1857 — The Indian Great Rebellion
- 1877 — Queen Victoria declared Empress of India
- 1885 — Indian National Congress (INC) formed

Source 2

The pink areas on the two maps show the Indian territory controlled by the British East India Company in **a** 1765 and **b** 1805.

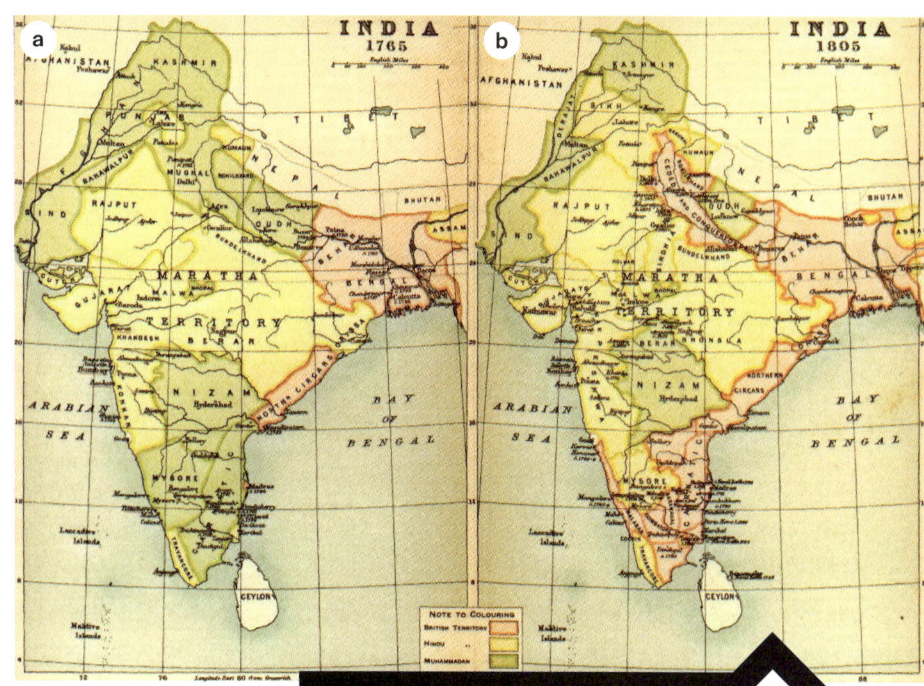

However, by 1750 the Mughal Empire was in decline. Expensive wars, high taxes and conflict over succession had resulted in the emperor, based in Delhi, having little control. Rulers of local provinces sought to increase their own territory and power. The Marathas, Hindu warlords from Maharashtra, successfully created their own empire across large areas of India. Meanwhile, European countries had been trading with India since the mid-16th century, but their power and influence began to grow.

The British East India Company

The British East India Company, a massive international trading and shipping business, entered India in 1600 and began to influence local politics in the region. The company helped to undermine the Mughal Empire, making separate deals with local rulers.

In 1756, as part of the Seven Years' War, the British East India Company was able to drive its French rivals out of India. A year later, the company took military control of the Bengal region in the Battle of Plassey. In 1765 Lord Robert Clive, a high-ranking member of the company, declared himself Governor of Bengal.

The company expanded its control over the next decade, taking over more provinces and cities. It introduced British law, replacing *maharajahs* (local rulers) with its own governors, as part of the Company **Raj** (a Hindi word meaning 'rule'). After he took office in 1774, Warren Hastings, the first Governor-General of India, tried to make the company become a political institution, not just a commercial one.

Following financial difficulties and allegations of corruption, in 1833 the British Parliament turned the Company into a managing agency for the British government of India. This forced it to focus on governing. By 1857, unrest and resentment at the British East India Company had grown to the point of rebellion.

Learning ladder H3.2

Show what you know

1. Who was the first Mughal ruler of India?
2. List the reasons for the decline of the Mughal Empire.
3. What impact did Warren Hastings have on India?
4. Draw a timeline of events to show how India became part of the British Empire.

Continuity and change

Step 1: I can describe continuity and change

5. Describe the changes in the British East India Company's territory between 1765 and 1805.

Step 2: I can explain why something did or did not change

6. Explain why Babur did not make changes to local religious practices during his rule.

Step 3: I can explain patterns of continuity and change

7. Explain how the power of the Mughal Empire declined.

Step 4: I can analyse patterns of continuity and change

8. Imagine you are an official with the British East India Company in 1775. Write about your experiences in India and the impact that the company has had on India.

Continuity and change, page 244

What impact did British rule have on India?

The presence of British rule had many economic, social and political impacts, both positive and negative. While the British Empire was motivated primarily by economic factors, it was also driven by the belief that the English were the superior race.

Land ownership changes

Before British colonisation, the Indian people had the right to use the land under collective ownership, while the ruling ***zamindars*** collected tax from peasant farmers. However, the Company Raj introduced a system of private property in 1793, which gave *zamindars* private property rights. This changed communities, as the rights to land had passed from one generation to the next under the Mughal Empire. Wealthier *zamindars* became powerful as they increased their land holdings, while peasant farmers lost their collective rights to land.

The higher land taxes demanded by the company meant that little improvement was made to land, and many *zamindars* turned to new crops that would generate greater profit. Farmland previously used for food production was instead used for growing cash crops such as cotton, **indigo** and tea. This, coupled with poor weather, caused a series of devastating famines in 1876–77 and 1899–1900.

Agriculture

Opium and tea were two crops that became essential in British India. While Britain bought many goods from China, the Chinese had little demand for British goods. In order to change this trading imbalance, the Company Raj began to sell Indian opium to China. This highly addictive drug was very lucrative, and the company's monopoly on opium contributed to as much as 20 per cent of British India's income.

There was an incredibly high demand for tea in England. Prior to British colonisation, no tea trade existed in India, although wild tea grew in Assam. In 1833, the British East India Company began commercial tea growing in this region; by the early 20th century, Assam had become the largest tea-producing region in the world. Darjeeling in West Bengal also became an important tea-growing area, with Darjeeling tea known as the 'champagne of teas'. Much of the tea grown in this region was actually cultivated from plants smuggled out of China.

Source 1

An 1850 photograph of workers weighing harvested tea in Assam

Destruction of the textile economy

Unlike previous rulers, who had made few changes to the economic structure, the British changed India's traditional trading practices. Many rural economies relied on making and selling **handicrafts**, especially textiles and clothing, for nobles and *maharajahs*. The British East India Company introduced mechanised looms and encouraged *maharajahs* to buy British manufactured textiles. As a result, there was no longer a demand for locally produced, hand-woven cotton garments, and the Indian textile market suffered.

The disruption caused by agricultural changes also altered rural economic patterns, and the introduction of railways disrupted traditional Indian industries. Silk and woollen textile production suffered the same fate as cotton-weaving, as it became cheaper and politically safer to buy British fabrics. Formerly a world-leading exporter of clothes, India became an exporter of raw cotton.

The introduction of import duties and restrictions on Indian goods, along with administration costs, further served to weaken the Indian economy.

Source 2

India's weavers were famed across the world for their light muslins. Production of these hand-made textiles declined following the arrival of the British.

'Britain's rise for 200 years was financed by its depredations in India. In fact, Britain's industrial revolution was actually premised upon the de-industrialisation of India.

'The handloom weavers, for example, famed across the world whose products were exported around the world, Britain came right in. There were actually these weavers making fine muslin as light as woven wear, it was said, and Britain came right in, smashed their thumbs, broke their looms, imposed tariffs and duties on their cloth and products and started, of course, taking their raw material from India and shipping back manufactured cloth flooding the world's markets with what became the products of the dark and satanic mills of the Victoria (sic) in England.'

Source 3

Extract from a 2015 speech by Indian politician Shashi Tharoor, arguing that Britain owed India reparations to make up for 200 years of colonial rule.

Religion

India has been a land of diverse religious beliefs for thousands of years, and is the place where four of the world's major religions – Hinduism, Buddhism, Sikhism and Jainism – were founded. While Hinduism was the dominant religion at the time of British colonisation, many other faiths also thrived in India – particularly Islam, as the Mughals were Muslims.

The British did not bring Christianity to India; missionaries had visited India since the time of the Mughals. In fact, the British East India Company actually banned Christian missionaries from coming to India! The company had a policy of 'religious neutrality', but not for moral or ethical reasons – its leaders simply wanted to avoid religious protests and uprisings. Fearing that an increased Christian presence would cause unrest, they banned missionaries and instead supported Hindu festivals.

Britain's rulers disapproved of this, and by the 1830s the company was forced to stop promoting Hinduism and to let missionaries in. Christianity started to develop a larger presence within India, which, as predicted, contributed to growing unrest within the Hindu community.

The Great Rebellion

The domination of India by the British East India Company resulted in previously independent kingdoms falling under company control. In 1833, the company lost its trade monopoly, which led it to increase the size of its army to protect its interests. The company employed professional Indian soldiers, known as *sepoys*. By 1852, the army of *sepoys* numbered 233 000.

Source 4

An 1859 illustration by George Francklin Anderson of *sepoy* troops during the failed rebellion. The British labelled this uprising 'The Sepoy Mutiny'.

However, the lack of respect given to Indian soldiers, and the British belief that their civilisation was superior, contributed to growing feelings of resentment among the Indian soldiers. In May 1857, the *sepoy* forces started a rebellion.

The immediate cause of the uprising was the introduction of the new Lee Enfield rifle. These rifles were loaded with paper cartridges that contained a bullet and gunpowder, and were sealed with grease to keep the powder dry. In order to load the rifle, the end of the greased cartridge had to be bitten off. While there is little evidence to support the claim, a rumour spread that the grease was made from a mix of pig and cow fat. This caused offence to both Muslim and Hindu *sepoys*.

The longer-term causes of the rebellion were the policies used by the company to increase its power and territory in India, such as the 'doctrine of lapse'. This allowed them to annex land if a Hindu ruler had no natural heir and replace the traditional rulers with their own officials. Hindus were also concerned about the spread of Christianity in India.

The uprising spread across India. Both elites and commoners joined the rebellion, united in their dislike of British rule, as did landlords who had been left impoverished by the company's economic policies. The British government sent troops and, amid atrocities on both sides, eventually defeated the *sepoys* in November 1858.

The British Raj

In June 1858, during the uprising, the British government took over control of India from the company, replacing the Company Raj with the British Raj. This meant that Britain now ruled India directly without the British East India Company's involvement. Some regions remained under the control of native princes, but these 'princely states' were fundamentally vassal states of the Raj.

In November 1858, Queen Victoria made promises to rule India for the good of its people. Indian subjects would be protected under British law and the land rights of native princes would be acknowledged. The Viceroy of India was to be responsible for both British India and Princely India, while the Governor-General ruled over British India.

In order to further strengthen political ties, Benjamin Disraeli, the British Prime Minister, and Lord Lytton, the Viceroy of India, declared Queen Victoria to be Empress of India in January 1877. Britain's ability to control India made it powerful, both economically and politically. Prime Minister Disraeli called India 'the brightest jewel in the crown'.

Learning ladder H3.3

Show what you know

1 How were peasants affected by changes to the agricultural economy?

2 Why was Queen Victoria declared Empress of India in 1877?

3 Describe how land ownership changed under the Company Raj.

4 Why did the *sepoys* rebel in 1857?

Historical significance

Step 1: I can recognise historical significance

5 Put these events on a scale from least important to most important:

Company Raj introduces private property rights

Introduction of mechanised looms to the Indian cotton industry

Queen Victoria becomes Empress of India

The Great Rebellion

Step 2: I can explain historical significance

6 Explain the significance of the Company Raj introducing private property rights.

Step 3: I can apply a theory of significance

7 Use Partington's theory (page 251) to explain why the introduction of the tea trade to India was significant.

Step 4: I can analyse historical significance

8 What was the significance of the Great Rebellion? Provide evidence to support your answer.

Historical significance, page 251

What was life like under the British Raj?

British rule in India brought technological and cultural change. Social reforms often conflicted with traditional Indian practices and began to change Indian society. However, some reforms were also supported by Indian citizens.

Britons abroad

The British population of India was small; there were never more than 100 000 Britons living there at one time. The wealthiest of them – politicians, businessmen and highly educated civil servants – lived lives of luxury on palatial estates with small armies of servants. However, the majority of Britons were workers, soldiers or Raj employees; they lived simpler lives, but still enjoyed a more privileged status than most indigenous Indians.

The British built leisure facilities to make India more enjoyable for them, such as racetracks, private clubs, golf courses and even ski resorts. Hill stations were built in the cooler mountainous areas. Originally built by the British East India Company for civil servants and their families, these served as a retreat from the hot, overcrowded cities. They also reinforced the belief in British superiority; for example, no Indians were allowed on the mall (the main street) in Shimla during the day.

Source 1

The Civil Secretariat building in Shimla, in the Himalayas, was the capital of the Raj during the hot summer months.

Source 2

An 1880 illustration of the clothing styles worn by different Hindu castes

Indians at home

The British held the first census of India in 1871; by 1881, there were estimated to be 255 million people living in the country. The majority lived in rural areas, working on farms held by *zamindars*. Others lived in urban centres, many of them working for British inhabitants or as employees of the Raj.

Exposure to Western ideas did not cause Indians to jettison their culture, as liberal reformers had anticipated. Rather, they reinvented and reinvigorated it, synthesising the differing cultures and creating new approaches. Western ideas and social models were adapted by a people operating within their own powerful traditions.

The legal system

The Raj implemented a system of Anglo–Hindu law, primarily based on British rather than Indian laws. A penal code was introduced in 1861, and other laws followed. One of them, passed in 1872, sanctioned inter-caste and inter-communal marriages. This challenged the **caste** system that existed within India.

The caste system, which divided Indian Hindus into four groups, was seen as contrary to the ideas of liberty and equality that Britain brought to India. Individuals such as reformer Raja Mohan Roy, began to criticise the rigid caste system and social practices such as *sati* (the ritual burning of a widow at her husband's funeral). This practice was banned in 1829.

Other changes to laws, such as the *Hindu Widows Remarriage Act* in 1856, also helped to improve the status of women in Indian society. Female **infanticide** was banned, and attempts were made to prevent child marriage (although this did not happen until the *Sharda Act* of 1926).

However, there was significant inequality within the legal system. Poor people found it difficult to access the law courts, and Europeans received preferential sentences and decisions compared with Indians.

Education

British politician Thomas Macaulay introduced a national system of education to India, something that didn't even exist in Britain at the time. A British supremacist, Macaulay argued that an education that stressed the value of British rule would result in Indians who were more English in their outlook, and that students should be taught in English.

Although schools were established across India, only 5 per cent of the Indian population could read by the end of the 19th century. In 1857 the first universities in India were founded in Bombay, Bengal and Madras. The children of wealthier Indians attended English-speaking universities in the hope that they would be able to enter the British civil service.

Education resulted in a new Indian middle class who, rather than being 'English in taste, in opinions, in morals and in intellect' as Macaulay proposed, began to challenge British rule in India.

New technologies

Prompted by the need for faster transportation of cotton, Lord Dalhousie (Governor-General of India from 1848 to 1856) pushed for the development of railways in India. The railways were intended to meet the Raj's economic and administrative needs; travel between cities by Indians was a by-product. The first railway opened in India in 1851, while the first passenger train ran from Bombay to Thane in April 1853 and carried about 400 passengers. Most of the passengers travelled in third class, as only about 10 per cent could afford first- or second-class travel. By 1890, India's largest cities were linked by 9000 miles of rail; by 1904, 28 000 miles of track had been laid.

Newspapers and print media were also popular in India at this time. By 1885 there were almost 100 English-language newspapers published within India, with a circulation of nearly 60 000 copies each week. There were also many Indian newspapers and journals, in native languages and with small print runs, circulating thanks to the rail network and an efficient postal system.

The first recorded game of cricket to be played in India was in 1721, when sailors from the British East India Company played in western India. However, it wasn't until 100 years later that the sport became popular in India, as British colonists brought the game with them. The first Indian community to adopt cricket was the Parsi community of Bombay (now Mumbai), who established the Oriental Cricket Club in 1848. Cricket's popularity grew exponentially from that date, to the point where it is now considered the unofficial national sport of India.

Source 3

The Loop at 'Agony Point' on the Darjeeling Himalayan Railway

Source 4

A hand-coloured photograph of soldiers of different Indian cavalry units, 1901

The military

The British Raj abolished the company armies and created a new British Indian Army. *Sepoys* were banned from some divisions and could not become officers. Most regiments consisted of soldiers from a single religious group – Hindus, Sikhs, Christians or Muslims.

During World War I, approximately 1.2 million Indians served in the Indian and British Army – more than the combined numbers of Australians, New Zealanders, Canadians and South Africans. Of these men, around 74 000 were killed during the fighting.

Indian soldiers served across all theatres of war. These included 700 000 *sepoys*, the majority of whom were Muslim; they served in Mesopotamia, fighting against fellow Muslims in defence of the British Empire. Five thousand Indian soldiers lost their lives at Gallipoli. Indian soldiers won numerous medals, including nine Victoria Crosses – the highest accolade that can be won in the British armed forces in the battlefield.

The names of 412 Indian soldiers who have no known graves are inscribed on the Menin Gate in Belgium. However, for those soldiers that survived World War I, their heroism was ignored or denigrated back at home because Indian nationalists felt they had fought in the service of a foreign master.

Learning ladder H3.4

Show what you know

1 Create a concept map to show the changes to India under the British Raj.

2 Why did the British build racecourses and ski resorts in India?

3 How many Indian soldiers fought for the British Empire in World War I?

4 How did changes in infrastructure and technologies enable Indians to challenge British rule?

Continuity and change

Step 1: I can describe continuity and change

5 Describe the legal changes that occurred in India.

Step 2: I can explain why something did or did not change

6 Why did forcing the middle classes to learn English fail to make them 'English in opinions'?

Step 3: I can explain patterns of continuity and change

7 To what extent did legal changes affect Indian society?

Step 4: I can analyse patterns of continuity and change

8 What evidence is there that education in India changed under British rule?

Continuity and change, page 244

How did nationalist movements gain influence in India?

Concepts such as liberty, equality and self-government resonated with Indians who felt oppressed by the policies of the British Empire. Indian identity was promoted by individuals who revived a sense of pride in Indian culture and heritage. Nationalist groups led political and sometimes violent campaigns against British rule.

Communication and unity

The British introduced new communications technologies and platforms to India, such as newspapers, railways and national and international telegraph networks. However, the colonialists of the Raj did not foresee that these new concepts would actually help **nationalist** movements gaining influence across India.

India was (and still is) a vast country, containing many different cultures and languages. The lack of a single common language caused communication issues throughout India's history and, during the Raj, it made it harder for those who opposed British rule to coordinate with allies or connect with supporters. This changed once the British introduced the national education system. Indians who received a Western education now had a means of communicating across the country, not only thanks to new technologies but because they spoke a common language – English.

Similarly, while the British introduced railways and telegraphs for their own convenience, Indian citizens used these technologies to exchange information. This helped to support a growing sense of Indian unity. Across the country, groups formed to discuss political issues and express their unhappiness with British rule.

The Indian Congress Party

The Indian Congress Party was one of the most prominent nationalist groups. In 1885 it established the Indian National Congress (INC) and began lobbying for Indian involvement in government. However, the prevailing British attitude was that Indians were unfit to run their own country.

Indian nationalists promoted the idea of *swaraj*, or self-rule. Individuals such as Dadabhai Naoroji, who became the first Indian member of parliament in 1892, argued that the British ruled in order to further their interests, rather than for the good of Indians.

The early nationalists were highly critical of the economic exploitation that had occurred under the Raj, and sought constitutional changes so that India could be governed by Indians but remain part of the British Empire. They favoured protest and petition as methods of trying to effect change in India. However, following the famine of 1899–1900, the INC adopted more direct methods in their attempts to achieve *swaraj*. Leaders advocated **boycotting** British goods and passive resistance.

Bal Gangadhar Tilak

Bal Gangadhar Tilak was a key independence activist who wanted to spread the idea of nationalism beyond those elite Indians who had received a British education. The British authorities banned protests, so Tilak used Hindu religious festivals such as Ganesh Chaturthi (a celebration of the god Ganesh) as opportunities to promote Indian nationalism to large numbers of people.

Tilak split with the INC in 1907, as he believed that Indians should use whatever means necessary to achieve self-rule, whereas moderates in the party believed in using only non-violent methods. In July 1908, following his public support of two men who had failed in their attempt to assassinate the chief presidency magistrate, Tilak was imprisoned by the British government for **sedition**. British officials realised how Tilak was using festivals to gain influence and labelled him 'the father of Indian unrest' in 1910.

Source 1

A modern-day Ganesh Chaturthi festival. Bal Gangadhar Tilak popularised this and other festivals in the late 19th century as a platform for anti-British protests.

In 1916, Tilak formed a radical party, the All India Home Rule League, in conjunction with British activist Annie Besant and fellow Indian activist G.S. Khaparde. The League promoted Home Rule – the promise of an India governed by Indians, not the British – and had more than 30 000 members by 1917. It promoted the idea of home rule through education and propaganda.

While the Home Rule Movement ultimately failed to reach the masses, it created links between urban and rural areas that were later used by Mahatma Gandhi, whose leadership was able to propel the movement forward.

Source 2

Bal Gangadhar Tilak

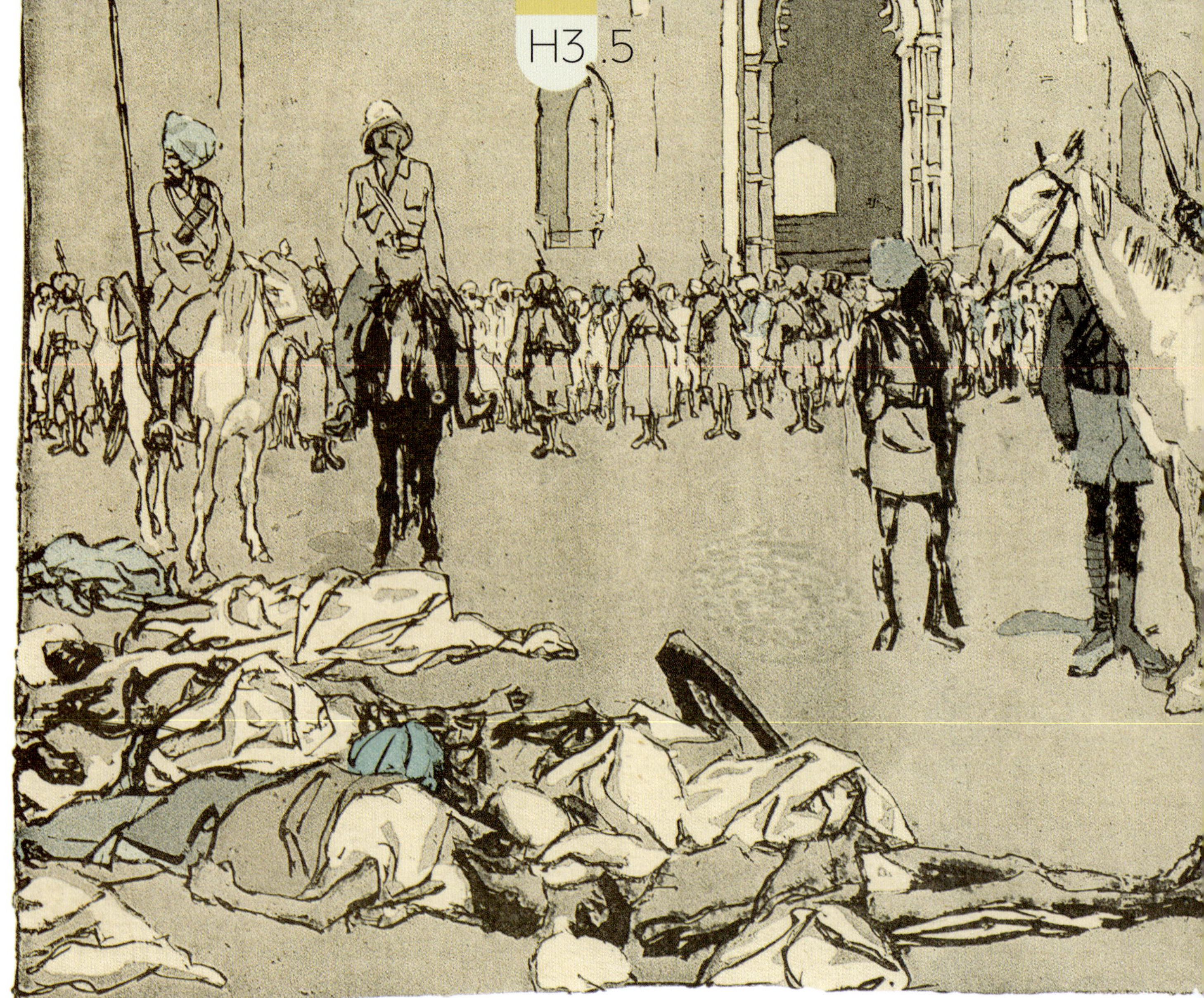

Post-war protests

Following the end of World War I, protests in favour of Indian self-rule became more frequent. It was felt that Indian sacrifice during the war should be rewarded. Britain felt differently and, despite high expectations in India, clamped down on protests. In 1919, British Parliament passed the *Rowlatt Act*, a law that gave authorities the power to crack down on what they considered to be **subversive** activities.

An example of the *Rowlatt Act* in action was the Amritsar Massacre of 1919. A crowd of over 10 000 people had gathered at Jallianwala Bagh for a protest meeting. General Reginald Dyer, commander of the local British Indian Army, ordered his troops to open fire without warning on the peaceful crowd. Over 400 men, women and children were killed and 1200 were wounded.

While some British leaders were horrified by the massacre, others applauded Dyer for taking decisive action. Rather than stifling dissent, the brutality of the killings and lack of remorse from the British only increased the sense of anger and nationalist sentiment in the Indian community.

Mahatma Gandhi

One of the most important figures in the campaign for Indian independence made his political debut during World War I, but would not rise to global prominence for another decade

Mohandas Karamchand Gandhi was an Indian lawyer living in South Africa, where he became known as a civil rights activist for his work helping Indian **expatriates**. The Indian community in South Africa gave him the **honorific** *Mahatma* (a Sanskrit word meaning 'the great souled one'), which stayed with him throughout his life.

Source 3

A 1920 illustration of the 1919 Amritsar Massacre. General Dyer's troops fired on the Indian protesters for 10 minutes, and only stopped because they ran out of bullets.

Gandhi returned to India in 1915 and became involved in the nationalist movement. While he supported Britain's war effort, and the deployment of Indian soldiers, he was critical of inequality and injustice against Indians living under the Raj. Gandhi promoted a form of civil disobedience called *satyagraha* ('truth force'). This non-violent protest movement involved boycotting British goods and institutions, among other methods.

Gandhi went on to lead the INC in the 1920s and became the most prominent figure in the Indian independence movement, as well as one of the most recognised people in the world.

Source 4

Mahatma Gandhi

Learning ladder H3.5

Show what you know

1 When did the Indian National Congress form?

2 Who was Bal Gangadhar Tilak and how did he encourage nationalism?

3 Which protest methods were most effective in the Indian Home Rule campaign?

4 How did nationalism in India gain momentum and spread across the country?

Cause and effect

Step 1: I can recognise a cause and an effect

5 Which of these actions caused the British to imprison Tilak?

His split with the INC

His use of religious festivals to promote Indian nationalism

His support of a failed assassination attempt

The foundation of the All Indian Home Rule League

Step 2: I can determine causes and effects

6 What caused the Amritsar Massacre?

Step 3: I can explain causes and effects

7 Explain why Indian nationalists wanted political change.

Step 4: I can analyse causes and effects

8 What were the most important factors in the campaign for Indian self-rule?

Cause and effect, page 247

PART II: CHINA

How did foreign powers interact with China?

China developed relationships with countries such as Britain and Russia, but imposed limits and taxes on foreign trade. The Western nations, in particular Britain, destabilised the Qing dynasty and provoked China into wars over opium.

Source 1

A 1758 portrait of Emperor Qianlong by Giuseppe Castiglione, depicting him as a monarch riding into battle – a theme from European painting traditions

China had been governed by successive **dynasties** for thousands of years, ever since the mythical Xia dynasty allegedly claimed power around 2070 BCE. The Qing (pronounced 'Ch'ing' – meaning pure) dynasty took power in 1636, and ruled China until the early 20th century. The Qing promoted Confucian principles in order to organise Chinese society.

Emperor Qianlong

In the middle of the 17th century, China was a vast and powerful nation. Emperor Qianlong ruled over 250 million people in a wealthy empire that included Mongolia, Xinjiang and Tibet. There were also **vassal** kingdoms that paid tribute to the emperor, such as Annam (Vietnam), Cambodia, Japan, Korea and Thailand. Emperor Qianlong, the absolute ruler of this empire, was the richest person in the world.

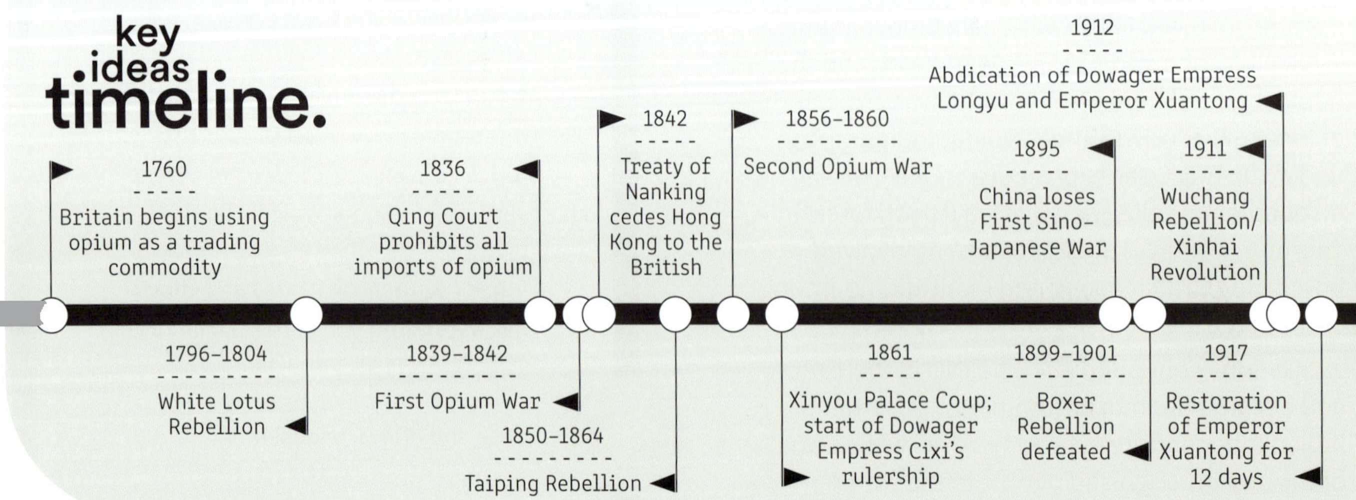

As a ruler, Emperor Qianlong was a scholar who tried to embody Confucian values. He loved beautiful objects and artworks, and collected works from throughout China. He also admired the works of European artists and even recruited the Italian artist Giuseppe Castiglione to serve as a painter in his court. However, Qianlong was less friendly towards European diplomats, and implemented business and bureaucratic reforms that favoured China.

Christianity

Reports by explorers inspired Christian missionaries to go out into 'heathen' countries. In 1750, French Catholic missionary Jean Joseph Marie Amiot persuaded Emperor Qianlong to allow Christian missionaries to come to China, effectively opening up the empire to the West.

Initially, foreign missionaries were only allowed in the port cities of Canton (modern-day Guangzhou) and Macau. Over time, their access was extended to Shanghai and other ports. The Christian missions in the port cities introduced Western languages and customs to China, and helped foreign traders deal with Chinese officials.

Trade

As the Chinese emperors feared foreign traders may exploit their country, only the port of Canton was opened to foreign trade in 1684. In 1711, the British East India Company established a trading post in Macau.

Trade was heavily regulated. Foreign traders were effectively locked inside a compound, known as a *hong* or a 'factory', and they could not venture out without a Chinese escort. All foreign traders had to have a Chinese merchant act as their agent, collecting customs duties and taxes. Traders protested at the taxes, shipping tolls (called 'chops') and prices for goods, but persisted in trading, because it was lucrative.

Britain became China's largest foreign trading partner. Tea was incredibly popular in England, and British companies also bought luxuries such as silk and porcelain. However, Chinese markets were uninterested in British goods, and Britain had a huge trade deficit with China. This lasted until British traders developed a new strategy – they promoted illegal opium to the people, then used that demand to spark war.

Source 2

The Western compound in Canton. Each foreign country had a 'factory' in the compound and a Chinese merchant agent responsible for them. [*The Hongs of Canton*, artist unknown, c. 1805]

Learning ladder H3.6

Show what you know

1 Which ports were opened initially to foreign traders?

2 What did the Chinese emperor fear might happen if China allowed trade with European merchants?

3 Source 2: How many different flags and countries are represented at the Western compound?

Cause and effect

Step 1: I can recognise a cause and an effect

4 What limitations were imposed on Europeans attempting to trade with China?

Step 2: I can determine causes and effects

5 What caused the British trade deficit with China?

Step 3: I can explain causes and effects

6 How did Britain address the trade deficit?

Step 4: I can analyse causes and effects

7 Source 1: What skills and knowledge other than religion did missionaries bring to China? Why would the emperor wish to be depicted in the style of Western rulers?

Cause and effect, page 247

What were the Opium Wars?

Foreign traders had to pay for Chinese goods in silver, rather than in foreign currency. The lack of desire for Western goods in China resulted in large debts to the Chinese and difficulties in obtaining enough silver to pay them. To counter this imbalance, the British East India Company grew opium in India and smuggled it into China.

Opium

Opium is a narcotic extracted from poppy plants. While it had been used for centuries as a painkiller, it is also an addictive drug. Opium was already known in China for its medical properties. During the 16th and 17th centuries, Portuguese and Dutch traders combined opium with smokable tobacco and supplied a growing Chinese market. Smokable opium was officially banned in China in 1637, but its use spread despite being illegal.

Opium use was illegal in Britain, but it was legal for the British to sell it to other countries, and Indian smokable opium was a popular trading commodity. The British East India Company developed a monopoly over the opium trade through Bengal. Using 'private traders' to circumvent official bans and edicts, they bribed key officials and smuggled opium into China. With the money from these illicit sales, the company now had enough silver to purchase goods at the official ports.

Source 1

A Hong Kong opium den in 1880. Opium became legal after China lost the Second Opium War.

A huge demand for smokable opium soon grew within China. 'Opium dens', houses where men could stay and smoke opium all day, sprang up in almost every Chinese town and city. The rise of opium addiction caused significant damage to China's society and economy. Earlier emperors had banned its importation and distribution, while Emperor Jiaqing added a ban to smoking opium in 1796, but these edicts had little effect.

The First Opium War (1839–1842)

By 1838, the British were selling almost 1300 tonnes of opium per year to China. The trade was so lucrative that it paid for the British Raj in India. Because of this, the British government was determined that the opium trade must continue.

In China, Special Imperial Commissioner Lin Zexu was appointed by the Emperor to eradicate the opium trade. Lin led the arrests of more than 1500 addicts and the confiscation of tens of millions of pipes. He also wrote a letter to Queen Victoria, which was later published in the London *Times,* urging her to put a stop to the opium trade. The letter never reached her.

In 1839, Lin confiscated 1000 tonnes of opium from British ships in Canton. In response, the British Superintendent of Trade declared the British government **guarantor** of the confiscated opium. This meant that traders could hand over their opium to Chinese officials, confident that they would be compensated by the British government. When Lin destroyed the opium, this gave Britain an excuse to declare war and send naval forces to China.

The war continued until 1842, and was primarily a naval conflict around port cities and major rivers. While the Chinese defending forces vastly outnumbered the invading army, the British Royal Navy's long-range weapons were far superior to the **antiquated** cannons, muskets and martial-arts weapons of the Chinese.

The invaders also had the *Nemesis,* a **paddle steamer** far more advanced than any Chinese boat. Its flat bottom allowed it to traverse rivers, regardless of wind or tide, and its iron armour resisted the Chinese cannons.

After British forces occupied major trading ports, and were poised to capture the city of Jiangning (Nanjing), Chinese commanders admitted defeat.

The defeat led to the signing of the Treaty of Nanking (1842). The Articles of the treaty:

- allowed Consuls appointed by Queen Victoria to collect dues, taxes and tariffs
- opened the cities of Amoy, Foochow, Ningpo and Shanghai to foreigners
- **indemnified** any Chinese people who worked on behalf of foreign traders, meaning they could not be punished for helping the British
- ceded the island of Hong Kong to the British, making it part of their empire; it soon became the centre of British activity in China.

A year later, the Treaty of the Bogue (1843) granted British citizens further rights, such as the right to be tried in British courts if they committed crimes in China. It also granted Britain 'most favoured nation' status for trading with China. These and other treaties with European and Asian countries (such as Japan) became known in Chinese history as the 'unequal treaties'.

The Second Opium War (1856–1860)

In the years after the first Opium War, opium remained illegal in China, but the British East India Company continued to smuggle it into the country and make huge profits. The British Empire grew during this period, and the British government was determined to gain more power within Asia.

In October 1856, Chinese forces in Canton seized a cargo ship called *The Arrow* on suspicion of piracy. *The Arrow* was a Chinese ship, but it had been sold to British traders and flew a British flag. The British demanded the crew be released without charge; when Qing officials didn't comply, a British warship attacked sites on the Pearl River. Locals in Canton set fire to the foreign trade warehouses, which was answered by military force – a second Opium War had begun.

A large number of British forces were deployed in India, so a coalition of British, US and French forces attacked China. Once again, superior technology and experience in modern warfare prevailed. Although they were outnumbered 10 to 1, the Western infantry and naval forces bombed river ports and routed the Chinese army.

Qing officials were forced to sign the Treaty of Tianjin (1858). Under the terms of this treaty:

- more ports were opened for Western use
- trading rights were granted to all Western nations, including the United States and Russia
- the Kowloon Peninsula near Hong Kong was ceded to Britain
- opium was legalised.

The treaty was signed in 1858 but Chinese military resistance continued, which led to retaliation on the part of the Western allies. Western forces advanced through China to Beijing. They stopped short of entering the Forbidden City, but destroyed part of the Summer Palace instead.

Source 2

This 1843 painting by Edward Duncan shows the *Nemesis* (on the right) destroying Chinese junks (ships) during the First Opium War.

At the end of the war, more trading rights and land were ceded to Britain. Traders and Christian missions gained access to the interior of the country, not just the port cities, and the foreigners offered Chinese people opportunities and passage on their boats. The Western powers established foreign **envoys** in Beijing, and China's isolationist policy was effectively finished.

Source 3

An illustration of Qing officials taking down the British flag on *The Arrow*, the trigger for the Second Opium War

Learning ladder H3.7

Show what you know

1 Who initially introduced smokable opium to China?

2 How did the Chinese emperors attempt to address the challenges of smokable opium? Were their attempts successful?

3 How did the British East India Company negate the emperors' measures?

4 Why was the Qing army ill-equipped to repel the foreign armies?

Historical significance

Step 1: I can recognise historical significance

5 What aspects of the treaties signed after the Opium Wars made them 'unequal treaties'?

Step 2: I can explain historical significance

6 What did the Chinese government lose with each treaty?

Step 3: I can apply a theory of significance

7 Briefly outline, in dot-points, the causes of the First and Second Opium Wars. Highlight major and minor causes in different colours.

Step 4: I can analyse historical significance

8 How did their advanced technology enable British forces to prevail over Chinese forces in both wars?

HOW TO

Historical significance, page 251

How did China change politically in the 19th century?

After the reign of Emperor Qianlong, the power of the Qing dynasty began to wane. In the aftermath of the Opium Wars, China owed massive amounts of money to Britain, France and the USA. The Western powers now had much greater influence over China's economy and infrastructure and could demand political concessions from the country's leaders.

The decline of the Qing dynasty

Emperor Qianlong officially stepped down as emperor in 1796, but in truth he maintained control over China until his death in 1799.

While his reign seemed like one of peace and prosperity, Qianlong had turned a blind eye to extensive corruption in the imperial bureaucracy. During his rule there had also been a period of huge population growth in China, which led to shortages of land, resources and food for almost 300 million peasants.

His son, Emperor Jiaqing, inherited these problems, as did the emperors that followed. While the Qing emperors attempted to address these issues, they had limited levels of success. A number of rebellions and popular uprisings, such as the White Lotus Rebellion (1796–1804) and the Eight Trigrams Uprising (1813), tested the dynasty's ability to control Chinese society. Although these rebellions were suppressed, anti-Qing sentiment began to increase around the country.

Western interference also damaged the power of the Qing dynasty. The opium trade shifted the balance of power and trade, making the British wealthy and the Chinese poorer. Widespread opium addiction damaged the fabric of Chinese society.

Outlawing the drug did little to affect the trade, instead making the imperial government look ineffectual. Ultimately, this culminated in the Opium Wars, which left the Chinese imperial court deeply in debt and greatly reduced in political power.

The Century of Humiliation

The period beginning with the First Opium War has come to be called the 'Century of Humiliation' in some sections of Chinese society. This was a period of continued loss and embarrassment for the Qing dynasty.

A number of significant Chinese losses and setbacks occurred during this time, including:

- defeat in the First Opium War
- the 'unequal treaties' that made massive trade concessions to Britain and other Western countries
- the Taiping Rebellion (1850–1864)
- defeat in the Second Opium War
- defeat in the Sino–French War (1884–1885), in which France took control over the Tonkin region that would later become modern Vietnam
- defeat in the First Sino–Japanese War (1894–1895), in which Japan took control of Korea.

Source 1

This 1898 French political cartoon from *Le Petit Journal*, shows foreign powers (Britain, Germany, Russia, France and Japan) dividing China among themselves. Acceptable at the time, it's extremely offensive by modern standards.

This was also a period in which foreign nations negotiated their own 'spheres of influence' within China, often dealing with regional powerbrokers rather than the Qing bureaucracy. Multiple foreign enclaves were established within China, and these enclaves followed their own laws rather than Chinese law.

Finally, it was also a period of natural disasters, including floods and droughts. A famine in northern China lasted from 1876 to 1879, and between 9 and 13 million people died. The famine also caused great economic distress, which led to further civil unrest.

These events raised difficult questions over whether the Qing dynasty had lost the *Tianming*, or 'Mandate from Heaven'.

Source 2

The signing of the unequal treaties forced China to make massive trade concessions to Western nations.

Reign of the Dowager Empress

While officially only acting as **regent** on behalf of underage male emperors, the true ruler of China in the second half of the 19th century was **Dowager** Empress Cixi, who ran the Qing court from 1861 until her death in 1908.

The woman who became the Dowager Empress was named Yehenara. She entered Emperor Xianfeng's court as one of his **concubines** in 1851, and gave birth to his only son in 1856. When Xianfeng died in 1860, he nominated their five-year-old son as Emperor Tongzhi and eight prince regents to rule until he had come of age. Lady Yehenara orchestrated the Xinyou Palace Coup in 1861, in which all eight regents were killed or forced to commit suicide. Her ally Prince Gong became sole regent, and she assumed the title Dowager Empress Cixi.

Empress Cixi began to rule 'behind the curtain' as the real decision-maker, with Prince-Regent Gong as her agent in the *Junchichu*, or Grand Council. In 1873, Emperor Tongzhi was put in charge of state affairs, but in name only, and in 1874 he died. Empress Cixi nominated her three-year-old nephew to become heir, Emperor Guangxu, and she continued to rule China until officially retiring in 1889.

Source 3

Detail from a 1906 painting of Dowager Empress Cixi by Dutch painter Herbert Vos

Source 4

The official portrait of Empress Dowager Cixi, taken by the court photographer in 1895

Source 5

Queen's Road in Hong Kong, 1902. Ceded to Britain after the First Opium War, Hong Kong was controlled by a British governor rather than the Qing dynasty of Empress Cixi.

Her retirement did not last long, however. Emperor Guangxu signed edicts to implement a reform agenda, such as building more railways and forming a national budget. Empress Cixi was not happy with the rapid pace of change. She arrested the emperor within the palace, squashed the reforms and did not allow Guangxu to reign. She took back control over China and remained in power until 1908; she died one day after the death of Emperor Guangxu (under possibly suspicious circumstances).

Known for her conservatism, cultivation of palace intrigues and the murder of a number of people (including one of Emperor Guangxu's favourite concubines), Cixi has often been portrayed as a **despot**. Such over-simplifications obscure her brilliance as a tactician who overcame traditional male power structures. Cixi did advocate for some modernisation, such as developing railways, but not if they ran through sensitive cultural places. She refused to introduce widespread industrialisation at any cost, as it would crush cottage industries such as spinning, but she also removed medieval punishments from the law codes.

She trod a difficult path. Each decision she made was complicated by maintaining a lopsided relationship with foreign powers, multiple internal rebellions and trying to develop the empire from within the Forbidden City. Critiques of her as a sinful poisoner and instigator of palace intrigues overlook the extent of the challenges and crises that she had to overcome.

Learning ladder H3.8

Show what you know

1 Which problems from Emperor Qianlong's reign were inherited by later emperors?

2 How did the Chinese people undermine the emperor and the vast bureaucracy?

3 How did natural disasters add to the problems faced by the Chinese emperors and their administrations?

4 Define 'sphere of influence' and 'enclave'.

Source analysis

Step 1: I can list specific features of a source

5 Refer to Source 1 and describe what is happening in the cartoon.

Step 2: I can find themes in a source

6 Refer to Sources 3 and 4. What does the adoption of photography show about Dowager Empress Cixi's approach to technology and modernisation?

Step 3: I can use the origin of a source to explain its creator's purpose

7 What or who is being made fun of in Source 1? Do the publication details influence your interpretation?

Step 4: I can analyse a source

8 Sources 3 and 4 are both portraits of Dowager Empress Cixi. Which one do you think she would have preferred? Explain your answer.

Source analysis, page 240

How did Chinese technology change during the 19th century?

For most Chinese peasants, life during the tumultuous 19th century wasn't very different to that of their grandparents. China did not industrialise until relatively late, and the technology of daily life changed very little. Things were different in the port cities, where Western inventions and ideas began to dominate society.

Source 1

Bronze coins from the Qing dynasty stamped with the name of Emperor Qianlong

Market towns and trade

The Qing dynasty promoted a primarily agricultural economy, in which small farmers and market gardeners were given incentives to develop land for agricultural use. Emperor Qianlong's reign saw the construction of weirs and canals. These opened up more land to farming, which led to a proliferation of market towns. In these towns, the money that changed hands for goods became the taxes paid to Qing officials, so taxation drove **monetisation** and economic growth.

Markets were initially held in central towns (usually a day's journey away) on certain days of the week. Under the Qing dynasty, sections of towns became permanent markets and merchant classes emerged to sell goods. Trade networks were sufficiently organised for goods to travel freely across provinces. Medicines from the north went south; cotton went from the periphery of the empire to be made into fabric in the centre. A system of banks called the *Piaohao* would take deposits in one location, issue a remittance certificate and allow the redemption of goods in other towns and cities.

One of the fastest ways to modernise a country is to monetise it, which happened during the Qing dynasty. Early Chinese coins were 'cast' by pouring metal into a mould, which meant coins were often smelted by the people to make utensils. Emperor Qianlong devised a plan to make coins out of alloys, which made them more valuable as coins than as raw metals. He also offered incentives for mints to buy back utensils for metal to make coins when metal shortages occurred. Copper coins (called *wen*) had a hole in the centre, so they could be put on strings and tied to a belt. The value of a silver ingot (or *tael*) varied, from 1000 wen (before 1820) to 2000 wen (after 1840).

New crops were introduced from Europe and the United States, such as corn, chilli, pepper, peanuts, potatoes and tomatoes. These allowed farmers to plant beyond their staple crops of rice and other vegetables, and expanded the repertoire of recipes and menus that can still be found today.

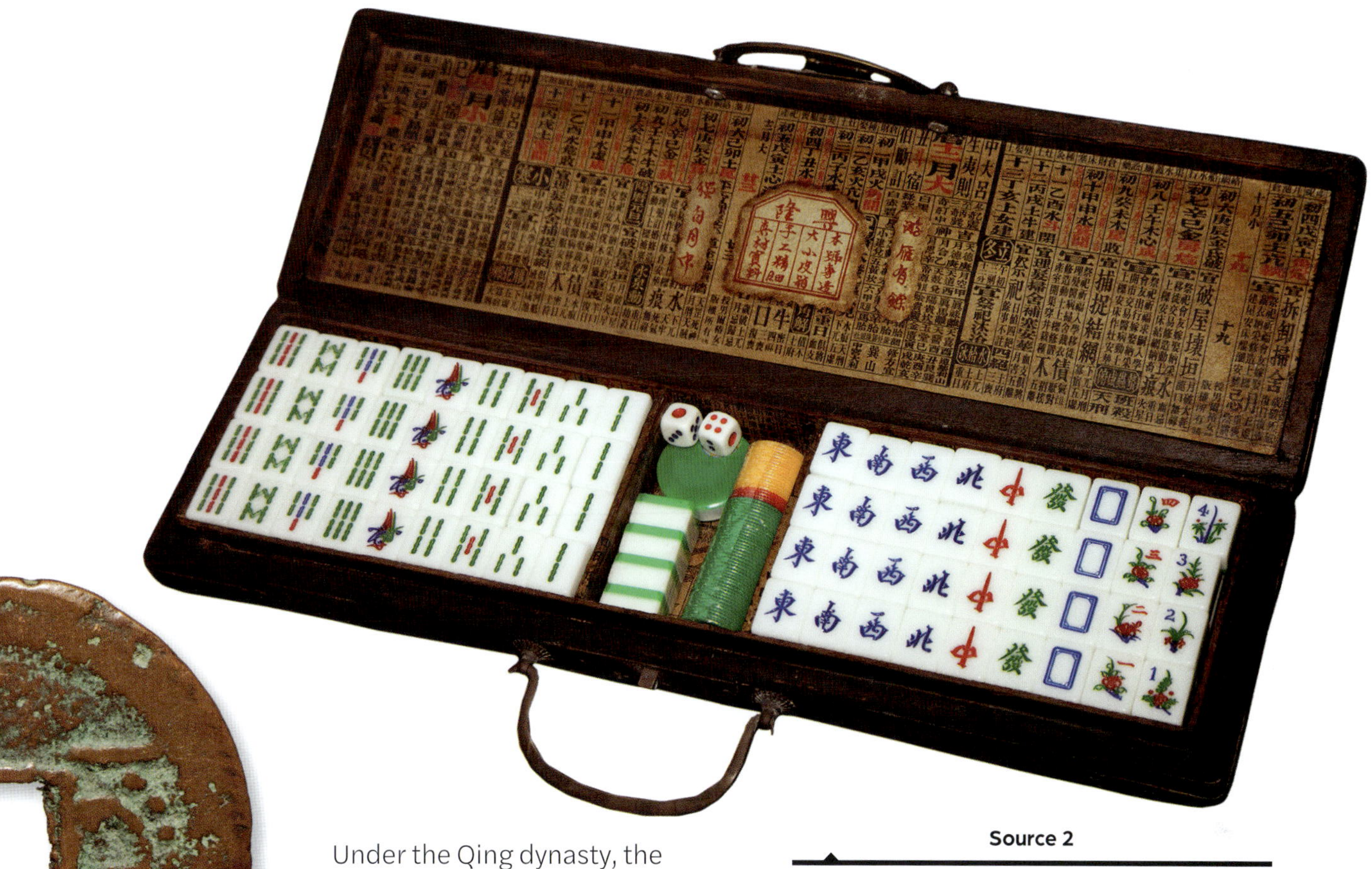

Source 2

A full mahjong set could be very expensive. The tiles were crafted from bone, often backed with bamboo, and were often carried in ornate wood or metal cases.

Under the Qing dynasty, the population increased fourfold. The increase in mouths to feed was supported by increasing the amount of fertile land. This was made possible by managing water resources and offering incentives to farmers. Embracing these reforms meant that food production could increase; however, droughts and warfare (which prevented seed planting) still affected food security. Such pressures often led to insurrections and rebellions when people were short of food.

The wealthy and the Westernised

Life was less difficult for the wealthy nobles of the court and for the officials of the Chinese imperial bureaucracy. The Qing dynasty introduced a variety of new noble titles and positions, meaning that the Qing court became larger and busier. These nobles had administrative roles to play in the imperial bureaucracy and had to contribute portions of their wealth to the emperor's coffers. Many nobles developed their own networks of power in regional areas and used these to negotiate deals with Western traders to make more money.

One way nobles passed the time was by playing *mahjong*, a tile-laying game. The exact origins of *mahjong* are unknown; it likely developed from a variety of games, including a card game called *ya pei* and another game called *madiao*. No matter its origin, *mahjong* became a popular pastime among those who could afford a set of playing pieces.

After the Opium Wars, a third way of life developed for some Chinese citizens. The British influence in port cities such as Shanghai, not to mention direct British control in Hong Kong, meant that Western dress, education and culture became much more common. Some Chinese locals were forced to adopt Western ways in order to find work; others chose to adopt those ways in order to gain social advantages. Western styles of dress, such as suits, ties and hats, became common, and some men cut off their plaits and grew their hair in Western styles. Hong Kong in particular became dominated by British culture, with the Chinese locals segregated from their new English rulers. The east portion of Hong Kong was mostly occupied by the British, who built racecourses, mansions and polo fields – and barracks to house large numbers of British soldiers.

Tea and other exports

Tea was so popular as a stimulant and a medicine that it became the national drink of China, and was taxed during the Tang dynasty (618–907 CE). Tea houses existed in every city and town, and were popular places to meet, gossip and socialise for peasants and nobles alike. Unhappily for the Qing dynasty, tea houses were also places where the social classes mixed, providing opportunities for everyone to share their discontent with the current political order.

Tea was also one of China's primary exports, especially to Britain, where it became incredibly popular. The popularity of tea, along with items such as porcelain tea sets, was one of the factors that led to Britain's massive trade deficit with China, which Britain ultimately addressed through the sale of opium and the Opium Wars. The British East India Company also worked to break China's monopoly over tea. One of the company's agents, botanist Robert Fortune, stole tea plants and seedlings during trips to China, bringing them to India's Darjeeling region to develop India's own tea industry.

Chinese artisans had been making porcelain and ceramics for thousands of years using various techniques and materials. Under the Qing dynasty, there was an emphasis on developing more colours, rather than just the blue and white designs popular during the Ming dynasty. Thanks in part to its connection to tea, porcelain became another major Chinese export – to the point where it became known as 'china' in many parts of the British Empire. Many tea services (cups, plates, bowls and tea pots) were decorated, treasured, collected and often handed down by successive generations. China (porcelain) also diversified into figurines, vases, statuettes and other art objects, which also became popular and valuable.

Silk was another luxury export, not just to the British but around the world. For 6000 years, silk had been part of China's export economy; it's what gave the 'Silk Road' trading network its name. Sericulture (silk production) had been a closely guarded secret all that time. Silk was usually sold as fabric, rather than Chinese clothing, so it could be used to create clothing in the styles popular in other countries. Within China, only nobles had been allowed to wear silk. Under Qing rule peasants were also permitted to wear it. Silk was also used for decoration, fishing and making bows.

Industrialisation

The Industrial Revolution that changed Britain, the United States and other nations did not occur in China at the same time. The controls that Emperor Qianlong and his successors placed on foreign traders also prevented many foreign innovations and inventions from entering China. During an audience with foreign ambassadors Qianlong said, 'I set no value on strange objects and ingenious, and have no use for your country's manufactures'. This demonstrated his unwillingness to allow foreign technology into his country.

Resistance to industrialisation also affected the militarisation of the Qing dynasty. The most advanced weapons in China at this time were gunpowder-fired cannons, which had been introduced during the previous Ming dynasty. The Qing emperors did not allow their armies to use guns or rifles, preferring traditional weapons, such as bows and arrows, swords and spears. These weapons failed when they faced the technologically superior munitions of the British during the First Opium War.

Source 3

This valuable Dragon jar is from the Ming dynasty and was created c. 1403–1424.

Source 4

The grand opening of the Woosung Road, China's first railway line, in 1876. It was dismantled in 1877. [From *The Graphic* magazine, 23 December 1876]

After the Opium Wars, more foreign technology was imported into China, and a desire for industrialisation began to grow. There was a demand not only for foreign weaponry and munitions, but also for steam power, mining equipment and better agricultural technology. The Qing authorities remained reluctant to allow new technology, so Chinese officials and traders often worked around them, dealing directly with foreigners and leaving out the bureaucracy.

Trains are a good example of the slow process of industrialisation in China. The first Chinese railway was built by the British in 1876 and operated near Shanghai, but it was closed and then destroyed by Qing authorities in 1877. Emperor Guangxu wanted to implement reforms and build more railways, but the conservative Dowager Empress Cixi deposed him and cancelled the reforms. By 1894, only around 480 kilometres of railway tracks had been built throughout China – far less than existed in England, a much smaller country.

Learning ladder H3.9

Show what you know

1 What changes did monetisation bring to Chinese society?

2 What were some of the agricultural benefits of trade with foreign empires and nations?

3 How did Britain manage to challenge the Chinese monopoly over tea?

4 How did high-ranking government officials increase their power in regional centres and ports?

Historical interpretations

Step 1. I can recognise that the past has been represented in different ways

5 Which national game developed during this time?

Step 2: I can describe historical interpretations

6 What does Emperor Qianlong's public statement about European inventions indicate about his public approach to foreign influence?

Step 3: I can explain historical interpretations

7 How did Emperor Qianlong's and others' approaches to modernisation affect the militarisation of China and the outcomes of conflicts as a result?

Step 4: I can analyse historical interpretations

8 How did the trade deficit on tea, porcelain and silk affect relations between Western European nations and the Qing Empire?

Historical interpretation, page 255

How did rebellions and revolutions change China?

Growing discontent with the rule of the Qing dynasty led to a number of rebellions and uprisings. In order to defeat them, the Qing brutally suppressed their own citizens, which weakened them politically. The 1911 revolution ended the reign of the Qing and the dynastic system, creating an entirely new government.

The White Lotus Rebellion (1796–1804)

The *Bai Lian Jiao* (White Lotus Society) was a **populist** Chinese secret society, based on a mixture of Taoist and Buddhist ideas, with a long history that stretched back to the Song dynasty. Secret societies were fairly common during this period; people from the lower classes, who bore the real hardships of agricultural failures and government corruption, joined such societies in order to vent their discontent.

The White Lotus Society didn't have much support until 1796, when high levels of immigration into Sichuan province from other provinces upset the local populace. Inadequate resources and high taxation angered the locals, who joined the White Lotus Society to express that anger. The people rallied beneath a banner that was openly anti-Qing and called for the return of the Ming dynasty.

Military forces struggled to suppress the growing rebellion. Local nobility had to recruit and pay exorbitant fees to mercenaries and bribe White Lotus members to change sides; the cost of fighting the rebellion ran into 100–200 million *taels* of silver. The mercenaries also required training, as they struggled to counter the White Lotus's **guerrilla** tactics of attacking and then disappearing.

The White Lotus Society members were eventually routed in 1804, after the deaths of around 100 000 rebels. While the rebellion was suppressed, the damage had been done to the Qing dynasty. Poor people had shown their dissent openly to their rulers, which inspired others to do the same.

Source 1

The White Lotus Society. [Handscroll; ink on paper, unidentified artist, c. 1368–1644.]

Source 2

Wu Youru's 1886 painting *Regaining Jinling* from the book, *A scene of the Taiping Rebellion, 1850–1864*.

The Taiping Rebellion (1850–1864)

A change of reign occurred in 1850, from Emperor Daoguang to Emperor Xianfeng; in the same year, a famine and other economic hardships affected the lower levels of society. During this period, a group called the *Bai Shangdi Hui* (God Worshipping Society) instigated a four-year rebellion that led to the occupation of the city of Jiangning and the death of 20–70 million people.

It began with a man named Hong Xiuquan. Fusing Christian ideas with Chinese thought, he convinced a large number of followers that he was either a **messiah** or the emperor of a future dynasty, with a mandate to vanquish the Qing. In 1850, Hong amassed approximately two million followers into military units and began to purchase weapons and supplies. Initial victories were seen as a sign from Heaven, and Hong declared the 'Taiping Heavenly Kingdom' as a new state.

Hong captured the major cities of Wuchan and Anqing but was met by Qing military forces at Beijing. The forces of the *Bai Shangdi Hui* forced Emperor Xianfeng and his troops to retreat during the attack on Tianjing. They then occupied the city of Nanjing in 1853; however, corruption set in among Hong's officials and advisers.

After the Treaty of Tianjin was signed following the Second Opium War, the British helped Emperor Xianfeng to crush the Taiping Rebellion as a show of good faith. The army besieged Nanjing in May 1862, and it fell in July 1864, a few weeks after Hong Xiuquan died of food poisoning.

Source 3

Hong Xiuquan, self-proclaimed 'Brother of Jesus' and leader of the Taiping Rebellion

Source 4

Qing troops escort and protect foreigners at the conclusion of the Boxer Rebellion. [Illustration by Oswaldo Tofani, *Le Petit Journal*, 15 July 1900]

The Boxer Rebellion (1899–1901)

In the north of China, secret societies stirred up nationalist discontent and began training rebellion troops. The largest of these societies was the Society of the Righteous and Harmonious Fists, or *I-ho ch'uan* ('The Boxers'). The Boxers called for a return to the 'traditional source' of Chinese warfare, such as spirituality and martial arts prowess; some even claimed (or truly believed) that they were immune to swords and bullets. Their war cry – 'Support the Qing, death to all foreigners' – united a people fed up with defeat after defeat.

The Boxers began their rebellion with attacks on Christian missionaries in 1899 and the destruction of railways and telegraphs in 1900. These attacks earned the love of the Chinese people but angered Western governments, who implored Emperor Guangxu to suppress the Boxers. Instead, Empress Dowager Cixi actually supported the Boxer campaign because of its popularity.

To protect their diplomatic missions in Beijing from the Boxers, and the Qing army that supported them, the Western powers sent more than 50 000 troops to China. An Eight-Nation Alliance was formed, made up of Austria–Hungary, Britain, France, Germany, Italy, Japan, Russia and the United States of America. (The British forces included Indian *sepoys* and Australian colonial militias.)

On 12 August 1900, Boxers and Qing troops attacked the diplomatic missions; on 14 August, Eight-Nation Alliance forces divided and attacked each gate of Beijing. Over the following months, tens of thousands of Chinese people (mostly civilians) were killed. Empress Cixi changed course, and ordered the Qing Army to assist the Western forces and arrest the Boxers. However, it was too late and, as the allied forces invaded, she left the Forbidden City in disguise. Meanwhile thieves sacked the city, taking palace antiques by the cartload.

Most of Chinese society had unified to reject the foreigners, and they had lost. The result was another 'unequal treaty': The Boxer Protocol of 1901. Under the terms of the treaty, the Chinese government had to pay 450 million *taels* of silver to the Eight-Nation Alliance countries and participation in anti-foreigner secret societies became a capital offence.

The Xinhai Revolution (1911)

Annoyed at government corruption and the Qing dynasty's inability to prevent invasions, many people in China felt that their country was in decline. A young leader called Sun Yat-sen organised his own revolution in 1895. It was suppressed, but from this experience he learned to use secret society networks across China and formed his own league, the *Tongmenghui*.

When the imperial court ordered the suppression of the Wuchang Rebellion in October 1911, the Qing Army refused to obey. The rebellion grew into a full-blown revolution, named the Xinhai Revolution; the army seized power from the Qing rulers and declared a new Chinese **republic**.

Sun Yat-sen returned from exile to become the first provisional president of the republic. Without armed forces to support him, Sun gave power to General Yuan Shikai, who recommended that the Empress Dowager Longyu, Prince Regent and three-year-old Emperor Xuantong abdicate, which they did on 12 February 1912.

Elections in February 1913 saw the election of the Kuomintang, or Chinese Nationalist Party, which remains active to this day in Taiwan. Extraordinary events saw Yuan Shikai declared as the Hongxian Emperor for 83 days before being removed from office; he died in 1916. Emperor Xuantong was reinstated for 12 days in 1917 by a coup, which was defeated by other Chinese republican forces.

The **abdication** of Emperor Xuantong brought an end to the Qing dynasty, as well as the institution of imperial rule of China, which had lasted for more than two millennia.

Source 5

Dr Sun Yat-sen became the Republic of China's first provisional president.

Learning ladder H3.10

Show what you know

1 How did the Boxers enlist followers?

2 Which institutions did the Xinhai Revolution bring to an end?

3 Name three reasons why the White Lotus Society's support increased after 1776.

Cause and effect

Step 1: I can recognise a cause and an effect

4 Source 5: What does Sun Yat-sen's style of dress indicate about his education and ideas?

Step 2: I can determine causes and effects

5 What long-term problems led to the White Lotus Rebellion?

Step 3: I can explain causes and effects

6 List the factors that inspired the White Lotus Rebellion, Taiping Heavenly Kingdom, Boxer Rebellion and Xinhai Revolution. Explain how each uprising was similar and different in terms of causes.

Step 4: I can analyse causes and effects

7 In what ways were the Chinese government's responses to the uprisings both adequate and inadequate?

Cause and effect, page 247

H3 .11

PART III: JAPAN

Why did the Tokugawa shogunate close Japan's borders?

The Tokugawa shogun, Japan's effective ruler, outlawed travel in and out of Japan in 1635. This prevented foreign ideas and religion from influencing Japanese society, and reduced the power and wealth of the shogun's rivals.

In 1600, Tokugawa Ieyasu became the nation's military leader, or shogun, beginning a dynasty that would last for almost three centuries.

While the emperor was Japan's official ruler, real power lay with the shogun, who oversaw national matters, including trade and defence. For the most part, peace reigned in Japan, although conflict persisted with the indigenous Ainu people of Ezo (modern-day Hokkaido), who clashed with Japanese settler colonies on the south of the island. Provincial lords, known as *daimyo*, retained their own *samurai* and controlled their lands in exchange for tribute and loyalty to the shogun.

Sakoku

In 1635, the shogunate implanted a policy of *sakoku* ('closed country'), which severely limited Japanese contact and trade with other countries.

Before *sakoku*, international trade and travel were largely unrestricted. Japanese sailors traded throughout Asia, and official envoys visited nations in Asia, South America and Europe. Large numbers of foreign traders lived in Japan, and foreign pirate ships were active in the seas surrounding the island.

Sakoku was a response to growing concern about the threat of these foreign traders and pirates. Christian missionaries were also seen as a danger to Japanese religion and *kokutai*,

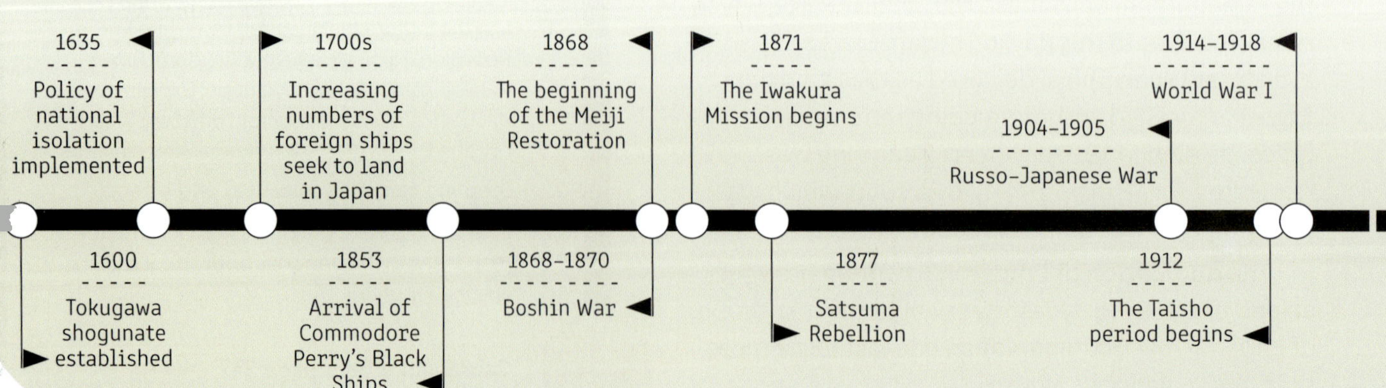

18th-century Japan

Source 1

This map shows Japan's proximity to China (the Qing Empire) and the Korean (Joseon) Peninsula (independent but under the Chinese tributary system).

the national essence; the Catholic Portuguese and Spanish were seen as especially zealous in this regard.

Under *sakoku*, the shogun officially forbade all foreigners from entering Japan or Japanese waters. Japanese citizens who returned from abroad were to be executed. For more than 200 years, Japan would remain isolated from the rest of the world.

Exceptions to the policy

Two exceptions to the *sakoku* restrictions existed in Nagasaki: a small community of Chinese traders could operate and Dutch traders were allowed on Dejima, an artificial island in Nagasaki harbour, although they could not set foot on Japanese soil.

Controlling trade in this way not only prevented creeping foreign influence, but also stopped regional *daimyo* from growing too wealthy by trading with outsiders. In turn, this restricted their ability to raise large armies and present any kind of threat to the Tokugawa dynasty.

Through official channels Japan maintained *tongsinsa*, or goodwill embassies, which facilitated trade and contact between Joseon Korea and Imperial China, along with the Ryukyu Islands to the south. Silver proved a particularly valuable export, and Chinese goods, developments and technology flowed into Japan in return.

Learning ladder H3.11

Show what you know

1 What was the *sakoku* policy?

2 Suggest why the shogun wanted Japan to remain isolated.

3 Outline the role of the *tongsinsa*.

Cause and effect

Step 1: I can recognise a cause and an effect

4 Why was Christianity seen as a threat to national unity in Japan?

Step 2: I can determine causes and effects

5 The Dutch were the only Westerners allowed to trade in Japan at this time. Why was this the case?

Step 3: I can explain causes and effects

6 What was the effect of *sakoku* on foreign trade in Japan?

Step 4: I can analyse causes and effects

7 To what extent did *sakoku* also strengthen the shogun's position against the *daimyo*?

Cause and effect, page 247

What was life like during the Tokugawa shogunate?

Japan during the Tokugawa shogunate operated under a feudal system. Everyone belonged to a particular class, with which came specific obligations and expectations. Japan's closed borders prevented foreign influences from changing society.

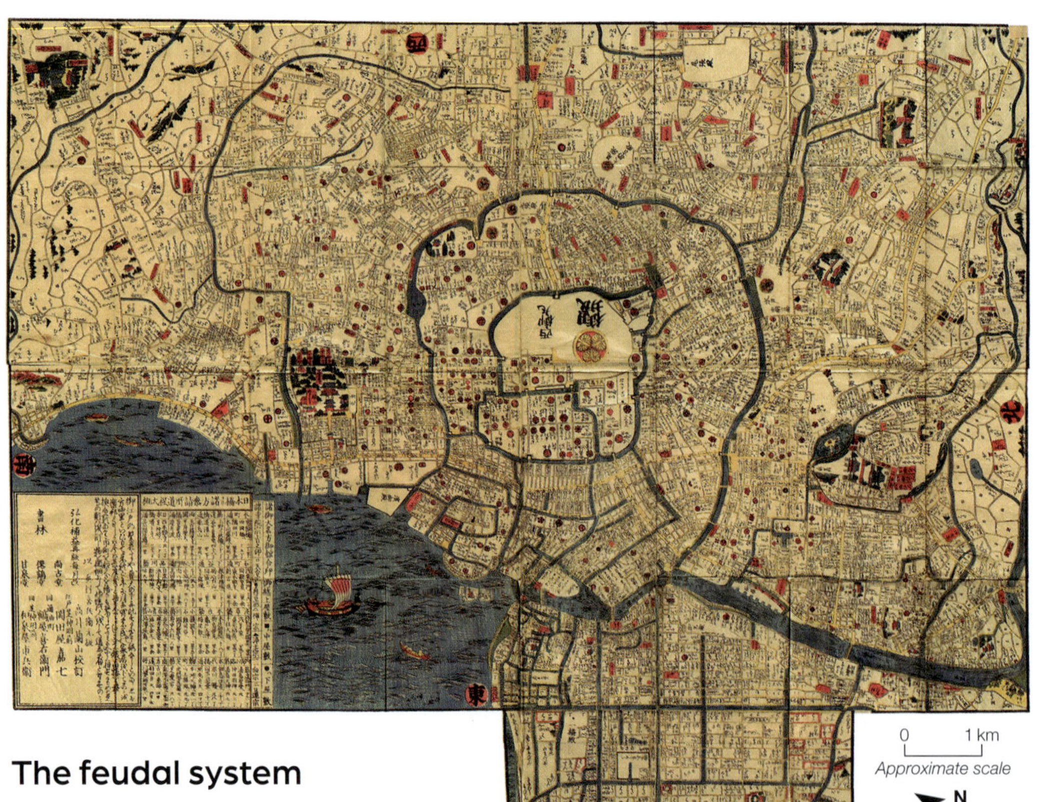

Source 1

A map of the city of Edo (modern-day Tokyo) from the 1840s. Edo was the seat of the shogun's power and the location of his palace.

The feudal system

The Japanese emperor stood at the very top of the system and was considered to be a descendant of Amaterasu, the sun goddess. During the Tokugawa shogunate, he was considered to be a spiritual head who held little real power or influence. His court was supported by *kuge*: nobles, aristocrats and bureaucrats.

Real power lay with the shogun, the emperor's leading general, who lived in Edo and ran the nation's domestic welfare and foreign affairs. The Tokugawa family held this role for centuries, primarily because of their large wealth and landholdings across key trade routes.

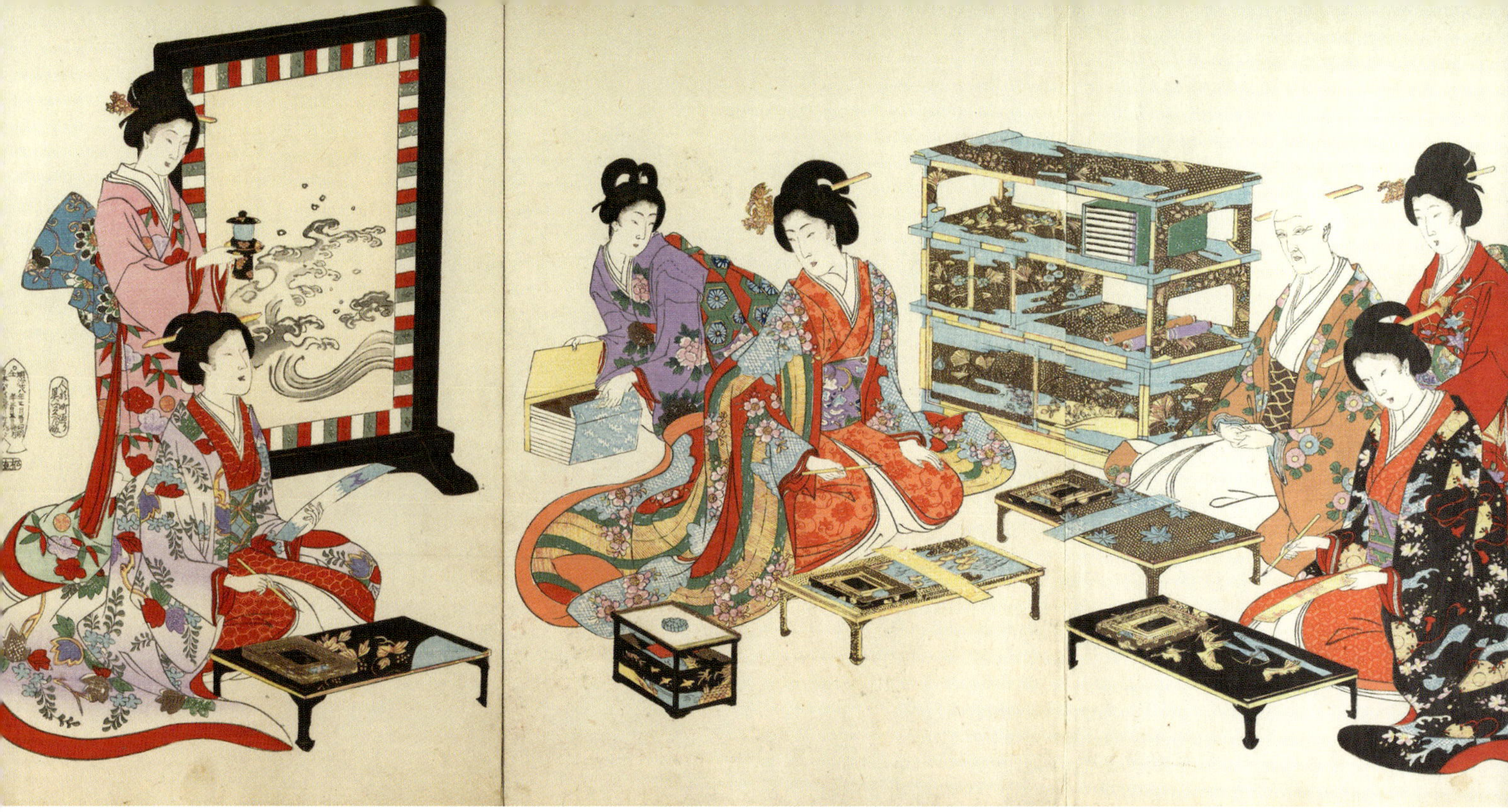

The nation was divided into approximately 250 *han* (domains), each ruled by a local hereditary lord (*daimyo*), who retained a private army and controlled laws and taxes. In order to ensure the *daimyo* could not become wealthy and pose any potential threat, the shogun spied on them and placed controls on their behaviour. He demanded they regularly upgrade roads and seawalls, regularly required them to live in Edo and prohibited alliances between *daimyo*, including those created through marriage.

The samurai were the warriors who made up the army of each *daimyo*. They commanded significant respect and lower classes were expected to show deference to them. They carried swords and lived according to the *bushido* (warrior code). By 1750, Japan had been relatively peaceful for more than a century, and the role of the samurai had become more administrative. Some had fallen on hard times and taken to gambling and other vices.

Everyone else was broadly classified according to the *shinokosho* (the four divisions of society): *shi* (samurai), *no* (farmers/peasants), *ko* (artisans/craftsmen) and *sho* (merchants/traders). Confucian traditions valued the role of farmers, who produced goods essential to society, over that of craftsmen. Merchants were considered of less value, as they generated wealth without producing goods.

Source 2

A 19th-century depiction of women in the Ooku quarters of Edo Castle. These women included the wives and concubines of current and former shoguns; they were not allowed to have relationships with other men. [Hashimoto Chikanobu, *Ukiyo-e depiction of the Ōoku,* 2 March 1895]

Some groups fell outside the feudal system, notably the *eta* (those who worked in 'tainted' industries such as leather-working, undertaking or animal slaughter) and *hinin* (people in indentured labour, ex-convicts, beggars and vagrants). These groups were considered untouchable and discrimination against them was widespread.

The role of women

Life for women in Tokugawa Japan was very different from that of men and depended greatly on their social status. They were expected to adhere to the Confucian tradition of the *Three Obediences*:

- as a maiden daughter
- as a chaste wife
- as a dedicated widow.

Marriages were generally arranged by parents and women held few rights. Female illiteracy was widespread; a woman could not own property and could be killed by her own husband if she were perceived to be lazy or unfaithful. Some were retained in entertainment and service, such as the *Ooku* women of Edo Castle.

冨嶽三十六景
神奈川沖
浪裏

Limited change

Through contact with Imperial China, new ideas, developments and technologies entered Japan, although the *sakoku* restrictions meant that the nation remained less open to change. *Rangaku* (Dutch studies) was a notable exception; books and texts obtained from Dutch traders at Dejima introduced new ideas around science, astronomy, medicine, languages and the natural world.

As swampland was filled and marshes drained, Edo grew during this period to become one of the largest cities in the world, although the crowded conditions and widespread use of wood created a risk of fires. Devastating fires proved a regular occurrence throughout the 16th and 17th centuries. As a result, thatched roofs were banned in favour of tiles and fire-prevention laws enacted.

As urban centres grew, lifestyles began to change for many. The rise of *ukiyo* ('floating worlds', or urban culture) saw a growth in new forms of entertainment, art (such as *ukiyo-e*) and business, and a gradual blurring of many of the older social hierarchies.

Source 3

The Great Wave off Kanagawa, 1831, by Katsushika Hokusai is one of the most important *ukiyo-e* artworks. *Ukiyo-e* was a genre of Japanese art, mostly paintings and woodblock prints, that emerged from the growing urban culture.

Learning ladder H3.12

Show what you know

1. What was the *shinokosho*?
2. Why were farmers considered to be more important than merchants?
3. Who were the *hinin* and *eta*? Why might they have suffered discrimination?
4. Explain how the emperor kept the *daimyo* under control.

Source analysis

Step 1: I can list specific features of a source

5. What is pictured in Source 3? Does this problem still exist in modern times in Japan?

Step 2: I can find themes in a source

6. Source 2 portrays a very traditional scene in the Imperial palace. What do the characters in the scene have in common? Suggest why.

Step 3: I can use the origin of a source to explain its creator's purpose

7. The author of the 1840s Edo map (Source 1) could not have seen the city from above. Why do you think he or she chose to represent it this way?

Step 4: I can analyse a source

8. To what extent do you feel that the lives of women in Tokugawa Japan are accurately reflected in Source 2? Use evidence from the source to support your response.

HOW TO

Source analysis, page 240

What caused Japan to re-engage with the West?

The arrival of US forces in 1853 marked a significant turning point in Japan's political outlook. It ended the *sakoku* period and began a growing internal push to modernise Japan. The Western powers clamoured to help accelerate this process.

Commodore Perry's Black Ships

By the mid-19th century, the USA was building its own empire of trade and vassal states, and was looking for ways to take power from its European rivals. Japan was seen as a strong potential trading partner within Asia, and the USA had made multiple unsuccessful attempts to establish diplomatic and trading ties with the isolated nation.

In 1853, the US government sent Commodore Matthew Perry, commanding four steam-powered warships, to present a letter from President Filmore to the Japanese Emperor. Perry's fleet of 'black ships' sailed around Japan, intimidating local *daimyos* who had never seen such advanced technology before. Arriving at the shogun's palace in Edo, Perry showed off the fleet's military power, firing cannons and making threats of force before delivering the President's letter to the shogun's aide. When the fleet left, Perry made it clear that the USA demanded Japan re-open its borders, and promised to return in a year's time for their reply.

Shogun Ieyoshi died a few days after Perry left, and his successor was in poor health, so it was left to the court's Council of Elders to decide how to respond to the threat. However, they were paralysed by indecision, unsure how to handle this new threat.

Perry returned in 1854 after only six months, this time with a fleet of 10 steamships and 1600 men. The Elders gave in to almost all of the US demands, and the Convention of Kanagawa was signed. This treaty opened the ports of Shimoda and Hakodate, established a US embassy and gave the USA preferential trading rights. It effectively ended the policy of *sakoku* and forced Japan to open its borders to the rest of the world.

The Meiji Restoration

The capitulation cost the shogunate a great deal of power and respect, and its control over Japanese society slipped. In late 1867, Shogun Yoshinobu was forced to resign, allowing Emperor Meiji to take back political control of the country. For the first time in centuries, Japan's emperor was a ruler rather than a figurehead. The period from 1868 is thus called the Meiji Restoration, and was a period of enormous change.

Japan began to define its own future. The emperor moved the capital city from Kyoto to Edo, which was renamed Tokyo, and the feudal domains ended in favour of a national government. In 1889, a new constitution created the imperial *diet* (parliament), sidelining the emperor and putting decision-making powers into the hands of the *genro*, older oligarchic advisors.

Renewed trade with the West brought modernisation and prosperity. Industrial development and foreign investment brought greater wealth, particularly to the *zaibutsu*, large business conglomerates that dominated industry. Japan's overseas colonies delivered raw materials,

Source 1

Detail from a depiction of Commodore Perry's flagship by Tsukioka Yoshitoshi, 1876. The Japanese were unfamiliar with industrial technology, and Perry's ships were far more powerful than the shogun's military forces.

such as coal and iron. These were particularly valuable for the growing Japanese military, which greatly increased Japan's presence in Manchuria, a region of north-eastern China.

Military expansion

The modernisation of Japan's military was a top priority during the Meiji Restoration. The Sino–Japanese war (1894–1895) resulted in Japan's victory over Qing dynasty China, which established Japan as a regional power and handed them control of Korea and some Chinese territories. The 1905 Eulsa Treaty formalised the Japanese sphere of influence on the Korean Peninsula, depriving it of independence.

Growing ties with the West, such as the Anglo–Japanese Alliance (1902), helped Japan to modernise; they also laid the foundations for acquiring Germany's colonies in the Asia–Pacific at the end of World War I. Growing tension between Japan and Russia over China brought the two nations into conflict. The Russo–Japanese War (1904–1905) was a comprehensive victory for Japan, the first victory by an Asian power over a European power in modern times.

Learning ladder H3.13

Show what you know

1 Identify three ways Japan began to change during the Meiji Restoration.

2 Why did the new government make it a priority to improve the military?

3 What was the outcome of the Russo–Japanese War?

4 Japan became increasingly influential in China and the Korean Peninsula at this time. How do you think people living there felt about this? Why?

Historical significance

Step 1: I can recognise historical significance

5 What do you think were the most important changes in Japan at this time? Why?

Step 2: I can explain historical significance

6 Why was the arrival of Commodore Perry's Black Ships significant?

Step 3: I can apply a theory of significance

7 Why was the Meiji period also known as the 'Meiji Restoration'? What was 'restored'?

Step 4: I can analyse historical significance

8 Japan's victory in the Russo–Japanese war changed the way that the world saw Japan. Why?

Historical significance, page 251

How did Japan modernise during the Taisho period?

The Taisho period began in 1912 and marked significant social change, particularly for women, following the end of World War I in 1918. Sometimes referred to as Japan's 'Jazz Age', a wave of prosperity, liberalism and intellectualism swept through urban centres.

The Meiji period ended with the death of Emperor Meiji. He was succeeded by his son, Prince Yoshihito, who declared his reign would be known as the Taisho ('great righteousness') period. To begin with, however, the Taisho period was one of upheaval and conflict. Japan had three different prime ministers between 1912 and 1913 because of conflicts between the civilian government, the military and the imperial court. Not long after the government stabilised, Japan entered World War I, allying with the United Kingdom against its enemies China and Germany (see Chapter H4).

Social change

In the post-war world, things changed dramatically for Japan, both at home and internationally. The country was granted a permanent seat on the Council of the League of Nations (the precursor to the modern United Nations) and was recognised as a world power.

In Japan's urban centres, a thriving film, music and literary culture rose to prominence, as did consumerism. *Moga* ('modern girl') and *mobo* ('modern boy') trends – heavily influenced by the West – became popular, and challenged traditional roles, fashions and expectations. People started to embrace the idea of the country becoming a democracy.

However, life was different in rural areas: spiralling inflation, rising national debt and increased military spending drove up the price of rice and led to violent clashes in July 1918. The *Kome Sodo* ('Rice Riots')

Source 1

A *moga*, or 'modern girl' – young people in Japan's urban centres were sophisticates heavily influenced by Western culture. [Kobayakawa Kiyoshi, *Tipsy*, 1930, woodcut print]

Source 2

Leaders of the Iwakura Mission to London (1872). The mission group spent two years travelling the world, gaining knowledge that could be used to modernise Japan.

indicated growing discontent, and martial law soon followed. Landless farmers, riots and union activity saw an increased interest in socialism and Marxism.

To quell discontent and prevent the spread of dangerous ideas, the *Peace Preservation Act of 1925* banned anything that could be perceived as dangerous to *kokutai*, or 'the national essence'. Tightening government controls, growing military influence and concern over the liberal direction of urban areas increased national and regional tensions.

Technological change

Before the Meiji period, Japan had been a medieval society. A great industrial revolution began during the 1870s. Railroads, shipping, gas lighting, textile manufacturing and banking reforms were widespread. Japan went from having 26 steamships in 1873 to more than 1500 by 1913; and from 29 kilometres of train tracks in 1872 to more than 11 000 kilometres by 1914.

The incredible speed of industrialisation was primarily due to the government's policy of *o-yatoi gaikokujin* (hired foreigners). Under this Meiji-era policy, up to 3000 foreign experts were brought into Japan to improve education, the sciences, engineering and the military. Japanese students were also sent to Europe and North America to acquire knowledge under an initiative known as the Iwakura Mission. The skills these students brought back shaped the advances of the Taisho period.

Learning ladder H3.14

Show what you know

1 What were *moga* and *mobo*?

2 How was the Taisho period different to the Meiji period?

3 What led to the Rice Riots?

4 What was the Iwakura Mission?

Continuity and change

Step 1: I can describe continuity and change

5 Think back to the *sakoku* period. Identify three things that had changed by the Taisho period and three that had not.

Step 2: I can explain why something did or did not change

6 Imagine you are an Australian newspaper reporter in the 1920s. Many of your readers still think of Japan as isolated and backwards. Write a brief article explaining how and why Japan has modernised.

Step 3: I can explain patterns of continuity and change

7 Describe how Japan changed between the Meiji and Taisho periods.

Step 4: I can analyse patterns of continuity and change

8 In 1750, Japan was an isolated nation; by 1918, it considered itself a great power. Rank the factors that contributed to this change.

Continuity and change, page 244

How do governments in Asia work today?

Asia is the world's largest continent, and is home to 48 different countries, some of which did not exist during the Age of Imperialism. These nations operate under many different forms of government.

Different types of government

Many different political systems are followed around the world, and every independent nation has their own government. The nations of Asia are governed in many ways, but four forms of government are common.

1 **Democracy**: a political system that allows each individual to participate. There are several different forms of democracy. Most democracies are **republics** – states where the people elect representatives to form and manage the government. However, some democracies are **constitutional monarchies**.

2 **Monarchy**: when one person inherits the position of head of state, and is the final word in government.

3 **Communism**: when a nation is run by an authoritarian government featuring a planned economy with equally shared resources.

4 **Dictatorship**: when a nation is run by an authoritarian government and a single individual rules the country and makes all of the decisions.

Countries of East Asia and their systems of government, 2020

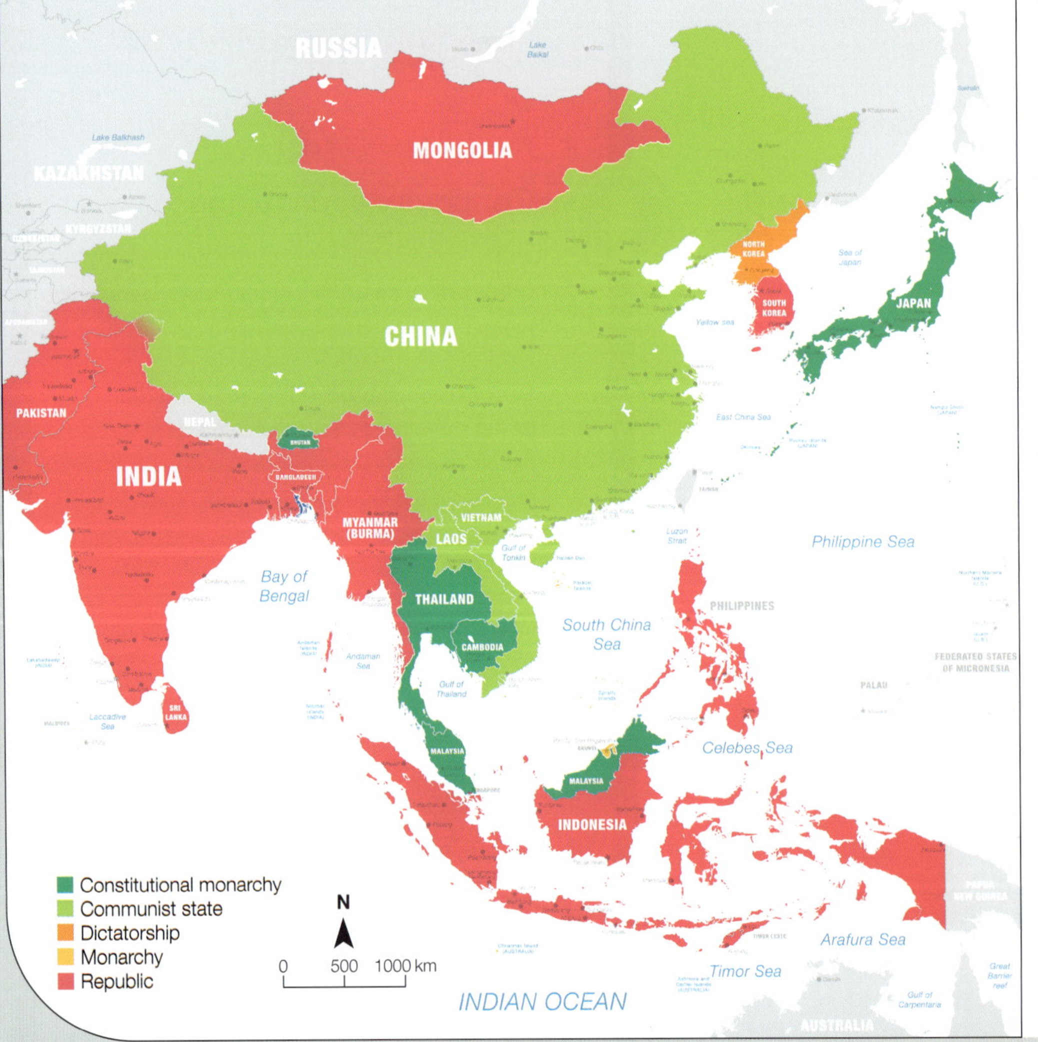

Source 1

Types of government across Asia

Source 2

A member of the world's largest democracy, India, proudly shows the mark of indelible ink that proves she has voted.

India: Democracy (republic)

Like Australia, India's government is modelled on the British Westminster System (see pages 124–25). The head of state and commander-in-chief of the armed forces is the president, elected by members of the Parliament of India and the state governments. The Prime Minister is appointed by the president and is the leader of the majority party in Parliament, elected by citizens over 18. The Prime Minister is responsible for legislation.

China: Communist state

Although China is sometimes referred to as a republic, the main power lies with the Communist Party. Elections are held for the National People's Congress, but the Communist Party is the only political party to vote for. The Prime Minister is the head of the majority party in the National People's Congress, the Communist Party.

Brunei: Monarchy

Brunei is an absolute monarchy or sultanate, where the Sultan of Brunei is both the head of state and the head of government. Brunei has a legislative council with 36 appointed members, who act as advisors to the Sultan.

Japan: Democracy (constitutional monarchy)

Like Australia, Japan is a constitutional monarchy – a democracy that also has a monarch. The Emperor of Japan is the head of the Imperial Family and the head of state, but the position is mainly ceremonial. The true power in Japan's government lies with the Prime Minister and Cabinet of Ministers, who lead the legislative branch of elected government. The Prime Minister is appointed by the emperor, but must have the support of the House of Representatives to remain in power. The Prime Minister leads the Cabinet and can appoint and dismiss ministers.

North Korea: Dictatorship

North Korea is controlled by Kim Jong-un and his family. He is the supreme leader of North Korea, and leader of the Workers' Party of Korea. North Korea is an authoritarian state in which all production and public services in the country are controlled by the state. North Korea prioritises its military and has an army of 1.2 million people, the fourth largest in the world.

Learning ladder H3.15

Civics and citizenship

Step 1: I can identify topics about society

1 What different roles are performed by the President and the Prime Minister in India?

Step 2: I can describe societal issues

2 Why can China be described as both a republic and a Communist state?

Step 3: I can explain issues in society

3 Rank the different forms of government according to the level of citizen participation.

Step 4: I can explain different points of view

4 Explain the difference between the monarchies that rule Japan and Brunei.

Step 5: I can analyse issues in society

5 Research the government of North Korea and its activities. Explain why the President of the United States claimed it was important to meet with the supreme leader of North Korea to discuss defence issues.

Masterclass

Learning ladder

Work at the level that is right for you or level-up for a learning challenge!

Source 1

An American cartoon from 1888 showing 'John Bull' (representing England) as an Imperial Octopus with its hands grabbing or interfering in various regions.

Source 2

A British political cartoon from 1858, satirising the passage of the *India Bill*, which transferred control of India to the English Crown.

Source 3

Adachi (Shōsai) Ginkō, *View of the Issuance of the State Constitution in the State Chamber of the New Imperial Palace*, 2 March 1889 (Meiji 22)

Step 1

a **I can list specific features of a source**

Source 2: Identify the features of the cartoon and provide a brief outline of the event pictured.

b **I can describe continuity and change**

Describe two examples of how India changed under British rule, and one example of how it did not.

c **I can recognise a cause and an effect**

What was the British East India Trading Company? How did its presence affect India?

d **I can recognise historical significance**

What was the Meiji Restoration?

e **I can recognise that the past has been represented in different ways**

Consider Source 1. How might imperialism have been seen by different people?

Step 2

a **I can find themes in a source**

Which elements or features in Source 3 suggest that Japan was becoming more 'Western'?

b **I can explain why something did or did not change**

Why did Japan decide to implement a policy of *sakoku*?

c **I can determine causes and effects**

Why did the British introduce opium to China? What effect did this have?

d **I can explain historical significance**

What type of government does India have today? Why?

e **I can describe historical interpretations**

How are the leaders of the British government portrayed in Source 2?

Step 3

a **I can use the origin of a source to explain its creator's purpose**

Source 1 was created by an American artist. How might Americans have felt about the British Empire? Suggest why.

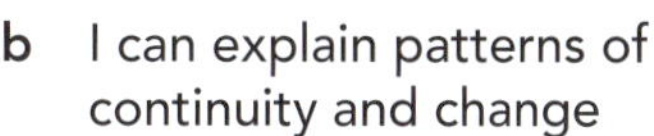

b **I can explain patterns of continuity and change**

What similarities in trade and European imperialism occurred across India, China and Japan?

c **I can explain causes and effects**

The coming of Perry's Black Ships can be considered a turning point in Japanese history. What changed after their arrival?

d **I can apply a theory of significance**

Rebellions and revolts triggered enormous change in China. Which of these was the most significant? Why?

e **I can explain historical interpretations**

Why was the Meiji Constitution presented in Source 3 in a westernised *ukiyo-e* style?

Step 4

a **I can analyse a source**

To what extent does Source 1 tell the story of British imperialism?

b **I can analyse patterns of continuity and change**

In what ways did India, China and Japan change as a result of European imperialism?

c **I can analyse causes and effects**

What events lead to the White Lotus Rebellion in China?

d **I can analyse historical significance**

Describe why Japan's victory in the Russo–Japanese War was so significant.

e **I can analyse historical interpretations**

Source 2 shows Britain's leaders treating control of India as sport. To what extent might this reflect attitudes within Britain regarding India?

Masterclass

Step 5

a I can evaluate a source

Source 1 suggests that Britain sought to interfere in the sovereignty of other nations. To what extent would you agree? Why?

b I can evaluate patterns of continuity and change

With reference to either China, India or Japan, discuss the following statement: 'While Western imperialism triggered significant changes across Asia, many nations were able to retain their own identity'.

c I can evaluate causes and effects

To what extent do you agree with the statement: 'Imperialism accelerated the modernisation of Asia'? Refer to specific events in your response.

d I can evaluate historical significance

Discuss how the Indian Rebellion of 1857, also known as the Great Rebellion, might be considered a major turning point for the British in India.

e I can evaluate historical interpretations

To what extent do you think Source 1 represents British colonial intentions? What legacies of British rule remain in its former colonies?

Historical writing

1 Structure

Imagine you are given the essay topic, '*Sakoku* was implemented to protect Japan from change, but it actually played a major role in transforming the nation'. Write an essay plan for this topic. Include at least three main paragraphs.

2 Draft

Using the drafting and vocabulary suggestions on page 262, draft a 600–800 word essay (at least 30–40 sentences) responding to the topic.

3 Edit and proofread

Use the editing and proofreading tips on page 263, to help edit and proofread your draft.

Historical research

4 Organise and present information

Imagine you are completing a research project on 'China: Fall of an Empire'. Write a contents page for this project. There should be an introduction, a conclusion, at least four main sections and many subsections. Number your chapters.

Capstone

How can I understand Asia in the Age of Imperialism?

In this chapter, you have learnt a lot about Asia in the Age of Imperialism. Now you can put your new knowledge and understanding together for the capstone project to show what you know and what you think.

In the world of building, a capstone is an element that finishes off an arch or tops off a building or wall. That is what the capstone project will offer you, too: a chance to top off and bring together your learning in interesting, critical and creative ways. You can complete this project yourself, or your teacher can make it a class task or a homework task.

mea.digital/GHV9_H3

Scan this QR code to find the capstone project online.

Australia at war

H4

World War I

WHY DID WORLD WAR I BECOME A GLOBAL WAR?

page 204

cause and effect

page 200

WHAT WERE THE **SHORT-TERM CAUSES** OF WORLD WAR I?

source analysis

page 216

WHY DID THE **UNITED STATES** JOIN THE WAR?

historical interpretations

page 222

HOW DID AUSTRALIAN SOLDIERS SERVE ON **THE WESTERN FRONT?**

How can we understand World War I?

World War I was a global conflict involving many nations. It changed the course of modern history and touched the lives of millions. Understanding its scale and impact is a broad task – one that begins well before the war itself.

Source 1

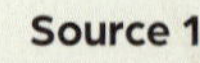

The gun used by Gavrilo Princip to assassinate Franz Ferdinand

learning ladder

Step	Source analysis	Continuity and change	Cause and effect
 step 5	**I can evaluate a source** I can present a judgement on the usefulness of a source based on its strengths, weaknesses and limitations. I can determine whether information is missing about the event or person the source refers to.	**I can evaluate patterns of continuity and change** I answer the question 'So what?' about patterns of continuity and change. I weigh up different aspects and debate the importance of continuity or change.	**I can evaluate causes and effects** I answer the question 'So what?' about cause and effect. I weigh up different things and debate the importance of a cause or an effect.
 step 4	**I can analyse a source** I can use my own knowledge to determine the reliability of a source and can explain whether it shows a one-sided view.	**I can analyse patterns of continuity and change** I can look deeper into patterns of continuity and change and determine the factors that contribute to them.	**I can analyse causes and effects** I don't just see a cause or an effect as one thing. I can determine the factors that make up causes and effects.
 step 3	**I can use the origin of a source to explain its creator's purpose** I combine knowledge of when and where a source was created to answer the question, 'Why was it created?'.	**I can explain patterns of continuity and change** I can see beyond individual examples of continuity and change between historical periods and explain broader patterns.	**I can explain causes and effects** I can answer 'How?' or 'Why?' a cause led to an effect in World War I.
 step 2	**I can find themes in a source** I look more closely at a source and find more than just features. I find themes and patterns in a source.	**I can explain why something did or did not change** I can give a reason for why something changed or why it stayed the same.	**I can determine causes and effects** Applying what I have learnt about World War I, I can describe what the cause or effect of an event was.
 step 1	**I can list specific features of a source** I can look at a World War I source and list the details I can see in it.	**I can describe continuity and change** I recognise what has stayed the same and what has changed from before World War I until now.	**I can recognise a cause and an effect** From a supplied list, I can recognise things that were causes or effects of each other in World War I.

Warm up

Source 2

The assassination of Franz Ferdinand was the 'spark' that ignited the Great War. This image by Achille Beltrame is from Italian newspaper *La Domenica del Corriere*, 12 July 1914.

Historical significance	Historical interpretations
I can evaluate historical significance I answer the question 'So what?' about things that are supposedly important in the history of World War I. I weigh up factors against one another and can cast doubt on how important things are.	**I can evaluate historical interpretations** I can weigh up the different historical interpretations that have been formed. I debate and challenge the interpretations that have been presented.
I can analyse historical significance I can separate out the various factors that make something historically important in the history of World War I.	**I can analyse historical interpretations** I can determine the factors that have led to why a historical interpretation has been formed.
I can apply a theory of significance I know a theory of significance. I use it to rank importance of changes, causes, effects and events in the history of World War I.	**I can explain historical interpretations** I can answer 'Why?' or 'How?' there are different interpretations of people and events in the past.
I can explain historical significance I answer the question 'Why?' about what was important in World War I.	**I can describe historical interpretations** I can provide different examples to show how people and events in the past have been interpreted.
I can recognise historical significance When shown a list of facts about World War I, I can work out which are important.	**I can recognise that the past has been represented in different ways** I can identify different views of people and events in the past.

Source analysis

1 Consider Source 2:
  a Identify clues that suggest a chaotic scene is taking place.
  b Describe the differences between the various people in the image.
  c Where is this image from? What other pieces of information could help us understand its context?

Continuity and change

2 Franz Ferdinand was assassinated in 1914. List 10 things in your life today that did not exist in 1914.

Cause and effect

3 World War I was Australia's first major military involvement. How do you think a political assassination could have led to Australia's participation?

Historical significance

4 The legend of the Anzacs is part of Australian folklore. How are they remembered today? How important is this legend to our nation?

Historical interpretations

5 In what ways do you think World War I is remembered in different countries and communities? Discuss your ideas as a class.

PART I: THE ROAD TO WAR

What was Australia like at the beginning of World War I?

1901 marked the birth of a federated Australia, as the separate colonies came together. A decade later, Australia was a young, confident nation, enriched by the gold rushes of the 19th century. It was also a conservative country that often looked to Britain as the 'mother country' to emulate.

Source 1

The Boer War (1899–1902) was a conflict between Britain and Dutch colonies in South Africa. The Australian colonies sent troops to support the British military.

key ideas timeline.

1882 – Germany, Austria-Hungary and Italy form the Triple Alliance

1901 – Australia becomes a federation

1902 – Japan forms an alliance with the United Kingdom

1907 – The UK, France and Russia form the Triple Entente

28 June 1914 – Franz Ferdinand and his wife Sophie are assassinated in Sarajevo, Bosnia

23–30 July 1914 – Austria-Hungary blames Serbia for the assassination and declares war

Political position

The Australian Commonwealth began on 1 January 1901, with Edmund Barton as its first elected prime minister. The uniting of the Australian colonies had been born out of hope, but also from a fear of isolation as the last imperial outpost on the fringe of the British Empire. By banding together, the new nation hoped to take control of its own destiny (see page 120).

Although it was an independent nation, Australia was still part of the British Empire, and did not make its own decisions on foreign policy. The economy used British currency until 1910, and the Australian government had strong ties to its British counterpart. The Australian military supported the British Empire in conflicts such as the Boer War in South Africa and the Boxer Rebellion in China. Australian soldiers followed a tradition of military service inherited from the British.

Technology

The period before World War I was one of rapid, extensive industrialisation, with great advances made in science and technology. The world was becoming increasingly interconnected, with communication via radio technology and faster travel via steamship. Telegraph wires criss-crossed the planet and postal services ran two deliveries a day to keep up with the large numbers of letters people sent all over the world.

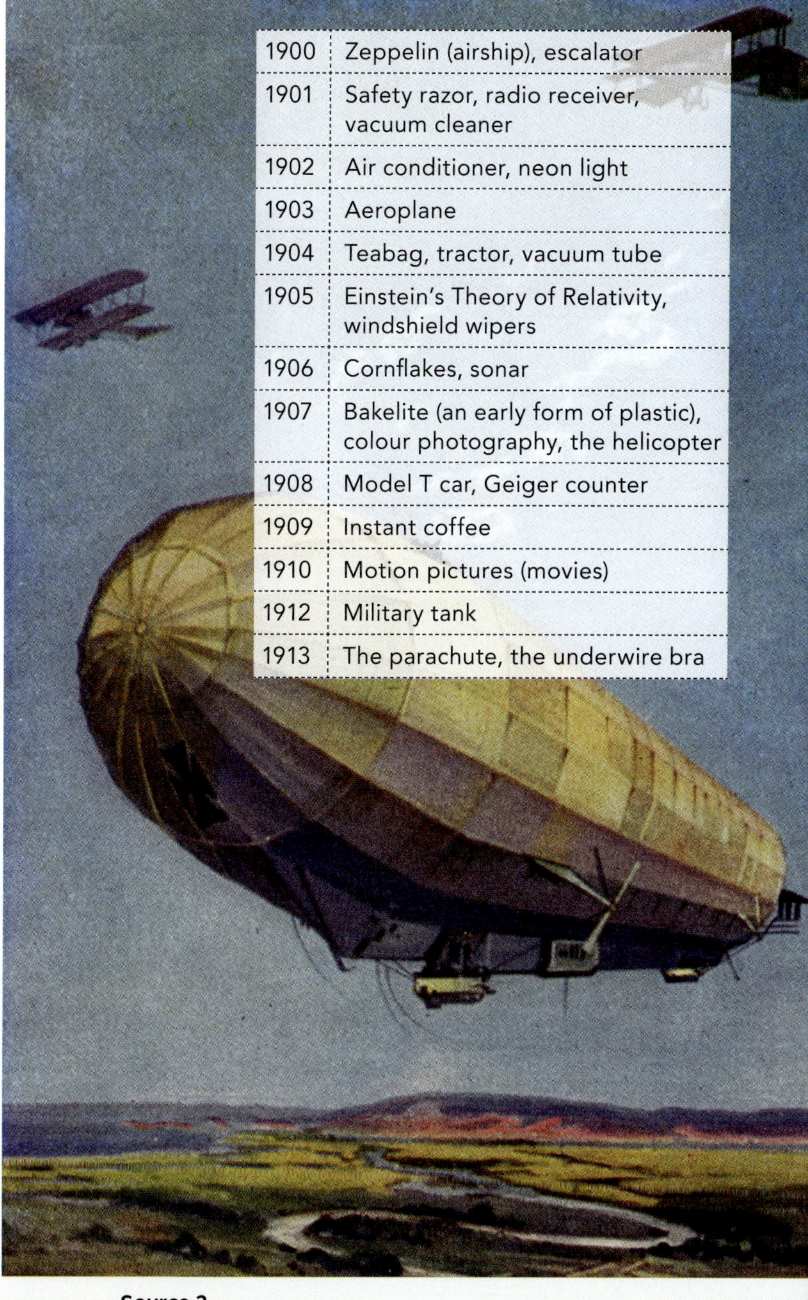

1900	Zeppelin (airship), escalator
1901	Safety razor, radio receiver, vacuum cleaner
1902	Air conditioner, neon light
1903	Aeroplane
1904	Teabag, tractor, vacuum tube
1905	Einstein's Theory of Relativity, windshield wipers
1906	Cornflakes, sonar
1907	Bakelite (an early form of plastic), colour photography, the helicopter
1908	Model T car, Geiger counter
1909	Instant coffee
1910	Motion pictures (movies)
1912	Military tank
1913	The parachute, the underwire bra

Source 2

Major inventions before World War I

1 August 1914 – Germany declares war on Russia in support of its ally Austria-Hungary

3 August 1914 – Australia offers a military force of 20 000 to the UK government

3–6 August 1914
- Germany declares war on France and invades Belgium.
- The UK declares war on Germany.
- This means members of the British Empire, such as Australia, are also at war.
- Austria-Hungary declares war on Russia.

12 August 1914 – Austria-Hungary invades Serbia; Britain and France declare war on Austria-Hungary

17 August 1914 – Russia invades Germany

1 November 1914 – The Ottoman Empire attacks Russia, entering the war on the side of the Triple Alliance

Source 3

Melbourne was Australia's cultural and shopping capital in 1914, and boasted electrical lights and trains.

Australia's major cities enjoyed many of these technological advances. Horse-drawn carriages were still the most common modes of transport, but by 1909 cars had become part of the hustle and bustle of traffic. Electric streetlights were first installed in 1894 in Melbourne, and in Sydney in 1904. Rural towns and properties still operated much as they had during the 19th century, although more people could now access the growing national rail network. Australia was still a long way from Europe – it took five to six weeks to travel there by steamship, much longer by sail – but news of international events reached Australia quickly through the telegraph, particularly in urban areas.

Society

The new Australia was something of a 'social laboratory'. Women – except those of Indigenous, Pacific Islander, African or Asian heritage – had gained the right to vote in federal elections and to stand for parliament. A series of laws had enshrined the right for workers to receive a decent minimum wage, collective bargaining rights and fair conditions in the workplace, transforming Australia into a 'working man's paradise'.

Despite these innovations, Australian society was conservative and slow to change. People still looked to Britain as the example of what Australian society should be, and followed British fashion, art and culture. Australian society was also very hostile to non-British people, including some European cultures. The White Australia Policy (see page 126) restricted immigration opportunities for non-British foreigners, while First Nations Peoples still being subjected to racist treatment such as children being separated from their families.

Most people lived with large families in small houses; they worked close to factories or shops and had little to no opportunity for travel or free time. While more middle-class men started to go to university, most Australians only completed a primary school education before entering the workforce.

Source 4

Australian fashions closely followed those popular in Britain.

Source 5

Most of the working classes in the cities lived in cottages close to the ports and factories where they worked. This photograph was taken in Port Melbourne in 1906.

Learning ladder H4.1

Show what you know

1 List five ways daily life in Australia began to change at the start of the 20th century.

2 Why did most Australians still have a strong feeling of loyalty towards Britain even after federation?

3 Which of the inventions listed in Source 2 might have had the greatest impact on day-to-day life at that time? Why?

Source analysis

Step 1: I can list specific features of a source

4 Review the sources in this section and identify features or elements that can still be found in Australia today.

Step 2: I can find themes in a source

5 Australia in the 19th and 20th centuries held strong links to Britain. What examples can you find in the sources to demonstrate this?

Step 3: I can use the origin of a source to explain its creator's purpose

6 Source 3 shows Swanston Street, Melbourne in 1914. Why do you think this picture might have been widely circulated? What impression does it give of Melbourne at the time?

Step 4: I can analyse a source

7 Identify other elements of Australia at this time that feature in Sources 1–5. How might they connect or relate to to each other?

HOW TO Source analysis, page 240

Which countries dominated Europe?

Australia was a minor nation in 1914, far from the world's cultural, political and military centre – Europe. This was the age of empires; Germany, Russia, Austria-Hungary and the Ottoman Empire were autocracies, ruled by royalty. France was a democratic republic, while England and Italy were constitutional monarchies.

Austria-Hungary

The Austro-Hungarian Empire was a multicultural powerhouse. It was home to Austrians, Bosnians, Croatians, Czechs, Germans, Hungarians, Italians, Poles, Romanians, Russians, Serbians, Slovaks, Slovenians, Ukrainians and Jewish people, all of whom had their own languages and distinct cultures. The Empire had a large military, but it lacked the latest technology.

Germany

Germany became a unified country in 1871, with a booming industrial economy in need of raw materials, resources and markets. In 1888, Wilhelm II became Kaiser (Emperor) of Germany. He was determined that Germany would have equal standing among the other European superpowers, and broke off ties with its traditional ally, Russia. Germany had invested heavily in improving its military and navy, which were modern, well equipped and well trained.

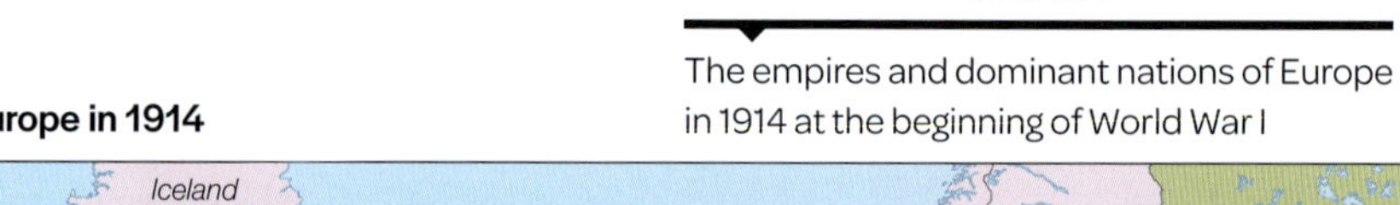

Source 1

The empires and dominant nations of Europe in 1914 at the beginning of World War I

Europe in 1914

Source: Matilda Education Australia

France

France was a **democracy** with a large colonial empire, mostly in north-west Africa and south-east Asia. Its army was not at full strength in 1914, and was concentrated along the border with Germany, a long-time enemy. The French army was not modern; it used **conscripts** and generally suffered from low morale.

Britain

Britain was a **constitutional monarchy**, and had been for hundreds of years. 'The sun never set' on the British Empire, as nearly a quarter of the world had been colonised by the British. Large amounts of resources flowed from the colonies to Britain. In 1914, the British Navy was the most advanced in the world. Britain was increasingly concerned about the growing economic, political and naval strength of Germany.

Russia

Russia was ruled by the **autocratic** Tsar Nicholas II. Russia's huge landmass and large population made it a powerful country, but its economy was undeveloped and based on agriculture. Its army was enormous, but it was badly equipped with outdated weapons.

Ottoman Empire

By 1913, the once-mighty Ottoman Empire had lost most of its European territories in the Balkan Wars. It was ruled by the autocratic Sultan Mehmed V. The Ottoman Empire's army had undergone a series of reforms and modernisations, with some help from Germany.

Serbia

Serbia was recovering from the first and second Balkan wars and had tense relations with its neighbour Austria–Hungary. By 1914, Serbia was a strategic political power in the region.

Japan

Through industrialisation and reform, Japan had become an important regional power. The 1905 victory in the Russo–Japanese conflict marked the first time an Asian nation had defeated a European one in modern history and heralded Japan's growing ambitions and capabilities.

China

By the late 19th century, the Qing dynasty had become weak and European powers moved in (see pages 162–65). France, Germany, Russia, England and Japan all controlled areas within China, which prompted a rise in anti-foreigner sentiment.

Learning ladder H4.2

Show what you know

1 Who were the great powers of Europe at the beginning of 1914?

2 China had a long and prosperous history. Why did this begin to change during the 19th century?

3 Why do you think the Austro-Hungarian Empire may have lacked the unity of many of its fellow empires?

Historical significance

Step 1: I can recognise historical significance

4 In your opinion, which of the listed empires were the largest and most influential?

Step 2: I can explain historical significance

5 Why were France and Germany considered historical enemies?

Step 3: I can apply a theory of significance

6 The Japanese defeat of Russia signalled both the power of Japan's navy and Russia's difficulties. Explain how.

Step 4: I can analyse historical significance

7 Imagine yourself as a citizen in one of these great empires. What do you observe around you? What is the feeling among everyday citizens? Prepare 3–4 short journal entries about your daily life and share them with the class.

Historical significance, page 251

What were the long-term causes of World War I?

Wars are caused by a combination of long-term, slow moving societal causes, suddenly triggered by short-term events. The long-term factors behind the outbreak of World War I included the key countries' militarism, imperialism and nationalism, as well as their various contemporary alliances.

The short-term event that triggered World War I was the assassination of Franz Ferdinand, heir to the Austrian throne, and the subsequent political crisis the shooting caused. The long-term causes are more complex. You can understand these causes – militarism, alliances, nationalism and imperialsim – using the acronym MANIAC.

Militarism

Militarism is the idea that a country should maintain a strong army to defend or promote its national interests. At the turn of the 20th century, the major powers tried to stay ahead of the others by investing in their military and navy.

An 'arms race' broke out between Germany and Britain. Wary of Germany's growing navy, Britain built the HMS *Dreadnought* in 1906, the largest, fastest and most modern battleship of its time. Soon Germany started to build its own dreadnoughts.

Alliances

In the years before World War I, European nations formed strategic military agreements (alliances) with each other. The most important of these were the Triple Alliance (known as the Central Powers) and the Triple Entente (known as the Allies or Allied Powers).

The Triple Alliance

The Triple Alliance was an agreement between Germany, Austria-Hungary and Italy. These nations promised to support each other if one was attacked. However, when Austria-Hungary declared war in July 1914, Italy claimed to be neutral and did not enter the war.

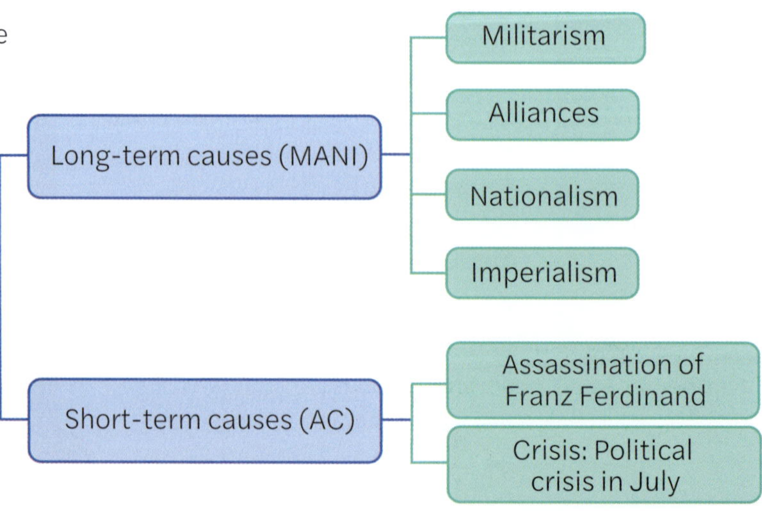

Source 1

The acronym MANIAC summarises the causes of World War I.

Source 2

Powerful new ships, such as the HMS *Dreadnought*, could easily outgun and outrun older vessels.

The Triple Entente

The word **entente** is French for 'friendly understanding'. In 1904, France and England formed an alliance called 'the Entente Cordiale'. Russia joined the agreement in 1907, and it became the Triple Entente. These agreements brought the countries together and solved arguments about territories in Asia and Africa.

Nationalism

Nationalism is a feeling of extreme loyalty to your nation–state, its people and culture, and involves promoting these interests ahead of other nations. The nation–state was relatively new to Europe. Austria–Hungary had become a nation in 1867, Italy unified in 1870 and Germany unified in 1871. Romania, Bulgaria and Serbia became sovereign nations in the late 1800s when the Ottoman Empire weakened its hold on eastern Europe.

By 1914, many of Europe's young nations were looking for a national identity to bind their peoples together. People were encouraged to be loyal to their own nation, so when war broke out, many men enthusiastically enlisted to support and defend their country.

Imperialism

Imperialism is the expansion of a country's power and influence by colonising other countries, often by force (see Chapters H2 and H3). By 1900, France, Britain and Germany had large overseas empires. The size of their empires was a badge of honour for the major European powers, as well as an important engine to keep their economies going.

Learning ladder H4.3

Show what you know

1 Which nations made up the Triple Alliance and the Triple Entente?

2 Use the sources in this section to suggest why the road to war may have been inevitable.

3 Suggest why nationalism might have encouraged young nations, such as Australia, to join the war.

Cause and effect

Step 1: I can recognise a cause and an effect

4 Define the term 'arms race'. What problems might such an event create?

Step 2: I can determine causes and effects

5 What were considered to be the four main long-term causes of World War I?

Step 3: I can explain causes and effects

6 Explain how each of the following factors contributed to the outbreak of war: militarism, alliances, nationalism, imperialism.

Step 4: I can analyse causes and effects

7 Select one of the four long-term causes. Using a guiding question, create a mind map to determine what factors influenced it. A sample guiding question for militarism could be: *What factors lead to an arms race in Europe?*

Cause and effect, page 247

What were the short-term causes of World War I?

The assassination of the Austro–Hungarian Archduke Franz Ferdinand in Sarajevo triggered a political crisis in the Balkans. When Austria–Hungary declared war on Serbia following Ferdinand's assassination, the tensions on the existing chains of alliances escalated, ultimately leading to a global war.

Tensions in the Balkans

The Balkans, a region in south-eastern Europe, had been ruled by different empires for hundreds of years. The most recent power in the area was the Ottoman Empire, but it had started to crumble in the late 1800s. This caused the different nationalities in the Balkans to start declaring independence, often through war. Other empires, such as Russia and the Austro-Hungarian Empire, saw this as an opportunity to exert their influence.

In mid-1914, Serbia was a growing power in the Balkans. It was also dealing with ethnic and political tensions in the territories it annexed during the Balkan Wars.

Source 1

Franz Ferdinand and his family in 1910. Ferdinand and his wife Sophie used their trip to Sarajevo to celebrate their 14th wedding anniversary.

Ethnic diversity of the Balkans region

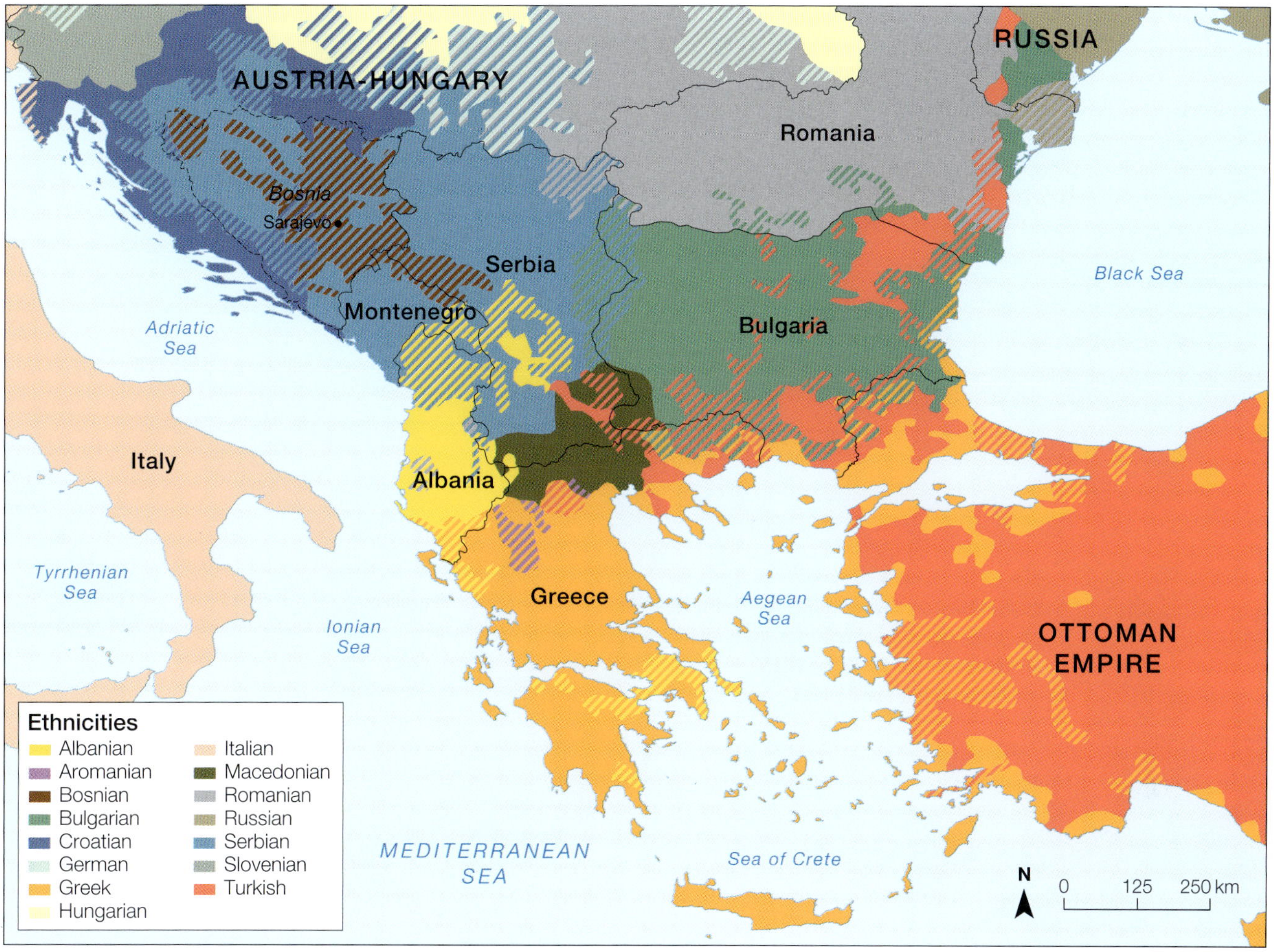

Ethnicities: Albanian, Aromanian, Bosnian, Bulgarian, Croatian, German, Greek, Hungarian, Italian, Macedonian, Romanian, Russian, Serbian, Slovenian, Turkish

Source 2

Source: Matilda Education Australia

The Balkans is one of the most ethnically and culturally diverse regions of Europe.

The shot heard around the world

Archduke Franz Ferdinand was the 55-year-old heir to the Austro-Hungarian throne and Inspector-General of the Austro-Hungarian army. In June 1914, he and his wife Sophie went to Bosnia to inspect the army's training and manoeuvres. This small country had recently been claimed by the Austro–Hungarian Empire as part of its empire. Serbian ministers warned the Austrian government that it might be dangerous to visit, but the Austrians did not take the warnings seriously.

On 28 June 1914, in the Bosnian city of Sarajevo, Franz Ferdinand and his wife were shot and killed. The assassin, Gavrilo Princip, was a Bosnian-Serb nationalist and a member of the Black Hand, an organisation that wanted to unite all Serbians into one nation. The assassination – called 'the shot heard round the world' in the newspapers of the day – led to a diplomatic crisis.

(REUTER'S SPECIAL CABLES.)

ASSASSINATED

DEATH OF AUSTRIA'S HEIR

THREE DETERMINED AT-TEMPTS

BOMBS AND REVOLVERS

HUSBAND AND WIFE KILLED WITH TWO SHOTS.

VIENNA, Sunday night.

Details of the assassination of the Archduke Francis Ferdinand of Austria and his morganatic wife, the Duchess of Hohenburg, show that the crime was a very determined one.

The Archduke, as Inspector-General of the land and sea forces, had made the voyage from Trieste to Metscovick in the Austrian Dreadnought, "Veribus Unibus," and afterwards proceeded to Mostar, where, in an address to the local authorities, he spoke partly in German, but mostly in Serbo-Croatian.

Source 3

The assassination of Archduke Franz Ferdinand made headlines around the word, including in Australia. This story appeared in the *Ballarat Evening Echo*, 29 June 1914.

Source 4

Austrian and Serbian forces in conflict during the first days of the invasion of Serbia, near the bridge over the Sava River. [Illustration by Achille Beltrame, from *La Domenica del Corriere*, 9 August 1914]

Weather Forecast:
Cloudy Tonight and Wednesday

The Washington Times

HOME EDITION

NUMBER 8244. WASHINGTON, TUESDAY EVENING, JULY 28, 1914. PRICE ONE CENT.

AUSTRIA HAS CHOSEN WAR

TYPICAL SERVIAN SOLDIERS AND THEIR ANTIQUATED EQUIPMENT

PHOTOS BY UNDERWOOD & UNDERWOOD

At the left is shown a detachent of Serbs ready for action. These men have seen service in both the Balkan wars and have demonstrated their courage and discipline.

At the right is shown a group of artillery officers placing an old-time field piece in position, as a protection in one of the border towns against invasion. Servia's artillery has not kept pace with modern tendencies in military equipment.

MEDIATION REJECTED, EXCEPT TO PREVENT SPREAD OF CONFLICT

Occupation of Belgrade Unofficially Reported—Servians Said to Have Withdrawn Without Contest—England Told Events Have Gone Too Far to Permit Turning Back.

LONDON, July 28—Austria today formally declared war against Servia, according to Vienna dispatches received here.

It is understood that Belgrade has already been occupied by the Austrians.

This announcement of war quickly followed the

Source 5

Austria-Hungary declared war on Serbia when it failed to meet only one of the 10 conditions of the Austro-Hungarian ultimatum.

Crisis

Austria-Hungary blamed Serbia for the assassination and issued a harsh ultimatum – 10 demands that would have politically embarrassed Serbia's rulers and drastically reduced its power in the Balkans. Serbia complied with all but one of the 10 demands – it refused to allow Austro-Hungarian agents to manage the assassination investigation. Because Serbia did not bow to all its demands, Austria-Hungary declared war on Serbia on 28 July and invaded the country two weeks later.

Austria-Hungary used Serbia's minor refusal as an excuse to declare war and try to expand its power in the Balkans. Austro-Hungarian leaders felt confident because Germany had declared it would support its ally with a 'blank cheque', meaning it would support Austria-Hungary both militarily and politically, without placing limits on the amount of assistance.

The first response to the declaration of war came from Russia, which backed its ally Serbia. As hostilities escalated, more allies entered the battle on both sides. Within four months, all the great powers of Europe were at war.

Learning ladder H4.4

Show what you know

1 In which part of Europe are the Balkans located?

2 At what point did Russia enter the war?

3 Why did Austria-Hungary choose to declare war on Serbia?

Cause and effect

Step 1: I can recognise a cause and an effect

4 Whose assassination angered the Austro-Hungarian Empire?

Step 2: I can determine causes and effects

5 How did the assassination of one man draw so many nations into conflict?

Step 3: I can explain causes and effects

6 Using Source 2, suggest why so many nations and empires had an interest in controlling the Balkans.

Step 4: I can analyse causes and effects

7 Using the information in the text as well as your own knowledge, explain why Australia entered World War I.

Cause and effect, page 247

Why did World War I become a global war?

The world was divided into a complex network of alliances long before the assassination of Franz Ferdinand. This meant the dispute soon escalated into a global war, in which Australian soldiers played active combat roles.

A global war

A local European quarrel quickly turned into a global war. The world became divided between the Allies (the Triple Entente, their colonies, territories and independent nations) and the Central Powers (Germany, Austria-Hungary and their allies). Not all countries sent troops, but many provided resources such as coal, oil, iron and food.

Although the war began in Europe, conflicts broke out across the world. Battles were fought in the colonies of the major empires, such as in German East Africa (modern-day Burundi, Rwanda and Tanzania) and Shandong in China. In India, the German navy bombed British oil storage tanks in Madras (modern-day Chennai) , while the coast of Chile was the site of several naval battles.

Countries involved in World War I

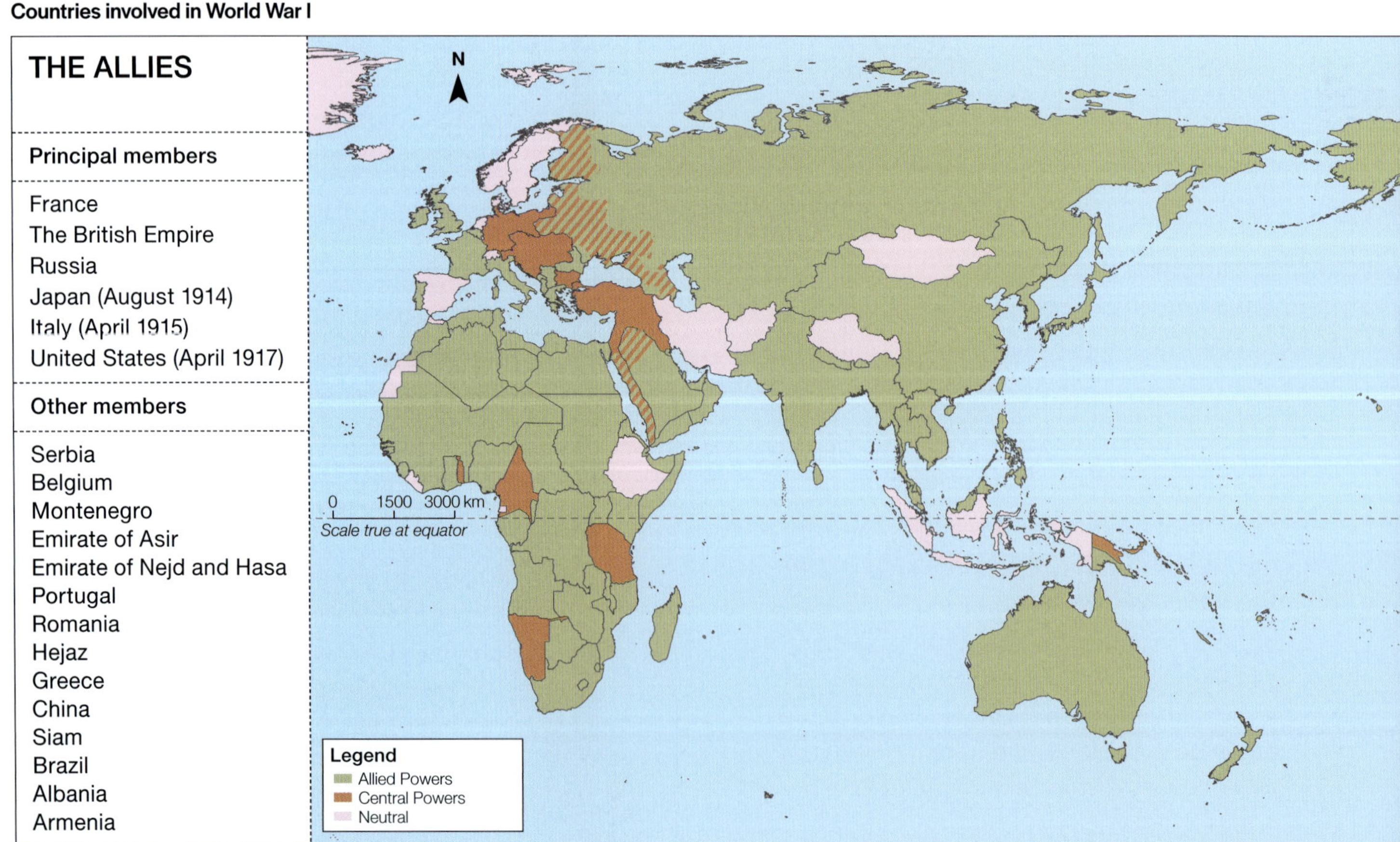

Nations in conflict

As the war continued, more nations took sides. The USA didn't join the war until April 1917, but proved to be a key participant on the Allied side. Other nations played smaller roles, only providing resources or information – enough so that, if their side won, they might be rewarded. A few countries had little choice, as they were colonies or 'client states' of more powerful empires that ordered them to become involved.

International armies

Britain deployed units from across its Empire. Soldiers from Canada, New Zealand and Australia were used in active combat. Men from the West Indies were not allowed to fight alongside white troops, and were mainly used in support roles.

More than one million Muslim, Sikh and Hindu soldiers from India and Gurkhas from Nepal volunteered for the British Indian Army.

Source 1

What began as a conflict between European powers soon drew in most of the world.

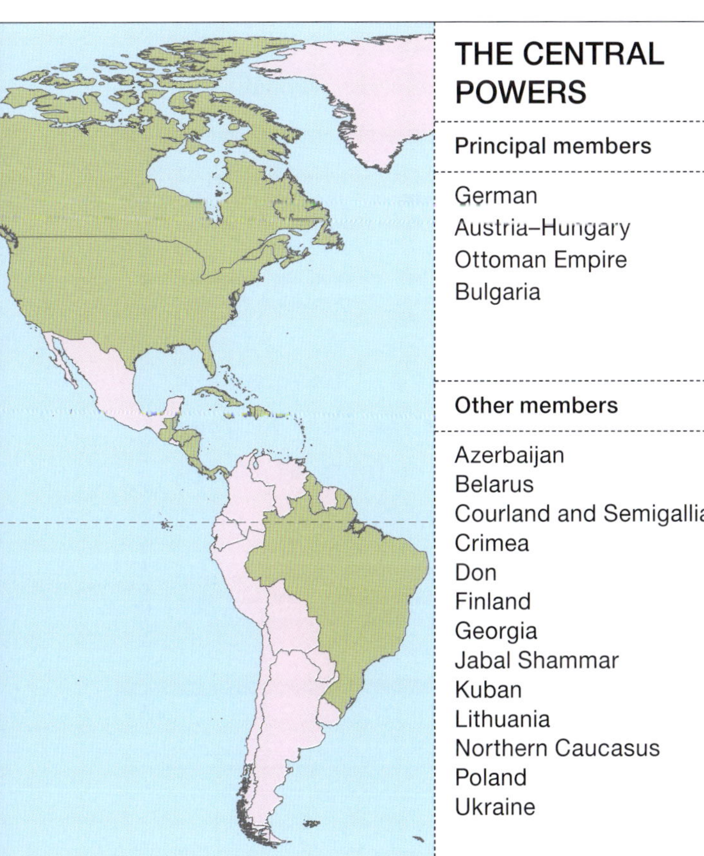

Source: Matilda Education Australia

At the Battle of Gallipoli, about 16 000 troops from the Indian subcontinent fought alongside British, French and Anzac soldiers.

While the British colonial troops consisted of volunteers, the French colonial army included some men who were forced to fight. Soldiers from Morocco, Tunisia, Algeria, Madagascar and Somalia were conscripted by the French, and they too were mainly used in non-combat roles.

Learning ladder **H4.5**

Show what you know

1 Why did a European conflict draw in other nations across the world?

2 Other than sending soldiers, how did some nations support the war effort?

3 Suggest the difference between volunteer soldiers and conscripts.

Continuity and change

Step 1: I can describe continuity and change

4 Compare a modern map of Europe with one from 1914. Which nations have remained the same? Which have changed or disappeared completely?

Step 2: I can explain why something did or did not change

5 From your understanding of this section, explain why life in countries outside of Europe was affected by the war.

Step 3: I can explain patterns of continuity and change

6 Some citizens in colonial nations, such as Madagascar, New Zealand and Belarus, volunteered to fight. Suggest why they might have done so.

Step 4: I can analyse patterns of continuity and change

7 Research some of the countries that remained uninvolved. What elements do these locations share in common? How might they have been indirectly affected by the war?

Continuity and change, page 244

What was Australia's response to the war?

During the early 20th century, Australia's ties to Britain were very strong. When Britain entered the war, Australia also joined by extension. Spurred by propaganda, large numbers of men and women enlisted to fight or support the troops. However, as the war continued, attitudes towards the conflict became much less positive.

Initial response

When Britain declared war on Germany on 4 August 1914, Australia considered itself at war with Germany too. Reactions to the war varied; some people believed that it would be 'over by Christmas', while others were more sceptical and concerned. Most Australians believed that it was right to enter the war and support 'Mother England' in its hour of need.

Australia's combined state armies were small, and by law these Commonwealth Military Forces could not fight outside of Australia. Therefore, when war broke out, the Australian Imperial Force (AIF) was created. This was a volunteer 'expeditionary force', which meant these men could fight overseas. The calls for enlistment started straight away.

> 'Whatever happens, Australia is part of the Empire right to the full. Remember that when the Empire is at war, so is Australia at war. That being so, you will see how grave is the situation. So far as the defences go here and now in Australia, I want to make it quite clear that all our resources in Australia are in the Empire and for the Empire and for the preservation and security of the Empire.'

Source 1

Extract from Prime Minister Joseph Cook's speech, given in Horsham, Victoria, on 1 August 1914

Australia's population was around 4 million people in 1914, which meant that there were about 820 000 men of fighting age (between 19 and 38). Australia offered 20 000 soldiers to the British government; the offer was accepted immediately.

AUSTRALIA'S OFFER ACCEPTED

TWENTY THOUSAND MEN WANTED

PRIME MINISTER RECEIVES MESSAGE

Official acceptance by the Imperial Government of the offer of the Commonwealth Government to provide a force of 20,000 men was announced by the Prime Minister today.

Mr Cook said that he had received the following message, through the Governor-General, from the Secretary of State for the Colonies:—

"His Majesty's Government gratefully accepts the offer of your Ministers to send a force of 20,000 men to this country."

The Prime Minister added: "We propose to send them at the earliest possible moment." He said that he could not say at this stage what the composition of the force would be.

Source 2

The ties between Britain and Australia remained strong, and Australian troops were sent to support the British abroad, as shown in this clipping from the *Melbourne Herald*, 7 August 1914.

Reasons for enlisting

Thousands of Australian men rushed to enlist for a variety of reasons.

- Many Australians still held strong familial and cultural ties to England, so fighting to protect the 'Mother Country' felt important.
- Australians were proud of their new nation; they wanted to represent their country overseas and show Europeans what Australians could do.
- Australia, a mostly rural nation, was very isolated. Travelling to Europe or the Middle East was an exciting idea for many young men.

Source 3

Sport is often seen as a metaphor for war. Good sportsmen – men of bravery and action – were believed to make good soldiers.

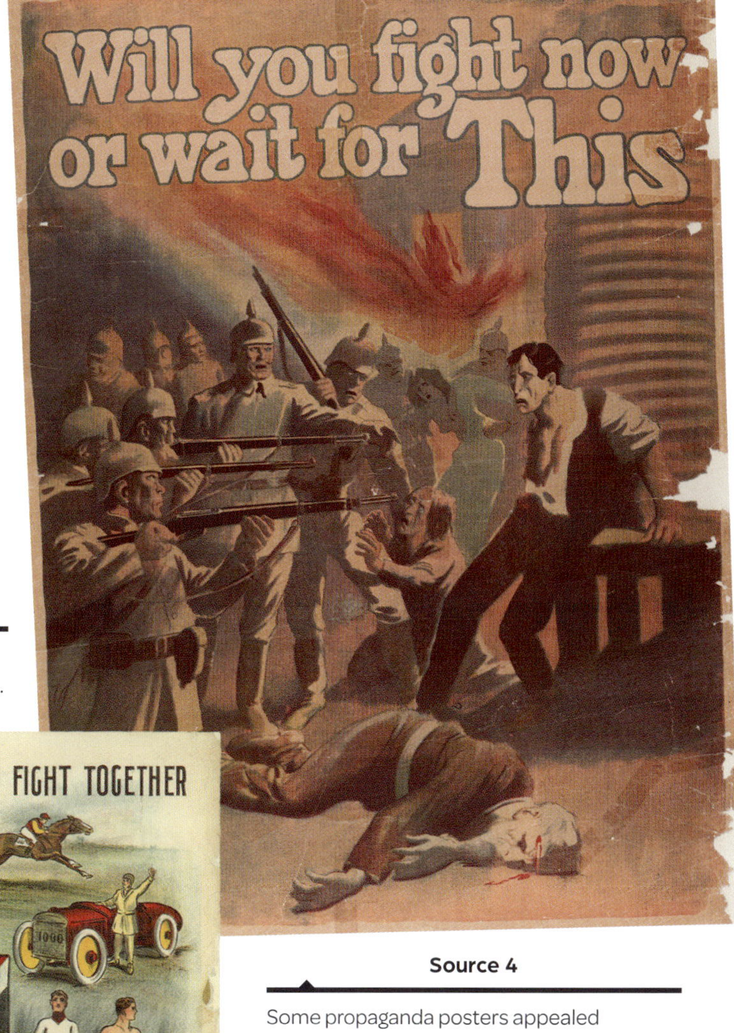

Source 4

Some propaganda posters appealed to a sense of duty and moral obligation. This poster is from 1918.

- There was a lot of social pressure. Army propaganda posters were everywhere and enlistment became the main topic of conversation. Eligible men who didn't enlist might find a white feather in their letterbox, suggesting that they were cowards.
- Football teams, cricket teams and groups from small towns would often all sign up together – keen to share the adventure.

Women signed up as nurses, with more than 2200 serving overseas. They also filled vital roles on the **home front**, doing the work of men who had gone abroad.

Source 5

Propaganda came in many forms, such as this handkerchief from 1915, which is printed with patriotic images and poetry.

Physical requirements

In 1914, about 33 per cent of volunteers were rejected because of age, height and health restrictions. As the war went on, these restrictions were relaxed to take in more men.

Period	Age requirement	Minimum height requirement
August 1914	19–38 years	5 ft 6 in (170 cm)
June 1915	18–45 years	5 ft 2 in (158 cm)
April 1917	18–45 years	5 ft (152 cm)

Use of propaganda

Propaganda is information that is used to influence an audience, often by producing an emotional response. Societies and governments have produced propaganda throughout history, in whatever formats were appropriate for their era.

World War I saw a massive increase in the production and variety of propaganda on all sides of the conflict. British and Australian propaganda sent a message that fighting the war against an 'evil' enemy was a moral imperative. Posters, pamphlets, newsreels and official speeches were all used by the Allies as propaganda. By modern standards, many propaganda images and messages of the time are racist and offensive.

The conscription debate

After the initial excitement, the number of recruits declined as the war went on and casualty rates increased. As news filtered back to Australia of the brutal nature of trench warfare, these numbers declined further.

By 1916 there was a shortage of men volunteering to enlist, and the Australian government was not able to provide the troops needed by Britain. Prime Minister Billy Hughes proposed that Australia create an army based on **conscription**, meaning that eligible men would be forced to enter the army for a certain period.

Army enlistments by year, 1914–1917

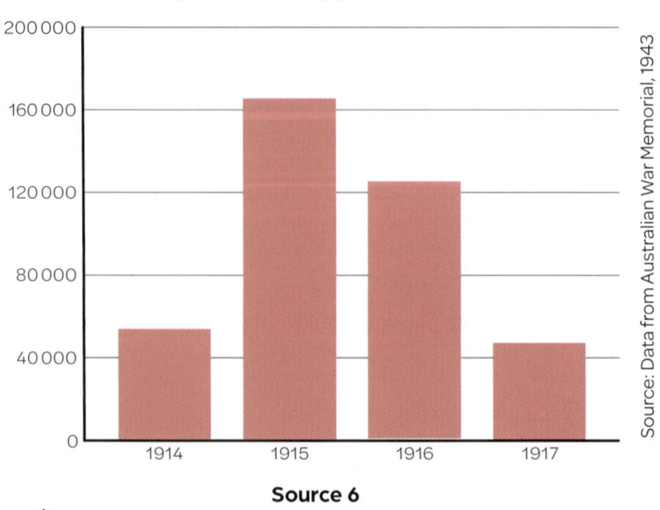

Source 6

After an initial rush in 1915, Australia simply could not meet British calls for more troops.

The government had the power to introduce conscription laws, but Hughes needed to demonstrate public support in order to pass legislation in the Senate. Hughes conducted two **plebiscites** during the war to try to bring in conscription. Like the 2017 marriage equality plebiscite, held a century later, these were non-binding national votes. (At the time, they were often incorrectly referred to as referendums.)

The first plebiscite to introduce conscription was held on 28 October 1916. It was narrowly rejected – 51 per cent of Australians voted against conscription. When Hughes was re-elected in 1917, he called for another plebiscite on conscription to solve the problem. On 20 December 1917 the plebiscite was rejected again, this time with a larger majority (54 per cent). The government did not try to implement conscription again during World War I.

The conscription debates were passionate and heated on both sides. Those in favour argued that Australia had a moral duty to support England, and that if the war was lost, Australia might also be invaded by Germany. There were various arguments against conscription; some were against war completely, while others supported the war effort but believed it was wrong to force men to become soldiers. The union movement argued against conscription because jobs would be taken by women or foreigners while men were on the frontlines.

Some arguments were along religious lines. Daniel Mannix, the Catholic Archbishop of Melbourne, declared that he was opposed to conscription. The Protestant majority then accused Australian Irish-Catholics of being anti-war and anti-empire.

Source 7

Propaganda was used by both sides during the conscription plebiscites.

Source: Australian War Memorial ARTV10140

Learning ladder H4.6

Show what you know

1 Many young Australians wanted to enlist. What were their main motivations?

2 Enlistment requirements became more relaxed as the war progressed. Suggest why.

3 Define the term 'conscription'. Why might it have been necessary?

Source analysis

Step 1: I can list specific features of a source

4 What kind of features in Sources 3–5 identify them as Australian sources?

Step 2: I can find themes in a source

5 What common themes emerged across recruitment posters and other war propaganda?

Step 3: I can use the origin of a source to explain its creator's purpose

6 Propaganda posters were commonly used to encourage enlistment. Suggest two different reasons for why posters were often used.

Step 4: I can analyse a source

7 What view of war and service did the posters provide? How was this different to the reality of war? Why might there be a difference between them?

Source analysis, page 240

PART II: THE COURSE OF THE WAR

What was trench warfare?

Trench warfare was a new fighting technique that was used heavily during World War I. The romantic ideas of sword fighting and cavalry of old armies were gone. The promises of glory and excitement that drew volunteers were replaced by the reality of life in the trenches, where boredom, disease and trauma proved worse than combat itself.

Trench warfare

In **trench** warfare, soldiers on the frontlines dug and occupied extensive trenches on the battlefield. Within the trenches, the soldiers were protected against enemy artillery. However, to take territory, the soldiers had to advance out of the trenches, where they could easily be shot.

The Western Front (page 222) was the main region of trench warfare. Both sides created extensive networks of trenches, along with underground tunnels and **foxholes**. These were surrounded by barbed wire, mines, traps and obstacles. The deadly zone between the opposing trenches was called No Man's Land.

Life in the trenches

Life on the frontline was a confusing mix of monotony and sudden action. Raids on the enemy's trenches were conducted at night; during the day, officers kept their soldiers busy with a strict routine of cleaning, training, repairing and building trenches.

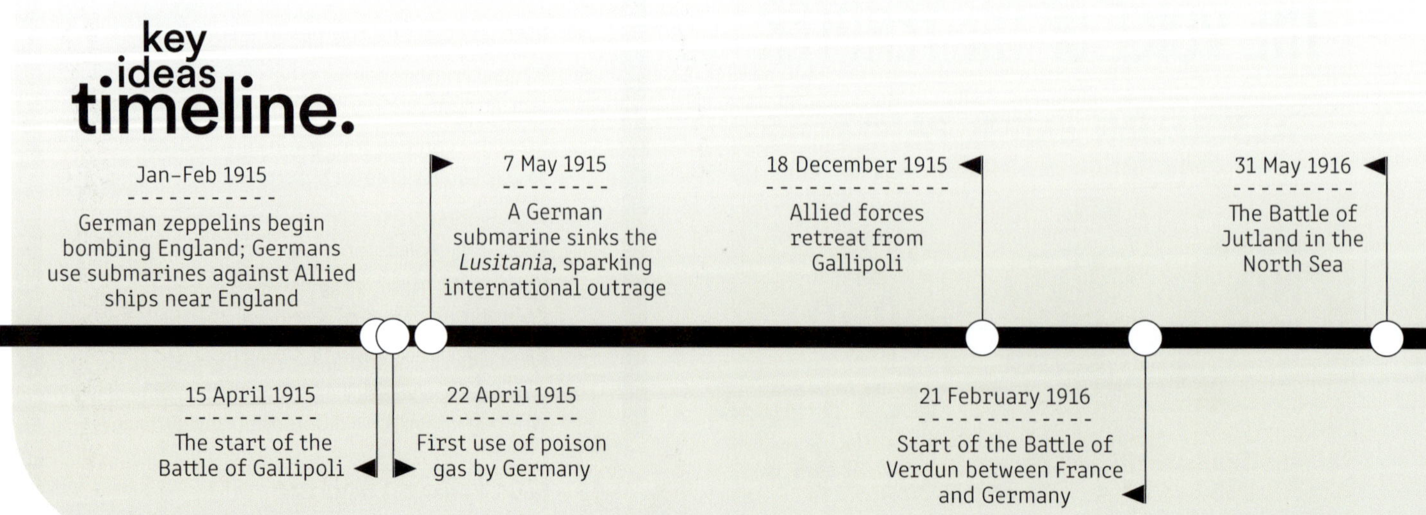

A typical day in the trenches

5 am	'Stand-to' (high alert for enemy attack): half an hour before daylight
5.30 am	Rum ration
6 am	Stand-to half an hour after daylight
7 am	Breakfast
After 8 am	Wash self and weapons; tidy trench
Noon	Lunch
Afternoon	Sleep and downtime (for every 10 men, one still on duty)
5 pm	Dinner
6 pm	Stand-to half an hour before dusk
6.30 pm	Stand-down half an hour after dusk
Overnight	Patrols: digging trenches, placing barbed wire, getting stores, night watch, some time for rest

Source 1

A sentry keeps watch in a British Army trench on the Western Front (1916).

Soldiers usually spent about five days per month in frontline trenches – five days of bombardment, knee-deep in freezing water and surrounded by corpses, rats and other vermin. After a short rest period away from the trenches altogether, soldiers would move to the supply and support trenches, and then back to the frontlines for their next stint.

1 July 1916 – The Battle of the Somme begins

15 September 1916 – The British use tanks on the battlefield for the first time

19 January 1917 – The British intercept the Zimmerman Telegram

8 March 1917 – The Russian Revolution begins

Show what you know

1. Why were soldiers 'stuck' in trenches?
2. What dangers or hazards did soldiers in trenches face?
3. Explain why trench warfare contributed to the high casualty rate on the Western Front.
4. Write a letter home to your parents, describing your life as a soldier in the trenches.

Historical significance

Step 1: I can recognise historical significance

5. Define the term 'trench warfare'.

Step 2: I can explain historical significance

6. Why was trench warfare so different to previous conflicts?

Step 3: I can apply a theory of significance

7. Prepare a sign for new arrivals to the trenches. List things they need to do, as well as dangers to watch out for. Which ones are the most important?

Step 4: I can analyse historical significance

8. Soldiers in the trenches were constantly 'on edge'. Decades after the war, many still had nightmares and other mental health difficulties. Suggest why.

HOW TO

Historical significance, page 251

Why was World War I known as the 'machine-age war'?

World War I was the first 'machine-age' war of the modern era. New weapons, such as the machine-gun, came to typify this new kind of warfare. World War I was also the first war fought in the air, as well as on the seas and on land.

War on land

World War I was the first modern war, and nowhere was this more devastatingly clear than in the development of **artillery**, large calibre guns such as howitzers and railway guns, along with other weaponry such as automatic machine guns, mortars and grenades.

Artillery was the most destructive weapon on the Western Front. Shells were filled with shrapnel – small bits of iron, nails or pellets. Even if a soldier wasn't killed by the impact of a shell, he could be killed or seriously wounded by shrapnel that flew around at incredibly high speeds. On the Western Front, big battles would often start with an enormous artillery barrage.

After the guns fell silent, soldiers would 'go over the top' (get out of the trenches) and advance through the treacherous No Man's Land (the deadly zone between the two opposing trenches). This area was riddled with bomb craters and barbed wire. Early in World War I, millions of kilometres of barbed wire were rolled out in No Man's Land in order to slow advancing troops. Men would get caught in knots of barbed wire, unable to escape while bullets rained down on them.

Tanks also represented a new challenge on the battlefield – heavily armoured, terrifying and capable of causing significant damage. Early tanks were slow and risked getting stuck, particularly in trenches, but they became increasingly deadly as designs improved.

Source 1

Powerful new weapons, such as grenades, were a hallmark of the war.

Source 2

Soldiers carried masks to protect themselves from poison gas attacks. These Australian troops have gas masks that connect to small respirator boxes on their chests.

Other new weapons on the battlefield included flamethrowers and poison gas. Volatile and leaking fuel, flamethrowers were often as dangerous for the person wielding them as they were for the soldiers subjected to the flames. Mustard gas (so-called because of its smell) was fired at enemy lines in special shells. The dangerous vapour flowed through the trenches, burning the eyes and lungs of soldiers who could not escape it. Sometimes the wind would blow the gas back over the soldiers that had just fired the gas shells. Gas killed many soldiers and left survivors with lifelong injuries; its use was banned by many post-war treaties.

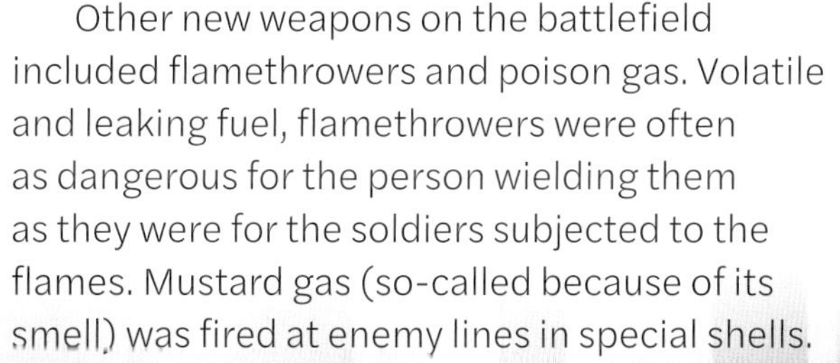

Source 3

French 'Saint-Chamond' tanks. Both sides fielded a variety of tank designs over the course of the war.

War in the air

The first aeroplane was invented in 1903 by American brothers Wilbur and Orville Wright. Just over a decade later, the technology of flight had improved to the point where powered aircraft became vehicles of war.

At the start of World War I, planes were only used for reconnaissance. Pilots would fly over enemy territory and sketch or photograph enemy positions. They did not go into battle; German and Allied pilots would greet each other in the air and fly on.

This soon changed. By 1915, planes were faster, could fly for longer and were more agile, which meant they could be used for air combat. Initially, pilots would shoot at each other with pistols, or drop bombs by hand on enemy positions. The invention of the 'interrupter' gear in mid-1915 allowed planes to mount machine guns; pilots could now shoot at planes in front of them without hitting their own propeller.

Skilled pilots often became famous. Manfred von Richthofen, the 'Red Baron', was a hero in Germany and was seen by some Britons as a 'noble enemy'. Films and books made life as a pilot seem romantic and exciting. In reality, planes were unreliable and air battles were deadly. In 1915, the average 'life expectancy' for an Allied pilot was just 11 days.

Planes were not the only air vehicles used in the war. Germany used zeppelins – airships that used hydrogen for lift – to conduct reconnaissance over the sea and for bombing raids on Britain, which killed many civilians.

Source 4

The plane of Manfred von Richthofen, Germany's infamous 'Red Baron', 1918

War at sea

The naval arms race between Britain and Germany led to the development of armoured dreadnoughts and battleships. However, World War I actually involved less naval warfare than the major conflicts of previous centuries. Battleships only came into direct conflict once, during the Battle of Jutland in June 1916, in which 14 British and 11 German ships sank.

As the frontlines of the conflict were on land, ships were primarily used to transport troops and supplies, or to attack enemy transport ships. British ships implemented a naval **blockade**, which stopped all supplies reaching Germany via the North Sea. This had a major impact on the German people; hundreds of thousands died of starvation and disease. The situation was made worse by bad harvests and mismanagement of existing food supplies.

The Germans responded with their newest invention – the submarine. While both sides developed submarines, and Britain had far more than Germany, the German U-boats were faster and more advanced. Germany's policy of 'unrestricted submarine warfare' meant their submarines would attack any ship they suspected of aiding the Allies, including civilian ships.

Source 5

A German U-boat (submarine) and its crew. U-boats armed with torpedos might attack any ship they encountered, whether or not it was a military vessel.

Source 6

A German Pickelhaube (spiked helmet), which was worn by infantrymen (left); a British 'Type A' helmet, which was worn by field medics (right). Helmets were the only form of protection available to individual soldiers.

Soldiers and their weapons

Because vehicles and artillery dominated the battlefield, commanders considered individual soldiers to be less important. Armies deployed large numbers of soldiers onto Western Front battlefields, but most were cut down by artillery or machine-gun fire long before they could come into contact with the enemy.

Most soldiers on both sides were armed with bolt-action rifles; these could only fire one shot at a time, but had a magazine of 5–6 rounds. Often these were tipped with **bayonets** for soldiers to use on the rare occasion they came to blows with the enemy. Armour was almost non-existent, apart from metal or leather helmets that might occasionally deflect a bullet.

World War I was also the last time that horse-mounted cavalry played a major role in war. While cavalry units on both sides had some successes early in the war, horses were no match for machine guns or tanks, and they were soon removed from battle.

Learning ladder H4.8

Show what you know

1 List three different technologies that changed the nature of war.

2 World War I was the first war fought on land, sea and in the air. Explain how.

3 Conduct further research into mustard gas. What were its effects? How did soldiers take precautions?

4 Who was the 'Red Baron'? Why might he have been both feared and admired?

Cause and effect

Step 1: I can recognise a cause and an effect

5 What was 'No Man's Land'? Why was it so deadly?

Step 2: I can determine causes and effects

6 Outline the positives and negatives of the new technologies emerging at this time.

Step 3: I can explain causes and effects

7 Explain how changes in weapons technology led to a sharp increase in fatalities during World War I.

Step 4: I can analyse causes and effects

8 Select one type of new technology, such as tanks or artillery, used in World War I, and outline how it affected the war.

Cause and effect, page 247

Why did the United States join the war?

When the war began, President Woodrow Wilson declared that the USA would remain neutral. American opinion began to change when Germany's policy of 'unrestricted submarine warfare' placed American ships in the firing line. The revelation of the Zimmerman Telegram cemented American hostility towards the Central Powers.

Initial neutrality

At the start of the 20th century, the USA was a fast-growing nation with an even faster growing economy. Every year, thousands of Europeans migrated to America in search of new opportunities. America mostly kept itself out of global and European politics, in a policy called **isolationism**. When war broke out, President Woodrow Wilson declared that America would be neutral and 'impartial in thought as well as in action'. Brazil, Chile and Venezuela also announced their neutrality.

Sinking of the *Lusitania*

While the USA preferred to stay out of the war, attitudes started to change when Germany announced its policy of 'unrestricted submarine warfare' and US ships came into the firing line.

A key event was the attack on the British civilian ship RMS *Lusitania*, in May 1915. The Germans suspected the ship was smuggling arms, and sunk it with a torpedo. Of the 1128 people who died, over 100 were American. This event fuelled strong anti-German sentiment that continued to build over the next two years.

"All the News That's Fit to Print."

The New York Times.

NEW YORK, SATURDAY, MAY 8, 1915.—TWENTY-FOUR PAGES.

LUSITANIA SUNK BY A SUBMARINE, PROBABLY 1,000 DEAD;
TWICE TORPEDOED OFF IRISH COAST; SINKS IN 15 MINUTES;
AMERICANS ABOARD INCLUDED VANDERBILT AND FROHMAN;
WASHINGTON BELIEVES THAT A GRAVE CRISIS IS AT HAND

SHOCKS THE PRESIDENT

Washington Deeply Stirred by Disaster and Fears a Crisis.

BULLETINS AT WHITE HOUSE

Wilson Reads Them Closely, but Is Silent on the Nation's Course.

HINTS OF CONGRESS CALL

Loss of Lusitania Recalls Firm Tone of Our First Warning to Germany.

CAPITAL FULL OF RUMORS

Reports That Liner Was to be Sunk Were Heard Before Actual News Came.

Special to The New York Times.

WASHINGTON, May 7.—Never since that April day, three years ago, when word came that the Titanic had gone down, has Washington been so stirred as it is tonight over the sinking of the Lusitania. The early reports told that there had been no loss of life, but the relief that these advices caused gave way to the greatest concern late this evening when it became known that there had been many deaths. Although they are profoundly reticent, officials realize that this tragedy, involving the loss of American citizens, is likely to bring about a crisis in the international relations of the United States.

It is pointed out that the sinking

The Lost Cunard Steamship Lusitania

X Where the First Torpedo Struck. XX Where the Second Torpedo Struck.

SOME DEAD TAKEN ASHORE

Several Hundred Survivors at Queenstown and Kinsale.

STEWARD TELLS OF DISASTER

One Torpedo Crashes Into the Doomed Liner's Bow, Another Into the Engine Room.

SHIP LISTS OVER TO PORT

Makes It Impossible to Lower Many Boats, So Hundreds Must Have Gone Down.

ATTACKED IN BROAD DAY

Passengers at Luncheon—Warning Had Been Given by Germans Before the Ship Left New York.

LONDON, Saturday, May 8.—The Cunard liner Lusitania which sailed out of New York last Saturday with 1,918 souls aboard, lies at the bottom of the ocean off the Irish coast.

She was sunk by a German submarine, which sent two torpedoes crashed into her side at 2:30 o'clock yesterday afternoon while the passengers, seemingly confident that the great, swift vessel could elude the German underwater craft, were having luncheon.

The great inrush of water caused the liner to list heavily

Cunard Office Here Besieged for News; Fate of 1,918 on Lusitania Long in Doubt

Fate of Most of the Well-Known Passengers Still in Doubt—Story of Disaster Long Unconfirmed While

Roosevelt Calls It Piracy; Says That We Must Act.

Special to The New York Times.

SYRACUSE, N. Y., May 7.—Ex-President Roosevelt

Meagre List of Saved Received in New York

Those whose rescue was reported to New York by

Loss of the Lusitania Fills London With Horror and Utter Amazement

News Held Back for Hours—Anxious Crowds Wait All Night at Steamship Offices for Word of

Source 1

The sinking of the *Lusitania* was a pivotal moment for the USA. This front page is from *The New York Times*, 8 May 1915.

The Zimmerman Telegram

Another incident that turned the American people against Germany was the discovery of a secret telegram in January 1917. This coded message was sent by German foreign minister Arthur Zimmerman to the German ambassador to Mexico. It promised that if Mexico joined the war on the German side, and Germany won, Mexico would regain Texas, New Mexico and Arizona from the United States.

British agents intercepted and decoded the message; they supplied it to the American government, which released it to the media on 28 February 1917. Popular opinion about the war immediately changed, with calls for America to defend itself against German and Mexican aggression.

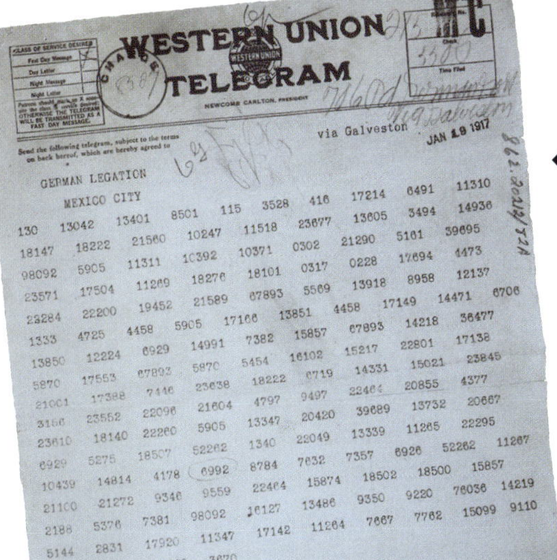

WESTERN UNION TELEGRAM

NEWCOMB CARLTON, PRESIDENT

Send the following telegram, subject to the terms on back hereof, which are hereby agreed to

via Galveston JAN 19 1917

GERMAN LEGATION

MEXICO CITY

130 13042 13401 8501 115 3528 416 17214 6491 11310
18147 18222 21560 10247 11518 23677 13605 3494 14936
98092 5905 11311 10392 10371 0302 21290 5161 39695
23571 17504 11269 18276 18101 0317 0228 17694 4473
23284 22200 19452 21589 67893 5569 13918 8958 12137
1333 4725 4458 5905 17166 13851 4458 17149 14471 6706
13850 12224 6929 14991 7382 15857 67893 14218 36477
5870 17553 67893 5870 5454 16102 15217 22801 17138
21001 17388 7446 23638 18222 6719 14331 15021 23845
3156 23552 22096 21604 4797 9497 22464 20855 4377
23610 18140 22260 5905 13347 20420 39689 13732 20667
6929 5275 18507 52262 1340 22049 13339 11265 22295
10439 14814 4178 6992 8784 7632 7357 6926 52262 11267
21100 21272 9346 9559 22464 15874 18502 18500 15857
2188 5376 7381 98092 16127 13486 9350 9220 76036 14219
5144 2831 17920 11347 17142 11264 7667 7762 15099 9110
10482 97556 3569 3670

BERNSTORFF.

Source 3

The coded version of the Zimmerman Telegram. It was intercepted and decoded by British intelligence.

The USA enters the war

In April 1917, President Woodrow Wilson told Congress that 'the world must be made safe for democracy'. Soon after, the United States declared war on Germany. It took time for the USA to mobilise its army, so it wasn't until June 1917 that 14 000 US troops landed in France to join the combat.

The arrival of well-supplied and fresh American troops was an enormous boost to the morale of the Allied soldiers on the Western Front, with more than 10 000 arriving each day. In total, the USA drafted more than 4 million men. About 50 000 of those would pay with their lives.

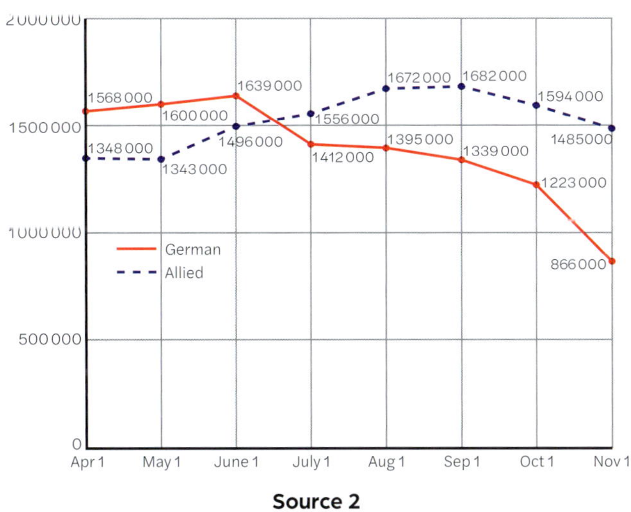

Source 2

As German troop numbers declined, Allied numbers rose, boosted by the United States entering the war. This graph shows the rifle strength of Allied and German armies on the Western Front in 1918.

Source: Leonard P. Ayers, ed. *The war with Germany: a statistical summary*

Learning ladder H4.9

Show what you know

1. Why was the USA initially determined not to enter the war?
2. What was the sinking of the *Lusitania*? Why did it change public opinion in the United States?
3. How did the Zimmerman Telegram attempt to persuade Mexico to ally with Germany?
4. Public support for isolationism was strong in 1914. Define this term and explain, using evidence, why it was supported within the United States.

Source analysis

Step 1: I can list specific features of a source

5. Source 1: Identify the key words/facts listed in the headline. How might these have made readers feel?

Step 2: I can find themes in a source

6. What does Source 2 show about the effect of the USA entering the war?

Step 3: I can use the origin of a source to explain its creator's purpose

7. The coded Zimmerman Telegram (Source 3) was made public after it had been decoded by British intelligence. Suggest why.

Step 4: I can analyse a source

8. Consider how Source 1 and Source 3 might have played a key role in swaying public opinion in the USA. What does this suggest about American society at this time?

Source analysis, page 240

Where did Australian soldiers serve?

As the global conflict escalated, Australian soldiers were drawn into battle in many different settings. With waves of volunteers drawn from across the nation, the Australians were keen to do themselves and their country proud.

Initial deployment

Australia's first active participation in combat actually occurred in Rabaul, a township in what is now Papua New Guinea. From the late 19th century, Germany had built a presence in the South Pacific with colonies in New Guinea, New Britain, the Solomon Islands, Palau, Nauru, Micronesia and the Marshall Islands.

The battle cruiser HMAS *Australia* of the Royal Australian Navy was sent to capture German New Guinea in September 1914. On 11 September, the Australian Naval and Military Expeditionary Force attacked a wireless radio station. Six Australians lost their lives in that attack – the first Australian casualties of the war. By 21 September, all German forces in the colony surrendered. It remained under Australian occupation until the end of the war.

Source 1

Regions where Australian servicemen and women were involved abroad in World War I

Postings of Australian forces in World War I

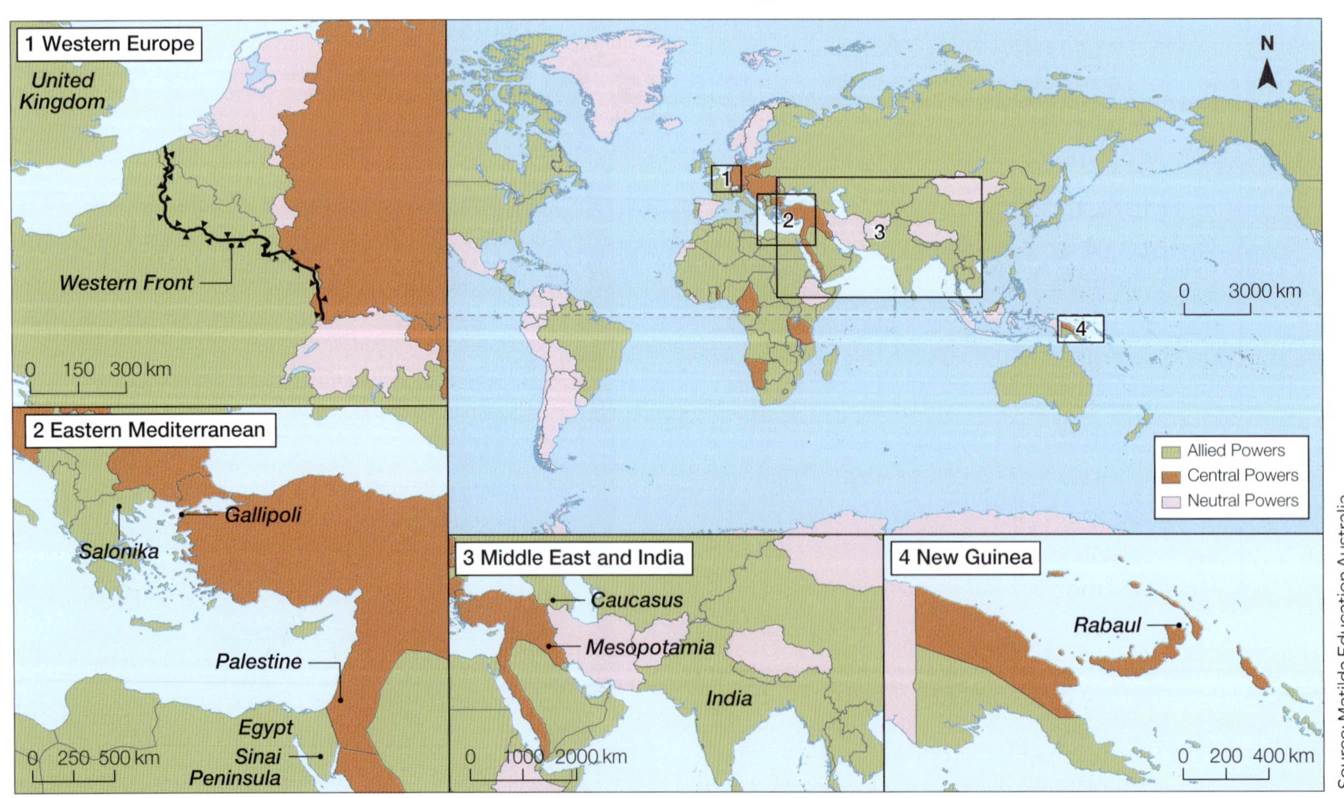

Source 2

Australian troops training in Egypt, 1914. Several soldiers smuggled kangaroos into Egypt, where they were mascots and pets to the troops. [Photo from Australian War Memorial C02588]

Formation of the Australian and New Zealand Army Corps

The Australian Imperial Force was a volunteer army. By September 1914, it consisted of 20 000 men, organised into a Light Horse Brigade and a 1st Infantry Division. By the end of the year they were in Egypt, training and getting to know the 10 000 soldiers from New Zealand who had joined them.

Together, these soldiers were known as the Australian and New Zealand Army Corps (Anzac). Despite this title, the Corps also included a small number of British and Indian units.

Deployment

By November 1914, the Anzacs had completed their training in Egypt, which had previously been part of the Ottoman Empire but was declared a British protectorate once the war began. From here, they travelled to Gallipoli and then on to the Western Front.

Between 1916 and 1918, the Anzacs also participated in battles within both Egypt and Palestine – a period known as the Sinai and Palestine Campaign – in support of British forces fighting against Turkey. Australian troops also supported the British in the Mesopotamian Campaign. Australian nurses served in Salonika and India, and the Australian Flying Corps flew in both France and the Middle East.

In total, 416 809 Australian men enlisted in World War I, of whom 331 781 served overseas. Of these, 61 720 died during the war and 137 013 were wounded.

Learning ladder H4.10

Show what you know

1 Other than Gallipoli and the Western Front, in which other regions did Australian soldiers serve?

2 Australian troops often fought alongside men of other nations, including New Zealand, India, Nepal and Canada. Suggest why.

3 Why was there a German presence in the South Pacific at the start of the 20th century? What happened to the Germans living there after their surrender?

Cause and effect

Step 1: I can recognise a cause and an effect

4 Why was Egypt a suitable training base for the AIF?

Step 2: I can determine causes and effects

5 Why was Australia's first active participation in the war at Rabaul (in modern-day Papua New Guinea)?

Step 3: I can explain causes and effects

6 Using Source 1, explain why Australians served in the locations indicated. Which opponents did they face?

Step 4: I can analyse causes and effects

7 To what extent was the German surrender in New Guinea likely to have affected morale in Australia? Compare the timing and speed of the campaign to the larger context of the war itself in your response.

Cause and effect, page 247

How did Australian soldiers serve at Gallipoli?

Australian soldiers played a key role in the Battle of Gallipoli. In one of the bloodiest battles of World War I, 8700 Australian men died in the fight to control the Dardanelles. Allied forces ultimately withdrew, and the Ottoman army declared victory.

The Gallipoli **peninsula** is a coastal region in Turkey. It runs along a narrow sea passage called 'the Dardanelles', which opens up to the Sea of Marmara, leading ultimately to the Black Sea. This was a key transport route as well as a communications link with Russia.

Source 1

Australian soldiers expected to be sent to Europe, but were instead deployed to Gallipoli. This photograph shows men of the Royal Naval Division and Australians in the same trench. One is using a 'sniperscope' and another a periscope.

When the Ottoman Empire joined the war, it closed the Dardanelles and the Bosporus Strait to ships from Russia, France and the UK. This meant that Russia could not easily receive supplies.To support its ally, the British Navy planned an attack to gain control of this strategic sea passage.

Landing of the Anzacs

After training in Egypt, the Anzac troops were transported to Gallipoli at short notice, landing before dawn on 25 April 1915. Anzac and Indian, British and French troops engaged with the defending Turks. They came under heavy fire from Turkish artillery and struggled to gain ground.

Despite the Turks holding the higher terrain, the Allies managed to establish a foothold; a few square kilometres of cliffs, gullies and beach. Here, they dug in.

A long campaign

The quick, decisive victory envisioned by the Allies did not occur. Fighting dragged on for months, as the Allies sought to gain more territory and the Turks attempted to drive them back.

As the death toll rose, so did the risk of diseases such as typhoid and dysentery; the corpses of soldiers, rotting in the hot sun, attracted swarms of flies. Occasional breaks in fighting allowed both

Source 2

The Dardanelles was a shipping passage of the utmost importance. Whoever controlled Gallipoli would control the Dardanelles.

sides to clear their dead and attend to the wounded soldiers in No Man's Land. Such breaks often followed attacks and counterattacks, such as the massive Turkish push on 18–19 May, in which 42 000 Turks attempted to rush the Allies. More than 10 000 were lost.

Anzac forces played a major role in two key battles.

- The Battle of Lone Pine: the Anzacs created a diversion to distract the Turks, allowing Allied troops to land at Suvla Bay and threaten the high ridges.
- The Battle of the Nek: Australian Light Horse troops attempted to storm a ridge and capture the Turkish positions. An early naval bombardment was meant to disrupt Turkish troops, but too much time elapsed between the bombing and the charge. Turkish troops returned to their trenches and bombarded the Australians with machine-gun fire.

Withdrawal

By November 1915, Allied leaders decided to withdraw and redeploy the remaining soldiers to the Western Front. Under cover of darkness, the Anzacs withdrew without further loss.

The Battle of Gallipoli was lost. Of the approximately 44 250 Allied troops who died during the campaign, 8700 were Australian. Approximately 100 000 Allies were injured. The Ottoman Empire lost approximately 86 700 men, with hundreds of thousands injured. Australia's experience at Gallipoli is seen by many as the nation's 'baptism of fire'.

Learning ladder H4.11

Show what you know

1. Which enemy did Australia face at Gallipoli?
2. Provide three reasons why the British had chosen to attack the Gallipoli peninsula.
3. List five challenges faced by Anzac soldiers at Gallipoli.
4. Conduct research about the experiences of Turkish soldiers at Gallipoli. How were their experiences both similar and different to those of Australians?

Historical significance

Step 1: I can recognise historical significance

5. How is the landing at Gallipoli remembered every year in Australia?

Step 2: I can explain historical significance

6. Was the Gallipoli campaign ultimately a success? Why or why not?

Step 3: I can apply a theory of significance

7. Why is this event considered to be so important to Australian history and to our national identity?

Step 4: I can analyse historical significance

8. How was the experience of Australian soldiers at Gallipoli different to those from Britain and New Zealand? How does this shape the way the campaign is commemorated in all three countries?

Historical significance, page 251

How did Australian soldiers serve on the Western Front?

The Western Front was the most renowned and deadliest battleground of the war. Close to half the Australians who fought there died.

Establishment of the Front

The Western Front was a stretch of land over 600 kilometres long, situated between the North Sea coast in the north of France and the German-Swiss border. Many of the worst and bloodiest battles of World War I were fought along this front.

The Western Front became a battleground early in the war, when Germany invaded Belgium on 3 August 1914. The German plan was to race through Belgium and conquer France within six weeks, and to then focus on fighting the Russians in the east. However, the small Belgian army slowed the advance long enough for the Allies to mobilise and meet the Germans in the north of France. Both sides began to dig in – literally, as they dug defensive trenches – for a long conflict.

Source 1

The Western Front saw Belgium and northern France turned into a series of trenches and devastated battlefields.

Location of the 1914–1918 Battlefields of the Western Front

Source: Matilda Education Australia

Source 2

Trench warfare was a relatively new phenomenon; very little was gained at enormous cost.

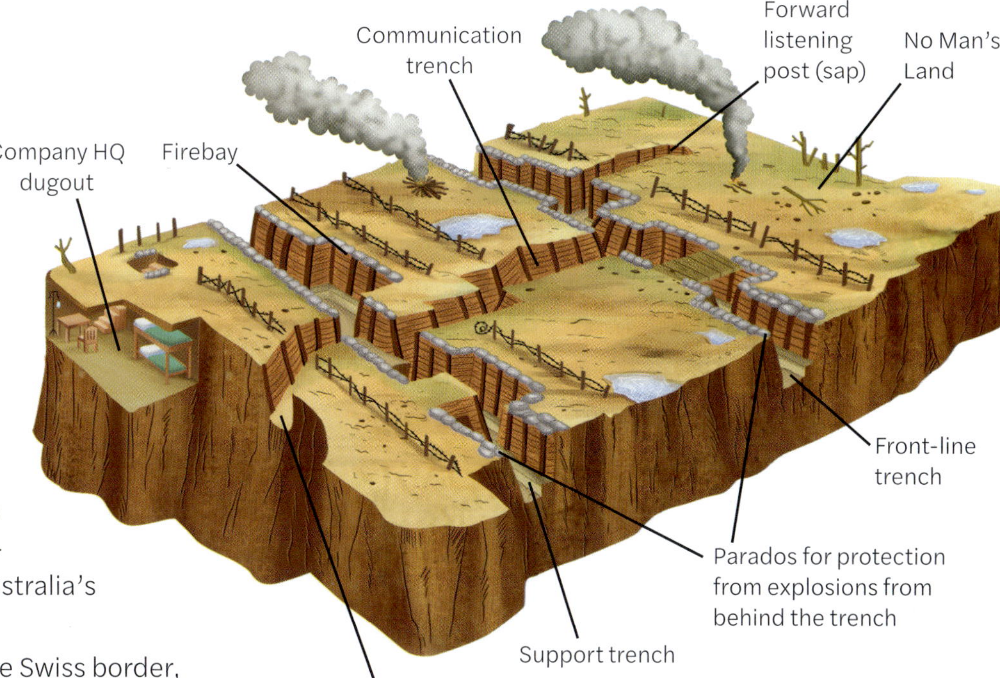

Stalemate

Hastily dug defensive trenches quickly developed into a sophisticated system of frontline and supply trenches. The Western Front ended up with 40 000 kilometres of trenches dug – enough to stretch around Australia's coastline 1.3 times.

From the North Sea to the Swiss border, the Western Front became the stage for four years of brutal and bloody battles, often for little gain and at a very high mortality rate.

The Allies had more soldiers on the Front, but the German trenches were better defended and protected, so neither side made much progress. This stalemate would not be broken until 1918, when new offensive weaponry was used in more effective ways, most notably by the Australian general John Monash.

Australians on the Western Front

Australian soldiers were sent to the Western Front in March 1916. They were involved in almost 30 battles, including at the Somme, Fromelles, Pozieres and Villers-Bretonneux. More than 295 000 Australians served on the Front, and around 46 000 soldiers died there. Another 132 000 were wounded.

Learning ladder H4.12

Show what you know

1 In your own words, outline what trench warfare involved.

2 Why were trenches used? Historically, such battles might have been fought on horseback. What changed?

3 Define the term 'stalemate'. Why do you think this happened on the Western Front?

Historical interpretations

Step 1: I can recognise that the past has been represented in different ways

4 Every Anzac Day is solemnly remembered, yet many veterans never wanted to speak of their experiences on the Western Front. Suggest why.

Step 2: I can describe historical interpretations

5 Military losses, such as those on the Western Front, were seen by some as the ultimate sacrifice, but by others as a senseless waste of young lives. Suggest why.

Step 3: I can explain historical interpretations

6 Casualties on the Western Front were particularly high. Explain how this might have influenced recruitment campaigns back in Australia.

Step 4: I can analyse historical interpretations

7 Today, some argue that the service of Australian soldiers on the Western Front remains central to the Anzac spirit and the birth of the nation. Explain how and why.

Historical interpretations, page 255

HOW TO

What was Australia's First Nations Peoples' experience of war?

When war began, approximately 90 000 First Nations Peoples lived in Australia. Although they did not have equal rights under Australian law until 1967, more than 1000 First Nations men served in World War I. A quarter of them gave their lives in service of a country that did not regard them as equals.

First Nations men enlist

In 1903, the *Defence Act* was passed by the federal parliament: it ruled that First Nations men could not enlist in the defence forces, even though they had served in the Boer War. The white colonial values that had directed immigration policy from 1901 now applied to Australia's armed forces.

In 1917, with demand for soldiers high and enlistment numbers falling, the *Defence Act* was amended. Volunteers from the First Nations could be accepted, provided they had one parent of European descent. Some men had actually enlisted before this, as recruiters often turned a blind eye to policy, particularly in rural and remote areas.

Why did these men volunteer to fight for a country that did not consider them equals? Many probably did so for the same reasons as other volunteers – loyalty to their friends and community, or a sense of duty. Others might have joined for financial reasons: privates were paid six shillings a day, which is more than they would have received working in Australia.

Source 1

An unidentified First Nations soldier from the 20th Battalion [Photo from Australian War Memorial P01703.001]

Source 2

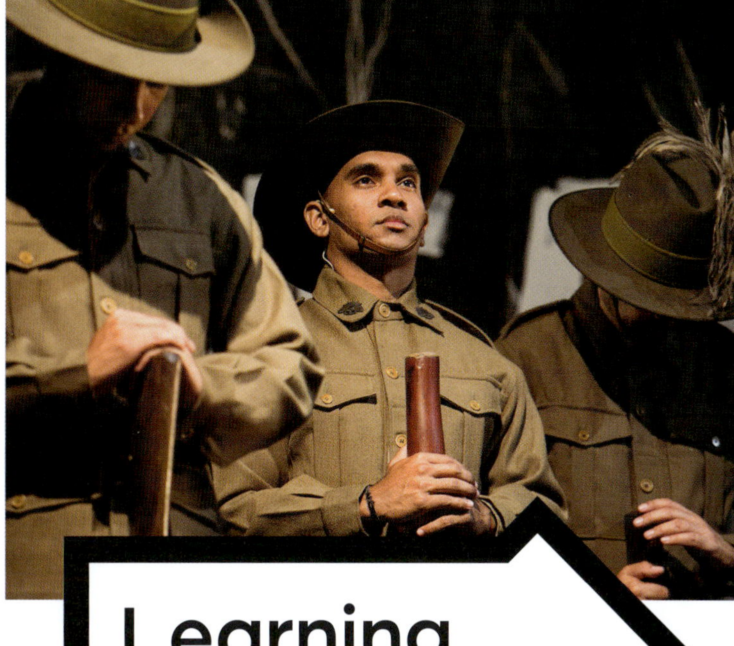

The play *Black Diggers*, from 2014, tells the stories of young First Nations men who volunteered to fight in World War I, and how the survivors were treated when they came home. [*Black Diggers* by Tom Wright (Queensland Theatre Company and Sydney Festival). Photography: Jamie Williams]

First Nations on the front lines

Approximately 1000–1300 First Nations soldiers served in World War I. At least 70 of them saw action at Gallipoli, and 13 lost their lives there. Information about these soldiers is difficult to uncover. Military records often did not record Indigenous status on the records of individual soldiers, describing them instead by their physical features.

Contemporary research suggests that around 100 First Nations men served in the Australian Light Horse Brigade. One Light Horse troop, which consisted of 26 First Nations soldiers, was active in the Sinai–Palestine Campaign.

After the war

When the war ended, many First Nations soldiers returned to their lives as second-class citizens. They were often not invited to join Anzac Day marches, and were prevented from entering Returned Servicemen's League (RSL) facilities. Some First Nations soldiers returned home to find that their children had been taken as part of the Stolen Generations (page 128) or their land seized from reserves to give to non-Indigenous returned servicemen as farm land.

As a minority group in their own homeland, the contribution of these soldiers has gone almost completely unrecognised. It has only been in recent years that their service has been celebrated and their stories told.

Men such as Douglas Grant, Harry Thorpe, Edmund Bilney, brothers Richard, Robert and George Kirby, Richard Martin and Chris Saunders are just some of the brave First Nations Anzacs. Marion Leane Smith, a First Nations woman from NSW, also served in World War I; she served in France in 1917 on an ambulance train, as part of the Canadian nursing division.

Source 3

Next time you see a $50 note, look at the church behind famous First Nations inventor David Unaipon. This church, located in Raukkan, is also a memorial to the 21 Ngarrindjeri Anzacs who signed up and fought on the Western Front.

Learning ladder H4.13

Show what you know

1 How many First Nations people joined the AIF during World War I?

2 Suggest what might have motivated some First Nations People to join the war effort.

3 List some of the challenges faced by First Nations servicemen, both during and after World War I.

4 Write a letter to the editor in 1918, making at least two arguments for the recognition of First Nations servicemen.

Continuity and change

Step 1: I can describe continuity and change

5 When did First Nations Peoples gain the right to join the Australian military?

Step 2: I can explain why something did or did not change

6 Indigenous Australians were not initially accepted into the Australian military. Explain why this changed as the war progressed.

Step 3: I can explain patterns of continuity and change

7 Outline how the changes to the *Defence Act* reflected the need for more soldiers.

Step 4: I can analyse patterns of continuity and change

8 Conduct research to compare the experiences of First Nations soldiers in World War I with those of New Zealand's Maori soldiers. How did their experiences differ?

HOW TO

Continuity and change, page 244

What was life like on the home front?

Australia's participation in the war affected the whole population, both directly and indirectly. There was widespread support in schools, charities and communities for the war effort, while others spoke out against the war. The loyalty of the German-Australian community was often questioned.

Source 1

Women could not join the Australian military until World War II, and could not operate in combat roles until 1990. Most Australian women who served in World War I were nurses.

Roles for women

At the beginning of the 20th century, most women worked in the home or in jobs such as domestic service, cooking, nursing or teaching. When the war broke out, many women wanted to support the war effort, so thousands of women volunteered for local organisations. They sent packages to the frontlines, raised funds to support soldiers, sent food to soldiers and volunteered at hospitals.

The only way women could serve in the army during World War I was to become a nurse, and more than 2200 women signed up with the Australian Army Nursing Service. The nurses had to be 25 years or older, and be unmarried or widowed. Many women lied to get through the recruitment process, so some 21-year-olds signed up, as well as married women. They served on transport and hospital ships, as well as in war zones. During the war, 46 Australian women lost their lives. These included nurses, munition workers, stewardesses and a doctor.

Not all women supported the war. Adela Pankhurst, daughter of famous British suffragette Emmeline Pankhurst, migrated to Australia in 1914. She took up an active role with the Women's Peace Army, a movement against war and conscription, alongside Melbourne suffragette Vida Goldstein. Both were leading proponents of peace and staunch advocates for women's rights.

Source 2

Children made socks and other supplies to send to soldiers on the frontline, part of a united push to support the war effort. [Photo from the Australian War Memorial H11581]

Tasks for children

For children, the war represented a significant change to daily life. Children were engaged in supporting the war effort in many ways, including:

- preparing clothing and other goods to send to soldiers
- taking on more household duties to cover for fathers on the frontlines and mothers at work
- fundraising
- collecting scrap metal for recycling
- undergoing cadet training: military training was compulsory for boys aged 12 and over
- enlisting to fight: some teenage boys lied about their age to join the army.

Schools played a key role in inspiring patriotism in children, through activities such as reading honour rolls of former students who had gone to war and emphasising the virtues of the Allies.

Patriotism

Patriotic rallies were commonplace in Australia during the war. Empire Day (24 May), Allies Day (19 November) and Anzac Day (25 April) were celebrated. The continued union between Britain and Australia was heavily emphasised, and the losses of Australian soldiers were considered to be for 'the greater good'. Propaganda portrayed the German soldiers as savage beasts, and Belgium was held up as an example of the need for Australia to act; one could not simply 'stand-by' while a bushfire raged in the small nation.

Associations such as the Belgian Relief Fund, the Travelling Kitchen Fund and the Blind Heroes Fund collected both donations and support, providing avenues for all Australians to help. War bonds and loans were used to fund the war effort, and 'loan drives' prompted citizens to contribute to each new issue of bonds.

However, not everyone felt patriotic. Some spoke out about the ongoing deaths of Australian men in far-off countries, while pacifists and 'conscientious objectors' rejected the very notion of conflict. Many Irish-Australians felt torn between loyalties, resenting the British for their occupation of Ireland at that time.

Source 3

The Molonglo Internment Camp held more than 150 families of German descent. They attended school, played tennis and interacted well with the local community. In 1919, after the war, they were deported to Europe. [Photo from the Australian War Memorial H17413]

Jobs and money

With manufacturing redirected to support the war effort and a freeze on wage growth in place, the cost of living rose sharply in Australia – it went up by 50 per cent between 1914 and 1918. Inflation increased, along with the price of food and rent; so did taxes, which were used to finance the war.

Growing inequality emerged, leading to serious strikes in 1916 and 1917, particularly involving seamen, miners, dock and transport workers. Strike-breakers – non-union workers employed to ensure production continued – were used to help end the strikes, which in turn led to further discontent.

Persecution of German-Australians

In 1914, Australia had a small German population of about 100 000 people, or 2 per cent of the population. Life changed drastically for these people when war broke out. All Germans were required to report to the local police station weekly, sometimes daily. The police would then fill out a secret report, stating whether this person could be trusted or if they were 'anti-British'.

Internment camps were established in some states, such as Holsworthy Army Barracks in NSW and Langwarrin Camp in Victoria. Many German and Austrian men were put into these camps for the duration of the war; they were joined by others who had been detained in Asia by the British. Some voluntarily went into camps to ensure that their wives and children would receive a government allowance. Women and children were kept at Molonglo in NSW.

In some towns, German-language schools and churches were closed, German food was renamed and German music was banned. Some Australians with German or Austrian heritage chose to anglicise their names. In addition, 42 place names were changed; for example, Blumberg in South Australia became Birdwood, while German Creek in NSW became Empire Bay.

Suspicions and restrictions

Many people in Australia were on high alert for spies and enemy agents, despite the war being thousands of kilometres away. A significant minority of Australians could trace their roots to continental Europe, particularly Germany. Despite being born in Australia, or having lived in Australia for some time, they attracted suspicion.

Suspicion also fell on trade unions, pacifists, socialists and anti-conscriptionists, who were seen as dangerous to the war effort or critical of the government. Many were investigated and sometimes even prosecuted. Passed in August 1914, the *War Precautions Act* gave the government wide-ranging powers to control all aspects of daily life. It remained in place until 1920, and aimed to:

- prevent acts of espionage
- prevent activity or communication that could jeopardise military operations
- prevent the spread of reports that might cause alarm
- prohibit foreign nationals and naturalised citizens from entering Australia, and allow the deportation of those already present
- prohibit foreign nationals and naturalised citizens from living in certain areas
- control the registration, movements, work and place of residence of foreign nationals and naturalised citizens in Australia
- prevent money or materials being shipped out of Australia without permission.

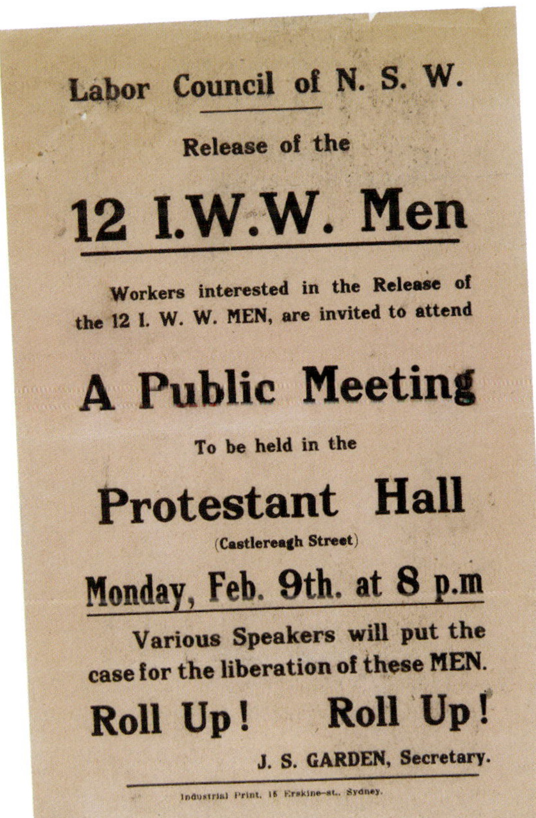

Source 4

In 1916, 12 men were arrested in Sydney under the *War Precautions Act*. They were members of the International Workers of the World, a radical labour organisation that was against the war. They were held for four years before being charged, and were finally released in 1920.

Learning ladder H4.14

Show what you know

1. Was all of Australia united behind the war effort? Why or why not?
2. What roles were traditionally undertaken by women in Australia at the beginning of the 20th century?
3. How did children support the war effort?
4. Define the term 'strike-breakers'. Why were they used and why were they despised?

Continuity and change

Step 1: I can describe continuity and change

5. What was the *War Precautions Act*?

Step 2: I can explain why something did or did not change

6. Imagine yourself as a government representative in 1915. Prepare a short article for the school community, explaining why the *War Precautions Act* is necessary.

Step 3: I can explain patterns of continuity and change

7. Provide examples of how daily life in Australia was changed by the *War Precautions Act*. For which groups would change have been the most significant?

Step 4: I can analyse patterns of continuity and change

8. Draw up a table with four columns: children, women, workers and German-Australians. In each column, identify ways in which the war both did and did not change the daily lives of those in that group.

HOW TO Continuity and change, page 244

economics + business

How can I make informed investments?

During World War I, Australians invested in war bonds to help support the war effort. Today there are many different investment options, with a range of risks and possible returns on the investment.

Investing in the war effort

Investing money means putting it into financial ventures such as shares, property or a business venture, with the expectation of making a profit.

During World War I, Australians were encouraged to invest in war **bonds**. By buying bonds, citizens effectively loaned money to the federal government, with the promise of being repaid with interest after the war. The marketing campaigns for war bonds aimed to convince people it was their patriotic duty to direct their personal finances towards the war effort.

Most countries involved in the war, on both sides, developed war loan programs to raise the funds needed to resource their expensive military campaigns. Money raised through bonds is estimated to have covered more than half of the cost of the war in Germany.

Source 1

World War I propaganda poster, calling on civilians to invest in war bonds to help raise money for the war in Europe

Investment and risk

All investments involve the risk of losing money. There's always the chance that an investment won't deliver the outcomes you want; that is, it may fall in value instead of grow. It's impossible to avoid all risks when you invest, but many investments have only a low risk. However, those with a greater potential for profit usually carry a greater risk. As a general rule, the lower the level of risk, the lower the return on investment will be.

Some investments, such as putting your money into bank **savings accounts** and term deposits (where your money is locked away for a set time), are low risk and government guaranteed for up to \$250 000. These low-risk options also provide a low return on investment, in the form of interest.

Investment specialists suggest that you will need to take some investment risk to achieve a healthier rate of return over time. Investing in real estate or the share market has the potential to earn higher returns, but it also comes with the potential for higher losses.

Source 2

Stocks (shares in a large business) become more valuable if the business is successful. Changing stock prices are listed and updated on the main board at the Australian Stock Exchange (ASX).

Investment options

Investing your money wisely takes a great deal of thought and research, and many people seek the advice of a financial consultant. Different investment options can also be combined into a managed fund, in which money from individual investors is pooled into a selected portfolio of investments such as shares, property or interest-earning investments.

Interest-earning investments

These investments involve lending money to a financial institution, company or government, and in return receiving payments of interest. Interest is calculated as a percentage rate, which might be a **fixed rate** or a **floating rate** that changes in response to the market.

Interest-earning investments also involve different time periods. Most banks give immediate or at-call access to your funds if you wish to withdraw your money. Others are fixed-term investments that last for a specified time. Term deposits are fixed-term investments; so are bonds.

The share market

Shares (also called stocks or equities) are units of ownership in a company. By investing in shares, you become a part-owner of a company, usually with the hope that it will increase in value. When a company earns a profit, they can pay a proportion of it as a dividend to its shareholders.

Shares offer potential for investors to grow their wealth over time. However, profiting from buying and selling shares requires skill and analysis.

Real estate

Real estate investment involves purchasing a house, apartment, factory or other building with the hope that it will increase in value. The owners can also receive rental payments as an investment return.

Superannuation

Superannuation is a way of saving for retirement. Employers make mandatory contributions to their employees' superannuation funds, and employees can add more as well. The fund's managers use that money to finance large investments, and the profits go back to the employees' savings. Superannuation is often taxed at a lower rate than other investments.

Learning ladder H4.15

Economics and business

Step 1: I can recognise economic information

1 What are you actually doing when you put your money into an interest-earning investment?

Step 2: I can describe economic issues

2 What happens when you become a shareholder? Refer to Source 2 in your response.

Step 3: I can explain issues in economic

3 How is return on investment related to risk?

Step 4: I can integrate different economic topics

4 Consider Source 1. What advantages did investors get when they put their money into war bonds during World War I?

Step 5: I can evaluate alternatives

5 Prepare your own risk analysis of investments in superannuation, real estate, shares and interest-earning investments. Rank the four types of investment according to risk and possible returns.

PART III: WAR'S END

How did World War I end?

The chief events that led to the end of the war played out on the Western Front. The German-initiated *Kaiserschlacht* of early 1918 proved to be the last crucial battle; the Allies defeated Germany, effectively signalling the end of the war.

Changing allies

In 1917, Russia's Tsar Nicholas II was forced to abdicate his throne under pressure from countrywide strikes and protests. In November 1917, Vladimir Lenin and the Communist Party came to power. One of the first acts of the new Communist government was to sign a peace treaty with Germany, allowing Russia to exit the war.

This was a major blow to France and England. It meant that Germany no longer had to fight Russia, and German soldiers could be sent from the Eastern Front to support the troops on the Western Front. The USA had joined the Allies in June 1917, and was sending more troops to the Western Front, but they would not arrive until the spring of 1918.

Source 1

After the 1917 Revolution, Vladimir Lenin and the Communist Party took control of Russia. They withdrew from the war in order to focus on internal Russian issues.

key ideas timeline.

- **6 April 1917** – USA declares war on Germany and US troops join the Allies in June
- **October 1917** – Revolution in Russia: the Communist Party takes over and withdraws from the war
- **March–July 1918** – The Ludendorff Offensive fails
- **11 November 1918** – Germany agrees to an armistice; the allies are victorious
- **1918–1919** – Almost 250 000 Australian troops return home; the 1918 flu pandemic sweeps the globe
- **1919–1920** – A number of treaties signed; the Treaty of Versailles marks the formal end of the war

The Ludendorff Offensive

General Ludendorff, the man in charge of the German army, knew that this was a critical time and planned a massive offensive for the early spring of 1918. This *Kaiserschlacht* ('the Emperor's battle') was meant to break the stalemate on the Western Front.

The Ludendorff Offensive was a gamble, but the odds were with the Germans. Still, they lost. A lack of resources, too much confidence placed on the Germans' ability to mount a strong attack, and tactical wins by the Allies, meant the offensive failed. The Germans had now effectively lost the war.

The armistice

On 1 October 1918, General Ludendorff suggested that an **armistice** was the only way forward. 'The High Command and the German army are finished', he said to his assembled staff officers. He placed the blame on the frontline soldiers, saying 'continually, units have proved themselves so unreliable that they have hurriedly had to be withdrawn from the front'.

Bolstered by the extra American troops, the Allies pushed back the German army. By November 1918, the German military commanders recommended to Kaiser Wilhelm II that an armistice was needed. On the 11 of November, at 11 o'clock in the morning, the armistice began. The news spread quickly across the world. The war was over.

Why did the Central Powers lose WWI?

- The failure of early attack, the Schlieffen plan, meant that Germany got bogged down in a stalemate on the Western Front.
- The Ludendorff Offensive of 1918 did not break the stalemate on the Western Front.
- The German allies were defeated. Bulgaria surrendered in September 1918, while Austria–Hungary devolved by October 1918.
- The British sea blockade of Germany had been effective; the Germans suffered from food shortages, made worse by bad harvests and bad government organisation.
- The policy of unrestricted submarine warfare brought the USA into the war.
- The USA's entry into the war in 1917 came at a crucial time, when the Allies were running low on materials and men.

Entire World Rejoices in Downfall of Autocratic Militarism

THE DAILY TELEGRAM

VOL. XXIX, NO. 118

LONG BEACH, CALIFORNIA, MONDAY AFTERNOON, NOVEMBER 11, 1918

TEN PAGES

PRICE 2 CENTS

WORLD WAR IS OVER

Beaten Germany Accepts Armistice Terms and Hostilities Cease on All Fronts

MONS TAKEN BY BRITISH AS NEWS REACHES ARMY | ALL CROWNED HEADS OF EMPIRE BUSY ABDICATING

Kaiser, Crown Prince and Hindenburg Seek Refuge in Holland

"IT'S OVER BOYS"

WAS WORD SENT ALONG LINES BY CYCLISTS

WHEN PEACE CAME

LONDON, Nov. 11.—It is officially announced that the armistice with Germany was signed at 5 a.m. today and that hostilities ceased at 11 a.m. Prime Minister Lloyd George announced that the cessation of hostilities was effective at 11 a.m. on all fronts of Europe.

WASHINGTON, Nov. 11.—It was shortly before 3 o'clock this morning when President Wilson, roused from a deep slumber, walked to the door of his bedroom and received from a white house attendant, the official State department message that the great war had come to an end.

GIVE UP "DIVINE RIGHTS"

ALL ROYALTY OF GERMAN EMPIRE HUNT COVER

GERMANY IS BRANDED NOW AS BEATEN NATION

Hun Acknowledges Superiority of Allies —To America Goes Glory of Turning Tide of War

HOLLAND'S PROBLEM

ARMED IMPERIALISM COMPLETELY ENDED BY ACCEPTED ARMISTICE

President Wilson, in Address to Joint Congress, Outlines the Conditions Accepted by Germany Which Puts Into Discard the Illicit Ambitions of Prussian Autocracy and Starts Peace Which Will Satisfy World Desires.

ALL DRAFTING

Source 2

The end of World War I made headlines around the world.

Learning ladder H4.16

Show what you know

1 Why did Russia withdraw from the war in 1917?

2 What was 'Ludendorff's gamble'? How might its success have changed everything?

3 Define the term 'armistice'.

4 Describe the difference between an armistice and a surrender.

Historical significance

Step 1: I can recognise historical significance

5 What happened at 11 am on 11 November 1918?

Step 2: I can explain historical significance

6 Why was the signing of the armistice such a historic moment?

Step 3: I can apply a theory of significance

7 As a senior advisor in the German military, write a letter to Kaiser Wilhelm II, outlining why an armistice is necessary. Use evidence to support your position.

Step 4: I can analyse historical significance

8 Many communities and nations had little to celebrate at the end of the war. Provide reasons why. To what extent did the end of the war resolve their problems?

Historical significance, page 255

How was peace maintained after the war?

World War I was an unprecedented event. It destroyed more human life and infrastructure than anything ever before. Dealing with the effects of this disaster was almost as big a challenge as fighting the war itself.

The aftermath of war

After four brutal years of slaughter and hardship, the Allies were looking for ways to settle the score. The armistice of 11 November ended the fighting, but loss and damage was widespread, particularly across continental Europe.

European political borders after World War I

Source: Matilda Education Australia

Source 1

The political borders of Europe shifted dramatically in only a decade. Compare this with the map of Europe in 1914 on page 196.

Trench warfare and new weapons technology had left significant environmental destruction, in addition to the huge death tolls, injuries and economic devastation. The empires of central and eastern Europe were broken, with new nations clawing for their own freedom amid widespread social unrest.

Treaties and agreements

To ensure that war would not return, US President Woodrow Wilson created a '14-point plan' to create peace. It was an idealistic document that focused on diplomacy, trade and self-governance to establish a new world order and 'make the world safe for democracy'.

A number of treaties were also signed. These assigned blame and consequences, redrew borders and ordered the Central Powers to provide **restitution** to the Allies. Five treaties were made with the members of the Central Powers:

1. The Treaty of Versailles (1919) – Germany
2. The Treaty of Saint-Germain (1919) – Austria
3. The Treaty of Neuilly (1920) – Bulgaria
4. The Treaty of Trianon (1920) – Hungary
5. The Treaty of Sevres (1920) – the Ottoman Empire

As a result of these treaties, new nations were formed, military capacities were restricted and further alliances forbidden. While they aimed to ensure peace across the continent, they also created discontent and anger in some nations.

Australia after the war

At war's end, more than a quarter of a million Australians needed to be returned to their country. But because of a lack of available ships, some took up to 18 months to get home.

Unfortunately, returning to a 'normal' life was almost impossible. There was widespread unemployment, so many soldiers couldn't find jobs. The government gave pensions to returning soldiers, nurses and war widows. This provided support, but was a massive drain on the Australian economy.

Health was also a major post-war problem. As some soldiers were exposed to the deadly influenza that swept through Europe in 1918–19, they needed to be quarantined upon their return home. Despite this, the disease entered the country and caused almost 12 000 deaths. Politically, though, Australia rose in stature and prominence as a result of its involvement in the war. Other countries now took the young nation much more seriously.

Source 3

The signing of the Treaty of Versailles. Australian Prime Minister Billy Hughes was present at the negotiations, and argued successfully for reparations for Australia. [Joseph Finnemore, *Key to the signing of the Treaty of Peace at Versailles*, 1919]

Source 2

Returning soldiers disembarking from a troopship at Port Melbourne, 1919

Learning ladder H4.17

Show what you know

1. How and why was World War I different from any other conflict before it?
2. What plan did Woodrow Wilson propose?
3. Briefly research the punishments or outcomes for each of the Central Powers. Show your findings in a table.
4. To what extent do you feel that each Central Powers nation was fairly punished?

Source analysis

Step 1: I can list specific features of a source

5. Outline the scene in Source 3. What features or people do you recognise?

Step 2: I can find themes in a source

6. Compare the map of pre-war Europe on page 196 with post-war Europe in Source 1. Create a list of new nations and note those that lost territory.

Step 3: I can use the origin of a source to explain its creator's purpose

7. Consider Source 3. Why would the presence of the Australian Prime Minister at Versailles be promoted both domestically and internationally?

Step 4: I can analyse a source

8. Describe the scene in Source 2. How might it differ to soldiers embarking for war years earlier?

HOW TO
Source analysis, page 240

Masterclass

Learning ladder

Work at the level that is right for you or level-up for a learning challenge!

Source 1

Trench warfare was brutal and unforgiving, but long hours of boredom were also commonplace. ['Studying French in the Trenches.' Cover, *The Literary Digest*, October 20, 1917.]

Source 2

'Bravo Belgium!' from *Punch Magazine*, 12 August 1914

Source 3

Gallipoli holds a special place in Australia folklore, famed as the birthplace of the Anzac legend. [Charles Dixon, *The landing at ANZAC, April 25 1915* (date unknown), oil on canvas]

Step 1

a I can list specific features of a source

Refer to Source 2. Identify the features of the cartoon and provide a brief outline of the event pictured.

b I can describe continuity and change

What was life like in Australia before the beginning of World War I? How did this change once war was declared?

c I can recognise a cause and an effect

Why was Archduke Franz Ferdinand assassinated? How did Austria-Hungary respond?

d I can recognise historical significance

World War I was seen as an important moment for the young Australian nation. Why?

e I can recognise that the past has been represented in different ways

Why might people have had different feelings about the war?

Step 2

a I can find themes in a source

What do you feel are the major themes of Source 3?

b I can explain why something did or did not change

Many young Australians rushed to enlist once war was declared. Explain why.

c I can determine causes and effects

Outline the four long-term causes of World War I.

d I can explain historical significance

Why was the Anzac legend 'born' at Gallipoli?

e I can describe historical interpretations

Describe how the landing at Gallipoli is portrayed in Source 3.

Step 3

a I can use the origin of a source to explain its creator's purpose

Why was Source 1 produced?

b I can explain patterns of continuity and change

What happened to Australian enlistments as the war progressed? Outline why.

c I can explain causes and effects

Explain the conscription debates. What were the outcomes and why did they develop?

d I can apply a theory of significance

Draft a response to the following statement: 'Technology played a key role in the war.' To what extent do you agree?

e I can explain historical interpretations

Explain why the debate over conscription is considered to be a time of bitter division within Australian society.

Step 4

a I can analyse a source

To what extent is Source 1 an accurate representation of trench warfare?

b I can analyse patterns of continuity and change

In what ways did the borders and nations of Europe change as a result of the war?

c I can analyse causes and effects

What events led to the entry of the USA into World War I?

d I can analyse historical significance

How might World War I be considered a turning point for women in Australian society?

e I can analyse historical interpretations

The way we understand and interpret the past can change over time, such as the debates around Australia Day. Prepare a brief paragraph explaining how this might be possible.

Masterclass

Step 5

a I can evaluate a source

How useful is Source 2 when describing the causes of World War I?

b I can evaluate patterns of continuity and change

How did the war change the world? Consider economic, political and social contexts in your response.

c I can evaluate causes and effects

How did trench warfare and new technology make World War I vastly different to any other conflict before it?

d I can evaluate historical significance

Draft a response to the following statement: 'World War I was a defining event for Australia, one that both changed and shaped it.'

e I can evaluate historical interpretations

To what extent does Source 2 support differing interpretations of the causes of World War I?

Historical writing

1 Structure

Imagine you are given the essay topic, 'World War I was a European war. Australia should never have been involved'. Write an essay plan for this topic. Include at least three main paragraphs.

2 Draft

Using the drafting and vocabulary suggestions on page 262, draft a 600–800 word essay (at least 30–40 sentences) responding to the topic.

3 Edit and proofread

Use the editing and proofreading tips on page 263, to help edit and proofread your draft.

Historical research

4 Organise and present information

Imagine you are completing a research project on 'Australian sacrifices in World War I'. Write a contents page for this project. There should be an introduction, a conclusion, at least four main sections and many subsections. Number your chapters.

Capstone

How can I understand World War I?

In this chapter, you have learned a lot about World War I. Now you can put your new knowledge and understanding together for the capstone project to show what you know and what you think.

In the world of building, a capstone is an element that finishes off an arch or tops off a building or wall. That is what the capstone project will offer you, too: a chance to top off and bring together your learning in interesting, critical and creative ways. You can complete this project yourself, or your teacher can make it a class task or a homework task.

mea.digital/GHV9_H4

Scan this QR code to find the capstone project online.

Source: Australian War Memorial E00019

History How-To

H5

History has its own set of skills to help us analyse and understand societies in the past and the key ideas, people and changes that shape the world we live in today. Historical skills are based around interpreting sources of evidence from the past, provoking debate and encouraging investigation.

Source analysis

Source analysis asks us to look at evidence and ask, 'How do we know what we know about the past?' A good source analysis interprets and makes meaning of the source:

- Who created it?
- When was it created?
- What was the author or creator's purpose?
- What is the historical context of the source?

I can list specific features of a source

When you list features of a source, they can be general or specific. For the general features of a source, when you list them, it is like you are providing a summarised version of what you see.

Listing specific, or detailed, features of a source is more like you are writing the long version of what you can see or interpret, describing as much as you can.

General features	Detailed or specific features
The most obvious features	Obvious and minor details
The most important things in the source	Everything in the source, whether or not you think it is 'important'
Using vague words: 'big', 'small', 'very', 'good' …	Using specific words and phrases, such as: 'three times bigger than …', 'small/big when compared to …', 'in the background/ foreground', 'useful for …'

I can find themes in a source

Often, a source contains a theme. The theme can help you uncover more meaning in a source. A theme is something that you might notice after recognising specific features.

Some examples of themes and how they might be shown are listed in this table.

Theme	How this might be shown in a source
Beauty	A statue of a handsome person with a muscular body and symmetrical facial features
Faith	A decoration on a vase showing people offering sacrifices to a god
Good vs bad	A statue of two figures fighting – one that looks like an angel and one that looks like a demon
Hierarchy	An image going from top to bottom, with gods on the top, then people, then animals, then rocks
Humanity vs nature	A building located in a natural place, dominating it; e.g. a temple on a mountain
Technology in society – good or bad	Good: a statue of a smiling person using a new farm implement Bad: a painting of a person using a machine to work looking unhappy
War or conflict	A decoration on a building depicting a large-scale battle

Source 1

A 19th-century woodcut of the smokestacks of steel factories in 1800s Sheffield. What themes do we see here?

If you think there is an underlying theme, after looking at all the details in a source, always give evidence *from the source* to back up your answer.

Not all sources have themes. If the source is a piece of technology, like a train, it might not have a 'theme' like the woodcut shown in Source 1. However, you could still comment on the type of society that produced it. For example, what materials did they use? Why did they need trains? When did they build railways?

step 3

I can use the origin of a source to explain its creator's purpose

Here are three ways you could explain a creator's purpose. You could:

1 use *who* created it to explain *why* it was made
2 use *when* it was created to explain *why* it was made
3 use both when it was created *and* who created it to explain why it was made.

The third explanation is the best. Here are some questions to ask of the source:

- What do you know about who created the source?
  - How old were they?
  - What gender were they?
  - What job did they do?
  - What position did they have in society? For example, were they powerful or powerless?
  - What beliefs did they have?
- What was going on at the time the source was made?
  - Were there any important events taking place?
  - What was going on politically?
  - What biases might people have had?
  - What was normal behaviour in society?
- Why was it produced?

The purpose of a source refers to what the source was originally made for. Don't get confused and think about what *we*, as historians, might use it for. Sources aren't usually created to leave records for historians.

Try to get into the head of the creator *at the time they were making the object*. For example, were they trying to:

- influence people?
- sell something?
- tell their version of events?
- make art? If so, who would enjoy it?
- make something practical, such as a tool?

Source 2

Rock art in the Northern Kimberley region, in the Gwion Gwion style, depicting a fishing hunt. [Raft Point Gallery, Kimberly region, Western Australia.]

The table below provides two examples of using the origin of a source to explain its purpose, using *The Communist Manifesto* and rock art by the First Nations Peoples of Australia.

Source	Origin: Who created the source	Origin: When the source was created	Why you think the source was created (its purpose)	Using the origin of the source to explain its purpose
Rock art	The Gwion Gwion people of the Northern Kimberley region	Between 20 000 and 50 000 BCE	To depict a fish hunt	The Gwion Gwion people were some of the most ancient First Nations Peoples of Australia. They used their art to show the hunting technologies they invented and the natural world around them.
The Communist Manifesto	Karl Marx and Friedrich Engels	1848 CE	To communicate the political theory of communism	Marx and Engels were German philosophers who were involved in the socialism movement. They wrote *The Communist Manifesto* to communicate their ideas about society, politics and economics.

step 4

I can analyse a source

Analysing involves breaking down something into its parts. If you can identify these different parts, and explain how together they make up the whole, you are analysing. If you can explain rules or theories that show how these parts are organised, you are analysing.

To analyse a source, consider the different elements of the source and see how they work together, and whether they produce an overall effect. We can break this into three parts:

1 Notice various elements of the source
2 Explain them
3 Show how they relate to each other, influence each other or are linked in some way to produce an overall effect.

As an example, we can analyse Source 3.

Source 3

A 1917 poster produced during the WWI conscription debate

1 There is a kangaroo, text and an army in silhouette.

2 a The kangaroo, a national symbol of Australia, is included to awaken nationalistic spirit.

b The text refers to Britain, which was still seen as the 'mother country'. The text also refers to a 'promise', making viewers feel they are obligated to help.

c The army in silhouette could be Australian soldiers on campaign in Europe.

3 The various elements all combine in an attempt to persuade the viewer, a citizen of Australia, to vote 'yes' in the upcoming plebiscite to implement conscription. Various persuasive techniques are used: appeals to nationalism, appeals to sense of duty, fear (the red foreboding colour) and even a direct command (note the lack of question mark at the end of the statement).

step 5 I can evaluate a source

Evaluation is a higher-order thinking skill. Evaluation can involve:

- assessing whether a statement or belief is true, such as 'does this image really show that working conditions during the Industrial Revolution were awful?'
- comparing different ideas, such as 'which source is a better example of pre-colonial First Nations life – rock art or shell middens?'
- judging between different things, such as 'were long- or short-term causes more important in causing the outbreak of World War I?'
- making a judgement about the value of something by asking, 'So what?'. Was a source a good or bad thing, or perhaps partially both? A historian may include both positives and negatives to form a balanced view.

An evaluation could involve one or more of these things.

Here is how Source 4 could be evaluated:

Some might ask whether this building represents the domination of the Indian people by the British. In it, we see both Indian and British architectural styles combined. On this reading, the building represents a successful merging of the two cultures.

On the other hand, the fact that this monument was built at all suggests domination. It is a major building in the Indian capital, built just to celebrate a short visit by the far-off monarch of Britain, an 'absentee landlord'. Indian nationalists might consider it negative, as it is a monument to the foreign power that ruled India for so many years.

A more measured approach might state that this gateway is a symbol of the modernity that the British influence brought to India, while still in some way respecting India's traditional culture.

This evaluation combined two of the possible elements of an evaluation: assessing whether a statement or belief is true, and making a judgement about the value of something.

Source 4

'The Gateway of India' in Mumbai, built by the British in 1924 to commemorate the 1911 royal visit

Continuity and change

One reason we study history is to see how life was different in the past. We can also see how an idea or a piece of technology evolved over time.

Continuity and change exist on a scale, between 'no change' and 'completely different', as shown below.

Continuity and change can exist at the same time: some things stay the same, while others change. Change can be fast or slow, happen gradually or in a burst.

Scale of change from least (on the left) to most (on the right)

No change at all (e.g. human DNA)	A bit different (e.g. food)	Quite different (e.g. attitudes to race, gender and sexuality)	Completely different (e.g. transportation technology)

step 1 I can describe continuity and change

Describing continuity and change means that you have to both recognise it and describe what you recognise. You should do this by writing descriptive sentences that use historical evidence to back up your claim.

Civilisation	Earlier time	Later time	Continuity between these times	Change between these times
First Nations Peoples of Australia	60 000 BCE: small number of First Nations People in the north	100 CE: First Nations Peoples had spread across the entire continent	First Nations Peoples lived in tribal groups	First Nations Peoples were spread across the whole of Australia
Imperial China	221 BCE: Qin Dynasty founded	1911: Wuchang Rebellion	China's economy is primarily agricultural	Imperial government and bureaucracy is overthrown

Source 5

A lion-dog statue from China's Forbidden City, which was built in 1420 and still stands today

step 2 I can explain why something did or did not change

Explanations require you to answer the question *why?* When explaining continuity and change, you need to:

- recognise it
- describe it
- know what caused it, or what effect it had.

step 3 I can explain patterns of continuity and change

Examples of patterns of continuity include:

- close family bonds
- the importance of food
- the impact of disease
- natural disasters.

Examples of patterns of change include:

- the improvement in technology
- the rise in population
- the spread of new ideas.

So you would need to explain *how* or *why* one of these things occurred. An example of explaining a pattern of continuity:

> Racism has been common in Australian society from the colonial era until the present. Why is this?
>
> In the 1700s, when Europeans first came to Australia, white people believed themselves to be superior in intellect and work ethic to other races. We now know that none of this is true. There are very few physical differences between 'races'; culture and upbringing account for other differences, not biology.
>
> Still, racism remains. Too many people are accustomed to seeing those that look different as a cause of fear, because they don't understand them.

step 4 I can analyse patterns of continuity and change

Analysing involves breaking a pattern down into its parts. If you can identify these different parts, and explain how together they make up the whole, you are analysing.

To analyse a pattern of continuity or change, you need to consider the different elements of the pattern and see how they interacted over time to produce an overall outcome.

An example of analysing a pattern of change is as follows:

> A major change in Australia's foreign policy since 1900 has been the weakening of ties with Britain.
>
> It began during World War I, when the mishandling of the Gallipoli campaign by British commanders led to defeat. Australians felt their military forces should be led by Australians. WWI also saw the end of Britain's dominance of the globe, replaced as global leader by the USA.
>
> After WWI, but especially after WWII, Britain began to integrate more closely with the rest of Europe and weakened ties with its former colonial possessions. At the same time, US culture became influential in Australia.
>
> Ultimately, the effect of these changes was that Australia became less politically entangled with Britain, even though the Queen remains Australia's head of state.

Source 6

Australian soldiers and American sailors socialise in Brisbane during World War II. [Photo from Australian War Memorial 008871]

I can evaluate patterns of continuity and change

Evaluation is a higher-order thinking skill. It involves remembering what you have learned and being able to use that information to make sense of something or make a judgement.

Evaluation can involve:

- assessing whether a theory or belief is true; for example, 'Is it true that all empires rise and fall?'
- comparing different ideas; for example, 'Which is a better job: working as a farmer or working as a miner?'
- judging between different things; for example, 'Some people think British rule was positive for the Indian people, while others think it was negative: who is right?'
- asking 'So what?'; for example, 'Was it a good or bad thing? Or was it both good and bad?'. A historian might include both positive and negative aspects to form a balanced view, as shown in this table. (See the key in the next column for an explanation of the colours.)

Each balanced view in the table contains four elements:

1. A statement about whether the situation was a continuity or a change.
2. A statement showing the positive aspect of it.
3. A statement showing the negative aspect of it.
4. A statement summarising the balance.

There is no 'right' answer when evaluating. Having a balanced view is not always the best answer, either. For example, it would be impossible to argue for the benefits of Hitler's policy to exterminate specific groups of people. However, some answers are better or worse than others. Better answers:

- use more historical evidence as examples in their evaluation
- use more logical reasoning – they show directly how beneficial the patterns of change were.

Situation	Positive thing	Negative thing	Balanced view
Continuity: The Japanese policy of *sakoku* forbade foreigners from entering the country	The policy protected the Japanese from foreign pirates and preserved their culture	Japanese technology was less advanced than that of industrialised nations	The policy of *sakoku* kept Japanese society isolated for more than 200 years. Foreign pirates were prevented from entering Japanese waters, and Japanese culture did not become diluted by Western ideas or religions. However, because it was not part of the Industrial Revolution, Japanese technology was quickly outmatched by that of Western nations. Ultimately, *sakoku* weakened Japan and left it vulnerable to the political and military manoeuvring of the United States and other nations.
Change: The restrictions on race in Australia's *Defence Act* were amended in 1917	Some First Nations men were able to earn a living as soldiers	First Nations veterans of WWI were treated badly at home	Because of the demands of WWI, the Australian government amended the 1903 *Defence Act* to allow some First Nations men to enlist. These soldiers earned better wages than they could at home, and were treated with respect. However, many died in battle, and the survivors were not given the same aid and benefits as white veterans. Overall, though, this was a positive step towards First Nations People becoming citizens and having the same rights as all other Australians.

Cause and effect

Cause and effect is visible every day. For example, when we open the fridge, the light comes on; when we wave our hand, the bus stops for us.

There are short-term and long-term causes. The short-term cause of the fridge light coming on is opening the door. The long-term cause is because we are hungry. Equally, there are short-term and long-term effects. After I wave my hand, the short-term effect is that the bus stops. The long-term effect is that I arrive at school on time.

Source 7

Nathaniel Dance-Holland, *Captain James Cook* (1775–1776), oil on canvas

Cause and effect requires understanding which events are linked and why. When we say things are linked by cause and effect, we say they have a *causal link*. This means that one thing *caused* the other.

Most things that happen have multiple causes and effects – some of which are more important than others. Two main types of causes are:

- historical actors: the individuals or groups involved; for example, Queen Victoria, Franz Ferdinand or Emperor Qianlong
- historical conditions: social, political, economic, cultural and environmental factors; for example, the invention of the steam engine or European political alliances in 1914.

However, just because one event happens after another, it doesn't always mean that the first event caused the second. For example, when a rooster crows in the morning, we don't think it makes the sun rise. You also need to be able to tell an acceptable story about why something caused its effect.

Events in history are not inevitable. When we study cause and effect, it can seem like things were always going to work out in a certain way. Yet, change a few conditions and things could have happened differently. If James Cook had not joined the Royal Navy in 1755, would Australia have been colonised in 1788? It is easy, with hindsight, to think our cause and effect explanations are perfect. But we need to be cautious when we make claims about cause and effect, as many events are unpredictable.

I can recognise a cause and an effect

Recognising cause and effect means correctly choosing from a list of possibilities. For example, which of these is most likely to be a cause of the Eureka Rebellion of 1854, when a group of gold miners erected a stockade in Ballarat to protest against the police?

A Gold license fees were abolished in 1855.

B The goldfields police extorted money from miners and did not protect them.

C The New South Wales government banned alcohol on the diggings.

D The Victorian economy thrived during the gold rush.

E The *Chinese Immigration Act* limited the number of Chinese immigrants allowed to enter Victoria.

The only one of these options that is linked to the conflict is B. Option A refers to a time *after* the Eureka Rebellion, so it can't be the cause. Option C refers to New South Wales, not Victoria, so it is not relevant. Option D and Option E just tell us facts that wouldn't necessarily lead to conflict.

For events to be causally linked:

- one event must come before the other
- you must be able to tell a believable story about why one event caused the other
- if possible, you should have some historical evidence that one event caused the other.

There is not necessarily a right or wrong answer, but there are better or worse answers. Better answers use more historical evidence, and have reasoning that is more logical.

I can determine causes and effects

Determining cause and effect means deciding what the cause or effect of something might be. Knowledge of the period will help with this.

Examples of historical causes include:

- conditions:
  - social
  - cultural
  - political
  - environmental
  - economic.
- actors:
  - individuals
  - groups.

If you suspect that two things are linked, see if one of the above items is the cause.

Type of cause	Example cause	Example effect
Social conditions	Industrialisation reduced the number of weaving jobs	The Luddites smashed looms and machines
Political conditions	Britain lost its American colony	Britain began a new program of establishing foreign colonies
Economic conditions	Famine in Northern China caused economic hardship	Chinese peasants rose up against the Qing dynasty
Cultural conditions	Japan's social hierarchy was set in stone	The feudal system remained in Japan for many centuries
Environmental conditions	Central Australia suffered a centuries-long megadrought	The First Nations Peoples invented sophisticated farming and hunting techniques
Individual actor	Gavrilo Princip wanted Bosnia to be part of Serbia	He assassinated Archduke Franz Ferdinand
Group actor	Britons imported the sport of cricket into India	Cricket became the most popular sport in India

step 3

I can explain causes and effects

Explaining cause and effect involves stating *how* or *why* a cause led to an effect.

Cause	Effect	*Explaining how* the cause led to the effect
The steam-powered train was invented	British people became more interested in politics and current events	The invention of steam-powered trains meant that goods and materials could now be moved faster and more cheaply around the country. This included newspapers, which were printed in London; they became larger and were circulated to rural towns and cities. This greatly improved communication, resulting in the spread of ideas, and people became more interested in politics.
Commodore Perry visited Japan in 1853 demanding trade	Japan opened up to the world and Westernised	Before 1853, Japan isolated itself from the world. When Commodore Perry from the USA came into Tokyo Bay with huge warships and demanded Japan trade with them, it forced Japan to open up to the rest of the world. The Japanese realised how less developed than other civilisations they were and began a modernisation program known as the Meiji Restoration.

step 4

I can analyse causes and effects

Analysing means the ability to break down something into its parts. If you can identify these different parts, and explain how together they make up the whole, you are analysing. If you can explain the rules or theories that show how these parts are organised, you are analysing.

The first step is being able to break something down into its parts. For example, what caused Britain to establish a colony in Australia?

We can break the cause into four parts:

1 Britain had colonies in America.
2 Britain suffered from high crime and unemployment rates.
3 The American colonies rebelled and became independent.
4 Britain needed a new source of income and a place to send convicts.

Breaking down the cause is the first part of the analysis. Now we can try to explain how these combined causes would have led to the colonisation of Australia. For example:

> The British Empire relied upon its American colonies as a source of income from trade, including the slave trade. At the same time, Britain was facing internal disruption from the Industrial Revolution, which led to many people becoming unemployed and some of those people turning to crime. When the American colonies rebelled and won the Revolutionary War, Britain lost both a key income source and a place to send its criminals and troublemakers.

Here we have linked Causes 1, 2 and 3 together.

Source 8

An 1851 illustration of a new steam-train design called 'Ariel's Girdle'. Trains revolutionised communication within Britain.

Another way to analyse cause and effect is to look at how strong the causal link is. Causes can have different strengths.

Small causal link	Large causal link
Gradual, minor, almost no part, short-lived, partly, partial, to some extent, small extent	Radical, powerfully, significant, important, to a great extent, considerable, main, crucial

Once you have decided how strong a cause was, here are some words you can use to describe it.

- If something had only a minor effect you could say:

  Only a small number of Australia's First Nations Peoples were affected by trade with Indonesian Makassars, and only to a limited extent.

- If something had a major effect you could say:

  All of Australia's First Nations Peoples were affected by European colonisation, which had a radical effect on all Indigenous groups.

step 5 I can evaluate causes and effects

Evaluation is a higher-order thinking skill. Evaluation can involve:

- assessing whether a theory or belief is true or not; for example, some historians think Britain's control over India was ultimately good for the Indian people. Is this true?
- comparing different ideas; for example, which is a more important effect of the colonisation of Australia by the British: the destruction of First Nations culture or the creation of the nation of Australia?
- judging between different things; for example, some people think Japan would have Westernised eventually anyway, others think the Americans were vital in opening up Japan to the world. Who is right?
- asking, 'So what?'. Were the causes or effects good or bad, or perhaps a bit of both? A historian may include both positives and negatives to form a balanced view.

Source 9

A McDonald's restaurant in Tokyo. Japan has embraced many Western brands and concepts. Would this have happened even if the USA had not forced Japan to open its borders?

Historical significance

How do we decide what is important to learn about from the past? How do we decide which events or time periods have historical significance?

We can use a model or theory to help us decide. A useful model is Geoffrey Partington's model of significance.

Partington's model states that you can determine historical significance by asking the following questions:

1 Importance: How important was it to people living at the time?
2 Depth: How deeply were people's lives affected?
3 Number: How many people were affected?
4 Time: For how long were they affected?
5 Relevance: How relevant is it to the present?

The tables below show some examples using Partington's model of significance.

Australia's conscription plebiscites	
Importance	They were very important, as conscription would have affected every family.
Depth	People were not affected deeply, because both plebiscites failed.
Number	More than 2 million people voted in each plebiscite; the population was around 4 million people at the time.
Time	The first plebiscite was in 1916 and the second in 1917.
Relevance	They are not very relevant to today.

Rise of *ukiyo-e* as an urban art form in Japan	
Importance	As most citizens lived in rural areas, it was not important to many people.
Depth	It was deep for the artists and citizens who found meaning in this art.
Number	Several million people lived in Japan's cities.
Time	The artform was in use from the 17th to the 19th century.
Relevance	It is quite relevant: *ukiyo-e* is an influential art form around the world.

I can recognise historical significance

Recognising historical significance means looking at a list of events or developments and deciding how important they are. (Significant means important; something worth noting.)

You should have some way of determining significance. You might ask: Was it important back then? Were people deeply affected? Did it affect a lot of people for a long time? Is it still relevant to modern times? The more you answer 'Yes' to these questions, the more significant the event was.

Historical significance is not a black and white issue, as it can be shown on a significance scale like the one below:

Least significant ◄ ► Most significant

James Cook's place of birth | The date Cook died | The Port Phillip colony is built | The federation of Australia

Source 10

More than 30 000 flag designs, such as this one, were considered in 1901. Only one of them was adopted as the Australian flag, and the rest are largely forgotten.

I can explain historical significance

Explaining historical significance means asking *how* or *why* something is important.

Here are some examples of significant and less significant events, based on Partington's model.

	More significant	**Less significant**
Importance	Federation was very important to Australia in 1901. All citizens were affected by it, as it united the different colonies together into a single self-governing nation.	The entries in the competition to select Australia's flag were not very important because there were so many of them, and only one of them became our flag.
Depth	The end of the *sakoku* policy was very significant as it affected everyone in Japan and many people in other countries.	The change in fashions that came with the Meiji Restoration, while important, was less significant, as clothing styles don't affect people as deeply as other aspects of life.
Number	The colonisation of Australia by Europeans is significant because it affected all First Nations Peoples at the time – between 300 000 and a million people.	The trading between Indonesian fishers and First Nations Australians is less significant, as it affected a much smaller number of people.
Time	The Industrial Revolution is very significant because it revolutionised technology and caused changes that still affect us 300 years later.	The invention of drones can't be considered historically significant yet because they have only been around for a few years.
Relevance	The 'Unequal Treaties' that reduced China's influence in Asia are still significant today, because China has gone to great efforts since then to regain its power and influence.	Traditional Chinese clothing is not that important today, because few people dress like that in modern China or elsewhere.

Good explanations of historical significance will discuss more than one of these elements.

Partington's model of significance is just an aid to your thinking. There are other things that could explain whether a historical event was significant. For example, a person might be important if they changed other people's ideas, or provided a good or bad example of how to live. An event might be important if it reveals underlying themes or patterns in history.

step 3 I can apply a theory of significance

Applying Partington's theory would mean looking at a set of events and ranking them against his categories. When applying his theory, use phrases such as those in the table below:

Importance	Issue A was more significant than Issue B because it was more important to people *at the time*.
Depth	Issue A is more important in history, because it affected people more *deeply* at the time than Issue B did.
Number	Issue A deserves the status of historical importance more than Issue B. Put simply, it affected more people.
Time	Issue B has been shown to be less important historically than Issue A, because it didn't last as long.
Relevance	Issue A is a more significant event than Issue B because it is still relevant to the present. It helps us to understand the modern world, whereas Issue B doesn't help as much.

step 4 I can analyse historical significance

Analysing means the ability to break down something into its parts. If you can identify these different parts, and explain how together they make up the whole, you are analysing. If you can explain the rules or theories that show how these parts are organised, you are analysing.

In the next paragraph, we will analyse the significance of the Opium Wars. Our breakdown of the topic provides us with three main points, not just one, and we can make a claim as to how each contributes to significance.

Key:

Main point

Claim

Evidence backing up claim

Summary statement

The Opium Wars of 1839–42 and 1856–60 were historically significant for several reasons. First, China was forced to allow foreign traders into the country, and to grant foreign powers rights to control their own enclaves. This ended China's cultural and political isolation. Secondly, the Wars showed that the technology of the industrialised nations was much more powerful and dangerous than Chinese technology. This provided an incentive for China to modernise and to import Western inventions and concepts. Finally, the Opium Wars left the Imperial court deeply in debt and greatly reduced in political power. This led to more populist uprisings and finally the Wuchang Rebellion. Therefore, the Opium Wars are significant because they ended isolation, pushed China to modernise and contributed to the end of the Chinese Empire.

Source 11

The key world leaders that signed the Treaty of Versailles after World War I – British Prime Minister Lloyd George, Italian Premier Vittorio Orlando, French Premier Georges Clemenceau and US President Woodrow Wilson – were all historically significant people.

Source 12

Signing of the Treaty of Nanking, at the end of the First Opium War. [John Platt, *Signing and Sealing of the Treaty of Nanking* (1846), coloured engraving.]

I can evaluate historical significance

Evaluating can mean asking, 'So what?'. In terms of evaluating historical significance, evaluating could include:

- questioning Partington's model of significance:
  - Perhaps the model suggests that Event A is more important than Event B, but you don't agree. Evaluating would involve you explaining why you think the model of significance doesn't give the right result in this instance.
  - Maybe you think some of the questions about significance are more important than others are. Perhaps you think that relevance to today is more important than how significant it was to people at the time.
- making a judgement call about the worth of important events:
  - Were these events 'good' or 'bad' in some way? When making an evaluation like this, make sure you define what 'good' or 'bad' means in this context.

Questioning the model of significance	Some historians think the number of people directly affected is crucial when determining historical significance, but this doesn't consider the long-term effects of events. Perhaps only 50 000 First Nations Peoples were directly affected by 'Aboriginal protection' policies in the early 20th century. However, the trauma and pain caused by these policies indirectly affected all First Nations Peoples, and contributed to transgenerational trauma that still continues today. Models of significance need to consider the ongoing effects of actions, not just the number of people affected at the time.
Judging the worth of an important event	How important was the Battle of Gallipoli in 1915? Some military historians might consider this battle to be only a minor event, much less significant to the outcome of World War I than the conflicts on the Western Front. However, many Australians see this battle as extremely important, a tragic 'baptism of fire' that helped establish Australian identity and that needs to be remembered. Whether an event is 'important' may be influenced by who is considering it and their own personal context.

Historical interpretations

The way we understand history is always changing. People take sources about the past and make meaning out of them, creating 'history' as opposed to 'facts'. *Interpreting* means to explain the meaning of information; different people will look at the same facts and interpret them differently. Studying **historical interpretations** is called **historiography**.

So why do people have different views? Some possible reasons include:

- they might have different political or economic views, such as favouring communism or capitalism
- as time passes, society changes, so people's view of the past changes too. For instance, overt racism is unacceptable now but was much more common 100 years ago
- new information may have been found
- historians might have different biases or emphases.

There are often debates between historians about the historiography of different events from the past.

step 1 I can recognise that the past has been represented in different ways

Recognising that the past has been represented in different ways involves understanding the difference between:

- historical facts (things that we know from evidence have actually happened)
- history (when we study and interpret these facts).

Different people interpret the same facts in different ways. When a sports team wins, their fans may think the game demonstrated their team's great skill. The losing team's fans might think the game featured incorrect refereeing, bad luck and unhelpful weather. When fans describe the game to others later, they will 'represent' it in different ways – the same facts, but different interpretations.

For example, one view of the Industrial Revolution is that it dramatically improved human wellbeing. Factories made vital products, so more people could afford clothing and bedding. Another is that the Industrial Revolution made human life worse. Work became more boring, as factory labour meant doing the same thing every day, and people were crowded into unhygienic cities to find work. The same facts about the Industrial Revolution have been represented in different ways: one positive, one negative.

Source 13

Were the textile factories of the Industrial Revolution a positive or a negative thing?

I can describe historical interpretations

Describing historical interpretations is difficult because you have to recognise and then describe them, not just notice them.

Once you have recognised one or more historical interpretations, you must then write descriptive sentences. Description involves writing at length, not just listing things or using dot points. It is also a good idea to use evidence from the interpretation to support your description.

For example, during World War I, the Anzacs fought at Gallipoli. There are different historical interpretations of this military campaign. Some thought it showed Australian soldiers to be brave and noble, yet down-to-earth. Others had a different view:

> '... I now write of the Dardanelles expedition ... It is undoubtedly one of the most terrible chapters in our history ...
>
> 'Some of the finest forces on the peninsula were used in this bloody battle [on August 21] ... They and other troops were dashed against the Turkish lines, and broken. They never had a chance of holding their positions when for one brief hour they pierced the Turks' first line; and the slaughter of fine youths was appalling ... to fling them without the element of surprise, against such trenches as the Turks make, was murder ...
>
> '... for the general staff, and I fear Hamilton, officers and men have nothing but contempt. They express it fearlessly ... What I want to say to you now very seriously is that the continuous and ghastly bungling over the Dardanelles enterprise was to be expected from such a General Staff as the British Army possesses, so far as I have seen it.'

Source 14

Extract from 1915 Gallipoli letter from Keith Arthur Murdoch to Prime Minister Andrew Fisher

Here is how this historical interpretation could be described:

Keith Murdoch's interpretation of the Anzac campaign was negative. He points out that the commanders who ordered the attack made an error, suggesting that they committed 'murder'. Murdoch also believes that this incident led Australian soldiers to have less respect for their commanding officers, having 'nothing but contempt' instead.

I can explain historical interpretations

Explanations require you to answer the questions 'Why?' or 'How?'. When explaining historical interpretations, you need to do one or more of these things:

- explain *why* what someone has written is a way of looking at historical facts, not just them listing those facts
- explain *why* a person interprets history in that way by referring to their political views, upbringing, biases or the new evidence they have access to
- explain *how* a person has interpreted historical facts.

Consider colonialism, which is the practice of taking over and controlling other lands. Britain was a major colonial power for many centuries. Some interpret this history as positive and some as negative. Source 15 contains one interpretation of colonialism, written in response to statements that Britain's influence was positive.

'Such responses demonstrate a profound ignorance about British imperial and colonial history, particularly about the impact of empire on not only the colonised but also the colonisers as well. But it is a state of denial about empire [...] To say that empire had "good bits" is to deny what empire entailed – namely the conquest, subjugation and exploitation of millions of people.

'It is to erase the tremendous structural and symbolic violence that empire unleashed. To praise Britain's role in abolishing the slave trade is only possible if we deny the various forms of economic, political, social and cultural violence that enabled the perpetuation of such a trade – in Britain and its empire – as well as the ongoing legacies of such forms of violence. To view empire as having "good" and "bad" bits also entails viewing the past in simplistic terms. And to claim students should only study the "good bits" of the past also begs the question: whose "good bits", exactly?'

Source 15

Extract from 2018 article, 'British Empire is still being whitewashed by the school curriculum' by historian Deana Heath

This historical interpretation could be explained like this:

> Historian Deana Heath is critical of British Imperialism and how it is taught in schools. She believes that a lot of history education only teaches the positive things about the British Empire, and not the negative things.
>
> In fact, she thinks splitting the Empire's actions into 'good' and 'bad' is too simplistic, because they must be seen together. For example, without the suppression of India's population, the modernisation of India would not have been possible.
>
> She writes like this because she is looking back into the past with a different ideology – that of equality and human rights – which was not as strong when the British Empire was at its peak.

step 4 I can analyse historical interpretations

Analysing is breaking something into its parts and explaining how they are linked or make up the whole. If you can explain the rules that show how these parts are organised, you are analysing.

When analysing historical interpretations, you could break down:

- the different reasons the writer decided to interpret the facts in that way
- the different parts or aspects of the interpretation
- a combination of the two.

The first step is always to break the interpretation down into its parts.

For example, in Australian history, some see the European colonists as 'settlers' and others see them as 'invaders'. We see this in the different ways of thinking about Australia Day. Some people call it 'Invasion Day' because it celebrates the day Europeans arrived and took land and resources from the First Nations Peoples. Here is one interpretation of Australian history.

'In this speech, Rudd turned his attention from history to the History Wars. He told us that he has no sympathy at all for those who have "refused to confront some hard truths about our past, as if our forebears were all men and women of absolute nobility, without spot or blemish". [...] he told us that he also has no sympathy for those who think we should only celebrate "renegades", "reformers" and "revolutionaries", while "neglecting" or even "deriding" the "explorers", "pioneers" and "entrepreneurs".'

Source 16

Analysis from 'The History Wars', Robert Manne, *The Monthly*, November 2009

Here is how you could analyse this historical interpretation:

In this text, the author argues that Kevin Rudd believes Australians need to accept the negative aspects of our history. Rudd is a left-wing politician, which means he is likely to be critical of traditional or conservative understandings of history. He is also well educated, and has learned about the violent events that happened in Australia's past. Rudd was also the politician who gave 'The National Apology to the Stolen Generations'; in this speech, he called for action to be taken to repair the damage caused. These three things together show why Rudd interpreted history and spoke the way he did.

Source 17

The British Empire established India's railway network, but also denied Indians their right to self-government. Bharatpur station, Rajasthan, India.

step 5 I can evaluate historical interpretations

Evaluation is a higher order thinking skill. Evaluation can include:

- assessing whether a theory or belief is true or not. Some historians think the Industrial Revolution could only have happened in Great Britain. Is this true?
- comparing different ideas. Which was the more significant factor in the abolition of slavery – the changing values of everyday people or the actions of a few passionate abolitionists?
- judging between different things. Some historians think Australia's contribution in World War I didn't have much impact, while others think Australia's support was vital. Who is right?
- asking 'So what?'. Were the causes or effects good or bad, or perhaps both? A historian may include both positives and negatives to form a balanced view.

Situation	Positive thing	Negative thing	Balanced view
British Imperialism in India	The British built railways and introduced a common language.	The British took India's natural resources and didn't let Indians rule themselves.	The British rule in India had benefits and disadvantages. On the one hand, the British introduced democracy, a common language and built thousands of kilometres of railway tracks. On the other hand, they also ruled over the Indians, denying them independence, and took many of their natural resources to enrich the British Empire.
Japan's modernisation after being confronted by US warships	Japan was able to copy technology from the West and modernise very quickly.	Japan's independence was challenged and some of its culture was lost as it copied the West.	When Japan modernised during the Meiji Restoration, parts of Japanese culture were lost as the country sought to imitate the West. However, the country benefitted from being able to copy Western technology and institutions, which allowed it to modernise much faster than other countries.

Historical writing

I can identify the writing purpose

If you are given a writing task, it will usually involve certain 'task words', such as *analyse, argue* and *compare*. These task words are explained below.

- *analyse*: look at the features of something, showing the relationships between the parts, how they're related and why they're important
- *argue*: make a case for or against something
- *compare*: discuss two things, emphasising what is the same and what is different between them
- *contrast*: discuss two things, emphasising what is different between them
- *describe*: write a detailed description of something, showing what something looks like, what it is for and how it works. Don't judge.
- *discuss*: write about something, talking about the arguments for and against and issue. Provide a balanced description, but make a judgement at the end.
- *evaluate*: make a judgment about something, but back it up with lots of evidence.
- *explain*: answer the question 'Why?' about something. Go into detail about the reasons for it, causes of it and effects of it.
- *justify*: provide reasons why a decision was or should be made, or why a conclusion was reached.
- *summarise*: briefly state the main points. Leave out the details.

After you know your purpose, figure out:

- *what kind of information you need to gather.* This relates to Stage 2 of the history-writing process: gathering information. Gather the right kind of information – but avoid gathering lots of irrelevant material.
- *how that information should be organised.* You will eventually need to write up any information you have gathered. How you do that – and what structure your writing takes – should be determined by the purpose of the writing.

Here is an example history-writing question, and how it can be tackled: 'Discuss how important Bal Gangadhar Tilak was in the political campaign for Indian independence.'

This is a 'discuss' type question, so it is asking us to write about the topic, discussing arguments for and against it. You should discuss both sides but also make a judgement.

Information needed to answer the question:

- details about the Tilak's political activities (but *not* details about every aspect of his life)
- details about the campaign for Indian independence. Gather information about the history of the campaign, especially the period before or during World War I.

How should that information be organised? A graphic organiser like the one below is a great way to structure your note-taking.

	General information	Evidence Tilak was important in the campaign	Evidence Tilak was not that important in the campaign
Tilak's political activities	○ ○ ○	○ ○ ○	○ ○ ○
The campaign for Indian independence	○ ○ ○	○ ○ ○	○ ○ ○

I can gather information

Good history writing involves providing lots of evidence. The more *relevant* information you use, the better. Relevant means the information is closely connected to what is being studied.

Gathering information will involve taking notes from historical sources. Academic historians look at many primary and secondary sources. For most school projects, you are likely to rely on secondary sources. Secondary sources provide a wide range of easily accessible information for young people. Textbooks and reference books are easy to obtain, relatively cheap, easy to read and contain pictures, facts, explanations and examples.

Follow these steps when taking notes for your history writing:

1. Purpose: *why* am I taking notes?
2. Organise: use a graphic organiser or codes
3. Skim-read the source. This is so:
   - you can look for topics, headings and so on
   - you *don't* have to read the entire source.
4. Find the *most* important information *for your purpose*:
   - rewrite it in your own words
   - write as briefly as possible
   - include keywords, and definitions of any words you don't know.

Source 18

A 1920s Japanese woodcut of a woman by Suzuki Harunobu, printed as a postcard, showing the art styles of the Taisho Period

I can organise information

Two ways to organise your information are to use a graphic organiser or use codes.

A graphic organiser is best for:

- when you know in advance the kind of information you will be taking notes about
- when the question you are answering has obvious parts to it that you can divide information into; for example, for and against.

Using codes is a process that involves:

- taking notes
- reading through your notes several times and seeing what patterns, themes or categories emerge
- making up a code for each pattern, theme or category; for example, 'W' for *war*; 'I' for *individual*; 'WWI' for *World War I*
- going through your notes and writing the code beside each point
- rewriting your notes in the code categories (This is much easier if you have taken notes electronically, because you can change their order without having to rewrite them.)
- using your notes in their coded categories to form the basis of your essay structure.

With either of these methods, don't forget to ask yourself which notes you should *not* use. You will always take notes that you thought were important but later realise don't actually matter. Get rid of them. Remember: the final written piece is what is most important, not your notes. Don't worry that you spent time writing those notes in the first place, because only your final piece of writing matters. Next time, try to take fewer irrelevant notes.

Here is an example of the process, with all the steps from note-taking to organising your notes.

Essay question: 'Did the Taisho Period bring stability and prosperity to Japan?'

1 Original notes

- Japan had three different Prime Ministers in 1912–1913 due to political infighting.
- Japan entered WWI as an ally of Britain against its enemies China and Germany.
- After WWI, Japan gained respect as a world power and annexed territories in the Pacific.
- Artistic, political and cultural movements thrived in urban centres.
- Women gained more rights and social status.
- Debts and food prices skyrocketed in rural areas.
- The *Kome Sodo* Rice Riots of 1918 led to martial law.
- The *Peace Preservation Act* of 1925 banned anything dangerous to 'the national essence'.
- Rapid industrialisation – Japan went from 29 km of train tracks in 1872 to more than 11 000 km by 1914.
- The policy of *o-yatoi gaikokujin* brought in foreign experts.

3 Notes not needed:

- Japan entered WWI as an ally of Britain against its enemies China and Germany.
- Rapid industrialisation – Japan went from 29 km of train tracks in 1872 to more than 11 000 km by 1914.

4 You could then put your notes into paragraphs:

- Taisho Period brought stability and prosperity: social change in cities, greater involvement in world affairs
- Taisho Period did not bring stability and prosperity: riots and food shortages in rural areas, increased government control

2 Put into a graphic organiser, the notes look like this:

	Taisho Period brought stability and prosperity	Taisho Period did not bring stability and prosperity
Social change	○ Artistic, political and cultural movements thrived in urban centres. ○ Women gained more rights and social status.	○ Debts and food prices skyrocketed in rural areas. ○ The Kome Sodo Rice Riots of 1918 led to martial law.
Political actions	○ After WWI, Japan gained respect as a world power and annexed territories in the Pacific. ○ The policy of *o-yatoi gaikokujin* brought in foreign experts.	○ Japan had three different Prime Ministers in 1912–1913 due to political infighting. ○ *The Peace Preservation Act* of 1925 banned anything dangerous to 'the national essence'.

step 4 I can structure a piece of writing

History essays should have an introduction, several body paragraphs and a conclusion.

When you are starting out writing essays, a paragraph structure that is easy to learn is **TEEL**. TEEL is an acronym for:

- **T**opic sentence
- **E**vidence
- **E**xplanation
- **L**ink.

These words are explained below. Every paragraph should have TEEL.

Introduction

The introduction should:

- show you understand what the question is asking
- say your overall response to the question
- introduce the main points.

Paragraphs using TEEL

Paragraphs using TEEL should include:

- Topic sentence: one sentence that summarises the whole paragraph
- Evidence: use *specific* examples, not general examples
- Explanation: how evidence supports your claim
- Link: at the end of the whole paragraph, link the main point in the paragraph back to the main question.

Conclusion

In your conclusion, make sure to:

- summarise your main points
- restate your response to the question.

step 5 I can write a draft

Drafting tips

- Focus on answering the question; don't just write everything you know about the subject.
- Don't worry about making mistakes when drafting – you will fix this later.
- Don't worry too much about punctuation, grammar or spelling when drafting.
- Start with the paragraphs. Draft the introduction and conclusion last.
- If you can, use a computer, as it makes it easier to edit and proofread your work later.
- Write the first draft quickly. Then edit and proofread slowly.

Sentence starters

Here are some sentence starters for introductions:

- This essay will discuss ...
- This essay will focus on ...
- The issue being focused on is ...

Other words you could use in sentence starters in place of *focused* are:

- described
- analysed
- evaluated
- explained
- explored
- justified
- outlined.

For conclusions, some starters include:

- In conclusion, ...
- In summary, ...
- It has been shown/demonstrated that ...
- Therefore/Thus/Hence, ...
- To summarise, ...

For comparing within your answer (when things are the same):

- By comparison, ...
- In the same way, ...
- Likewise, ...
- Similarly, ...

For comparing (when things are different):

- However, ...
- In contrast, ...
- On the other hand, ...
- Then again, ...

For adding more:

- Additionally, ...
- Also, ...
- First, ... Second, ... Third, ... Finally, ...
- Furthermore, ...
- In addition, ...
- Moreover, ...

For giving examples:

- For example, ...
- For instance, ...
- An illustration of this is ...
- As an example, ...
- ... such as ...

For showing effects:

- As a result, ...
- For this reason ...
- It can be seen that ...
- The evidence suggests ...
- The result of this is that ...
- These factors contribute to ...

Different ways to say 'caused':

- resulted in
- created
- lead to
- determined
- is attributed to
- meant that
- is dependent on
- forced
- made.

step 6

I can edit and proofread

The point of writing is to communicate – using words to pass ideas from you to another person. So, keep your writing clear and simple.

Editing means checking for meaning, to make sure your text answers the question and meets the task requirements. For example, does your writing need a bibliography? Does it need labelled pictures? Proofreading means checking the grammar, spelling and punctuation of your work. Always edit first, then proofread.

Editing tips

- Always use headings, unless told otherwise.
- Delete any words, sentences or paragraphs that do not help the piece of writing overall.
- Check what you have written against the requirements of the task. Ask yourself:
  - does my writing answer the question asked?
  - is it clear that I have done the full task?
  - is there an assessment schedule or rubric my writing will be marked against? Mark yourself against these criteria. Is there time to improve at least one aspect of what you have written?
- What is your worst paragraph? Why? What would it take to make it your best paragraph?

Proofreading tips

Proofreading is going back over your finished work and looking for errors.

- Don't try to fix every problem at once. Pick one thing to correct each time you proofread. For example, first look at spelling, then look at punctuation, then look at confused words, then look at making your vocabulary more interesting.
- Read your work aloud. Even better, record it, then play it back to yourself a bit later.
- Ask someone else to read your work aloud.
- Read sentences backwards to check for mistakes. This will help you pick up more errors.
- If you know you are a bad speller, don't trust your instincts. Check words you are unsure about in the dictionary.
- Spellcheckers are not perfect. A word can be spelled correctly but still be the wrong word, so don't rely on a spellchecker!
- Read a printed copy of your work, rather than reading on screen.

Common errors

Following is a list of common errors:

- only use apostrophes for shortening words and ownership, not for plurals
- write short sentences, preferably less than 25 words
- only use capital letters at the start of sentences and for proper nouns
- confusing 'your' and 'you're'
- confusing 'there', 'their' and 'they're'
- writing informally: don't use 'I' (unless told to), '&', 'etc.', 'e.g.', 'i.e.', 'wanna', 'heaps', 'stuff'
- could of / would of / should of are incorrect; replace with could have / would have / should have
- confusing 'to', 'too' and two'
- confusing 'much/many': much = for an item that can't be counted (e.g. water), many = individual item that can be counted
- then = something happening after something else; than = comparing
- subject–verb agreement. If the subject is a plural, the verb must be too, for example 'towels *are* in the closet'
- be careful about starting a sentence with 'and', 'but' or 'because'
- a full sentence should have a subject (doer), verb (action) and an object (the thing the verb is happening to)
- use the same tense (future, present, past) in the whole text
- avoid using boring words: very, good, bad, amazing, interesting, crazy, mad, extremely
- avoid 'passive' sentences. Instead of 'The warships were commanded by Commodore Perry', write 'Commodore Perry commanded the warships'.

Historical research

I can define the problem

To define the subject you will research, get some background information and build up a list of keywords.

Start by reading a simple Wikipedia page about your subject.

Get keywords for your topic. Think of different ways of saying your topic, or google 'synonyms for ...' and insert your search term.

I can decide what information to find and where to find it

What type of evidence do you want?

Include these kinds of words in your search:

- facts, examples, definitions, quotes, artefacts, images, data, statistics
- primary and secondary sources
- databases, links, archives, collections, references, research, museums, journals, graphs, tables, letters.

Where is your evidence?

There are many different types of websites to look at: scholarly works, databases, archives, reference sources and information pages.

How credible is the evidence?

Ask yourself the following questions:

- Is the content *relevant*? Is it useful for my purpose? Does it contain links to other relevant sources? Is it at an appropriate reading level?
- Is the source *reliable*? What type of source is it? (Published or official sources are better.) Who is the author? (Experts are better.) When was it published? (Newer is usually better.) In what way is the source biased?
- Is the source *true*? Is it backed up by other sources? Does it *sound* right? Does it fit in with other things you know?
- Does the source state where its information comes from? This means it is more likely to be credible (able to be believed).

I can find information

Online search strategies

Following are some search strategies:

- After you type in a search term, scan through the first page of results. If they are not relevant, change your search.
- Start with a wide search, then get more specific.
- Learn *from* your search. Change what you are searching for based on what you learn after you start searching.
- Be ready to stop a search if it is taking you in the wrong direction.

Tips for searching with Google

- Every word matters
- The order of the words matters
- Capitalisation doesn't matter
- Punctuation doesn't matter
- Specific search terms are better

- Use these capitalised terms to narrow your search: AND, OR, NOT.
- A search with 'filetype:' will find specific files. For example, 'Meiji Restoration filetype:ppt' will find PowerPoint files about the Meiji Restoration.
- A search with 'site:' will find things *within* a website. For example: 'World War One site: britishmuseum.org' will find WWI-related material from the British Museum website.
- Use Google's subject tabs to search by category, such as images, news and maps.
- Use a hyphen to exclude words and narrow your search. For example, adding a hyphen to the search 'Danish -pastry' will find information about Danish people and culture but exclude the term pastry, to avoid finding information on Danish pastries.
- Search for a range of numbers using two full stops between speech marks: '..' For example:
  - '2001..2004' searches between 2001 and 2004
  - '..2004' searches before 2004.
  - '2004..' searches after 2004.
- An asterisk acts as a wildcard. So, for example, 'teen*' will return results with any of the words *teen, teens, teenager* in them.
- Use exact phrase searching by putting speech marks around a search to find exact text.

step 4 I can extract information

This note-taking stage is the same as Step 2 in the Historical writing section. Read that section on page 260.

step 5 I can organise and present information

This stage is very similar to Step 3 in the Historical writing section. Read that section on page 260.

Research will be presented in a number of different ways, and will usually include some history writing. History writing is generally presented as text with perhaps some supporting pictures.

You should also edit and proofread your research, just as you do with your writing. Read Step 6 from the Historical writing section on page 263.

step 6 I can evaluate information

You can improve every time you conduct research by asking yourself these questions after you finish:

- What worked? What didn't work?
- How could I work smarter next time?
- Can I apply what I've learnt to other situations?
- How could I have improved:
  - the project?
  - the way I worked on my project?
  - the way I managed my time?

Glossary

abdicate/abdication to formally give up a position of power

annex the act of one state claiming sovereignty over another state's territory

antiquated old-fashioned or outdated

arithmetic change/growth change or growth that occurs at a constant rate over a set period of time; for example, 1, 2, 3, 4 ...

armistice an agreement to lay down arms and stop fighting

artillery large weapons capable of launching heavy munitions much further than traditional guns

assimilation replacing the language and culture of a non-dominant social group with those of the dominant social group

autocratic a type of government rule in which power is held by one person

autonomy self-government

bias to show preference for or prejudice towards something

bayonet a bladed weapon, generally a knife or sword, attached to the end of a rifle, allowing it to be used as a spear in close combat

blackbirding the process where Pacific Islanders were duped, kidnapped or coerced into boarding ships for Australia, only to end up working on the sugar and cotton plantations of Queensland for extremely low wages

blockade sealing off an area or region to prevent the flow of people and goods

bond a loan taken out by a company, where investors buy the bonds; in exchange for the **capital**, the company pays an annual interest rate on the bond

boycott to intentionally abstain from an activity, from using or buying a product, or from interacting with a person, group, state or country usually for moral, social or political reasons

capital wealth in the form of money or other assets that is available for investing

capitalism a political and economic system in which private individuals and companies own property and goods, rather than the government; the companies or individuals then compete to make a profit

caste a class structure that is determined by birth

colonise to settle in, and take control of, an area away from one's home territory

communism when a nation is run by an authoritarian government featuring a planned economy with equally shared resources

conciliator a person who mediates or negotiates between two groups in a dispute

concubine in 19th-century China, a woman who lived in the imperial palace whose job was to keep the emperor entertained; it was considered an honour to be an emperor's concubine

conscription a system where eligible people (usually young men) are forced to enter the army for a certain period

conscripts people who are conscripted into a military force

constitution the basic laws and principles of a nation or state

constitutional monarchy a democracy that also has a monarch; the monarch's role is mainly ceremonial, while true power lies with the government

democracy a political system that allows each individual to participate; most democracies are republics – states where the people elect representatives to form and manage the government

despot a ruler who wields absolute power

dictatorship when a nation is run by an authoritarian government where a single individual rules the country and makes all of the decisions

dispossession the act of depriving someone of land, property or other possessions

dowager a woman who holds a title or property from her deceased husband, especially the widow of a king or emperor

dynasties systems of rule where one family maintains power over many years by handing on the throne to an heir, usually the oldest son

economist a person who studies how we produce and consume products and how wealth is established and distributed

electorate a geographically defined area represented by a single member of parliament

emancipist a former convict who had completed their sentence or been granted a pardon

enclosure where land was divided into farms surrounded by fences or hedges

entente a French term meaning 'friendly understanding'

entrepreneur a person with the ability to organise the factors of production and transform them into a business

envoy a messenger or representative

expansionist a policy of territorial expansion (the act of becoming larger)

expatriate a person who lives outside their country of birth

exponential change/growth change or growth that occurs more and more rapidly over a set period of time; for example, 1, 2, 4, 8, 16, 32 ...

fixed rate the interest rate on a loan, where the rate does not increase or decrease during the fixed rate period of the loan

floating rate the interest rate on a loan, where the interest rate increases or decreases over the life of the loan according to market conditions; also known as a variable rate

foxhole a hole in the ground used by soldiers to provide cover from gunfire

genocide the deliberate killing of a large group of people because of their nationality or ethnic group. The United Nations defines five acts that are committed with intent to destroy the whole or a part of a national, ethnic, racial or religious group including killing, causing physical or mental harm, preventing births or transferring children.

globalisation when businesses and other organisations interact and integrate with businesses, organisations and people worldwide

global warming the unnatural rise in Earth's temperature linked to the increase of fossil fuel use since the Industrial Revolution

guarantor a financial term describing a person who promises to pay another borrower's debt if the borrower defaults on the loan (cannot keep up with loan repayments)

guerrilla a member of a small group taking part in impromptu fighting, typically against a larger more-organised force

handicrafts objects made by hand

historian a person who specialises in the study of history by using evidence to answer questions about the past

historiography the study of the methods of history and historical interpretations

home front the civilian population and daily activities of a nation whose armed forces are fighting overseas

honorific a title or word used to show respect

House of Representatives the lower house of the bicameral Australian Parliament; it is made up of 151 democratically elected members who act on behalf of their electorate to pass new laws or make amendments to existing laws

humanistic describing a system of thought with primary importance attached to human rather than religious matters

imperialism the policy or ideology of extending a country's power and influence, usually through colonisation or military force

indemnified compensated or cleared of blame

indigo a shrub with red or purple flowers, used as a source of indigo dye

industrialise to introduce industry to an area; to use machinery to do things previously done by hand

indigenous the original or earliest known inhabitants of an area in contrast to groups that have colonised more recently; also known as First Peoples, First Nations Peoples or aboriginal peoples

infanticide the killing of an infant

innovation the creation, development and implementation of a unique idea, product, process or service

isolationism a policy of not engaging in the affairs of other nations

manufacturing the making of products for use or sale using labour and/or machinery

mechanisation the use of machines to replace, wholly or in part, the work of humans and/or animals

member of parliament (MP) a member of the House of Representatives in the Australian Parliament, who acts on behalf of their electorate

Men's Business when a council of male Australian First Nations Elders comes together to initiate, teach and regulate spiritual and legal matters

mercantilism a system of profitable trading based on the idea that increasing exports would increase a nation's power

merchant a person or company involved in wholesale trade

messiah the holy saviour of a cause or people

microhistory a genre of history centred on small and focused research, such as interviewing an individual to gain a perspective on the big picture

militarism the idea that a country should maintain a strong army to defend or promote its national interests

millennium, millennia (plural) a period of one thousand years

moiety a system of kinship used by First Nations Peoples of Australia to divide the community into groups that reflect environmental, regional, matrilineal (through the mother), patrilineal (through the father) and generational factors

monarch a sovereign head of state; usually a king, queen or emperor

monarchy where one person inherits the position of head of state, and is the final word in government; a constitutional monarchy

monetisation earning revenue from something; e.g. a product or a service

nationalism/nationalist extreme loyalty to your nation-state, its people and culture, and involves promoting these interests ahead of other nations

oligarchy control of a country by a small group of people

outsource to contract work out to a person (or another company) that does not work in-house for the company requiring the work

paddle steamer a boat powered by a steam engine that turns large paddle wheels

peninsula a coastal landform projecting out into a large body of water, such as a sea or an ocean

phenomenon something that is observed to exist or happen

plebiscite a non-binding vote on a proposed law, generally used to gauge public opinion

political party an organisation that represents a particular group of people or a set of ideas, values or philosophies

populist/populism a political stance that emphasises the (morally good) 'people' against the (morally corrupt) 'elite'

primary source a source that was created or existed at the time under study, such as a book, a letter, an artefact or a building

privy a toilet located outside a house

Raj a Hindi word meaning 'rule' or 'government'; used to describe the British government in India

referendum, referenda (plural) a vote by the electorate on a specific political question posed by the government

regent a person appointed to temporarily administer a state because the monarch is currently too young

republic a state or region ruled or governed by the people or their elected representatives, independent of a monarch

restitution the return or replacement of something lost or stolen

revolution a radical change to an established system or process

savings account a personal bank account where money is deposited, sometimes with a limit on the number of withdrawals

secondary source a source created after the time under study, such as a textbook, website or documentary

sedition speaking out or inciting protest against a government or monarch

Senate the upper house of the bicameral Australian Parliament; it is made up of 76 democratically elected members who represent the views of their constituents, debate and vote on bills and scrutinise the work of the government

senator a member of the Australian Senate, elected to represent a state or territory

skin name a system of kinship used by First Nations Peoples of Australia that relates to a person's genetic matrilineal or patrilineal lines

socialism a political and economic system in which communities control businesses and industry for the benefit of their members

socialist someone who advocates for or believes in socialism

sovereignty the authority of a state or nation to govern itself independently of other nations

status quo a Latin phrase meaning 'the existing state of affairs'; usually relates to politics

stockade a defensive barrier or fence

subversive something that is intended to reverse the principals, morals or norms of a current system or system of thought

suffrage the right to vote in political elections

suffragette women who participated in direct or even violent protest to demand voting rights

suffragists women who advocated for the right to vote rather than engaging in direct protest

republic states where the people elect representatives to form and manage the government

temperance movement a social movement that promoted moderation in, and often complete abstinence from, the consumption of alcohol

totem a system of kinship used by First Nations Peoples of Australia that includes national, clan, family and personal factors

trade union an organisation that protects and promotes the rights of a specific group of workers

transgenerational trauma psychological pain that affects multiple generations

trench a narrow excavation that is generally deeper than it is wide; during World War I, multiple trenches were often connected in zigzag patterns and were used extensively to provide cover and shelter from gunfire

urbanisation a process where an increasing number of people relocate to cities from the countryside in search of work

vassal a person who or kingdom that has a mutual obligation to a lord or monarch, often including military service in return for privileges such as holding of land

Women's Business when a council of female Australian First Nations Elders comes together to initiate, teach and regulate spiritual and legal matters

zamindar in pre-colonial India, a landowner or landlord, who leased his land to tenant farmers and collected tax revenue

Index

Aboriginal and Torres Strait Islander Peoples *see* First Nations Peoples
Aboriginal Customary Law (ACL) 63, 67, 86, 99
aeroplanes 10, 11
Africa, slaves from 46
Age of Enlightenment Scientists 14, 24
Age of Imperialism 136, 138–9
  in China 141, 156–73
  in India 141, 142–55
  in Japan 141, 174–85
  timeline 140–1
Age of Reason 14, 24
agricultural changes
  China 166
  India 144
agricultural revolution 21–2, 87
air pollution 48–9
aircraft 214
airships 214
All India Home Rule League 153
alliances, pre-World War I 198–9
Allies 204, 234
  restitution from Central Powers 234
  Western Front 222–3, 232, 233
Allies Day 227
American Revolution 87
Amritsar Massacre 154
analysing 242–3, 245, 249–50, 257–8
Anglo-Japanese Alliance 181
anti-conscription campaign/anti-conscriptionists 209, 226, 229
anti-immigrant sentiment 113, 121
anti-slavery campaign 47
Anti-Transportation League 109
Anzac Day 227
Anzacs
  at Gallipoli, Turkey 220–1
  deployment 219
  formation 219
  on Western Front 223
Aotearoa (New Zealand), circumnavigated by Cook 85
arithmetic change 10
Arkwright, Richard 26, 27
armistice 233, 234
artificial intelligence 37
artillery 212, 215
Asia in the Age of Imperialism 135–84
  see also China; India; Japan
assimilation policy 108, 109, 128, 129
Australia
  8 Hours Movement 53
  after the war 235
  colonial period (1770–1901) 82–119
  Federation (1901–1914 and beyond) 120–31, 193
  First Nations Peoples *see* First Nations Peoples
  gold and industry 52, 110–15
  government in 124–5
  how it obtained its name 81
  immigration policy 113, 121, 126–7
  impact of Industrial Revolution on 50–3
  independent colonies 120–1
  industrialisation 50–1, 193
  as part of British Empire 193, 206
  perceived by British as *terra nullius* 70, 81, 86–7
  political position, pre-war 193
  pre-World War I 192–5
  railways 52, 121, 194
  rise in stature and prominence as a nation following the war 235
  as a sacred country (1750–1770) 58, 60–81
  society, pre-war 194–5
  strong ties with Britain 192, 193, 206
  suffragists 42
  technological changes, pre-war 193
  unemployment, post-war 235
  wool industry 52, 93
  working conditions 53, 194
  World War I *see* Australians at war (World War I)
Australia, HMAS (battle cruiser) 218
Australian and New Zealand Army Corps *see* Anzacs
Australian Army Nursing Service 226
Australian continent before colonisation 60–1
timeline 60–1
Australian flag 121
Australian Flying Corps 219
Australian Imperial Force (AIF) 206
Australian Light Horse troops 219, 221, 225
Australian Naval and Military Expeditionary Force 218
Australian nurses 219, 226, 235
Australian South Sea Islander community 127
Australians at war (World War I) 205
  after the war 235
  Anzacs 219, 220–1, 223
  casualties 219
  children's tasks during the war years 227
  conscription debate 208–9
  deployment 219
  in Egypt and Palestine 219
  enlistments and overseas service 219
  First Nations Peoples experiences of war 224–5
  in Gallipoli, Turkey 205, 220–1
  initial deployment 218–19
  initial response 206
  investing in the war effort 227, 230
  jobs and money during the war years 228
  life on the home front 226–9
  patriotism 227
  persecution of German–Australians 228
  propaganda 207, 208, 227, 230
  reasons for enlisting 207
  regions where Australian servicemen and women were involved 218
  returned soldiers 225, 235
  suspicions and restrictions at home 229
  Western Front 222–3
  women's roles 226
Australia's Constitution 123, 124–5
Austria, Treaty of Saint-Germain 234
Austria–Hungary 196, 199, 233
  assassination of Archduke Franz Ferdinand 198, 200, 201
  declares war on Serbia 198, 200, 202–3
  in Triple Alliance 198
autocracy 197
automation 36

B

Babur, Zahiruddin Muhammad 142
Bai Lian Jiao (White Lotus Society) 170
Bai Shangi Hui (God Worshipping Society) 171
balanced view 246, 258
Balkans region, tensions in 200–1, 203
Ballarat

Eureka Stockade and its aftermath 114–15
licence hunts 111, 115
miners form unions 114
murder and arson 115
soldiers sent to maintain public order 115
Ballarat Reform League 114
banking system 24
Barak, William (Wurundjeri man of the Woi wurrung Clan) 116–17
barbed wire 212
Barton, Edmund 193
Batman, John 103, 106
negotiates treaty with First Nations Peoples 104–5, 116
problems with the treaty 105
treaty negotiated with *Ngurungaeta* of tribes of the Kulin nations 104, 105
Battle of Gallipoli 205, 220–1
Battle of Lone Pine 221
Battle of the Nek 221
battleships 198, 199, 214
bayonets 215
Belgium 222, 227
Besant, Annie 153
bias in sources 4
bicameral parliaments 124
Billibellary 116
Black Wars (1824–1831), Van Diemen's Land 100–1, 103, 104
blackbirding 126–7
Blackstone, Sir William 87
Bligh. Governor William 95, 96
Boer War 192, 193, 224
boomerangs 75
Boonwurrung people 102, 103
Bourke, Governor Richard 103, 105, 107
Boxer Protocol (of 1901) 173
Boxer Rebellion (1899–1901) 172–3, 193
Eight-Nation Alliance countries send troops to China 172
rejection of foreigners 172, 173
support for Qing 172
boycotting (of British goods in India) 152
Britain
agricultural revolution 21–2
arm's race 198
Australia's strong ties to 192, 193
capitalism 24, 25–6
declares war on Germany 206
expansionist agenda 139
imperialism 23
Industrial Revolution 20–49
legal system 86
mercantilism 23
naval power 198, 199, 214
opium trade with China 144, 158–61, 167
population change 20
pre-World War I 197
reasons for establishing Australian colony 86–7
sea blockade of Germany 214, 233
trade and the Age of Imperialism 139
trade with China 157, 158–61, 168
transportation of convicts 86, 87
in Triple Entente 199
urbanisation 20
British
colonisation of Australia 60, 61
control in Hong Kong 141, 167
displace original Australian inhabitants 60
early contact with First Nations People 81
imperialism in India 141, 142–55
perceive Australia as *terra nullius* 70, 81, 86–7
secret instructions to Cook to search for New Holland and take possession 84, 87
British East India Company
Company Raj 143, 144, 146, 147
employs professional Indian soldiers (*sepoys*) 146–7
expands control in India 143, 146
and the Great Rebellion 146–7
loses its trade monopoly 146
smuggles opium into China 144, 158–9, 160
trading post in Macau 157
British Empire 139, 197
Australia as part of 193, 206
India as part of 142–3
and slave trade 46
and trade 24, 25
British imperialism 139
British India 147
British Indian Army 151, 205
British monarch 86, 96, 123
British Navy 197, 220
British Raj
creates British Indian Army 151
education 150
legal system 149
life for Britons in India 148
life for Indians at home 149
life under the 150–3
nationalist movements against 152–5
new technologies 150
and opium trade to China 159
takes over from Company Raj 147
British troops, World War I 219, 220
Brunei, as monarchy 185
Bulgaria 199, 233
Treaty of Neuilly 234
Bungaree 81
bush medicine 77
bush tucker 72
bushrangers
New South Wales 119
as outlaws or icons 118–19
Victoria 118

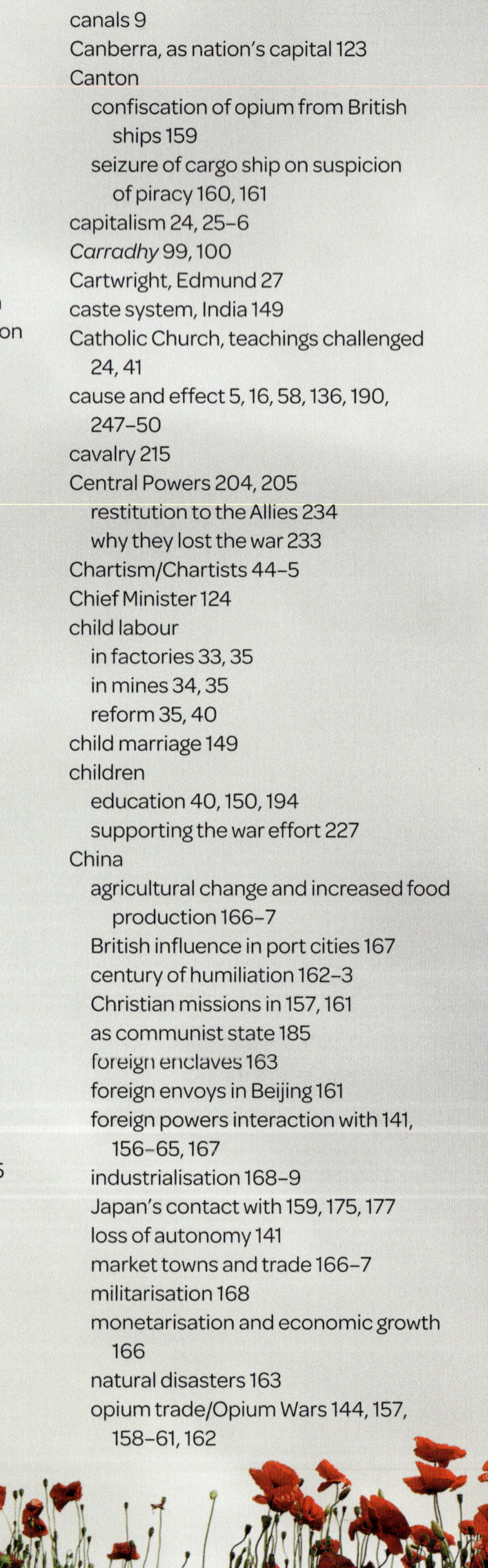

C

Canadian soldiers 205
canals 9
Canberra, as nation's capital 123
Canton
confiscation of opium from British ships 159
seizure of cargo ship on suspicion of piracy 160, 161
capitalism 24, 25–6
Carradhy 99, 100
Cartwright, Edmund 27
caste system, India 149
Catholic Church, teachings challenged 24, 41
cause and effect 5, 16, 58, 136, 190, 247–50
cavalry 215
Central Powers 204, 205
restitution to the Allies 234
why they lost the war 233
Chartism/Chartists 44–5
Chief Minister 124
child labour
in factories 33, 35
in mines 34, 35
reform 35, 40
child marriage 149
children
education 40, 150, 194
supporting the war effort 227
China
agricultural change and increased food production 166–7
British influence in port cities 167
century of humiliation 162–3
Christian missions in 157, 161
as communist state 185
foreign enclaves 163
foreign envoys in Beijing 161
foreign powers interaction with 141, 156–65, 167
industrialisation 168–9
Japan's contact with 159, 175, 177
loss of autonomy 141
market towns and trade 166–7
militarisation 168
monetarisation and economic growth 166
natural disasters 163
opium trade/Opium Wars 144, 157, 158–61, 162

political change in 19th century 162–5
pre-World War I 197
Qing dynasty 156–61, 162, 163, 166–7, 168, 170–3, 197
railways 169
rebellions and revolutions 162, 170–3, 193
reign of Dowager Empress Cixi 164–5, 169
tea and other exports 157, 168
technological change in the 19th century 116–19
timeline, 1760–1917 156
trade with Japan 175, 177
trade regulation 157
under Emperor Qianlong 156–7, 162, 166
'unequal treaties' and trade concessions with Western powers 159–61, 162, 163, 173
wealthy and the Westernised 167
Chinese immigrants
on the goldfields 112, 113
racism towards 113, 121
residency and mining licence fees 113, 114
Chinese Nationalist Party 173
Chinese republic 173
Chisholm, Caroline 93
Christian missions in Chinese port cities 157, 161
Christianity 146
civics + citizenship 124–5, 184–5
Cixi, Empress Dowager, reign of 164–5, 169, 172–3
Clive, Robert 143
coal-driven smoke from factories 48
coal/coal mines 29, 31, 32, 34, 35
Collins, Lieutenant-Governor 103
colonial Australia (1770–1901) 82–119
convicts 86, 87, 90–1
federation 120–3
free settlers 92–3
Frontier Wars 98–101
governors 95, 96
independent colonies in 1900 120–1
power of the military 94–5
reasons for establishment 88–9
representative government 97
timeline 82–3
see also New South Wales colony; Van Diemen's Land; Victorian colony
colonisation 12, 23
colonists
conflict with First Nations Peoples 98–101, 103, 104, 107
deaths from Frontier Wars 101
First Nations Peoples working with 116–17
gold rush and First Nations Peoples 110
in New South Wales 96
in Van Diemen's Land 100–1, 102
in Victoria 104, 110
see also free settlers
Commonwealth of Australia 123, 124, 193
Commonwealth Military Forces 206
communication 30, 152, 193, 194
communism 184, 185
Communist Party (China) 185
Communist Party (Russia) 232
Company Raj 143, 144
policy of 'religious neutrality' 146
replaced by British Raj 147
sells Indian opium to China 144
concubines 164
Confucianism 156, 157, 177
conscientious objectors 227
conscription debate, World War I 208–9
conscripts 197
Constitution of Australia 121, 123, 124–5
changes to 125
constitutional monarchy 123, 185, 197
continuity and change 4, 16, 58, 136, 190, 244–5
Convention of Kanagawa 180
convicts
crime and punishment 90
demographics 90
female 90–1
freed 90
transportation of 88, 89, 109
Cook, James
and Britain's secret agenda to take possession of New Holland 84, 87
circumnavigates Aotearoa (New Zealand) 85
contact with First Nations Peoples 85
reaches New Holland in the *Endeavour* 85
seizes New South Wales without consent of Indigenous inhabitants 85
sets sail for Tahiti 84, 85
Coranderrk reserve 116–17
closure and impact 117
cotton/cotton mills 23, 26–7, 32, 144
Creation Spirits 63, 65, 86, 103
creator's purpose 241–2
cricket 150
criminal justice system, Australia 109
Crompton, Samuel 27
crop rotation system 22
currency lads and lasses 91
cycle of prosperity 30

daimyos 174, 175, 177, 180
Dampier, William 81
Daoguang, Emperor 171
Dardanelles 220
Darwin, Charles 41
deforestation 50
democracy 184, 185, 197
describing continuity and change 244
despots 165
Dharuk language 67
dictation test 126
dictatorship 184, 185
diggers 111, 114–15
dingoes 72
diseases 38, 39, 46, 49, 235
dispossession 61
dreadnoughts (battleships) 198, 199
Dyer, General Reginald 154

E

Eastern Front 232
economic boom, from Victoria's gold rush 112
economic factors
as cause of the Industrial Revolution 20
for establishing Australian colony 89
economic thinking, advances in, impact on Industrial Revolution 24–6
economics + business 36–7, 230–1
editing 263
education
Australia 194
for children (Industrial Revolution) 40
India 150
for women (Industrial Revolution) 41
Egypt 219
8 Hours Movement 40, 53
Eight-Nation Alliance countries 172, 173
Eight Trigrams Uprising 162
Elders 63, 64, 85
electorates 125
emancipists 88
Empire Day 227
employment growth, Industrial Revolution 36
enclosure 21–2, 32
Engels, Friedrich 41
The Enlightenment 14, 24, 25
enlisting
numbers by year 208
physical requirements of volunteers 208
reasons for 207
Entente Cordiale 199
entrepreneurs 26
environmental effects of Industrial Revolution 48–9
equality 41
espionage 229
Eureka flag 115

- Eureka Rebellion 114–15
- Eureka Stockade 114, 115
- Europe
  - aftermath of war 234
  - influenza, 1918–1919 235
  - political borders after World War I 234
  - pre-World War I 196–7
  - trade with China 159
  - war in 13, 190–235
- European explorers, contact with First Nations Peoples 80
- European imperialism 138, 139
- European nations, search for southern continent 60, 80
- evaluation 243, 246, 250, 254, 258
- evolution, theory of 41
- exclusives 92
- executive branch (of government) 124
- expansion of empires 12–13, 23
- exponential change 10
- exports 37

F

- factories
  - advances in technology 26–7
  - child labour 33, 35
  - and welfare of workers 40
  - working conditions in 32
- fashion 194
- Fawkner, John Pascoe 103, 106
- features of a source 240
- Federal Council of Australia 121
- federal government 125
  - branches 124–5
- federal parliament 125
  - representing electorates 125–6
- federation 120–3, 193
  - and the Constitution of Australia 123
  - opposition to 122
  - Parkes' proposal for 121
  - referenda and voting rights 122–3
  - support for 121
  - timeline 120
- female convicts 88–9
- female infanticide 149
- Ferdinand, Archduke Franz, assassination 198, 200, 201, 204
- feudal system, Japan 176–7
- firestick farming 73
- First Fleet 89
- First Nations Peoples 13
  - Aboriginal Customary Law 63, 67, 86, 99
  - ancestral Creation Spirits 63, 65, 86, 103
  - attempts at 'civilising' and assimilation 108, 109, 128, 129
  - Australia as a sacred place 58, 60–81
  - bush tucker 72
  - as bushrangers 119
  - casualties 101
  - concepts of time 64–5
  - conflict with colonists 98–101, 103, 104, 107
  - contact with Cook 85
  - Coranderrk reserve 116–17
  - cultures 62–5
  - dating their occupation sites 71
  - dispossession by colonists and invaders 61
  - early conflicts 99
  - economic structures 68–9
  - employed as Native Police 108–9
  - enlist World War I 224
  - firestick farming 73
  - forced removal of children from their families 128–31
  - Frontier Wars 98–101, 103, 104, 107
  - gatherings 68, 70
  - genocide 61, 101, 117, 128
  - on the goldfields 111
  - health and wellbeing 69
  - hunting 73, 74, 75
  - ignored in federation debate 121
  - impact of gold rush on 110, 111
  - interactions with other cultures 78–81
  - kinship system 63
  - land management 73, 85, 87
  - languages and language groups 66–7
  - legal structures 62, 63
  - massacres 101
  - Men's and Women's Business 63, 65, 86
  - navigation techniques 76
  - of New South Wales 85, 98–101
  - political structures 62
  - population pre-1770 62
  - 'protection' of 129, 131
  - racism towards 129
  - resistance and violence 99, 100, 107
  - rock art 72
  - semi-permanent residences and food production 60, 70–1, 87
  - social structures 63
  - sovereign ownership of the land 65
  - Stolen Generations 128–31
  - technologies 74–7
  - toxic plant harvesting and bush medicine 77
  - trade and exchange 68
  - treatment of returned soldiers after the war 225
  - treaty with Batman 104–5
  - of Van Diemen's Land 100–1, 103, 104
  - of Victoria 102–3, 104–5, 108, 116–17, 128, 129
  - voting rights 123
  - women's equal status 88
  - work with colonists 116–17
  - World War I 224–5
- First Nations Peoples of Canada, apology to over Stolen Generations policy 131
- FitzRoy, Governor 111
- fixed rate interest 231
- flamethrowers 213
- Flinders, Matthew 81
- Flinders Island 100
- floating rate interest 231
- food production
  - agricultural revolution 22
  - China 166–7
  - First Nations Peoples 60, 70–1
- fossil fuels, burning of 48, 49
- foxholes 210
- France
  - Australian Flying Corps in 219
  - battlefields 222
  - in Triple Entente 199
  - *see also* Western Front
- free settlers 92–3
  - conflict with First Nations Peoples 99–101
- free thinkers 40–1
- French soldiers 205
- Frontier Wars 98–101
  - casualties 101
  - massacres 101

- Gallipoli, Turkey 205, 220–1
  - Anzac forces 220–1
  - and blocking of Mediterranean Sea passage to Russia 220
  - First Nations soldiers at 225
  - landing of the Anzacs 220
  - long campaign 220–1
  - Turkish forces 220, 221
  - withdrawal 221
- Gallipoli peninsula 220
- Gandhi, Mahatma 153, 154–5
  - as civil rights activist 154
  - prominent in Indian independence movement 155
  - promotes non-violent protest 155
- gathering information 260
- genocide 61, 101, 117, 128
- German–Australians, persecution of 228–9
- German New Guinea 218
- Germany 196
  - armistice 233
  - arm's race 198
  - Britain declares war on 206
  - British sea blockade of 214, 233

defeated by the Allies 232, 233
on the Eastern Front 232
invades Belgium 222
Kaiserschlacht 233
Ludendorff Offensive 233
nationalism 199
naval power 198, 214–15
policy of 'unrestricted submarine warfare' 214, 216, 233
Russian Communist government signs peace treaty with 232
supports Austria–Hungary 203
Treaty of Versailles 234
in Triple Alliance 198
United States declares war on 217
on the Western Front 222–3, 233
see also Central Powers
Ghan cameleers 113
Gipps, Governor George 108
global concerns, global conflicts 13
global war 204–5
globalisation 37
gold rush 52
impact on First Nations Peoples 110, 111
in Victoria 110–13
goldfields
Chinese immigrants on 112, 113, 114
cultural change and ethnic tension 113
Eureka Rebellion 114–15
gold licences 111, 113, 114, 115
hard mining life 111
politics 114
'traps' and diggers 111
Goldstein, Vida 226
Gong, Prince-Regent 164
Governor-General 123, 124
governments
in Asia 184–5
in Australia 124–5
Governor, Jimmy 119
great southern land, search for 60, 80
Green, John (Protector of Aboriginal Peoples in Victoria) 116, 117
grenades 212
Grose, Governor Francis 96
Guangxu, Emperor 164, 165, 169
guerrilla tactics 170

Hargreaves, James 26
Hartog, Dirk 80
Hastings, Warren 143
health and wellbeing, First Nations Peoples 69
Hickman, Jessie 119
highway robbery 118
Hindus/Hinduism 146, 147, 153
castes 149
historians
at work 3
what do they do? 2–3
historical interpretations 5, 17, 59, 136, 191, 255–8
historical research 4–5, 264–5
historical significance 5, 17, 59, 136, 191, 251–4
historical thinking skills 4–5
historical writing 259–63
historiography 255–8
Hoddle, Robert 107
home front, life on (World War I) 226–9
Hong Kong
British control in 141, 167
British culture in 167
ceded to the British 159
Hong Xiuquan 171
hot air balloons 11
Hotham, Lieutenant-Governor Charles 115
House of Representatives 125
Hughes, Billy 208, 209
humanism 14
Hungary
Treaty of Trianon 234
see also Austria–Hungary
Hunter, Governor John 96
hunting, First Nations Peoples 73, 74, 75
hunting boomerangs 75

I

Ieyasu Tokugawa 174
Ieyoshi, Shogun 180
immigration policy 113, 121, 126–7, 194
Immigration Restriction Act 1901 113, 126
imperialism 12, 23, 138–9
in Asia 140–1
as cause of World War I 199
what is it 138
see also China; India; Japan
independent colonies (Australia) 120–1
see also New South Wales colony; Van Diemen's Land; Victorian colony
India
agricultural changes 144
Australian nurses serve in 219
boycott of British goods 152
British East India Company 143, 144, 146–7
British imperialism in 141, 142–55
British Raj 147, 148–55
Britons living in 148
caste system 149
communications and unity 152
Company Raj 143, 144, 146, 147
cricket 150
as democracy (republic) 185
destruction of textile economy 145
education 150
Gandhi's influence 153, 154–5
Great Rebellion 146–7
how it became part of the British Empire 142–3
impact of British rule in 144–7
land ownership changes 144
legal system changes 149
military 146–7, 151
Mughal Empire 142–3
nationalist movements 152–5
newspapers 150
post-war protests and massacre 154
push for self-rule (*swaraj*) 152, 153, 154–5
railways 150, 152
religions 146, 147, 151
status of women 149
Indian Army 151
Indian Congress Party 152
Indian National Congress (INC) 152, 153
Indian soldiers, World War I 151, 205, 220
Indians, life for under British Raj 149
Indigenous peoples 13
see also First Nations Peoples
indigo 144
Indonesian traders, interactions with Australia's First Nations Peoples 79
Industrial Revolution 16–53
and advances in economic thinking 24–6
affect on the environment 48–9
in Australia 50–3
in Britain 20–49
and capitalism 24, 25–6
and entrepreneurs 26
free thinkers who challenged the status quo 40–1
groups who fought for social and political change 42–5
impact on living conditions 38–9
impact on working conditions 32–5, 39
political factors 23
and the slave trade 46–7
social and economic factors 20
sources of power 27
and technological advances 26–7
technological factors – the agricultural revolution 21–2, 87
and The Enlightenment 24, 25
timeline 18–19
transport 28–31
industrialisation 11, 26, 38
Australia 50–1, 193
China 168–9
inflation 228
influenza 235
innovation 20

intellectuals 14
interest-earning investments 231
internal combustion engine 31
internment camps 228
interpreting information 5
inventions, pre-World War I 193
investments
  options 231
  and risk 230
  war bonds 227, 230
Italy, in Triple Alliance 198
Iwakura Mission 183

Janszoon, Willem 80
Japan
  arts, business and urban growth 177
  Commodore Perry's 'black ships' demand opening of borders to US trade 141, 180
  concern over threat of foreign traders 174–5
  as democracy (constitutional monarchy) 185
  Edo (Tokyo) 179, 180
  enters World War I 182
  exceptions to *sakoku* restrictions 175, 177
  feudal system 176–7
  hires foreigners to improve education, sciences, engineering and the military 183
  industrialisation 180–1, 183
  limited change and introduction of ideas 177
  Meiji Restoration 180–1, 182
  military expansion 181
  modernisation during the Taisho period 182–3
  pre-World War I 197
  re-engagement with the West 180–1
  sakoku (closed borders) imposed by shogunate 174–5
  social change (post-war) 182–3
  social structure (feudal period) 176–7
  takes control of Korea 181
  technological change 183
  timeline, 1600–1918 174
  Tokugawa shogunate 174–80
  trade with China and Joseon Korea 175, 177
  women's role 177
Jiaqing, Emperor 159, 162
Johnston, Major George 95
judiciary 124

Kaiserschlacht 233
Kelly, Ned 118
Kelly gang 118
Khaparde, G.S. 153
King. Governor Philip Gidley 96, 103
kinship system 63
Kome Sodo ('Rice Riots'), Japan 182–3
Korea 175, 181
Kowloon Peninsula, ceded to Britain 160
Kulin nations 100, 102, 104, 105
Kuomintang 173

laissez-faire approach to economic management 25
Lalor, Peter 115
land management, First Nations Peoples 73, 85, 87
languages and language groups, First Nations Peoples 66–7
LaTrobe, Lieutenant-Governor Joseph 109, 111
Learning Ladder 6, 16–17, 58–9, 136–7, 190–1
legal system, British Raj 149
legislature 124
Lenin, Vladimir 232
Lin Zexu 159
living conditions (Industrial Revolution) 38–9
  reform 40
Luddites 45
Ludendorff Offensive 233
Lusitania, sinking of 216

Mabo Decision 65, 86
Macarthur, John 93, 94, 95
Macau 157
Macauley, Thomas 150
machine guns 212, 215
Maconochie, Alexander 108, 109
Macquarie, Governor Lachlan 96
Macrozamia cycad, toxin removal from seeds 77
maharajahs 143
mahjong 167
Makassans, contact with Australia's First Nations Peoples 79
MANIAC (causes of World War I) 198–9
manufacturing 31, 33, 36, 37
  in Australia 2, 50–1
Marxism 183
Matra, James Mario 88
Mehmed V, Sultan 197
Meiji, Emperor 180, 182
Meiji Restoration 180–1
Melbourne
  conflict with First Nations Peoples 107
  debate with Sydney as to nation's capital 123
  founding of 106–7
  laid out by Hoddle 107
  naming of 107
  slums 51
Melbourne Gaol 109
Members of Parliament (MPs) 125
Men's Business 63, 65, 86
mercantilism 23, 25
merchants 25
Mesopotamian Campaign 219
messiah 171
Mexico 217
Middle East, Australian Flying Corps in 219
militarisation
  China 168
  Japan 181
militarism, as cause of World War I 198
mines
  child labour 34
  working conditions 33, 35
  see also goldfields
Ming dynasty 168, 170
mining licences 111, 113, 114, 115
modern world, factors shaping the 10–14
moiety 63
Molonglo Internment Camp 228
monarchy 184, 185
Monash, John 223
monetarisation, China 166
Monroe Doctrine 139
motor vehicles 31
Mughal Empire 142–3
Muslims 146, 147, 151
mustard gas 213
Myall Creek massacre 101

Naarm 102
Naoroji, Dadabhai 152
national capital 123
nationalism 12
  as cause of World War I 199
nationalist movements, India 152–5
Native Police 108–9
native title 65
natural selection, theory of 41
naval blockades 214
navigation techniques, First Nations Peoples 76
Nemesis (paddle steamer) 159
New Holland 81
  Cook's search for 84, 87

New Lanark mill, Scotland, workplace and social reform 40
new nations, and ancient peoples 12–13
New South Wales colony
'black market' economy 95
bushrangers 119
convicts 88, 89, 90–1
First Nations Peoples 85, 98–101
free settlers 92–3
governors 95, 96
as penal settlement 86, 87, 94
power of the military 94, 95
representative government 97
Rum Rebellion 95
takes control of Victorian colony 106
New South Wales Corps, power of 94, 95
New Zealand soldiers 205
Newcomen, Thomas 27
newspapers 30, 150
Ngurungaeta 104, 116, 117
Nicholas II, Tsar 197, 232
No Man's Land 212, 221
North Korea, as dictatorship 185
NSW Parliament House 97
nurses, World War I 219, 226, 235

occupations, economic classification 36
oligarchy 94
online searching 264–5
opium
ban in China 159, 162
British sales to China 158, 159, 167
confiscation of 159
smuggled into China from Bengal 144, 158–9, 160
opium dens 159
Opium Wars 167
First (1839–1842) 159, 162
industrialisation following 169
leaves China in debt 162
Second (1856–1860) 160–1, 162, 171
optically stimulated luminescence 71
organising information 260–1
Ottoman Empire 197, 199, 200, 219, 220
Treaty of Sevres 234
see also Turkey
outsourcing 37
Owen, Robert 40

Pacific Islander Labourer Act 1901 127
Pacific Islanders working in Queensland
sugar cane industry 122, 126
deportation 127
pacifists 227, 229
Palawa people, Van Diemen's Land 100, 101, 103, 108
Palestine 219
Pankhurst, Adela 226
Parkes, Henry 120, 121
Parliament House, Canberra 123
parliamentary system 121, 123, 124
Paterson, Governor William 96
patriotism 227
patterns of continuity and change 245–6
Pemulway 100
pensions 235
Perry, Commodore Matthew 176
PEST factors, Industrial Revolution 20–3
Phillip, Governor Arthur 87, 96, 99, 103
pilots 214
Pitt, William 88
planes 214
poison gas 213
police force
New South Wales 119
Victoria 109, 111, 118
political and legal structures, First Nations Peoples 62
political change, groups who fought for 42–5
political factors, as cause of the Industrial Revolution 23
political parties 125
population change
Britain 20
post-colonisation Australia 62
porcelain and ceramics 157, 168
Port Phillip Bay 102, 103
and founding of Melbourne 106–7
Port Phillip Protectorate 100, 107, 108
attempts to 'civilise' First Nations Peoples 108, 109
Native Police 108–9
pre-World War I
Australia 192–5
European nations 195–6
see also World War I
Premiers 124
primary sources 4
Prime Minister 124, 185
Princely India 147
prison hulks 87
privies 38
proofreading 263
propaganda, World War I 207, 208, 227, 230
Protection Boards 129

Qianlong, Emperor 156–7, 162, 166
Qing Dynasty 156–61, 166–7, 168, 197
decline 162, 163
end of 173
rebellions and revolutions 162, 170–3
The Boxers support for 172
Queensland 120
Pacific Islanders working in 122, 126–7
unicameral government 124

Rabaul 218
racism 113, 129
see also White Australia Policy
radiocarbon dating 71
railways 30–1
Australia 52, 121, 194
China 169
India 150, 152
real estate, investment in 231
Rebecca Riots 27
rebellions and revolutions, China 162, 170–3, 193
reconnaissance 214
recruitment posters 207, 208
referendum 125
religions, India 146, 147, 151
representative government, NSW 97
restitution 234
returned soldiers 225, 235
returning boomerang 75
revolutions
what are they? 18–19
see also rebellions and revolutions
Richthofen, Manfred von ('Red Baron') 214
rifles 215
roads 28–9
Robinson, George Augustus 100, 108
Romania 199
Rowlatt Act 154
Rudd, Kevin 131
Rum Rebellion 95
Russia 197
and closure of Dardanelles 220
signs peace treaty with Germany and exits the war 232
supports Serbia 203
trading rights with China 160
in Triple Entente 199
Russo-Japanese War 181, 197

sakoku 174–5
Salonika 219
samurai 174, 177
Savery, Thomas 27
savings accounts 230
Schlieffen plan 233
search strategies 264–5
secondary sources 4

selective breeding 22
self-rule, India 152, 153
Amritsar Massacre of 1919 154
Gandhi's role 154–5
post-war protests in favour of 154
Senate 125
senators 125
sepoy soldiers
banned from some divisions of British Indian Army 151
British troops defeat 147
employed by British East India Company 146, 147
rebellion by 147
Serbia 197, 199, 200
assassination of Franz Ferdinand in 198, 200, 201
at war with Austria–Hungary 202–3
Seven Years' War 139, 143
shares 231
silk 145, 157, 168
Sinai and Palestine Campaign 219, 225
Sino–French War 162
Sino–Japanese War 162, 181
skin name 63
slave ships 46
slave trade 46–7
abolition movement 47
Smith, Adam, The Wealth of Nations 24, 25
smog 48, 49
social change
groups who fought for (Industrial Revolution) 42–5
Japan 182–3
social factors
as cause of the Industrial Revolution 20
for establishing Australian colony 89
social structure
First Nations Peoples 63
Japan 176–7
socialism/socialists 40, 41, 183, 229
society, Australia, pre-World War I 194–5
Society of the Righteous and Harmonious Fists ('The Boxers') 172–3
source analysis 4, 16, 58, 136, 190, 240–3
South Africa, Boer War 192, 193
South Australia 120
South Sea Island labourers 126, 127
Southern Cross 115
sovereignty 12
spinning machinery 26–7
squatters 92, 93
state parliaments 125
steam engines/steam power 10, 26, 27, 32, 50
steamships 31
Stolen Generations 128–31
and Aboriginal 'protection' 129, 131
broken families 129
official apology to 131
ongoing legacy 131
transgenerational trauma 130
stone tools 74
strikes 228
submarines 214–15
suffrage 42, 44
suffragists/suffragettes 42, 226
Sullivan Bay (near Sorrento) 103
Sun Yat-sen 173
superannuation 231
Sydney, debate with Melbourne as to nation's capital 123
Sydney Cove 88

Taiping Heavenly Kingdom Rebellion (1850–1864) 162, 171
Taisho period, Japan, modernisation 182–3
Taiwan 173
Tang dynasty 168
tanks 212, 213
Tasman, Abel 80
Tasmania 120
see also Van Diemen's Land
tea trade 144, 157, 168
technological advances 10–11
Australia, pre-World War I 193–4
Industrial Revolution 26–7
technological factors, as cause of the Industrial Revolution 21–2
telegraph 193, 194
temperance movement 42
terra nullius 70, 81, 86–7
territory parliaments 125
textile market decline, India 145
The Arrow (ship) 160, 161
themes in a source 240–1
theory of significance 253
thylacines 72
'Ticket of Leave' 90
Tilak, Bal Gangadhar 153
time, concept of (First Nations Peoples) 64–5
timelines 12–13
Age of Imperialism 140–1
Australian continent before colonisation 60–1
British in India 142
China, 1760–1917 156
colonial Australia (1770–1901) 82–3
federation and beyond 120
Industrial Revolution 18–19
Japan, 1600–1918 174
World War I 192–3, 210–11, 232
Tokugawa shogunate
feudal system 176–7
life during 176–7
why it closed Japan's borders 174–5
Tolpuddle Martyrs 43
Tonghzi, Emperor 164
Torres, Luis Váez de 80
Torres Strait Islander peoples, interactions with First Nations of northern Australia 78
totem 63
towns and cities, living conditions (Industrial Revolution) 38–9
trade 24, 25
and the Age of Imperialism 138, 139
with China 144, 157, 158–61, 168
and exchange, First Nations Peoples 68
in goods for slaves 46–7
with Japan 141, 170, 174–5, 177, 180
trade unions 43, 228, 229
transport
Australia, pre-World War I 194
during Industrial Revolution 28–31
transportation of convicts 88, 89, 109
'traps' (police) 111
Treaty of Nanking (1843) 159, 171
Articles 159
Treaty of Neuilly 234
Treaty of Saint-Germain 234
Treaty of Sevres 234
Treaty of the Bogue (1843) 159
Treaty of Tianjin (1858) 160
Treaty of Trianon 234
Treaty of Versailles 234
trench warfare 210, 222–3
life in the trenches 210–11
typical day in the trenches 211
Triple Alliance 198
Triple Entente 199, 204
Truganini 100–1
Turkey
Dardanelles 220
Gallipoli 220–1
Turkish troops, Gallipoli 220–1
turnpike trusts 29

U-boats 214–15
Underwood, Jacky 119
unemployment 235
United States
attitude towards Germany 216, 217
declares war on Germany 217
demands Japan open its borders to trade 141, 180
enters World War I 205, 216, 217, 233
as imperialist power 139
initial neutrality, World War I 216

isolationist policy 139, 216
and sinking of the civilian ship *Lusitania* 216
trading rights with China 160
and the Zimmerman Telegram 217
'unrestricted submarine warfare' (German policy) 214, 216, 233
urbanisation 20, 32, 38–9

Van Diemen's Land 80, 103, 106
Black Wars 100–1, 103, 104
colonists 100–1, 102
Viceroy of India 147
Victoria, Queen 147, 159
Victoria, 8 Hours Movement 53
Victorian Aboriginal Protection Act 1869 129
Victorian colony
Aboriginal 'protection' through forced child removal 129
Anti-Transportation League 109
Batman's treaty with First Nations Peoples 104–5
Bourke nullifies Batman treaty 105, 106
bushrangers 118
comes under NSW control 106
Coranderrk reserve 116–17
economic boom from gold 112
first colonisation attempts 103
First Nations Peoples 102–3, 108, 109, 116–17, 129
founding of Melbourne 106–7
from Naarm to Port Phillip Bay 102–3
gold rush 110–13
law and order 108–9
police force 109, 111, 118
policy of assimilation 129
reasons for settlement 103
Stolen Generation 129
see also Port Phillip Protectorate
Victorian Police 109
voting rights
federation referenda 122–3
First Nations Peoples 123
women 42, 122, 194

war bonds and loans 227, 230
War Precautions Act 229
war widows 235
water pollution 49, 51
Watt, James 27
welfare of workers 40
Western Australia 120
Western Front
Allies on 222–3, 232, 233
American troops on 217, 233
Australian soldiers on 223
establishment 222
German forces on 222–3, 233
Kaiserschlacht 233
Ludendorff Offensive 233
rifle strength of Allied and German armies 217
soldiers on 215
stalemate 223, 233
trench warfare 210, 211, 222–3
Westminster system 121, 124, 185
White Australia Policy 113, 126–7, 194
immigration restrictions 126
impact on Pacific Islanders working in Queensland 126–7
White Lotus Rebellion (1796–1804) 162, 170
wife-selling auctions 91
Wilhelm II, Kaiser 196, 233
Wilson, President Woodrow 216
'14-point plan' to create peace 234
Wollstonecraft, Mary 41
women
equality and education for 41
Indian society 149
single women, colonial Australia 93
Tokugawa Japan 177
voting rights 42, 122, 194
World War I 226
Women's Business 63, 86
Women's Peace Army 226
women's rights 226
women's suffrage movement 42, 226
Wonga, Simon (Wurundjeri man of the Woi wurrung Clan) 116
wool industry 52, 93
woomera 74
working conditions (Australia) 53, 194
working conditions (Industrial Revolution) 39
for children 33–5, 40
in factories 32
groups pushing for reform 43
in mines 33
reform 40
workplace change today 36–7
World War I 13, 190–235
aftermath of war 234
Allies 204, 222–3, 232, 233, 234
armistice 233, 234
Australia *see* Australians at war (World War I)
Britain declares war on Germany 206
causes (MANIAC) 198–9
Central Powers 204, 205, 233, 234
changing allies 232
course of the war 210–29
Eastern Front 232
end of 232–5
European borders after 234
European countries prior to 196–7
Gallipoli, Turkey 205, 220–1
as a global war 204–5
international armies 151, 205
Japan enters 182
long-term causes 198–9
as 'machine-age war' 212–15
nations in conflict 205
peace after the war 234–5
the road to war 192–209
short-term causes 198, 200–3
soldiers and their weapons 215
timelines 192–3, 210–11, 232
treaties and agreements after 234
trench warfare 210–11
United States enters 205, 216–17, 233
war in the air 214
war at sea 214–15
war on land 212–13
Western Front 210–11, 215, 217, 222–3, 232, 233
why the Central Powers lost the war 233
see also Germany
writing
editing and proofreading 263
gathering, organising and structuring 260–2
task words 258
Wuchang Rebellion (1911) 173
Wurundjeri People 102, 116, 117

Xia dynasty 156
Xianfeng, Emperor 164, 171
Xinhai Revolution (1911) 173
Xuantong, Emperor, abdication 173

Yoshihito, Prince (Taisho period) 182–3
Yoshinobu, Shogun 180
Yuan Shikai, General 173

zamindars 144
zeppelins 214
Zimmerman Telegram 217

Acknowledgements

The author and publisher are grateful to the following for permission to reproduce copyright material:

PHOTOGRAPHS: agefotostock/fototeca gilardi/Marka, **191**, /Heritage Image/The Print Collector, **28**, /Karl Johaentges, **i** (centre left), **58**; Alamy, /agefotostock, **18** (bottom left), /Album/British Library, **146–7**, /Alpha Stock, **127** (bottom), /Archivart, **52**, **230**, /Art World, **62**, /Arterra Picture Library/Collection Phillipe Clement, **212**, /Artokoloro, **80** (bottom right), /Auscape International Pty Ltd, **68**, /BNA Photographic, **80** (bottom left), /Charles Knox - Creative, **265**, /Chronicle **44–5**, **54**, **112** (top), **137**, **138–9**, **154–5**, **221**, **249**, /Emanuele Ciccomartino, **132** (bottom), /Classic Image, **23**, **48**, /Colport, **114** (bottom), /CPA Media Pte Ltd/Pictures From History, **163** (bottom), **164** (right), **165**, **171** (bottom), /Dinodia Photos, **i** (centre right),**135**, **155**, **153** (top), /EDU Vision, **183**, /Crystal Egan, **7** (top), /Everett Collection Historical, **213** (bottom), /Everrett Collection Inc, **157**, /FLHC12, **72** (bottom left), /Nigel Forder, **74** (bottom), /GL Archive, **61** (bottom), **118**, **247**, /Glasshouse Images/JT Vintage, **181**, /Granger Historical Picture Archive, **140** (top), /david hancock, **101**, /Heritage Image Partnership Ltd, **11** (bottom), /Hi-Story, **214–15**, /Hirarchivum Press, **160–1**, /Historic Collection, **100** (bottom right), **117** (top), /Historic Images, **116**, /Historical image collection by Bildagentur-online, **93**, /History and Art Collection, **95**, /hoch2w0, **36** (education and industry icons), /David Hodges, **167**, /Peter Horree, **14**, **182**, /incamerastock, **178–9**, **226**, /INTERFOTO, **11** (top), /Jemastock, **36–7** (robot icons), /Robert Kawka, **166–7**, /John R. Kreul/Independent Picture Service, **75** (top), /Viacheslav Iakobchuk, **37** (top left), /Lebrecht Music & Arts, **193**, /Lordprice Collection, **214**, /Natalia Lukiianova, **244**, /mauritius images GmbH, **115**, /MeijiShowa, **260**, /Niday Picture Library, **140** (bottom left), **171** (top), /North Wind Picture Archives, **241**, /PhotoBliss, **60** (bottom right), /Pictorial Press, **29** (top), **192**, /Pictorial Press Ltd, **90**, **104**, **163** (top), **200**, /Private Collection / AF Fotografie, **47**, /William Robinson, **61** (top), /Science History Images, **213** (top), /Science Photo Library/Simon Stone, **5**, /Shawshots, **211**, /Cigdem Simsek, **121**, /Kumar Sriskandan, **258**, /Jack Sullivan, **190**, **/**Go Australia - Neil Sutherland, **i** (bottom left), **189**, /The Archives, **81**, /The Artchives, **42** (bottom), /The Granger Collection, **24**, **186** (top left), /The Granger Historical Picture Archive, **186** (top right), /The Natural History Museum, **99**, /The Picture Art Collection, **18** (bottom right), **83** (top), **168**, /Marc Tielemans, **215**, /Penny Tweedie, **60** (bottom left), /Vernon Lewis Gallery/Stocktrek Images, **12–13**, /John Warburton-Lee Photography, **64–5**, /World History Archive, **43**, **140** (bottom right), **232**; Women working in the Roslyn Woollen Mill, Ref: MNZ-0704-1/4-F. Alexander Turnbull Library, Wellington, New Zealand, **255**; Auscape/Reg Morrison, **71** (top); Bridgeman Images/Clifford Possum & Tim Leura Tjapaltjarri, *Warlugulong*, 1976, AGNSW, © Estates of the artists licensed by Aboriginal Artists Agency Ltd, **59**, /© British Library Board. All Rights Reserved, **85** (top), /Commonwealth Club, London, UK, **88**, /Mitchell Library, State Library of NSW, **82–3**, /Private Collection/The Stapleton Collection, **46**, /Private Collection/© Look and Learn, **45**; Peter Campbell, **117** (bottom); John Morieson /Alex Cherney, **76**; Courtesy Frank Tung Foo, Chinese Museum of Chinese Australian History, FC127, **113**; Fairfax/Rick Stevens/SMH, **79**; Getty Images, /adoc-photos, **158–9**, /Anadolu Agency, **53** (bottom), /Auscape, **i** (centre left), **58**, **72** (top), **75** (bottom), **242** (top), /Bettmann, **19** (bottom), /Culture Club/Hulton Archive, **161**, /DEA PICTURE LIBRARY, **18** (bottom (centre), /DEA/A. DAGLI ORTI, **202**, /DEA/BIBLIOTECA AMBROSIANA, **169**, /DEA/M. SEEMULLER, **172**, /Christian Ender, **4**, /Fox Photos, **49**, /Manfred Gottschalk, **97** (top), **133**, /Heritage Images, **30–1**, **150**, /Hulton Archive, **29** (bottom), /ilbusca, **26** (top), /Mark Kolbe, **266–8**, /MPI, **216**, /INDRANIL MUKHERJEE, **153**, /Nastasic, **26** (bottom), /Nikada, **280**, /Tracey Nearmy, **78**, **124**, /PA Images, **253**, /Peter Zelei Images, **279**, /Photo 12, **25**, /Pool, **131** (top right), /Print Collector, **27** (top), **148**, **151**, /REDA&CO, **103**, /Jan Rzaczek / EyeEm, **i** (top right), **15**, /Frans Sellies, **40** (bottom), /Andrew Sheargold/Stringer, **131** (top left), /stock_colours, **269–78**, /Koukichi Takahashi / 500px, **250**, /Topical Press Agency, **237**, /PEDRO UGARTE, **18** (top), /ullstein bild Dtl., **173**, /United Archives, **144–5**, /Universal History Archive, **149**, /UniversalImagesGroup, **156**, /urbancow, **i** (bottom right), **239**; Karen Hughes, **3**; © Imperial War Museum (Q13427), **220**; iStock, /Arrlxx, **77**, /duncan1890, **38**, /FiledIMAGE, **72** (bottom right), /Grafissimo, **34**, /iShootPhotosLLC, **231**, /kokkai, **125**, /Leonsbox, **30** (bottom), /lovleah, **63** (background image), /nimis69, **i** (top left), **1**, /Phillip Openshaw, **20–1**, /petekaraci, **30** (ticket background), /SeanShot, **2**, /Dedy Setyawan, **36** (globe), /upstudio, **30** (top), /sara_winter, **i** (background), /AlexanderYershov, **9**, **56**,**134**, **188**, **238** (arch icon); Museum of Applied Arts and Sciences/Photograph 85/1284-2818, by Kerry & Co. Tyrrell Collection. Gift of Australian Consolidated Press under the Taxation Incentives for the Arts Scheme, 1985, **53** (top), /Licence 2000/128/l. Gift of the Royal Australian Historical Society, 19E1. Photo: Marinco Koidanovrkl, **114** (top); Metropolitan Museum of Art/Item 13.220.24, **170**, /Item JP3536, **177**, /Item JP3235, **186** (bottom), /21.175.334a, b, **187**, /Item 2001.561.1, **251**, /Item 29.110.55, **256**; Museums Victoria, collections.museumvictoria.com.au/items/765582, **195**; National Archives of Australia/A7342, Album 2, **235**

(bottom), /Removal of quadroon and octoroon children from the N.T. - Offers of Accommodation (Newspaper photograph of half caste Aboriginal children), A1, 1934/6800, **129**; Official U.S. Navy photo, NH 63596, courtesy of Naval History and Heritage Command, **199**; NSW State Archives/Darlinghurst Gaol; NRS 2138, Photographic Description Books, 1871-1914; 13150 Jessie McIntyre [3/6083] Digital ID: 2138_a006_a00603_6083000103r, **119**; Newspix/Jason Busch, **74** (top), /News Ltd, **112** (bottom); Queensland Police Museum, **108**; Shutterstock/Boris-B, **243**, /Everett Historical, **32–3**, /Fran Veale, **19** (top), /pingebat, **184**, /Roop_Dey, **142**; Ballarat District Orphan Asylum, Sovereign Hill Museums Association Collection, Accession Number 80.1629, **128–9**; State Library of Queensland, **130** (top); State Library of South Australia/B70647, **42** (top), /PRG 1664/2/1, **194** (bottom); State Library of Victoria/Item H2009.95/42, **194** (top), /F. Oswald Barnett Collection, **51** (top), **55**, /Charles Nettleton, **51** (bottom), /Washbourne, Thomas J, **110–11**; *Australia's coloured minority: its place in the community* by A.O. Neville, Introduction by A.P. Elkin, Neville, A.O., 1875-1954, State Library of Western Australia, **130** (bottom); Stewy, **40** (top); Photo © James Horan for Sydney Living Museums, **69**.

OTHER MATERIAL: Poster 'Union is Strength' Melbourne Punch (Periodical), Melbourne Punch. Ref: J-040-003. Alexander Turnbull Library, Wellington, New Zealand. /records/22845835, **120**; Illustration, 'Australian Spear-traits and their Derivations', D.S. Davidson, 1934, JPS, volume 43, page 72, **100** (bottom left); Charles Dixon, *The Landing at Anzac*, AAAC 898 617 NCWA Q388, Archives New Zealand, **236** (bottom); Cartoon 'Bravo, Belgium!', Punch, or the London charivari, 1914, British Library, **236** (top right); Carved shield, ©The Trustees of the British Museum. All rights reserved, **85** (bottom); F.G. Moon, *The signing and sealing of the Treaty of Nanking*, 1846, Anne S.K. Brown Military Collection, Brown University Library, **254**; Historical Map (1765 and 1805), *Imperial Gazetteer of India*, volume 26, Clarendon Press, Oxford, 1909, p.27,**143**; Extract from 'British Empire is still being whitewashed by the school curriculum - historian on why this must change' by Deana Heath, University of Liverpool. Originally published on *The Conversation*, theconversation.com/au, **257**; *H. I. M., the Empress Dowager of China,* Cixi (1835-1908), Harvard Art Museums/Fogg Museum, Bequest of Grenville L. Winthrop, © President and Fellows of Harvard College, 1943.162 (detail), https://hvrd.art/o/311922, **164** (left); National Library of Australia/Edward Backhouse, *A chain gang, convicts going to work near Sidney (i.e. Sydney)*, New South Wales, 1843, Adams Hamilton, London, nla.obj-138467409, **94**, /*The Chinese invasion, northern Queensland* [Melbourne], Illustrated Australian news, 1877, nla.obj-135884777, **126**, /*Captain Cook's landing at Botany, A.D. 1770*, 1872, nla.obj-135775020, **84**, /Extract, James Cook, John, Hutchinson, Samuel Wallis & Henry William Ferdinand Bolckow, *Journal of H.M.S. Endeavour*, 1768, nla.obj-228958440, p299, **69**, /Edward William Cooke, *Prison-ship in Portsmouth Harbour, convicts going aboard*, 1829, nla.obj-135934086, **89**, /Article 'Assassinated, Death of Austrias Heir' *The Evening Echo* (Ballarat, Vic: 1914 - 1918) 29 June 1914: 4 (2ed), nla.news-article241766044, **201** (bottom), /Joseph Finnemore, *Key to the signing of the Treaty of Peace at Versailles*, 1919, nla.obj-139642280, **235** (top), /Thomas Samuel Gill & Macartney & Galbraith, *Pensioners, Forrest Creek*, 1852, nla.obj-135675521, **111**, /Article 'Australia's Offer Accepted' 7 August, 1914, *The Herald* (Melbourne, Vic: 1861 - 1954), p14, nla.news-article242297753, **206**, /*Kanaka labourers on a Queensland pineapple plantation*, nla.obj-136808212, **127** (top), /William Knight, *Collins Street, town of Melbourne*, 1839, nla.obj-135240584, **107**, /Joseph Lycett, *Aborigines cooking and eating beached whales*, Newcastle, New South Wales, 1817, nla.obj-138499828, **70–1**, /Joseph Lycett, *Aborigines using fire to hunt kangaroos*, 1817, nla.pc-an2962715-s20, **73**, /Joseph Lycett, (1824), *View of the heads at the entrance to Port Jackson New South Wales*, 1824, nla.obj-135701554, **86**, /Henry Macbeth-Raeburn and Francis Wheatley, *The pioneer, in 1788 Captain Arthur Phillip R.N. proceeded from Botany Bay to Port Jackson*, 1936, nla.obj-136096089, **96**, /*Melbourne Punch* (Vic: 1855 - 1900), 10 May 1888, nla.news-page20443538, **121** (top), /*Portrait of Bennilong, a native of New Holland, who after experiencing for two years the luxuries of England, returned to his own country and resumed all his savage habits*, 1798, nla.obj-136020945, **97** (bottom); John Constable, *Wivenhoe Park*, Essex, 1816, Widener Collection, 1942.9.10, Courtesy National Gallery of Art, Washington, **17**; National Gallery of Victoria, Melbourne/Fred Kruger, *Victorian Aboriginals and war implements* c. 1883, albumen silver photograph, 13.2 x 20.0 cm (image), Gift of Mrs Beryl M. Curl, 1979 (PH230-1979) **132** (top), /Frederick McCubbin, *The Pioneer*, 1904, **92**; Graph, Esteban Ortiz-Ospina and Diana Beltekian, 'Trade and Globalization' published online at ourworldindata.org/trade-and-globalization, Data source: World Bank – World Development Indicators, CC BY 4.0, **37** (top right); Tom Roberts, *The Opening of the First Parliament of the Commonwealth of Australia*, 1901, Royal Collection Trust, © Her Majesty Queen Elizabeth II 2020, **122–3**; State Library of New South Wales/Samuel Calvert, *Fight between aborigines and mounted whites*, Mitchell Library, **98**, /Engraving from *The narrative of a voyage of discovery :*

performed in His Majesty's vessel the Lady Nelson, of sixty tons burthen, with sliding keels, in the years 1800, 1801 and 1802, to New South Wales, by James Grant, London 1803, **100** (top), /G. Wickham, *Howells' Mill*, Parramatta, 1849, Mitchell Library, **50**; State Library of Victoria/John Batman, *The Batman deed Melbourne*, 1835, **105**, /*Founding of Melbourne August 29 1835*, 1885, **106**, /*Australian Labor Party Anti-Conscription Campaign Committee*, Melbourne, Fraser & Jenkinson, 1916, **209** (bottom), /Poster 'Will you fight now or wait for This', Norman Lindsay, **207** (top), /Propaganda handkerchief, George Sylvestor Naylor, 1915, MS10024, **208**, /*Frescoes for the new houses of parliament. no. VI [picture]; La Trobe and the chieftains resist the landing of the convicts. Melbourne*, 1856, printed and published by Edgar Ray and Frederick Sinnett, **109**, /Poster 'Australia Has Promised Britain 50,000 More Men. Will YOU Help Us Keep That Promise', Syng printer, 1914, **242** (bottom), /Poster 'Enlist in the Sportsmen's 1000. Play up and play the game', Troedel & Cooper lithographers, 1915, **207** (bottom), /John Helder Wedge, *Plan of the Port Phillip District 1835*, **102**; Sydney Living Museums/Ticket of Leave, Thomas Beaton, 1840, Hyde Park Barracks Archaeology collection, HPB2003/23, **91**, (bottom), /Advertising flyer, Private collection, **229**; Map of Edo (Tokyo), 1844-1848 courtesy of the University of Texas Libraries, The University of Texas at Austin, **176**; Zimmerman telegram, 1917, Courtesy of the U.S. National Archives and Records Administration, **217** (top); Royal Proclamation, reproduced in *One hundred and twenty Canadian historical pictures, portraits, and documents, from the Dominion Archives and other sources: selected from the reproductions made for 'Canada and its provinces' and published solely for the subscribers to that work*. Shortt, Adam (1859-1931), Toronto, 1914, Virtual Museum of Canada, **141**; Cartoon, 'A court for King Cholera', Wellcome Collection. CC BY 4.0, **39**.

While every care has been taken to trace and acknowledge copyright, the publisher tenders their apologies for any accidental infringement where copyright has proved untraceable. They would be pleased to come to a suitable arrangement with the rightful owner in each case.

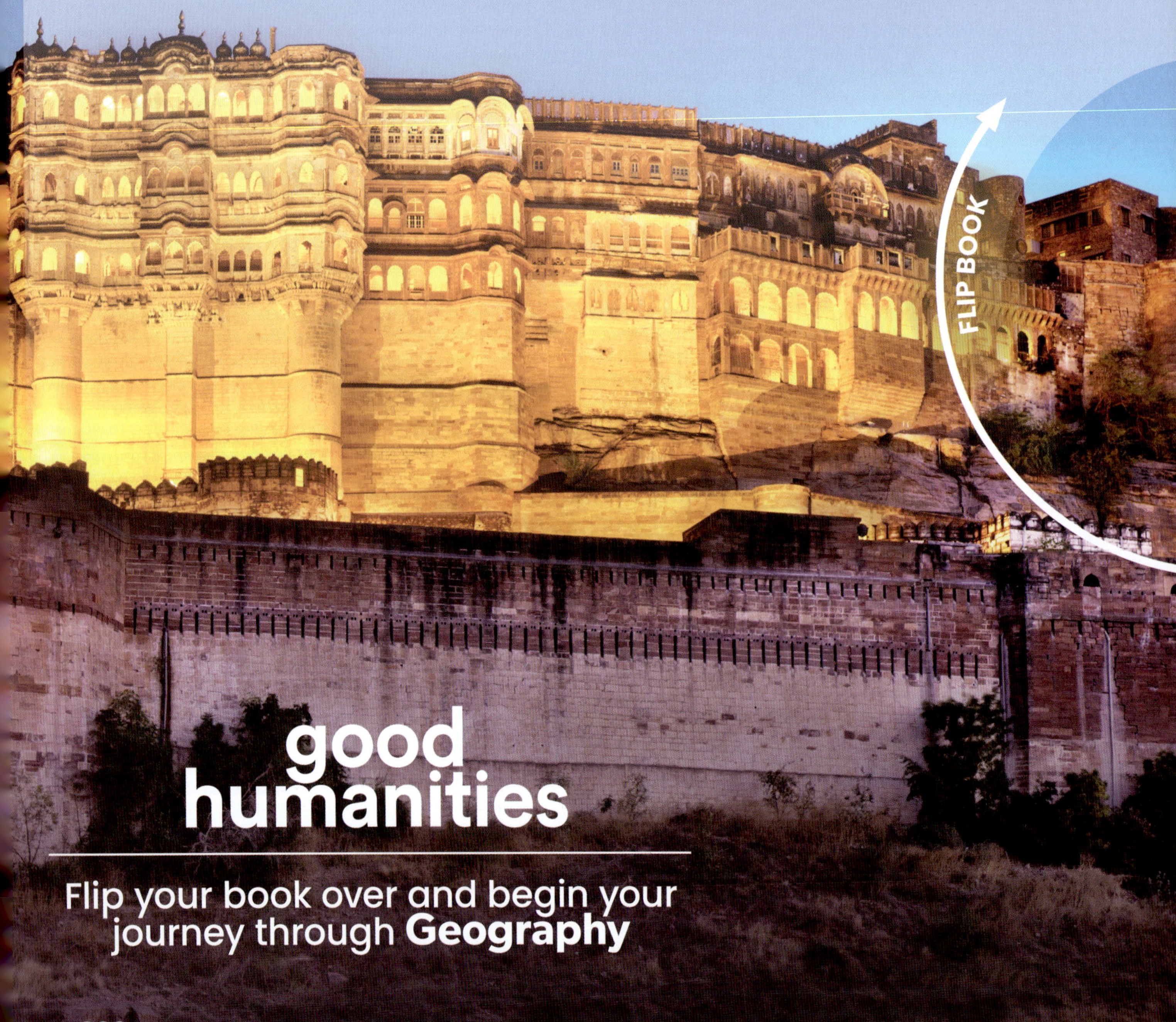

good humanities

Flip your book over and begin your journey through **Geography**

civics+ citizenship

history

geography

economics+ business

contents

G0 Introduction to Geography page 1

Biomes and food security

G1 A natural world page 11

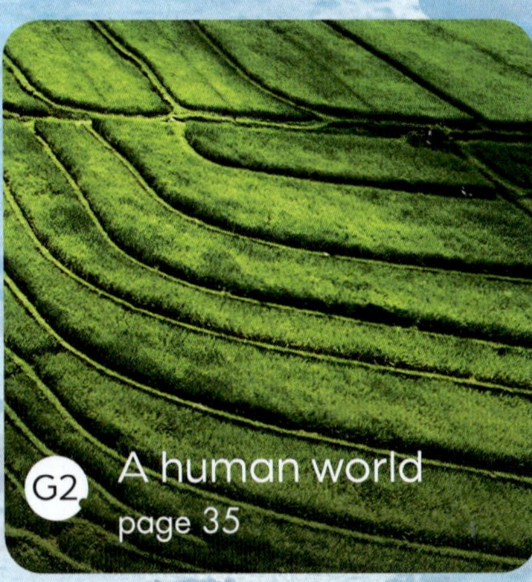
G2 A human world page 35

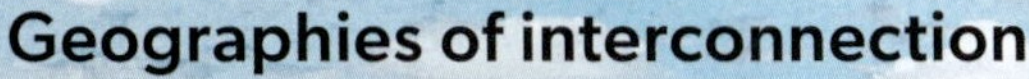

Geographies of interconnection

G3 Perceptions and places page 67

G4 Choices and changes page 95

Geography concepts + skills

G5 Fieldwork page 121

G6 Geo How-To page 131

Good Humanities 9
Victorian Curriculum
1st edition
Damien Green
Ben Lawless
Phillip O'Brien
Danielle O'Leary
Natalie Shephard
Ilja van Weringh
Aleryk Fricker

Publisher: Catherine Charles-Brown
Publishing director: Olive McRae
Copy editors: Kate McGregor, Linden George
Project editors: Kate McGregor, Isabelle Sinclair
Development editor: Patrick O'Duffy, Jana Gabriel
Proofreader: Kelly Robinson
Indexer: Max McMaster
Text and cover designer: Jo Groud
Typesetter: Paul Ryan
Illustrator: QBS Learning
Cartography: Wayne Murphy, MAPgraphics
Production controller: Sarah Blake
Digital production: Erin Dowling
Permissions researchers: Samantha Russell-Tulip,
Hannah Tatton
Cover images photographed by:
Geography: Photo by Getty Images/Wahyu Noviansyah/EyeEm
History: Photy by p1xels, artwork by Cam Scale
Spine: Photos by Getty Images/Dima Viunnyk (rock detail);
Stocksy United/Akela – from alp to alp

First published in 2021 by Matilda Education Australia, an imprint
of Meanwhile Education Pty Ltd
Melbourne, Australia
T: 1300 277 235
E: customersupport@matildaed.com.au
www.matildaeducation.com.au

National Library of Australia Cataloguing-in-Publication data
Author: Damien Green, Ben Lawless, Phillip O'Brien ,
Danielle O'Leary, Natalie Shephard, Ilja van Weringh
Title: Good Humanities 9 Victorian Curriculum
ISBN: 9781420247220

A catalogue record for this book is available from the National Library of Australia at www.nla.gov.au

Warning: It is recommended that Aboriginal and Torres Strait Islander peoples exercise caution when viewing this publication as it may contain names or images of deceased persons.

Printed in Malaysia by Vivar Printing
Sep-2024

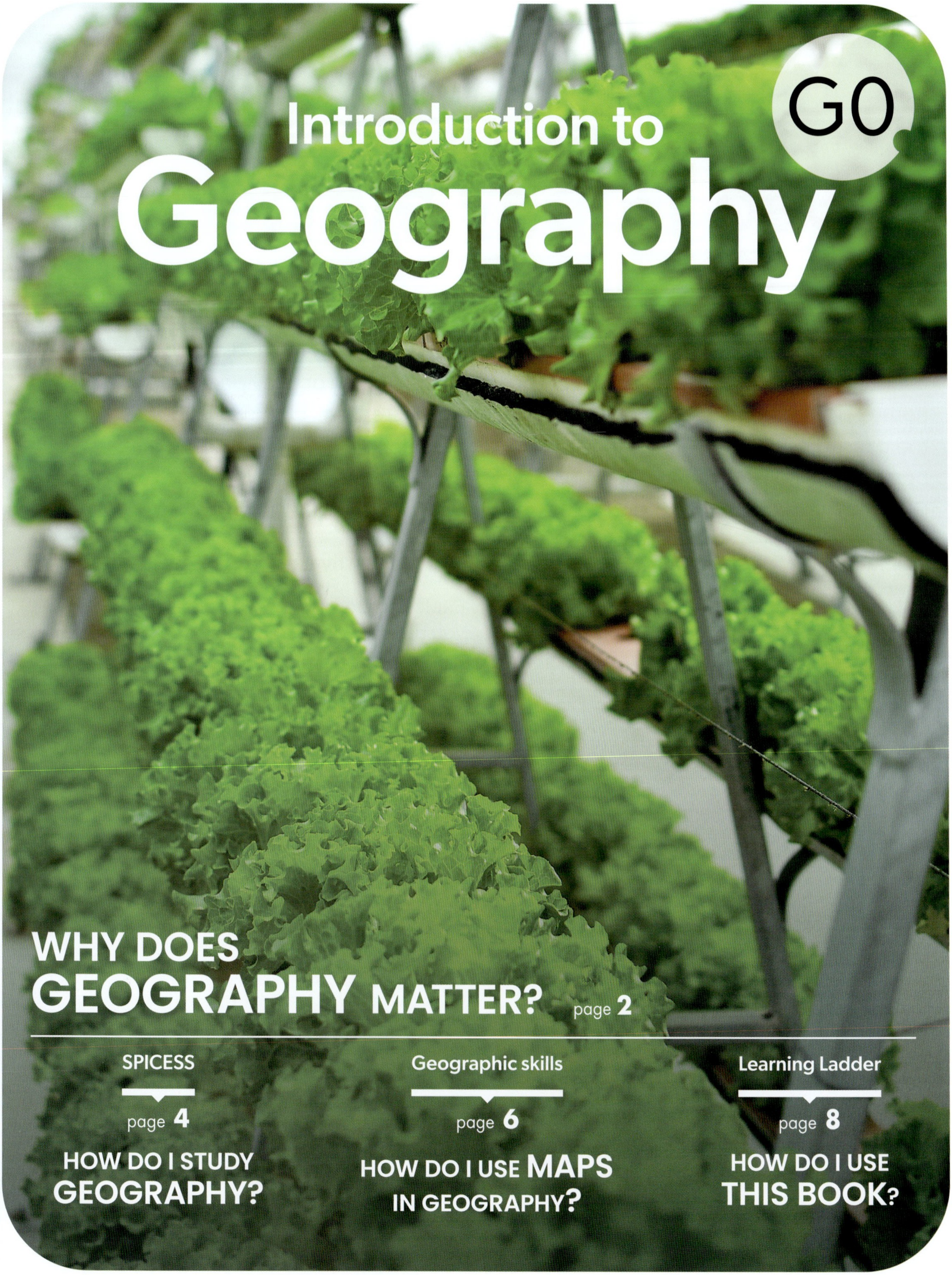

G0

Introduction to Geography

WHY DOES GEOGRAPHY MATTER?

page 2

SPICESS

page 4

HOW DO I STUDY GEOGRAPHY?

Geographic skills

page 6

HOW DO I USE MAPS IN GEOGRAPHY?

Learning Ladder

page 8

HOW DO I USE THIS BOOK?

Why does Geography matter?

Geography is the study of our world – its characteristics, patterns and changes over time. In Geography we focus on two main ideas: human activity and natural processes. We consider changes in the characteristics of places and the consequences of these changes, we look at spatial distributions and find patterns, we consider interconnections over time and at different scales. Geography is key to understanding the world around us and our role in it.

Thinking like a geographer

As a Year 9 Geography student, you have already studied and practised many geographic concepts and skills and you have a rich bank of geographic knowledge to draw from. You have explored how humans influence the *space* in which they live and, in return, how the natural *environment* influences how we as humans live. You may have practised 'thinking like a geographer' while you engaged with important issues, such as climate change and the construction of new coal mines in Australia, as well as positive initiatives such as the clean-up of the Great Pacific Garbage Patch.

This year you will continue to explore contemporary geographic issues using maps, images, graphs and other sources of data. You will explore spatial patterns. You will also conduct your own fieldwork to answer research questions about different phenomena and interconnections. You will consider how your own choices and actions can result in positive changes in our increasingly interconnected world.

Three reasons why Geography matters

1 Exploring Geography will help you develop powerful skills

Geography is an active and vibrant study where you will *learn by doing*. To help you complete your learning, you will continue to develop your geographic skills. These skills will remain with you and will help you in all aspects of your life. Some of these valuable new skills include reading maps, interpreting data, creating visual representations, navigating and understanding how other people experience the places in which they live.

2 Studying Geography lets you see the world (sometimes without leaving your desk!)

Do you want to visit the Sahara Desert? Are you concerned about the impact of climate change on the canals of Venice? Do you daydream about purchasing a round-the-world plane ticket?

The study of Geography allows you to access every corner of the world: either on a fieldtrip or at your desk using Geographic Information Systems (GIS), such as Esri or Google Earth. Through this process you will come to understand the *interconnected* nature of our world and the role we all play as custodians of our planet.

Source 1

A satellite image of Parliament House in Canberra.

2 Understanding Geography makes us informed, global citizens

Studying Geography helps us understand ourselves and our planet. There are two main areas of study in Geography: physical (or natural) geography and human geography. Physical geography is the study of the Earth – its landscapes, landforms and natural processes – some of which have been occurring over hundreds of thousands of years. Human geography is the study of how we interact with our natural world.

This year, you will investigate the Earth's environments and how these environments sustain life; how humans change these environments for food and fibre production; the impacts of globalisation and consumerism on people and places; and the role of tourism, trade and technology in our interconnected world. The study of Geography will empower you to be an active global citizen who makes informed decisions.

Learning ladder G0.1

1. Brainstorm all the skills and concepts you have learned in Geography so far. Discuss as a class.
2. Discuss the difference between geographical knowledge and geographical skills with a partner. Provide examples of each.
3. Consider how your Geography skills have helped you in different subject areas. Hint: Did you need to read maps on camp? Have you used your knowledge of processes in science?
4. 'Being able to think like a geographer today is more important than at any other time in human history.' Hold an informal class debate on this topic. Remember to respect all of your classmates as they express their opinions.

G0.2

How do I study Geography?

In Year 9 Geography you will focus on five geographical skills: spatial distributions and patterns, patterns and interconnections, changes and implications, communicating data and digital and spatial technologies. You will also practise the important skills of research and writing. On the next couple of pages, you can read an introduction to some of these skills before practising them in your first geographical study. The Geo How-To section on pages 131–151 will also give you step-by-step support when you begin to apply these skills.

Geographic concepts

Geographic concepts will help you to think like a geographer. The acronym **SPICESS** can be used as a prompt to help you remember the seven geographic concepts: space, place, interconnection, change, environment, scale and sustainability.

Using geographic language is important when writing your responses. Remember, you do not have to use the exact terms in your responses. Instead, try to use the concept to make your writing more geographical.

Space

In a geographical context, *space* does not refer to 'outer space'; instead, it refers to the ways in which we use, change and distribute things on Earth's surface. For example, this environment can be broken up into two key spaces: the natural mountain space and an agricultural space.

Environment

An *environment* can be defined by its **geographic characteristics**. Some environments are largely natural and are untouched by humans, such as coastlines, islands and forests. Other environments have undergone significant change and are largely unnatural, such as cities and other urban areas. Within environments we can observe processes, interconnections between **phenomena** and change over time. This is an example of both a natural and a human environment.

Change

Change refers to how a place is altered due to shifts in the environment or to meet the needs of humans. Change can be positive or negative, and can occur over short or long periods of time. In this place, we can observe a change over time from the natural mountainous landscape to extensive rice terraces and farmland.

Scale

Scale usually refers to the size of something. Scale can be literal, like a scale on a map showing you using quantitative data how big something is in real life. It can also be used as a qualitative word, such as when describing a region. For example, patterns can exist on a local, regional, national or global scale. This is an example of large-scale agriculture.

Sustainability

Sustainability is the concept of maintaining and preserving resources and environments for future generations. This could be through the use of **sustainable**, renewable energy, such as solar power or wind generated electricity. Small-scale farming can be sustainable; however, when we overuse Earth's resources we degrade its nutrients and alter natural processes.

Source 1

Terraced rice fields in Mù Cang Chải, Yên Bái Province, Vietnam

Interconnection

Interconnection is the idea that two things or phenomena are related, interact or are linked in some way. For example, there is a strong interconnection between the success of an agricultural harvest and the economy of the local region.

Place

The concept of *place* allows humans to identify the location or position of something within a space. We can identify place through describing the **relative** or **absolute location** of that area. This place is an agricultural field and rice terraces in Vietnam.

Learning ladder G0.2

1. Identify the place shown in Source 1 on page 3.
2. Locate an image of a natural landscape online. Annotate the image with the SPICESS terms, explaining how each term relates to the image.
3. Create a photo essay that illustrates the variation of natural landscapes around the world.
4. For each of the images collected for the photo essay in the previous question, discuss as a class how that place or environment may change over time.

How do I use maps in Geography?

Maps are vital in Geography to visualise spatial patterns and identify *interconnections* between phenomena. There are many different types of maps that all have slightly different uses.

Physical maps

Physical maps show the terrain or natural features of a location. For example, in the map of Africa in Source 1, you can see the large expanse of desert in the north of the continent and the forests around the Congo Basin.

Source 1

This physical map shows the natural features of Africa.

A physical map of Africa

Source: Matilda Education Australia

Cartograms

Cartograms are maps that have been distorted by data. For example, in the map in Source 2, countries that are larger in size are those that are the most popular with tourists, while narrow, squished looking countries are those that have far fewer visitors. The UK and Europe are large because they are very popular tourist destinations, while Australia is narrow because it is less popular.

Source 2

The cartogram is another way to represent data; for example, how popular particular countries are as tourist destinations.

Popularity of different destinations for tourists, 2017

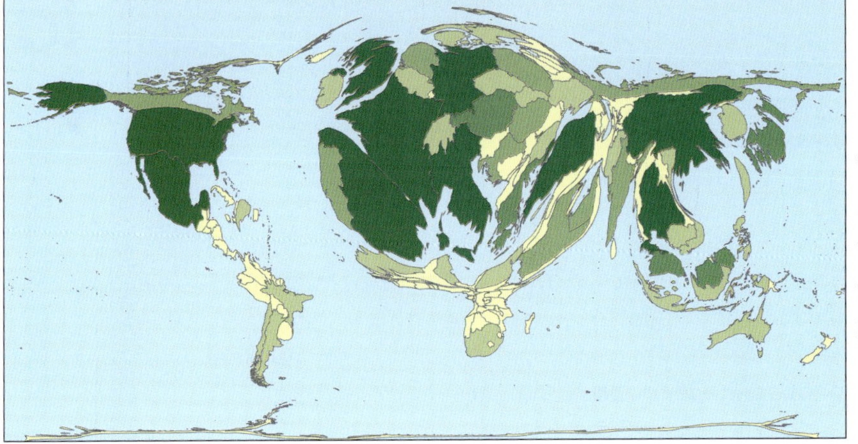

Source: Matilda Education Australia, UN World Tourism Organisation

Choropleth maps

Choropleth maps use darker and lighter shades of colour so that the reader can instantly see a pattern. The choropleth map in Source 3 uses darker shades to represent the areas with the highest areas of species richness and the lighter shades represent the areas with the lowest levels of species richness. It is best to use the same colour in different shades, such as light to dark green, to clearly show the pattern.

Bird species richness in North America

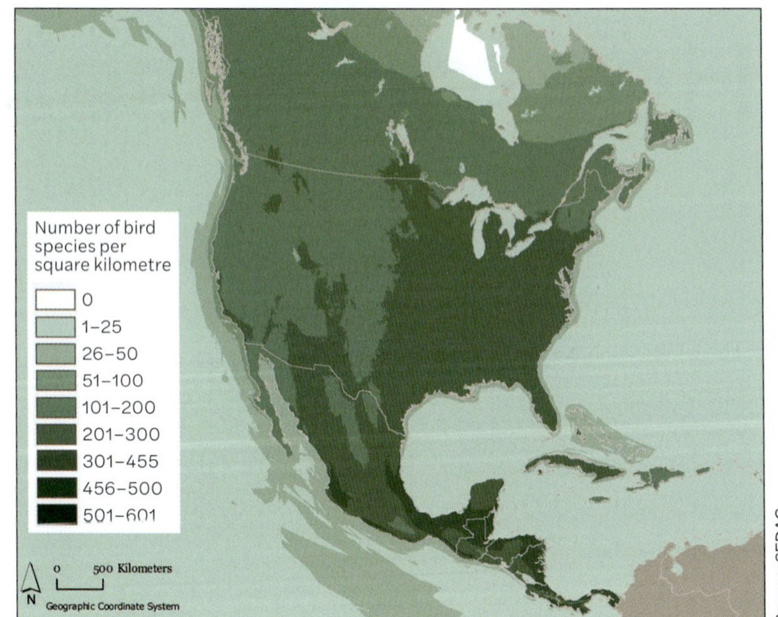

Source: SEDAC

Source 3

A choropleth map showing species richness (birds) in North America

Australia's climate zones and ocean currents

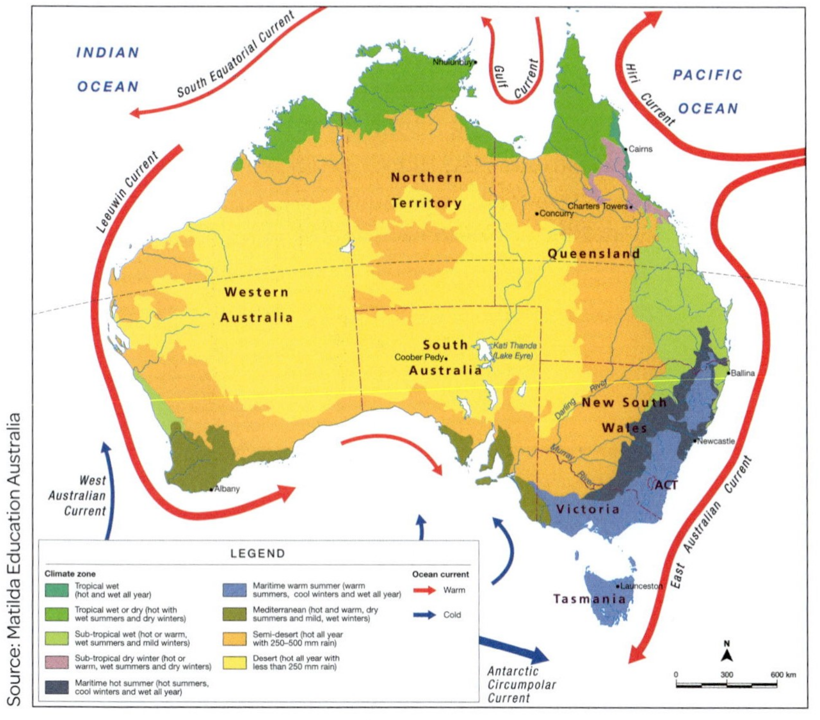

Source: Matilda Education Australia

Climate maps

In Geography, **climate** is important for determining land cover and how liveable a place is. While weather is what we observe day to day through our classroom window, climate is the average conditions (usually temperature and **precipitation**) within a region over a period of time. Climate is important when studying Geography as it gives us an indication of expected rainfall. In the map in Source 4 we can clearly see different climate zones on the continent of Australia.

Source 4

A climate map of Australia

Learning ladder G0.3

1. Suggest which map type would be best to display the following data:
   a. population distribution between regions
   b. location of forests in a country
   c. flow of rivers on a global scale
   d. differences in temperature across a country.
2. Locate a map showing the distribution of the world's human population. What map type is used to display this data? Why?
3. Source 2: Using your understanding of cartograms, list three places (continents or countries) that are not popular tourist destinations.

How do I use this book?

Good Humanities has been built to help you thrive as you move through the Level 9 Humanities curriculum and to demonstrate your progress in every lesson. The Geography section of this book explores two geographical topics: biomes and food security, and geographies of interconnection. You will also find a Fieldwork section and a Geo How-To skills section. The Geo How-To section is vital and you should refer to it often.

Climb the Learning ladder

Geography is a skills-based subject. This means that learning content alone is not enough to give you a geographical understanding. In order to be a geographer, you need to be able to write geographically, read a variety of data, interpret data, and conduct and communicate your own research.

Each chapter in this book begins with a Learning ladder. This is your 'plan of attack' for the skills you will practise in each chapter. It lists the five geographical skills you will be learning and five levels of progression for each of those skills. Read it from the bottom to the top. As you progress through the chapter, you will climb UP the Learning ladder.

Each skill described in the Learning ladder is a higher progression than the one before it. To be able to accomplish the higher-level skills, you need to be able to master the lower ones. Practising activities at all the levels will help you develop 'higher-order' skills, such as evaluating. This approach is called developmental learning and puts you in charge of your own learning progression.

Source 1

The Learning ladder helps you to take charge of your own learning!

learning ladder

Step	Spatial distributions and patterns	Patterns and interconnections	Changes and implications	Communicate data	Digital and spatial technologies
step 5	I can identify multiple spatial distributions and patterns	I can interpret causes of patterns and interconnections	I can interpret data to quantify predictions based on research	I can evaluate the success of research methods	I can draw conclusions from geographical information in digital and spatial technologies
step 4	I can use data to support exceptions to spatial distributions and patterns	I can use relevant sources to research further reasons for patterns and interconnections	I can make predictions and outline consequences of change over time	I can organise data collected according to relevance for a research question	I can manipulate data using digital and spatial technologies
step 3	I can describe spatial distributions and patterns	I can use data to support explanations of patterns and interconnections	I can explain the causes behind the change over time in a place	I can filter collected data	I can access and use spatial technology platforms such as GIS
step 2	I can use data to quantify spatial distributions and patterns	I can explain patterns and interconnections	I can describe how places have changed over time	I can successfully use data collection methods	I can construct paper maps using correct cartographic conventions
step 1	I can identify spatial distributions and patterns	I can provide short explanations for patterns and interconnections	I can identify that changes occur in the characteristics of places over time	I can list primary and secondary methods useful for my study	I can interpret different map types using cartographic conventions

Check your progress

Each chapter is divided into multiple sections. Each section is designed to cover one lesson, but sometimes your teacher might decide to spend more or less time on a particular section. A section is usually two pages long, but some are four pages. At the end of every section, you will find a block of questions to help you check your progress.

1 Show what you know

These questions ask you to look back at the content you have read and viewed and to show your understanding of it by listing, describing and explaining. Sometimes these questions will take you outside your classroom.

Learning ladder

These activities are linked to the Learning ladder. You can complete one of the questions or several of them, depending on your progress. Throughout a chapter you will complete at least one activity for each level of the Learning ladder.

Case studies

Throughout every chapter, you will discover a variety of case studies that ask you to think about local, national and global places, issues and events. A case study is an in-depth exploration of a single subject (the case) and is usually based on data or research. The local case studies are focused on Victorian contexts. The national case studies explore a case study from elsewhere in Australia, and the global case studies describe places and events in other countries. Many of these link to interactive sources so you can explore further using your eBook.

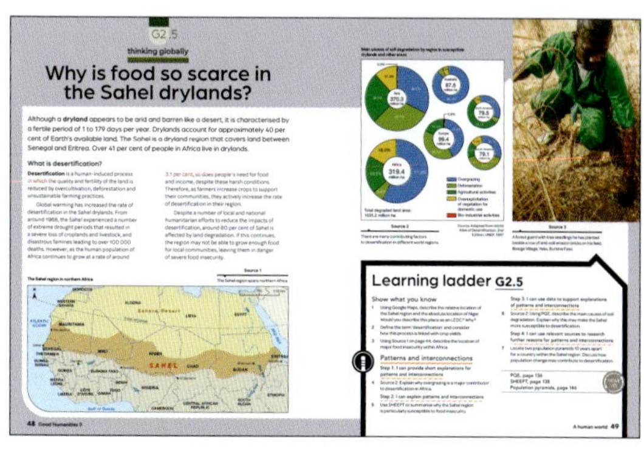

Source 3

Case studies are an important part of your Geography course.

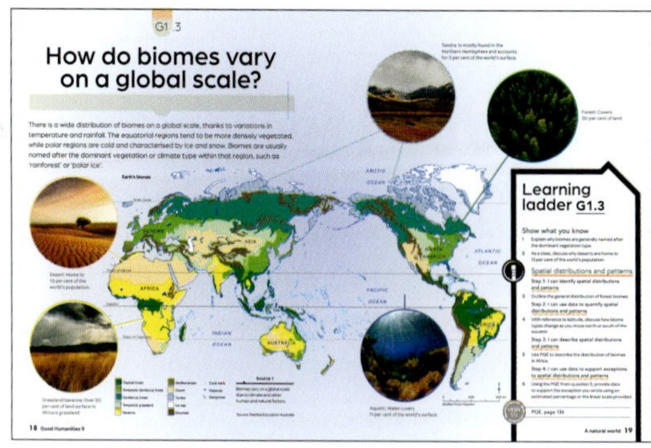

Source 2

Check your progress regularly. You can attempt one or several of the Learning ladder questions.

civics+ citizenship | economics+ business

The study of Humanities is more than just Geography and History – it is also the study of Civics and citizenship and Economics and business. In every chapter of both the Geography and History sections of this book, you will discover either a Civics and citizenship *or* an Economics and business lesson. School is busy and you have a lot to cover; designing a textbook where the important Civics and citizenship and Economics and business content is placed meaningfully next to relevant Geography or History lessons makes good sense and will help you connect your learning.

Geo How-To

In the middle of the book, you will find a guide to fieldwork and a skills chapter called Geo How-To. The How-To chapter explains how to perform each skill and the steps involved in writing and research. There are *lots* of examples. Refer to it often, especially when completing the Learning ladder questions and Masterclass activities.

The Fieldwork chapter of your textbook explains all the skills you need for hands-on research and gives you several suggested tasks.

Source 4

The Geo How-To section is your key to success – refer to it often!

Masterclass

At the end of each chapter is a review section, called the Masterclass. The questions here are organised by the steps on the Learning ladder. You can complete all of the questions *or* your teacher might direct you to complete just some of them, depending on your progress.

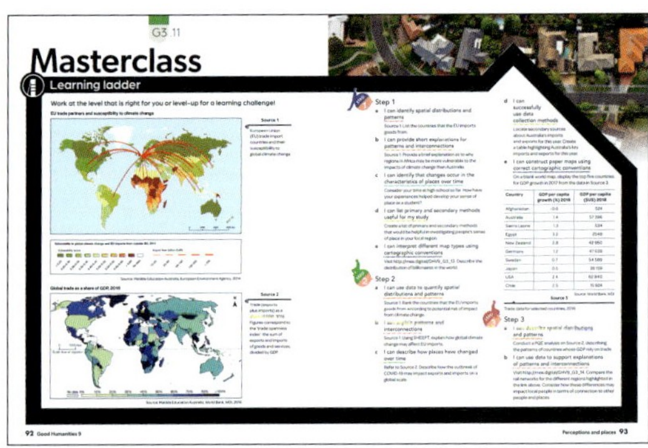

Source 5

The Masterclass is your opportunity to show your progress. Take charge of your own learning and see if you can extend yourself.

Capstone

After you complete a chapter, it's time to put your new knowledge and understanding together for the capstone project to show what you *know* and what you *think*. In the world of building, a capstone is an element that finishes off an arch, or tops off a building or wall. That is what the capstone project will offer you, too: a chance to top off and bring together your learning in interesting and creative ways. It will ask you to think critically, to use key concepts and to answer 'big picture' questions. The capstone project is accessible online; scan the QR code to find it quickly.

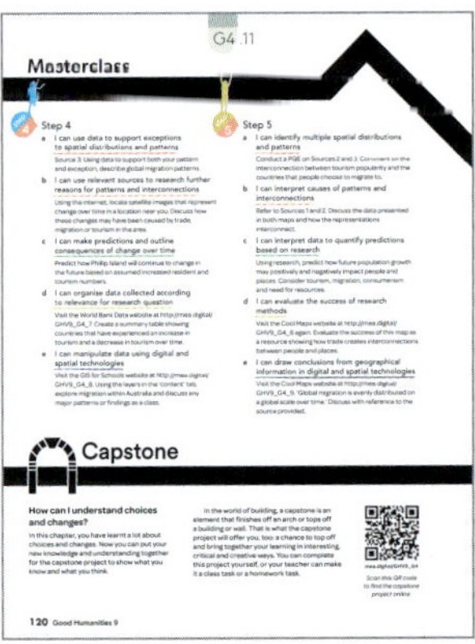

Source 6

The capstone project brings together the learning and understanding of each chapter. It provides an opportunity to engage in creative and critical thinking.

Learning ladder G0.4

1. What skills do you think are important in Geography? How are they reflected in the Learning ladder?
2. How can you use the Learning ladder to monitor your progress in Year 9 Geography?
3. As a class discuss the idea of 'monitoring your own progress'. Why is this important?
4. Read through the steps of the Geography Learning ladder and consider where you may already be up to for each skill, based on your prior learning.

G1

A natural world

WHY IS STUDYING BIOME PRODUCTIVITY IMPORTANT?

page 26

digital + spatial technologies

page 14

HOW DO WE CLASSIFY ENVIRONMENTS?

thinking nationally

page 22

WHY ARE AUSTRALIA'S BIOMES SO VARIED?

thinking globally

page 28

HOW DOES PRODUCTIVITY VARY IN NORTH AMERICA?

How can I understand our natural world?

Earth is a big planet! To help with their study of Earth's environments, geographers divide the planet into distinct areas known as **biomes**. Each biome has its own unique climates, soils, vegetation and levels of productivity. The distribution of biomes on Earth is influenced by a number of natural and human factors.

learning ladder

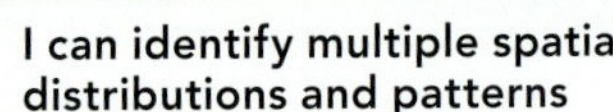

Step	Spatial distributions and patterns	Patterns and interconnections	Changes and implications
step 5	**I can identify multiple spatial distributions and patterns** I can take my PQE one step further to find links or relationships that exist in the natural world.	**I can interpret causes of patterns and interconnections** I can use multiple sources to find links or relationships that exist in the natural world and can explain 'Why?'.	**I can interpret data to quantify predictions based on research** I can use external data from research as evidence of the positive and negative impacts of a change I have predicted.
step 4	**I can use data to support exceptions to spatial distributions and patterns** I can use data to answer 'Why?' about the exceptions identified in a PQE analysis of our natural world.	**I can use relevant sources to research further reasons for patterns and interconnections** I can use sources other than this textbook to further research patterns I observe in the natural world.	**I can make predictions and outline consequences of change over time** I can use my knowledge of natural processes and world regions to make an educated guess of the positive and negative impacts of change in the natural world.
step 3	**I can describe spatial distributions and patterns** I can describe patterns, quantify them and point out exceptions (PQE) to describe our natural world.	**I can use data to support explanations of patterns and interconnections** I can use data from a map or graph to explain patterns I observe in the natural world.	**I can explain the causes behind the change over time in a place** I can use my knowledge of natural processes and world regions to explain why changes may occur over time in the natural world.
step 2	**I can use data to quantify spatial distributions and patterns** I can read data and use it to measure key trends on a map or graph about the natural world.	**I can explain patterns and interconnections** I can identify social, historical, economic, environmental, political and technological (SHEEPT) factors to help me explain the natural world.	**I can describe how places have changed over time** I can use specific examples to describe changes over time in the natural world.
step 1	**I can identify spatial distributions and patterns** I can find key trends on a map or graph about the natural world.	**I can provide short explanations for patterns and interconnections** I can write descriptions of patterns and interconnections that I find in the natural world.	**I can identify that changes occur in the characteristics of places over time** I can read information and answer questions about changes over time in the natural world.

Source 1

The tundra biome is distinctive because of its extremely cold climate and low rainfall. There are very few animal and plant species in this biome.

Warm up

I can evaluate the success of research methods
I can look back and comment on the data collection methods I used and evaluate how successful they were in helping me answer a research question about the natural world.

I can organise data collected according to relevance for a research question
I can review the data I have collected in the field and display it using graphs, tables, annotations and captions.

I can filter collected data
I can review my collected data and select the most relevant data to answer a research question about the natural world.

I can successfully use data collection methods
I can use primary and secondary data collection methods in the field and classroom to investigate the natural world.

I can list primary and secondary methods useful for my study
I can create a checklist of methods to investigate the natural world and categorise them as primary or secondary methods.

Communicate data

I can draw conclusions from geographical information in digital and spatial technologies
I can interpret and analyse patterns by using different layers and features on spatial technology platforms.

I can manipulate data using digital and spatial technologies
I can work with layers and other features on spatial technology platforms to further explore data and interconnections.

I can access and use spatial technology platforms such as GIS
I can use spatial technology platforms to explore data and find patterns.

I can construct paper maps using correct cartographic conventions
I can use a pencil, paper and ruler to construct a map that follows BOLTSS conventions.

I can interpret different map types using cartographic conventions
I understand data found in different types of maps and graphs and use the data to answer questions about the natural world.

Digital and spatial technologies

Spatial distributions and patterns

1 Describe the distinctive features of the biome shown in Source 1.

Patterns and interconnections

2 Identify how the characteristics of the biome shown in Source 1 might be interconnected with the adaptations of local animal and plant species.

Changes and implications

3 Predict changes that might occur in the biome in Source 1 over the following time periods:
a 6 months
b 75 years.

Communicate data

4 What are the limitations associated with fieldwork in the tundra biome?

Digital and spatial technologies

5 Use an outline map of the world to colour where you predict tundra biomes are found on Earth. Annotating your map, explain the reasoning behind your predictions.

How do we classify environments?

Earth can be divided into four main spheres: the atmosphere, hydrosphere, lithosphere and biosphere. The four spheres interact with each other as elements move within and between them.

Source 1

Earth's four spheres all interact with each other.

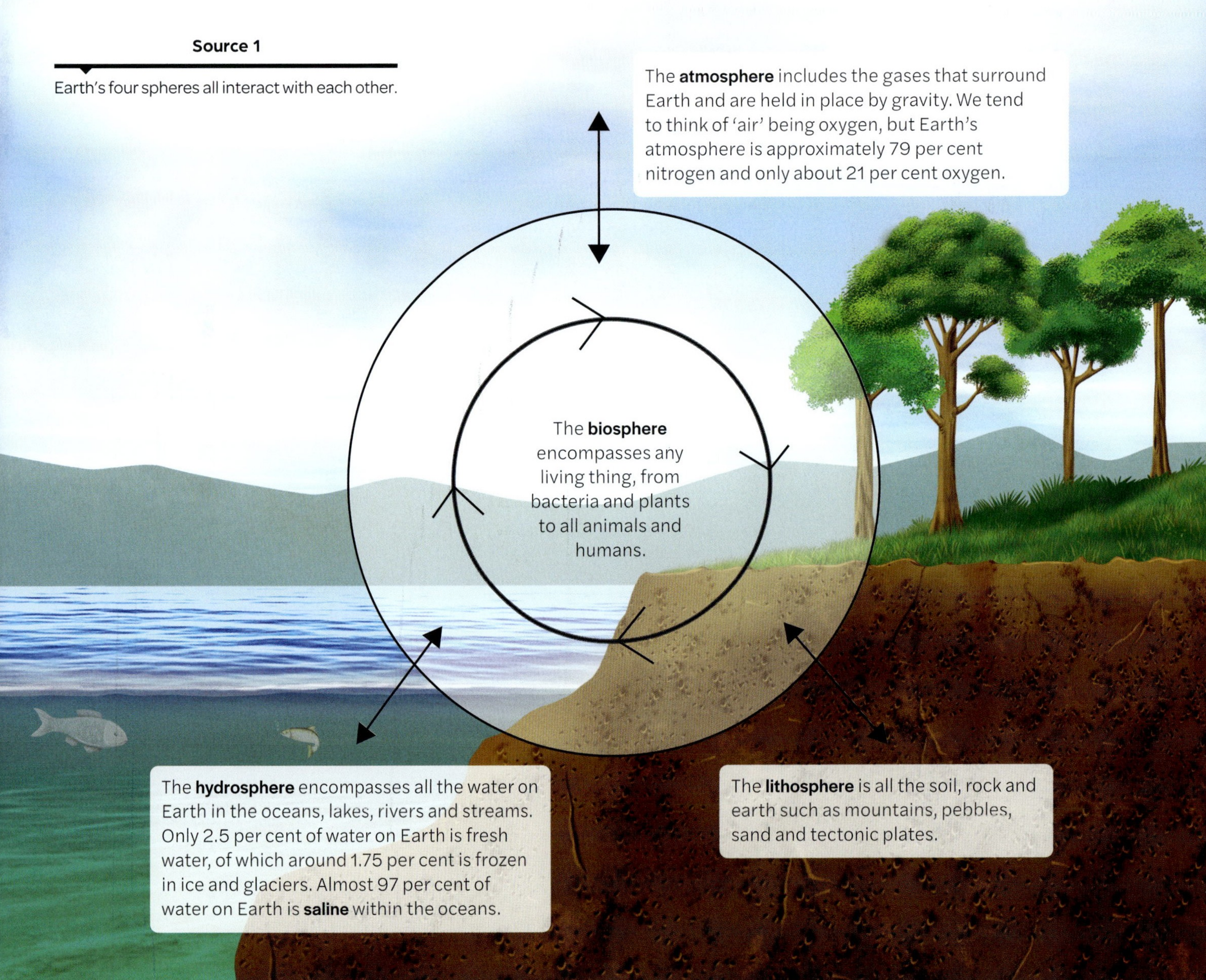

Source 2

Within Earth's biosphere there are many biomes. However, most geographers identify five main types of biome on Earth. Some geographers break up the forest biome into temperate and taiga/ boreal.

Biomes

The biosphere can be further broken up into sub-groups called biomes. Biomes are determined by the characteristics of places and **ecosystems**. Biomes vary depending on location, climate (average temperature and precipitation), soil type and human interaction.

There is no set number of biomes on Earth. Some geographers simply divide Earth into five main biomes – aquatic, grassland, forest, desert and tundra – while other geographers use a much more detailed set of biomes.

Temperature and rainfall are the main factors that determine biomes. High temperatures and low rainfall lead to deserts, while low temperatures and high rainfall create tundra.

Source 3

The effect of temperature and rainfall on biomes

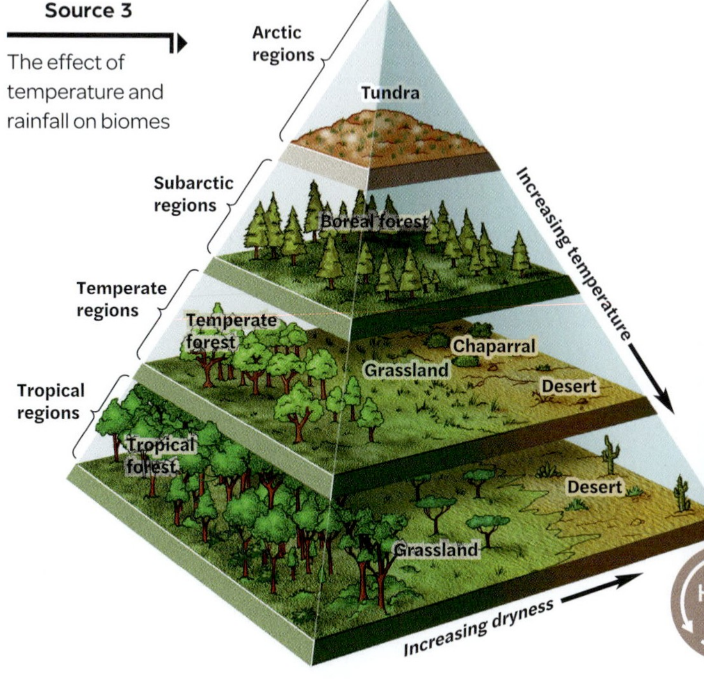

Learning ladder G1.1

Show what you know

1 What are the five types of biome that some geographers use to describe our planet?

2 Construct a four-way Venn diagram to illustrate how Earth's four spheres interact with each other.

Digital and spatial technologies

Step 1: I can interpret different map types using cartographic conventions

3 Consider Source 3. What information is shown here, and why is it useful in understanding about biomes?

Step 2: I can construct paper maps using correct cartographic conventions

4 Using a blank world map and correct conventions, show the distribution of the five global biomes.

Step 3: I can access and use spatial technology platforms such as GIS

5 Access http://mea.digital/GHV9_G1_1. to explore the world's biomes. Which are located in Australia?

Step 4: I can manipulate data using digital and spatial technologies

6 Access http://mea.digital/GHV9_G1_2. Identify the locations and biomes of the following latitudes and longitudes:

a (52.1° N, 19.4° E)

b (74.8° N, 41.4° W)

HOW TO BOLTSS, page 132

Is there a pattern to biome distribution?

The types of biome in an environment vary depending on where you are in the world. Biome distribution is influenced by natural and human factors. In general, there is a pattern to the way biomes are distributed on the planet, and we can use this information to make predictions. As the climate changes, the pattern of biome distribution may continue to change.

Solar radiation and the latitudinal effect

The Sun's rays are a critically important input for processes within biomes, such as **photosynthesis** in plants. Plants use photosynthesis to create energy to grow. Animals, in turn, feed on plants to support complex ecosystems.

On Earth, the **equator** (zero degrees latitude) receives most of the Sun's rays (see Source 1), while the poles (90 degrees north and south of the equator) receive the least. Therefore, the equator tends to be associated with biomes such as forests and grasslands, where lush vegetation requires vast amounts of sunlight to grow. As you move away from the equator, the reduced energy provided by the Sun results in lower vegetation density, and biomes such as savannahs, deserts and taigas (snow forests) are more common. At 90 degrees north and south of the equator, the poles are characterised by tundras, polar ice and snow.

Atmospheric circulation

Precipitation is just as important as the Sun's energy for plant growth and the development of ecosystems. Like the Sun's rays, precipitation can also vary depending on latitude. The **Hadley cell** (see Source 2) is the name for the circulation of air in the atmosphere near the equator. At zero degrees latitude the warm air rises and water is evaporated, leading to cloud formation and large amounts of rainfall. This precipitation also contributes to the dense vegetation found near the equator. At approximately 30 degrees north and south the air descends, meaning there is less precipitation, less vegetation and more dry biomes, such as deserts.

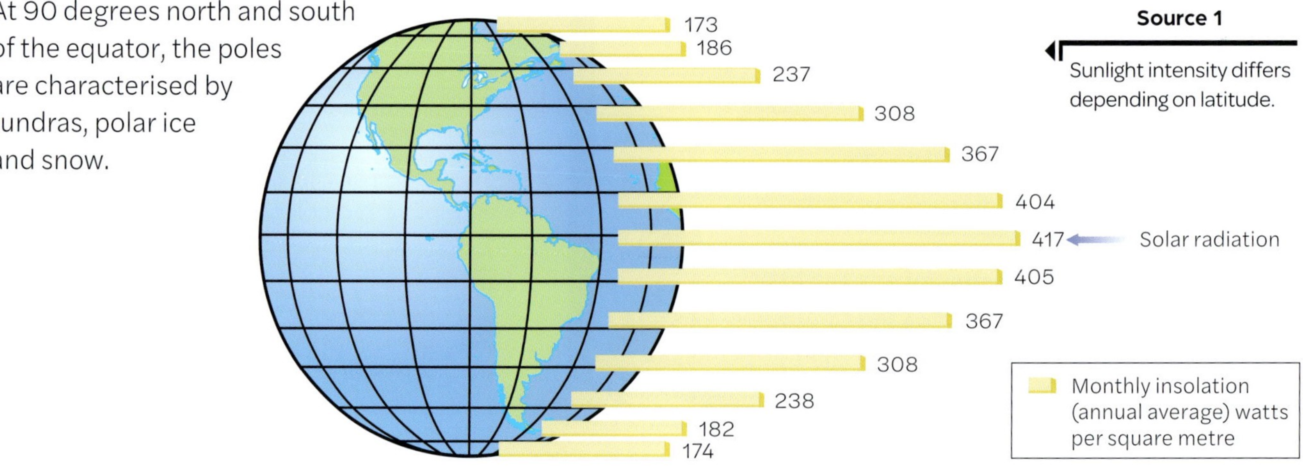

Source 1

Sunlight intensity differs depending on latitude.

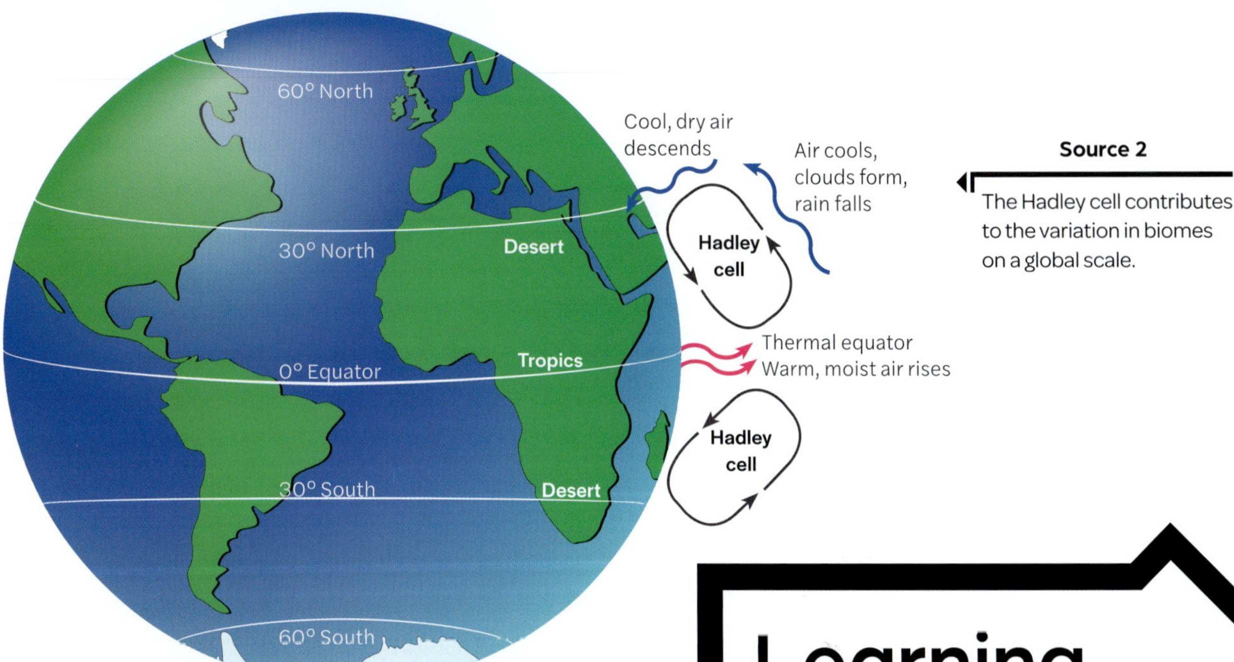

Source 2

The Hadley cell contributes to the variation in biomes on a global scale.

Source 3

A rainforest biome in northern Ecuador, near the equator

Source 4

Most desert biomes are found along the Tropics of Capricorn and Cancer.

Source 5

Glaciers in Greenland within the Arctic Circle

Learning ladder G1.2

Show what you know

1 Outline two natural factors that affect the distribution of biomes on a global scale.

2 What kind of biome would you expect to see near the Tropic of Cancer?

3 Explain the process of the Hadley cell and its impacts on environments in different regions.

Changes and implications

Step 1: I can identify that changes occur in the characteristics of places over time

4 Consider Sources 3–5. Brainstorm two ideas to explain how biomes might change over time in a region due to changes in natural processes.

Step 2: I can describe how places have changed over time

5 Identify three places that are currently undergoing environmental change, then locate an image to illustrate each change.

Step 3: I can explain the causes behind the change over time in a place

6 Using the images you collected in question 5, provide one reason for environmental change at each location selected. Consider climate, atmospheric circulation and human needs.

Step 4: I can make predictions and outline consequences of change over time

7 Evaluate how the human population's increased need for resources may play a role in altering the world's biomes over time.

HOW TO

Photo essays, page 143

How do biomes vary on a global scale?

There is a wide distribution of biomes on a global scale, thanks to variations in temperature and rainfall. The equatorial regions tend to be more densely vegetated, while polar regions are cold and characterised by ice and snow. Biomes are usually named after the dominant vegetation or climate type within that region, such as 'rainforest' or 'polar ice'.

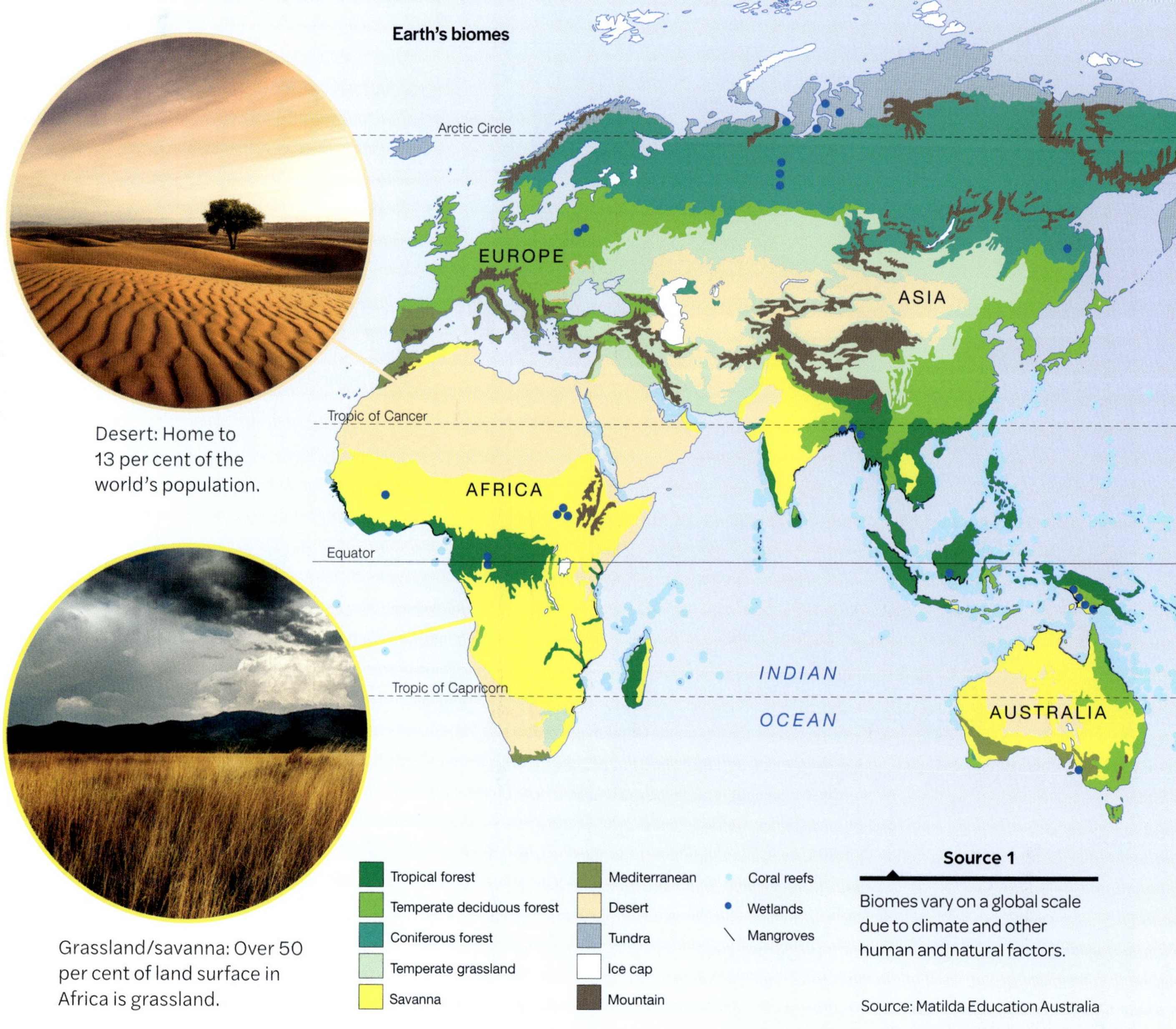

Desert: Home to 13 per cent of the world's population.

Grassland/savanna: Over 50 per cent of land surface in Africa is grassland.

Source 1

Biomes vary on a global scale due to climate and other human and natural factors.

Source: Matilda Education Australia

Tundra: Is mostly found in the Northern Hemisphere and accounts for 3 per cent of the world's surface.

Forest: Covers 30 per cent of land.

ARCTIC OCEAN

NORTH AMERICA

ATLANTIC OCEAN

PACIFIC OCEAN

SOUTH AMERICA

0 2000 4000 km

N

Modified Times Projection

Aquatic: Water covers 71 per cent of the world's surface.

Learning ladder G1.3

Show what you know

1 Explain why biomes are generally named after the dominant vegetation type.

2 As a class, discuss why deserts are home to 13 per cent of the world's population.

Spatial distributions and patterns

Step 1: I can identify spatial distributions and patterns

3 Outline the general distribution of forest biomes.

Step 2: I can use data to quantify spatial distributions and patterns

4 With reference to latitude, discuss how biome types change as you move north or south of the equator.

Step 3: I can describe spatial distributions and patterns

5 Use PQE to describe the distribution of biomes in Africa.

Step 4: I can use data to support exceptions to spatial distributions and patterns

6 Using the PQE from question 5, provide data to support the exception you wrote using an estimated percentage or the linear scale provided.

HOW TO

PQE, page 136

How do biomes vary?

Earth's biomes vary in many ways. Climate and soil are two distinctive features of a biome. These have an impact on the animals and plants that will thrive in a particular area and affect the productivity of the biome.

Distinctive climates

Unlike weather, which changes daily, climate is the average temperature and average precipitation in a region usually measured over a year. As the growth of plant life and success of animal communities largely depends on the availability of sunlight and water, the latitudinal effect and Hadley cell play a major role in determining the local climate and type of biomes present in any given place on Earth.

We then classify each biome based on the type of vegetation or local climate. For example, the term 'tropical' tends to be associated with warm average temperatures and high precipitation. People like to visit tropical locations as they tend to be warm holiday spots, with lush forests and extensive wildlife to explore.

Variation in soil

Soil type also varies depending on location and can be influenced by climate through **weathering**, **leaching** and **erosion**. Soil is a **non-renewable resource**; it can take up to 500 years to develop a new one-centimetre layer of fertile top soil. The soil in biomes that have a high density of deciduous vegetation tends to be some of the most fertile in the world. This is because seasonal falling leaves create a natural compost layer on the earth.

Source 1

Soil type can influence the growth of vegetation and therefore the type of biome that exists in that location.

Coniferous forest soil

Tropical rainforest soil

Desert soil

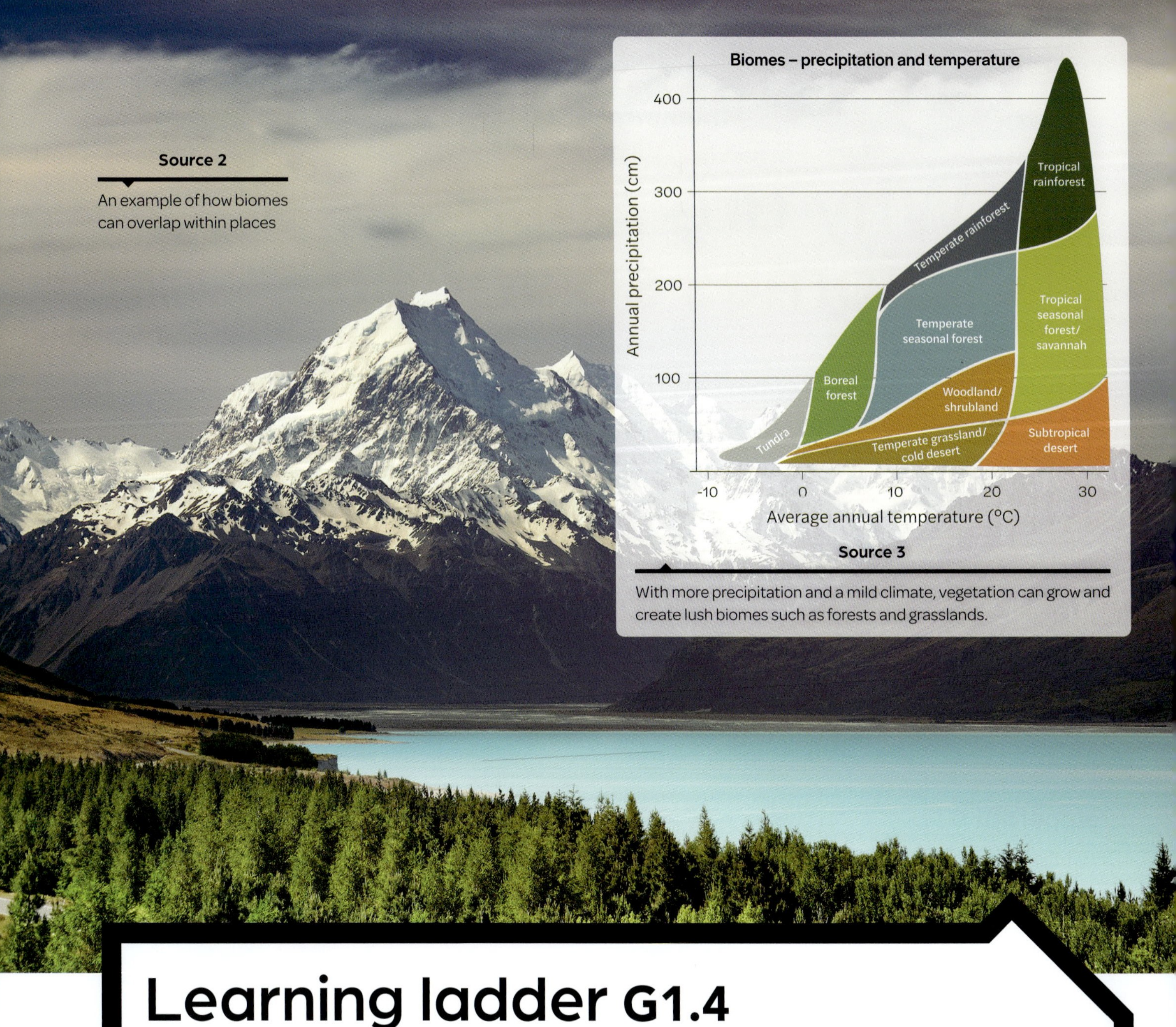

Source 2

An example of how biomes can overlap within places

Source 3

With more precipitation and a mild climate, vegetation can grow and create lush biomes such as forests and grasslands.

Learning ladder G1.4

Show what you know

1 Discuss why water and sunlight are vital to the animals and plants within a biome.

2 Explain why soil type is strongly interconnected with whatever biome is present in that region.

3 Create a Venn diagram comparing the concepts of 'weather' and 'climate'.

Communicate data

Step 1: I can list primary and secondary methods useful for my study

4 Imagine your school principal has asked you to collect data to determine what biome your school is in. Create a list of primary and secondary methods that would assist you to collect this information.

Step 2: I can successfully use data collection methods

5 Apply one of the methods you outlined in question 4 to gather some relevant data. Discuss your findings as a class.

Step 3: I can filter collected data

6 Outline what data has been collected in Source 3. How would this information be helpful to Geographers?

7 Access http://mea.digital/GHV9_G1_3 and use the information to identify:

a the hottest month in Melbourne

b the wettest month in Melbourne.

Step 4: I can organise data collected according to relevance for a research question

8 Climate is determined by collecting data on average temperatures and precipitation. Using the weblink from question 7, create a summary table that shows the average maximum temperatures and mean rainfall for Melbourne over a calendar year.

G1.5

thinking nationally

Why are Australia's biomes so varied?

Australia has a vast array of biomes. This is largely due to the size of Australia's landmass, which stretches from approximately 11 degrees to 44 degrees south of the equator. Australia's biomes are largely dependent on climate and human interaction.

Rainfall

Australia's rainfall is heavy in the tropical regions in the north of the country and in Tasmania, but much of the inland regions receive less that 200 millimetres of rain annually.

Along the coast of Australia, rainfall is between 800 and 1600 millimetres per year.

Temperature

Australia's temperature heavily affects its biomes. Nearly the whole country has an average daily maximum over 15 degrees Celsius, while the far north of Australia is regularly higher than 27 degrees.

Only along the Great Dividing Range and in Tasmania do average daily maximums reach less than 10 degrees.

Source 1

Biomes vary on a regional scale within Australia.

Biomes of Australia

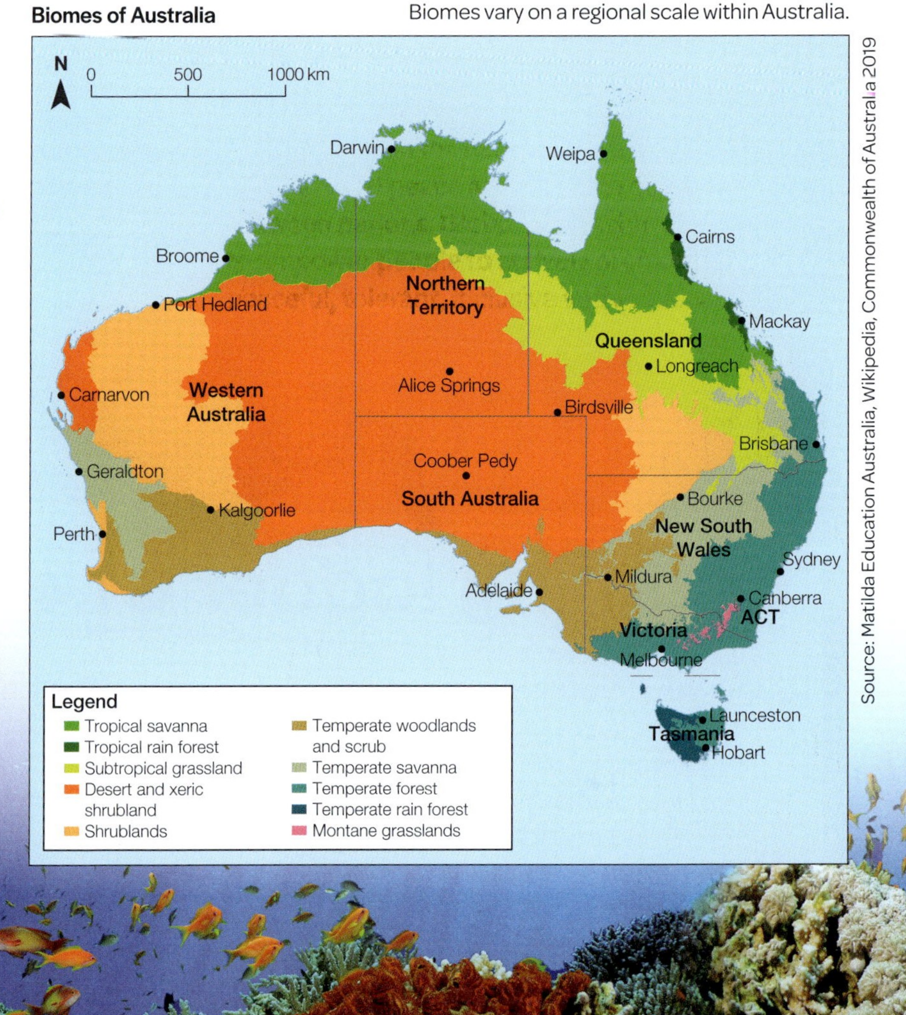

Source: Matilda Education Australia, Wikipedia, Commonwealth of Australia 2019

Source 2

Australia has a vast range of biomes, from desert to rainforest, because of its varied climate.

Learning ladder G1.5

Show what you know

1 Create a list of the biomes present in Australia.

2 Consider your local area. Which biome best describes your local environment and why?

3 Develop a photo essay that highlights the biomes you listed in question 1. Annotate each photo with an explanation of why that biome may occur in Australia (consider natural and human factors).

Patterns and interconnections

Step 1: I can provide short explanations for patterns and interconnections

4 Provide a brief explanation of Australia's biome distribution with reference to its climate.

Step 2: I can explain patterns and interconnections

5 Using SHEEPT, explain why the biome types in Australia differ from those in Antartica.

Step 3: I can use data to support explanations of patterns and interconnections

6 Using the scale on Source 1, calculate the width of the desert and shrub biome in the centre of Australia from Carnarvon to Birdsville.

Step 4: I can use relevant sources to research further reasons for patterns and distributions

7 Research the percentage of Australia that is classified as:

a desert

b tropical

c grassland.

HOW TO

SHEEPT, page 138

How does natural productivity vary between biomes?

Biomes can also be described based on how productive they are. The **productivity** of a biome is linked to the rate of energy exchange between the living things in the biome. Lush, tropical rainforest biomes have a higher productivity than tundra biomes with their extremely cold, dry climates.

Producers and primary productivity

Producers are organisms that use sunlight, water and carbon dioxide to photosynthesise and produce their own energy. Green plants, such as trees or aquatic algae, are typical examples of producers. Plants tend to grow best in wet, tropical regions that allow maximum access to the Sun's rays.

As plants produce their own energy, they are classified as **primary producers**. Plants use some of the energy they create to grow, and the rest is stored. When an animal consumes a plant, it gains that stored energy and uses it for its own growth. The amount of energy a plant uses minus the amount they store is referred to as **net primary productivity**. Net primary productivity is a **quantitative** measure of how efficient a biome is. The biomes with the highest net primary productivity are wetlands, tropical rainforests and coral reefs; while the biomes with the lowest net primary productivity are open oceans and deserts.

Source 1

Primary productivity is vital for the success of consumer species.

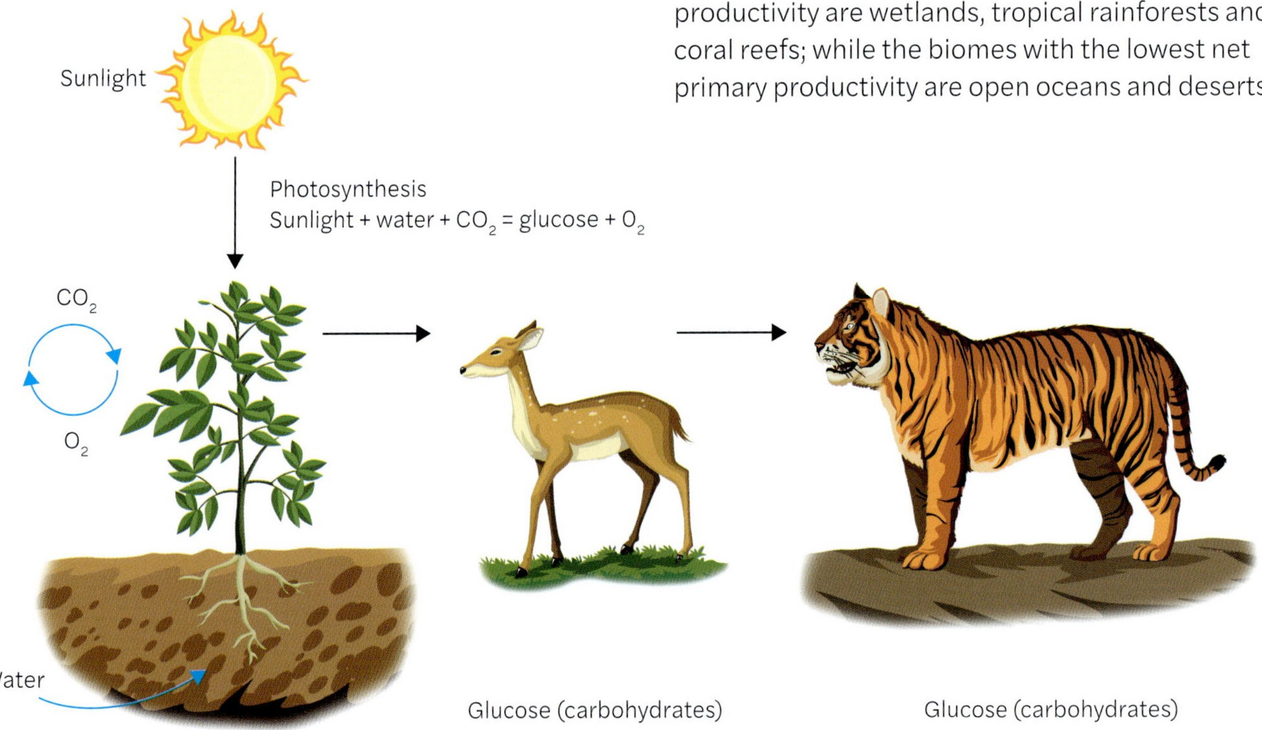

Biomes and productivity

Biome	Average net primary productivity (kcal/m²/year)
Estuaries, swamps and marshes	
Tropical rainforests	
Coral reefs	
Temperate deciduous forests	
Temperate grasslands	
Lakes and streams	
Rocky/sandy beaches	
Tundra	
Open oceans	
Deserts	

Source 2

Mangrove forests, swamps and marshes are some of the most productive in the world. Scientists use the amount of leaf litter that falls to the ground as a way of measuring productivity in these regions.

Source 3

Biomes vary in average net primary productivity

Learning ladder G1.6

Show what you know

1 Provide one example each of a primary producer and a consumer.

2 Observe the flow diagram in Source 1. Using your drawing and annotation skills, create another example of a primary producer, primary consumer and secondary consumer. Incorporate the following terms in your diagram: photosynthesis, producer, consumer, net primary productivity and ecosystem.

Digital and spatial technologies

Step 1: I can interpret different map types using cartographic conventions

3 Explain what data is being displayed in source 3.

Step 2: I can construct paper maps using correct cartographic conventions

4 Create an overlay map to compare the distributions of net primary productivity and forest biomes.

a On a blank world map, complete BOLTSS and highlight the regions of the world with highest net primary productivity.

b Lay tracing paper over your base map. Using the outline, annotate the main locations of forest biomes on a global scale.

c How are the distributions interconnected?

Step 3: I can access and use spatial technology platforms such as GIS

5 Visit the NASA Earth Observatory website at http://mea.digital/GHV9_G1_4 and search for 'net primary productivity'. Use the materials provided to comment on how net primary productivity changes between years on a global scale.

Step 4: I can manipulate data using digital and spatial technologies

6 Visit the NASA Earth Observatory website at http://mea.digital/GHV9_G1_4 and search for 'net primary productivity'. Use the materials provided to comment on how net primary productivity changes in a particular region between seasons.

BOLTSS, page 132
Overlay maps, page 148

Why is studying biome productivity important?

Highly productive biomes such as swamps, marshes, rainforests and coral reefs are vital for the success of many animal communities. For example, it is estimated that tropical forests are home to more than two-thirds of the world's known animal and plant species. These places also are important for humans as a source of food, timber and other resources that are traded around the world.

Interconnection between biomes and productivity

Biomes that have high net primary productivity also tend to have a high species richness. **Species richness** is the number of different species within a **community**. Humans and other animals rely on primary producers, such as plants, to provide food for herbivores, which in turn are eaten by carnivores (see page 24). Therefore, the net primary productivity of a biome has a direct influence over the diversity of species (the **biodiversity**) within a place. These interconnections can be seen in Source 1.

In South America, the Amazon Rainforest is home to a staggering 16 000 different species of trees. It is one of the most productive places in the world, and supports a wide range of animal life including dolphins, monkeys, jaguars and sloths (Source 3).

As the human population increases, we need more space for housing, transport and agriculture. Mass deforestation is currently occurring on a global scale. It is estimated that, over the last 30 years, humans have cut down or burned 1.3 million square kilometres of rainforest and replaced it with farmlands.

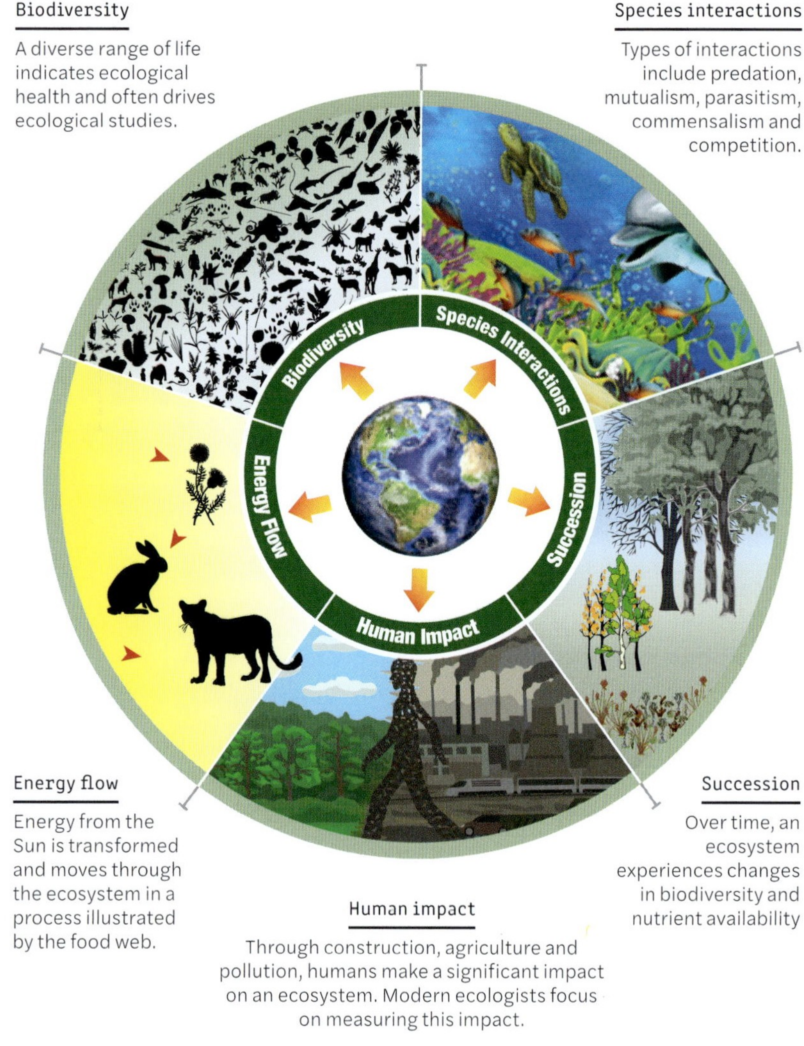

Source 1

Biomes contain strong interconnections between species richness, energy flows, human impacts and biodiversity.

Source: Carolina Biological Supply Company

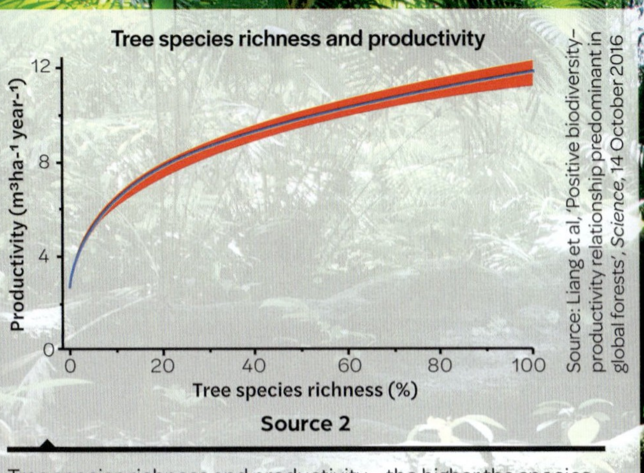

Source 2

Tree species richness and productivity – the higher the species richness (number of species in a given region), the higher the net primary productivity.

Removing primary producers such as trees not only decreases habitat and biomass from the environment but also contributes to our global carbon emissions. As these highly productive biomes are altered, we may see an increase in global temperatures, a reduction in precipitation and a decline in the efficiency of agriculture over time.

Species richness in North and South America

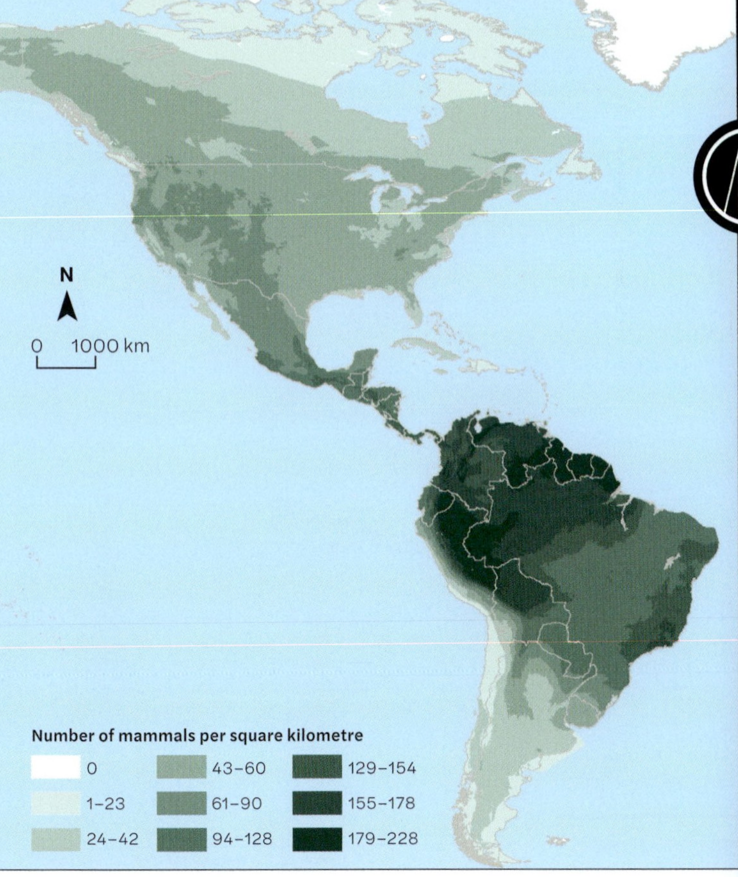

Source: SEDAC

Source 3

Species richness (mammals) per square kilometre in North and South America

Learning ladder G1.7

Show what you know

1 Define the term 'species richness' using examples from the spread.

2 Discuss why productivity and species richness are important to understand when learning about biomes.

3 Construct a block diagram of a forest biome. Annotate how humans, animals, vegetation, climate and soil type interconnect.

Patterns and interconnections

Step 1: I can provide short explanations for patterns and interconnections

4 Source 3: Provide a brief explanation of what this graph is showing.

Step 2: I can explain patterns and interconnections

5 Source 2: Explain why productivity increases with tree species richness.

Step 3: I can use data to support explanations of patterns and interconnections

6 Source 3: Using PQE, describe the distribution of mammals in North and South America. Using information from this chapter, explain why mammals may occur in these locations.

Step 4: I can use relevant sources to research further reasons for patterns and interconnections

7 Investigate how humans are affecting biomes worldwide. Create an infographic of at least 10 facts and four images to show this.

HOW TO

Block diagrams, page 142

How does productivity vary in North America?

North America is the third largest of the seven continents and its landmass covers 24 230 000 square kilometres, or 16.5 per cent of the available land on Earth. It contains 23 countries including Canada, the USA and Mexico.

Productivity and biome interconnection

North America is home to a vast range of biomes, because of its large land size, as shown in Source 3. The most northern areas of the continent are contained within the Arctic circle, while the southern most points are tropical and densely forested.

These variations across the continent directly influence the net primary productivity and species richness of different regions. When you compare Sources 1 and 3, you can see that the differences in numbers of breeding bird species around North America is strongly interconnected with biome type, climate zone and net primary productivity. For example, bird species richness increases in regions of tropical and consistent climates.

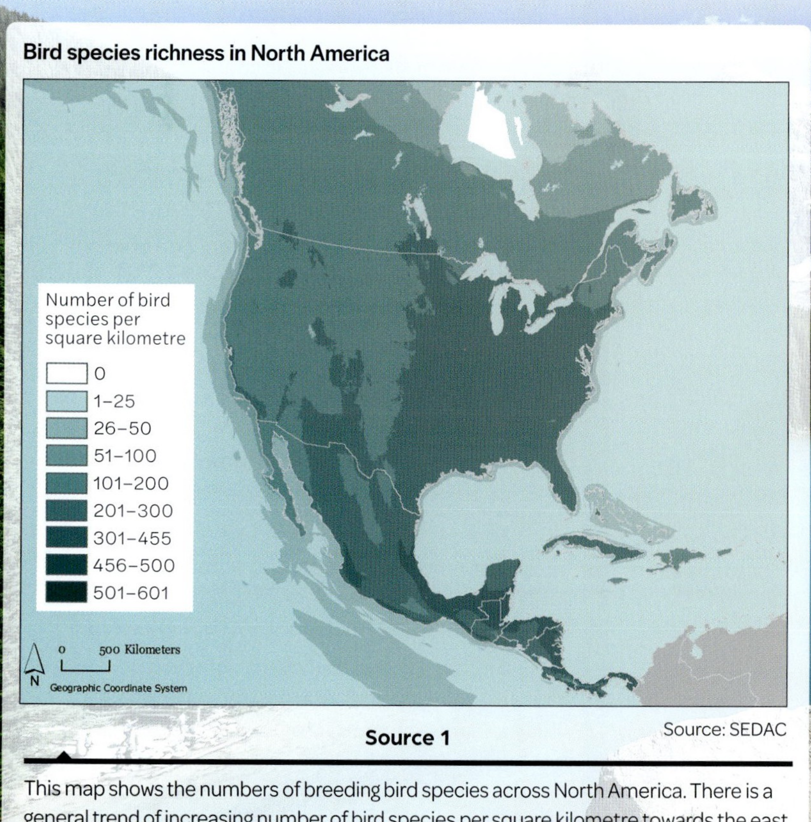

Source 1

This map shows the numbers of breeding bird species across North America. There is a general trend of increasing number of bird species per square kilometre towards the east and south coasts where we observe more tropical biomes.

Source 2

Forest and mountainous biomes in Colorado, USA

Climate zones of North America

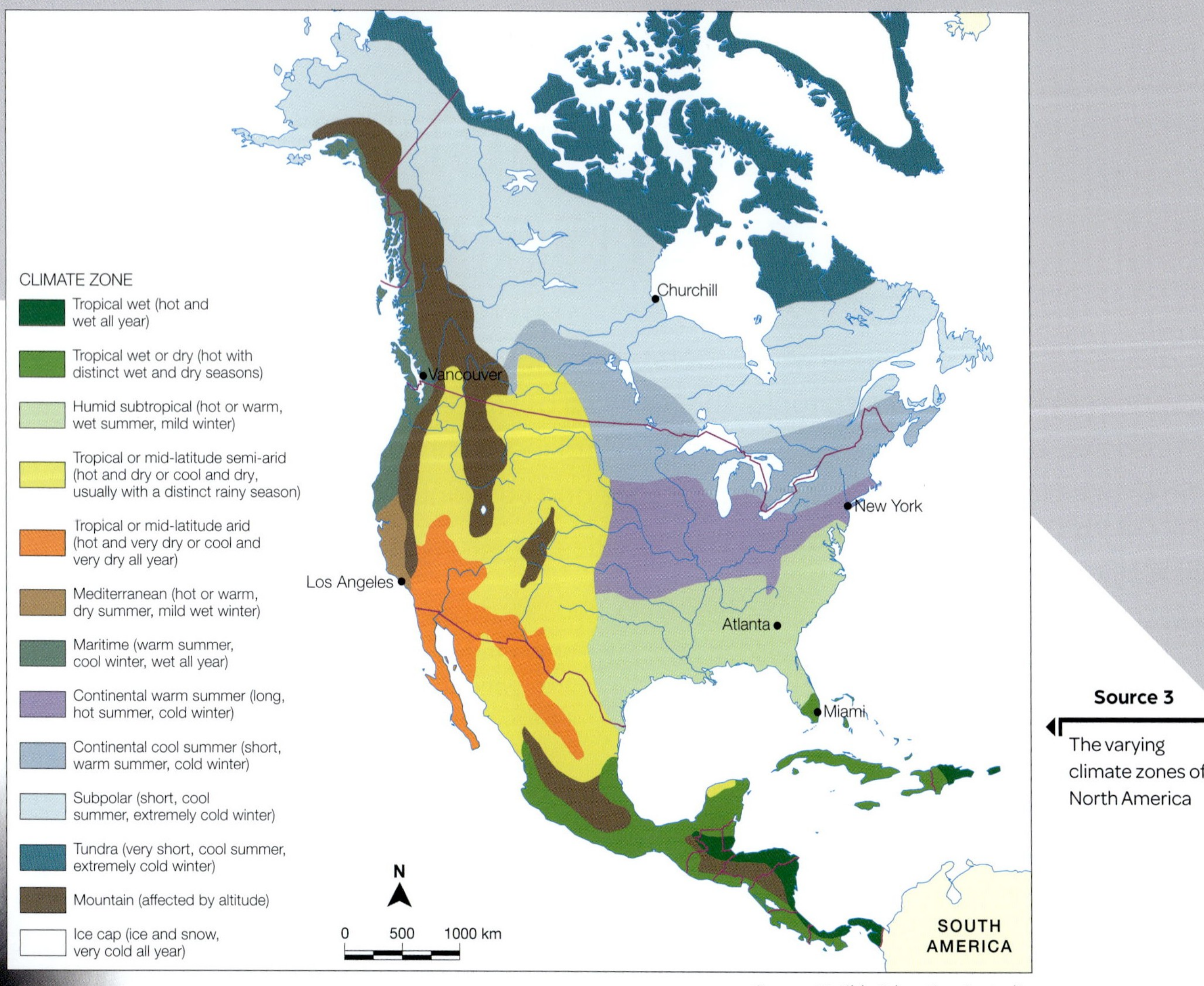

Source: Matilda Education Australia

Source 3

The varying climate zones of North America

Learning ladder G1.8

Show what you know

1 Outline the relative location of North America and find out the absolute location of Atlanta, USA.

2 Compare North America to Australia in terms of its variation in climate and biomes across the landmass.

3 Using Source 3, identify the major biomes in the following locations:

a Atlanta

b Churchill

c Los Angeles.

Communicate data

Step 1: I can list primary and secondary methods useful for my study

4 Identify two methods that would help you collect data on climate zones as pictured in Source 3.

Step 2: I can successfully use data collection methods

5 Using correct techniques, create a field sketch of your local environment. In the annotations, comment on how human actions have changed the natural biome of the region.

Step 3: I can filter collected data

6 Imagine you were conducting the study on breeding birds shown in Source 1. List five points of data you would need to collect in order to understand how breeding birds are affected by primary productivity.

Step 4: I can organise data collected according to relevance for a research question

7 Rank the list of data you suggested in question 6 from most important to least important.

Sketches and annotating, page 140

G1.9

civics + citizenship

What is global citizenship?

A global citizen is someone who understands that they are part of a wider world and wants to make that world a better place. Global citizenship involves people in many nations taking local action to make our planet fairer, safer, more tolerant and sustainable.

Global citizens

Global citizens are individuals who believe they have a civic responsibility, as human beings, to respect other humans and the planet Earth itself. **Global citizenship** is not a legal status, like Australian citizenship, but represents people who attempt to be **socially responsible** and act to benefit all societies across the world, not just their own. The concept of global citizenship could be summed up by the words of the American political philosopher Thomas Paine, who said in 1792, 'my country is the world, and my religion is to do good'.

Global citizenship education is promoted by the **United Nations (UN)**, an organisation formed in 1945 to maintain international peace and promote co-operation between nations. Global citizenship education aims to empower people to actively promote a more peaceful, tolerant, inclusive and sustainable world.

Source 1

Students at the School Strike for Climate demonstration

Source 2

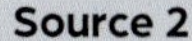

A global citizen is someone who takes responsibility and acts to make the planet a more tolerant, safe and sustainable place.

Global challenges

Global citizens need to address many important environmental and societal issues around the world, including:

- climate change
- degradation of environments
- gender inequality
- poverty.

To build a more peaceful and **sustainable** world, we all need to change the ways we coexist with the planet and each other. We need to think globally and act locally to shape a more tolerant and sustainable world. Participating in **awareness campaigns** about global issues, writing letters, signing petitions and attending demonstrations are all ways that you can be a responsible global citizen.

School Strike for Climate

The 2019–2020 bushfires once again sparked conversation between the media, parliament and lobby groups around global warming and our use of fossil fuels. School students across the country joined forces to hold a 'school strike' and attend climate change rallies. In Victoria, over 100 000 young people gathered to protest about the need for further action on climate change. Many held signs, with some reading 'One day I will vote, but today I march'.

Chrissy Downes, a Year 10 student from Glen Iris in Melbourne, said she had joined the School Strike for Climate demonstration because the bushfires 'currently sweeping our nation are undoubtedly influenced by the climate crisis'.

'We're doing this protest today as a solidarity sit down to show how much we care about those that are affected by the bushfires that are wreaking havoc across our country,' she said.

'This has been noted from firefighters to Indigenous leaders to scientists for years and it's time our policymakers and politicians both on a state, local and federal level give them a voice and listen to them,' she added.

It is important that the awareness raised by events such as the school strike is not lost, and that we continue trying to live more sustainably.

Learning ladder G1.9

Civics and citizenship

Step 1: I can identify topics about society

1 How is global citizenship different to citizenship in a nation?

Step 2: I can describe societal issues

2 Prepare a bulleted list that describes the aims of global citizenship.

Step 3: I can explain issues in society

3 Select one of the global issues listed in this spread, and explain why it is an issue that deserves attention.

Step 4: I can explain different points of view

4 Select one of the global challenges listed in this spread, and explain two different points of view about the issue.

Step 5: I can analyse issues in society

5 What was the purpose of the student climate change rally? What could students do during pandemic conditions to act as global citizens and raise awareness about human-induced climate change?

Masterclass

Learning ladder

Work at the level that is right for you or level-up for a learning challenge!

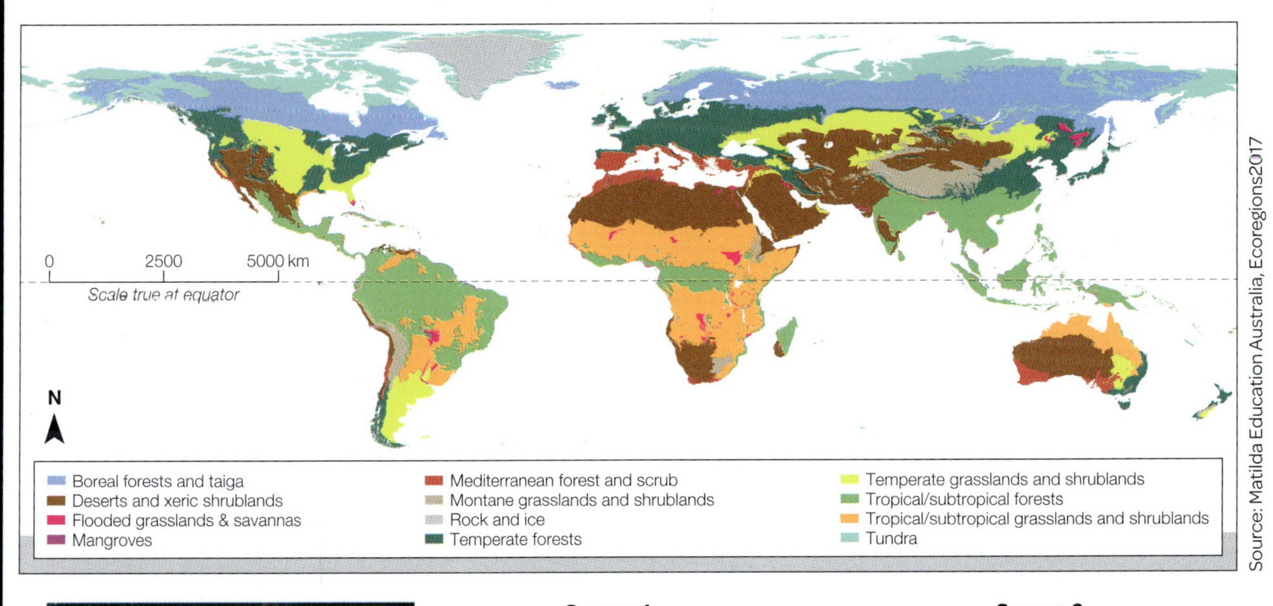

Source 1

Change over time in Cambodia: **a** 2000 and **b** 2015

Source 2

The distribution of biomes on a global scale

Source 3

Main biomes of the world

Change in species richness over time in North America

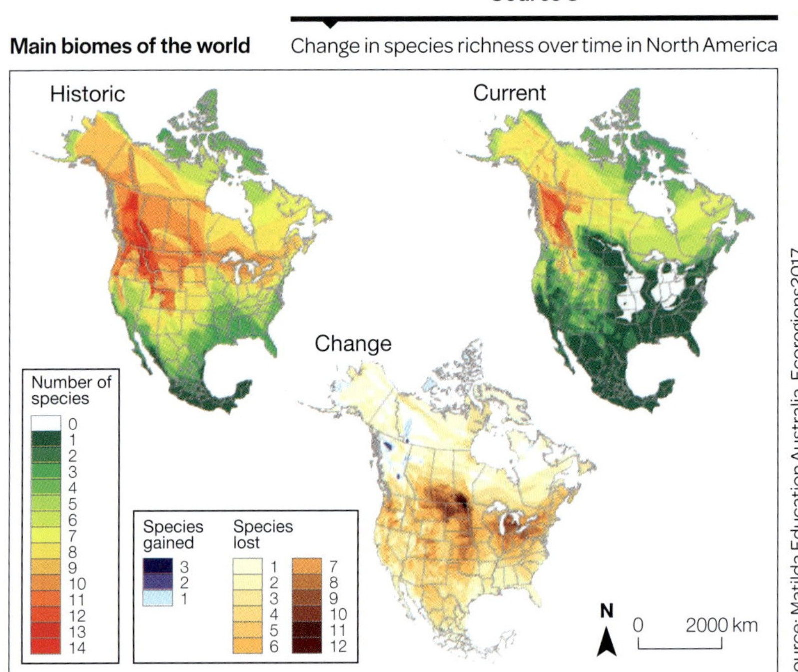

Source 4

Climate graph for Brisbane, Australia

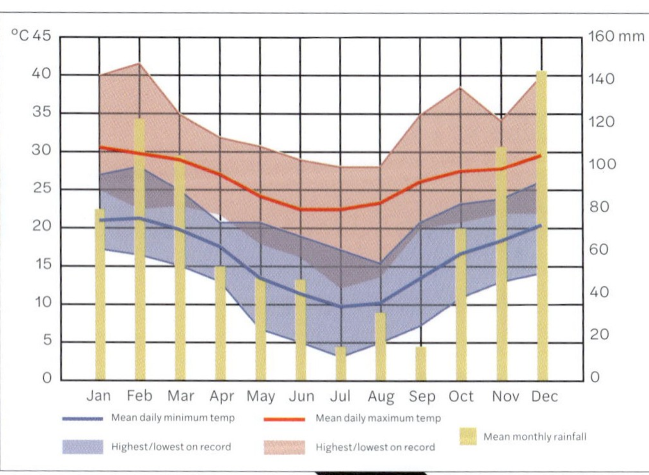

Source: Weatherzone

Step 1

a I can identify spatial distributions and patterns

Identify the primary locations of desert biomes in the world.

b I can provide short explanations for patterns and interconnections

Provide two explanations for the locations of desert biomes in the world.

c I can identify that changes occur in the characteristics of places over time

Refer to Source 4. Comment on how precipitation changes monthly in Brisbane, Australia.

d I can list primary and secondary methods useful for my study

Investigate the research question: 'How does climate affect biomes?' Create a list of two primary and two secondary methods you could use to collect data to answer this question.

e I can interpret different map types using cartographic conventions

Identify the map type used in Source 3.

Step 2

a I can use data to quantify spatial distributions and patterns

Refer to Source 4. Using data, describe how temperature changes monthly in Brisbane, Australia.

b I can explain patterns and interconnections

Source 3: Using SHEEPT, suggest the potential causes of changes to species richness in North America over time.

c I can describe how places have changed over time

Source 1: Using examples, describe the changes that have occurred in Cambodia over time.

d I can successfully use data collection methods

Explore climate data at the Bureau of Meteorology website and collect data that explains why Central Australia is dominated by desert biomes.

e I can construct paper maps using correct cartographic conventions

Using this data table and a blank map of Australia, create a choropleth map showing the difference in temperature between states.

7 January 2013	Average maximum
Queensland	33.0°C
NSW	34.5°C
Victoria	34.7°C
Tasmania	27.1°C
South Australia	38.9°C
Western Australia	38.0°C
Northern Territory	37.5°C

Step 3

a I can describe spatial distributions and patterns

Source 3: Use PQE to describe the change in species richness in North America over time.

b I can use data to support explanations of patterns and interconnections

Using data and examples from this chapter to support your response, explain which of the following biomes would have the highest levels of net primary productivity.

- Forest
- Desert
- Tundra

c I can explain the causes behind the change over time in a place

Source 1: Outline three possible causes for the change over time in Cambodia.

d I can filter collected data

Investigate the following research question: 'What natural factors determine the types of biomes present in a region?' Select which sources in this spread would provide the most appropriate evidence.

Masterclass

e I can access and use spatial technology platforms such as GIS

Access http://mea.digital/GHV9_G1_1. to explore the world's biomes. Record data on the following topics for five locations:

i climate ii animal species iii biomes present.

Step 4

a I can use data to support exceptions to spatial distributions and patterns

Source 4: Using PQE, describe the climate of Brisbane, quantifying both your pattern and exception.

b I can use relevant sources to research further reasons for patterns and interconnections

Source 4: Which biomes are present in Brisbane? Discuss the interconnection between these biomes and their climate.

c I can make predictions and outline consequences of change over time

Source 1: Suggest five consequences of the change over time in Cambodia, considering environmental, economic and social factors.

d I can organise data collected according to relevance for a research question

Access the Switch Zoo Animal Games at http://mea.digital/GHV9_G1_7. Navigate to the 'Build a biome' game. Build a series of biomes based on your knowledge of their characteristics and climate.

e I can manipulate data using digital and spatial technologies

Access the Our Environment website at http://mea.digital/GHV9_G1_8. Explore the different landscapes and habitats in New Zealand and create summary notes of your observations.

Step 5

a I can identify multiple spatial distributions and patterns

Research the web to find a climate graph of Melbourne. Compare it with the climate of Brisbane in Source 4 and, using PQE, identify any similarities or differences between the two cities.

b I can interpret causes of patterns and interconnections

Discuss the interconnection between the patterns of species richness in Source 3. What could be the reasons for the change?

c I can interpret data to quantify predictions based on research

Refer to Source 3. Conduct some research and, using quantitative data, predict the future pattern of species richness in North America.

d I can evaluate the success of research methods

Source 1: Evaluate the usefulness of satellite images in determining change over time to Cambodia's natural environment.

e I can draw conclusions from geographical information in digital and spatial technologies

Visit the VRO Agriculture website at http://mea.digital/GHV9_G1_9. Type 'primary production landscapes of Victoria with representative soil profiles' into the search bar. Use the information to discuss how natural factors, such as soil type, can influence biome type and land use in particular regions.

Capstone

How can I understand the natural world?

In this chapter, you have learnt a lot about the natural world. Now you can put your new knowledge and understanding together for the capstone project to show what you know and what you think.

In the world of building, a capstone is an element that finishes off an arch or tops off a building or wall. That is what the capstone project will offer you, too: a chance to top off and bring together your learning in interesting, critical and creative ways. You can complete this project yourself, or your teacher can make it a class task or a homework task.

mea.digital/GHV9_G1

Scan this QR code to find the capstone project online.

G2

A human world

HOW DO HUMANS CHANGE THE ENVIRONMENT?

page 38

spatial distributions and patterns
page 42
HOW DOES GLOBAL WARMING IMPACT PRODUCTIVITY?

thinking globally
page 46
WHY IS FOOD SO SCARCE IN THE SAHEL DRYLANDS?

economics + business
page 56
HOW DO WE DETERMINE THE VALUE OF RESOURCES?

How can I understand our human world?

Humans alter the biomes of the world for many reasons, including transport, tourism and safety. One of the main reasons is for the purposes of food and fibre production. When humans alter a biome to produce food, industrial materials and fibres, these alterations can have significant environmental impacts.

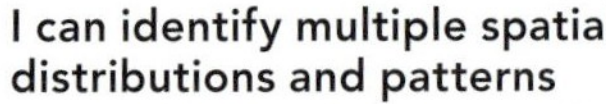

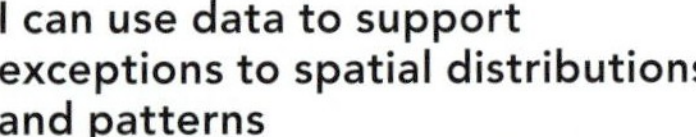

Step	Spatial distributions and patterns	Patterns and interconnections	Changes and implications
step 5	**I can identify multiple spatial distributions and patterns** I can take my PQE one step further to find links or relationships that exist in the human world.	**I can interpret causes of patterns and interconnections** I can use multiple sources to find links or relationships that exist in the human world and can explain 'Why?'.	**I can interpret data to quantify predictions based on research** I can use external data from research as evidence of the positive and negative impacts of a change I have predicted.
step 4	**I can use data to support exceptions to spatial distributions and patterns** I can use data to answer 'Why?' about the exceptions identified in a PQE analysis of the human world.	**I can use relevant sources to research further reasons for patterns and interconnections** I can use sources other than this textbook to further research patterns I observe in the human world.	**I can make predictions and outline consequences of change over time** I can use my knowledge of natural processes and world regions to make an educated guess of the positive and negative impacts of change in the human world.
step 3	**I can describe spatial distributions and patterns** I can describe patterns, quantify them and point out exceptions (PQE) to describe the human world.	**I can use data to support explanations of patterns and interconnections** I can use data from a map or graph to explain patterns I observe in the human world.	**I can explain the causes behind the change over time in a place** I can use my knowledge of natural processes and world regions to explain why changes may occur over time in the human world.
step 2	**I can use data to quantify spatial distributions and patterns** I can read data and use it to measure key trends on a map or graph about the human world.	**I can explain patterns and interconnections** I can identify social, historical, economic, environmental, political and technological (SHEEPT) factors to help me explain the human world.	**I can describe how places have changed over time** I can use specific examples to describe changes over time in the human world.
step 1	**I can identify spatial distributions and patterns** I can find key trends on a map or graph about the human world.	**I can provide short explanations for patterns and interconnections** I can write descriptions of patterns and interconnections that I find in the human world.	**I can identify that changes occur in the characteristics of places over time** I can read information and answer questions about changes over time in the human world.

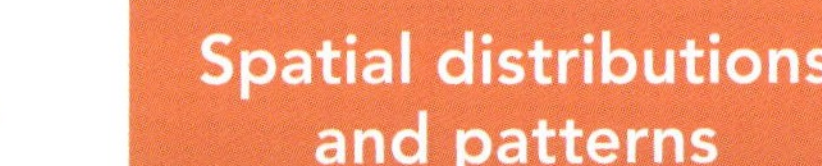

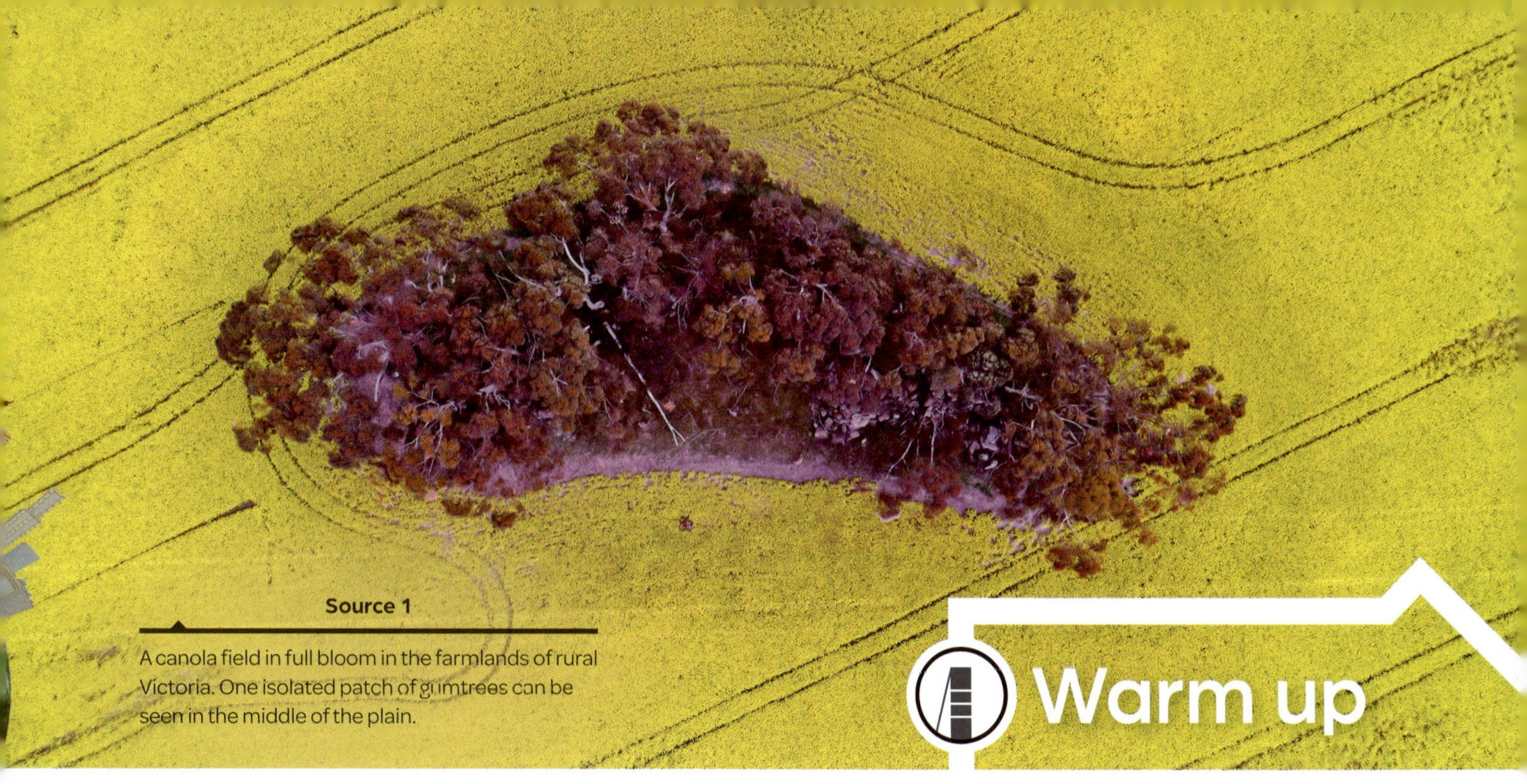

Source 1

A canola field in full bloom in the farmlands of rural Victoria. One isolated patch of gumtrees can be seen in the middle of the plain.

Warm up

Spatial distributions and patterns

1 Refer to Source 1. Describe the type of agriculture that is being undertaken in this region. What other vegetation is present?

Patterns and interconnections

2 What evidence of technology can you see in Source 1?

Changes and implications

3 Describe the biome in Source 1 before it was altered by humans to serve as a farm for food production.
   a What evidence do you have for this change?
   b Write down one positive and one negative impact of this change.

Communicate data

4 What type of photograph is shown in Source 1?

Digital and spatial technologies

5 Sketch a map of Source 1 and label any evidence of human alteration of this place.

I can evaluate the success of research methods
On reflection, I can look back and comment on the data collection methods I used and evaluate how successful they were in helping me answer a research question about the human world.

I can organise data collected according to relevance for a research question
I can review the data I have collected in the field and display it using graphs, tables, annotations and captions.

I can filter collected data
I can review my collected data and select the most relevant data to answer a research question about the human world.

I can successfully use data collection methods
I can use primary and secondary data collection methods in the field and classroom to investigate the human world.

I can list primary and secondary methods useful for my study
I can create a checklist of methods to investigate the human world and categorise them as primary or secondary methods.

Communicate data

I can draw conclusions from geographical information in digital and spatial technologies
I can interpret and analyse patterns by using different layers and features on spatial technology platforms.

I can manipulate data using digital and spatial technologies
I can work with layers and other features on spatial technology platforms to further explore data and interconnections.

I can access and use spatial technology platforms such as GIS
I can use spatial technology platforms to explore data and find patterns.

I can construct paper maps using correct cartographic conventions
I can use a pencil, paper and ruler to construct a map that follows BOLTSS conventions.

I can interpret different map types using cartographic conventions
I understand data found in different types of maps and graphs and use the data to answer questions about the human world.

Digital and spatial technologies

G2.1

How do humans change the environment?

The natural characteristics of biomes can influence the way humans use the land. We also alter the environment to suit our own needs. These changes can have both positive and negative outcomes for a region.

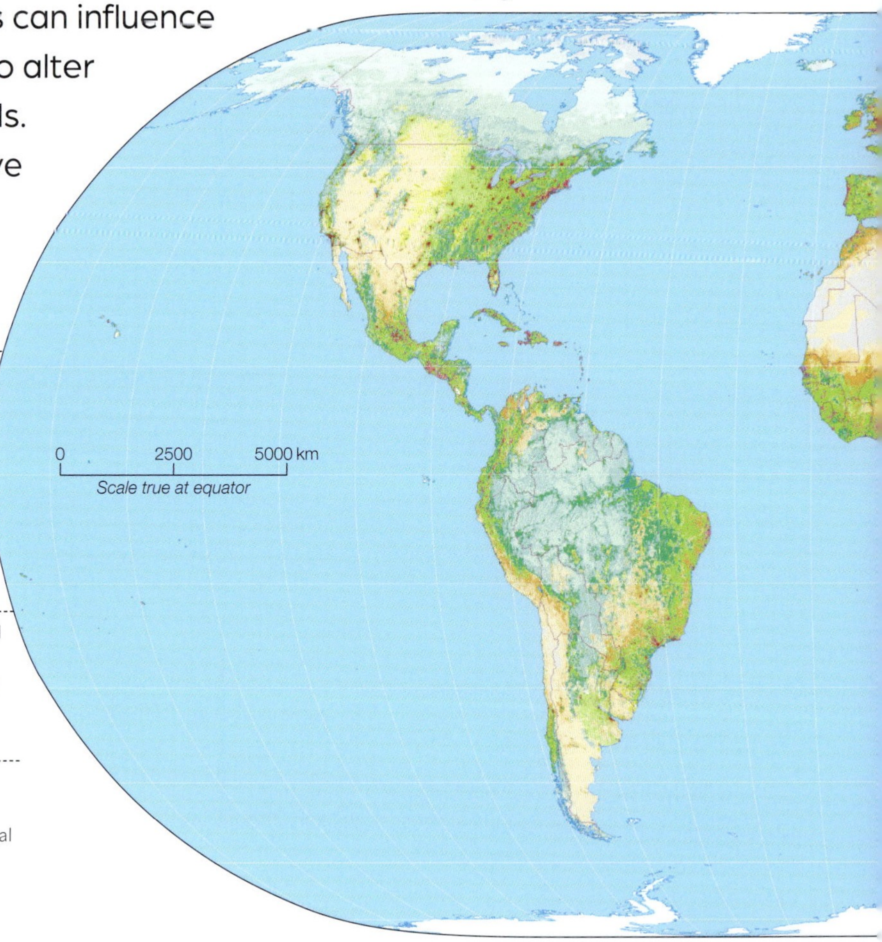

Source 1

A map showing three anthropogenic biomes of the world.

Anthropogenic biomes of the world

Urban		Dense
		Urban
Wildlands		Barren
		Sparse
		Wild
Forested		Populated
		Remote
Rangelands		Residential
		Populated
		Remote
Croplands		Residential irrigated
		Residential rainfed
		Populated irrigated
		Populated rainfed
		Remote
Villages		Rice
		Irrigated
		Cropped and pastoral
		Pastoral
		Rainfed
		Rainfed mosaic

0 2500 5000 km

Scale true at equator

Source 2

A fish kill site is an example of water degradation caused by human impact.

Water degradation

Aquatic environments are affected by surrounding human activities, including urbanisation and agriculture. When fertilisers and other products containing high levels of nutrients are washed into these natural environments, it causes algal blooms to form. These blooms attract bacteria that feed upon decaying matter. As the number of bacteria grows exponentially, the level of oxygen in the water decreases and this can cause large 'fish kills' within the region. The dead fish, which may have also been feeding on the toxic algae, also pose a risk to other consumers, such as aquatic birds and mammals.

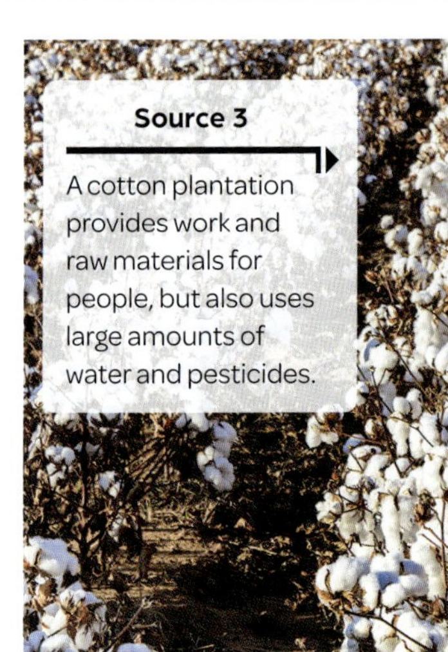

Source 3

A cotton plantation provides work and raw materials for people, but also uses large amounts of water and pesticides.

Source 4

Borneo's environment is being altered by palm oil plantations.

Land degradation

One of the biggest anthropogenic changes occurring in Borneo is **deforestation** to make way for palm oil plantations. Palm oil is a popular and versatile oil that is a component in more than 50 per cent of Australian supermarket products. As more plantations are created, Borneo loses more than 1.3 million hectares of forest a year, and the World Wide Fund for Nature estimates that less than 24 per cent of forest cover remains in 2020. These forests also provide habitat for many species that will become extinct without the food and shelter that the rainforest environment provides.

Source: Erle Ellis, UMBC

Industrialisation

Cotton is a versatile crop that can be used for materials and food. Currently, cotton production provides work for more than 7 per cent of people in **less economically developed countries (LEDCs)**. However, cotton requires a lot of water: 20 000 litres are needed for enough cotton to make a t-shirt and a pair of jeans. Of all crops, cotton consumes the largest amount of pesticides (11 per cent of all pesticides globally), which eventually make their way into the natural land and water systems.

Learning ladder G2.1

Show what you know

1 Outline how humans use the land. Classify each use as positive or negative for the environment.

2 Why do humans alter biomes, even though we understand their importance to ecosystems and communities?

3 Source 3: Locate historical and current satellite images of Borneo and discuss how the environment has changed over time because of human influence.

Communicate data

Step 1: I can list primary and secondary methods useful for my study

4 Imagine you plan to investigate the impact of the cotton industry on your local environment. (Remember, impacts can be both beneficial and harmful.) List one primary and two secondary methods for collecting appropriate data.

Step 2: I can successfully use data collection methods

5 Using secondary sources, collect some data that would help address the research topic outlined in question 4.

Step 3: I can filter collected data

6 Using the data you collected in question 5, create a poster showing the impacts of the cotton industry to educate others in your community.

Step 4: I can organise data collected according to relevance for a research question

7 As a class, create a data bank that highlights the impacts the cotton industry might have on your local region.

HOW TO

Satellite images, page 147

G2.2

Are freshwater biomes our most valuable?

Life depends on water. For humans, water is not only vital to our health, but also crucial for agriculture, sanitation, cooking and recreation. While Earth is known as the 'blue planet', only 2.5 per cent of the water on Earth is fresh. Of that 2.5 per cent, around 1.75 per cent is locked in glaciers and sea ice, so is not accessible to humans. The remaining fresh water comes from rivers and lakes, springs, purpose-built catchments such as dams and rainwater tanks, as well as bores and wells located underground.

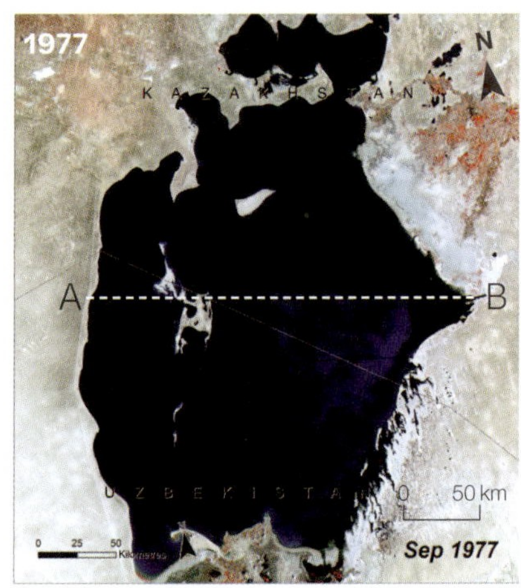

Source 1

The Aral Sea, in Central Asia, was once an expansive freshwater lake, but over time it has almost completely dried up because of drought, overuse and irrigation.

Source 2

The Ganges River in India is an example of a freshwater biome that has become polluted.

Freshwater biome degradation

Water pollution and degradation is a global issue. As the human population increases, so does our output of pollution and waste, as well as our overall reliance on fresh water sources. Key resources such as the Ganges River or the Aral Sea, which were once large freshwater systems, are now largely degraded, polluted or lost because of human use.

The Ganges River, in India, is a highly sacred, spiritual place. However, because of its many uses – from drinking, washing, cooking and transport, to burial ceremonies and human waste disposal – the river has become highly polluted. This has resulted in many health issues for those living alongside it.

Irrigation

In many dry regions, such as Australia, agricultural success relies on **irrigation**. Irrigation diverts water from rivers and lakes to crops, either directly, via channels, or indirectly, through pipelines. While irrigation helps to increase suitable growing conditions, which increases crop yields, it removes water from natural systems and therefore alters local biomes.

Source 3

Irrigation assisting crop growth in dry regions

The world's freshwater biomes

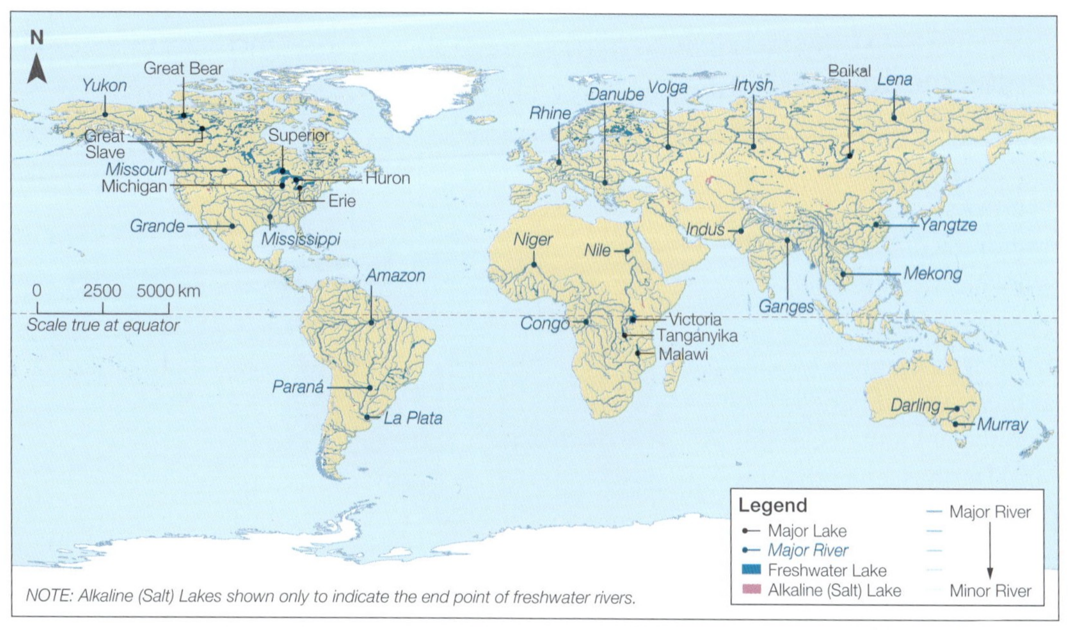

Source 4

Location of freshwater biomes on a global scale

Matilda Education Australia

Learning ladder G2.2

Show what you know

1 List two reasons fresh water is a scarce resource.

2 Outline how water access is interconnected with food crop yields.

3 Define the term 'degradation' using the terms 'pollution' and 'water'.

Patterns and interconnections

Step 1: I can provide short explanations for patterns and interconnections

4 Source 4: Suggest why access to fresh water is not evenly distributed on a global scale.

Step 2: I can explain and patterns and interconnections

5 Source 2: Using the SHEEPT factors, explain why the Ganges River has become so degraded over time.

Step 3: I can use data to support explanations patterns and interconnections

6 Source 1: Using the scale, measure the length of the transect line (from A–B and C–D) and calculate the change in the expanse of the Aral Sea from 1977 to 2015.

Step 4: I can use relevant sources to research further reasons for patterns and interconnections

7 Conduct some research to create a summary table that highlights how much water is used in the production of different crops such as cotton, nuts and wheat. Comment on how irrigation may play an important role in the success of these crops in Australia.

SHEEPT, page 138
Transects, page 149

How does global warming impact productivity?

Often the terms 'climate change' and 'global warming' are used interchangeably. However, these terms can actually refer to different processes on Earth.

Climate change versus global warming

The term **climate change** usually describes the global average temperature fluctuation over time due to a range of natural processes. Climate change has been occurring for hundreds of thousands of years and is responsible for the cycle of very cold glacial periods known as **ice ages**.

Global warming, on the other hand, is a human-induced process by which industrialisation has dramatically increased the amount of carbon dioxide in the atmosphere. This has led to the unnatural, exponential warming of the globe. Recent studies have revealed that levels of carbon dioxide in the atmosphere are at their highest in 400 000 years; this has led to an average global temperature increase of 0.8°C since 1880. Temperatures are predicted to increase by 2–6°C in the next century – 20 times faster than increases in historical records.

Influence of global warming on productivity

The impacts of global warming range from an increase in disastrous climatic events to altered agricultural production and ecosystem structures. Global warming is expected to change productivity on a global scale and this will have flow-on effects on agricultural yields, food security and poverty. Source 1 illustrates how changes to climate may decrease agricultural success in major farming regions throughout the world.

'A key culprit in climate change, carbon emissions, can also help agriculture by enhancing photosynthesis in many important (...) crops such as wheat, rice, and soybeans. The science, however, is far from certain on the benefits of carbon fertilisation.'

William R. Cline, *Global Warming and Agriculture, Finance and Development*, 2008.

Projected impact of climate change on agricultural yields

0 1500 3000 km
Scale true at equator

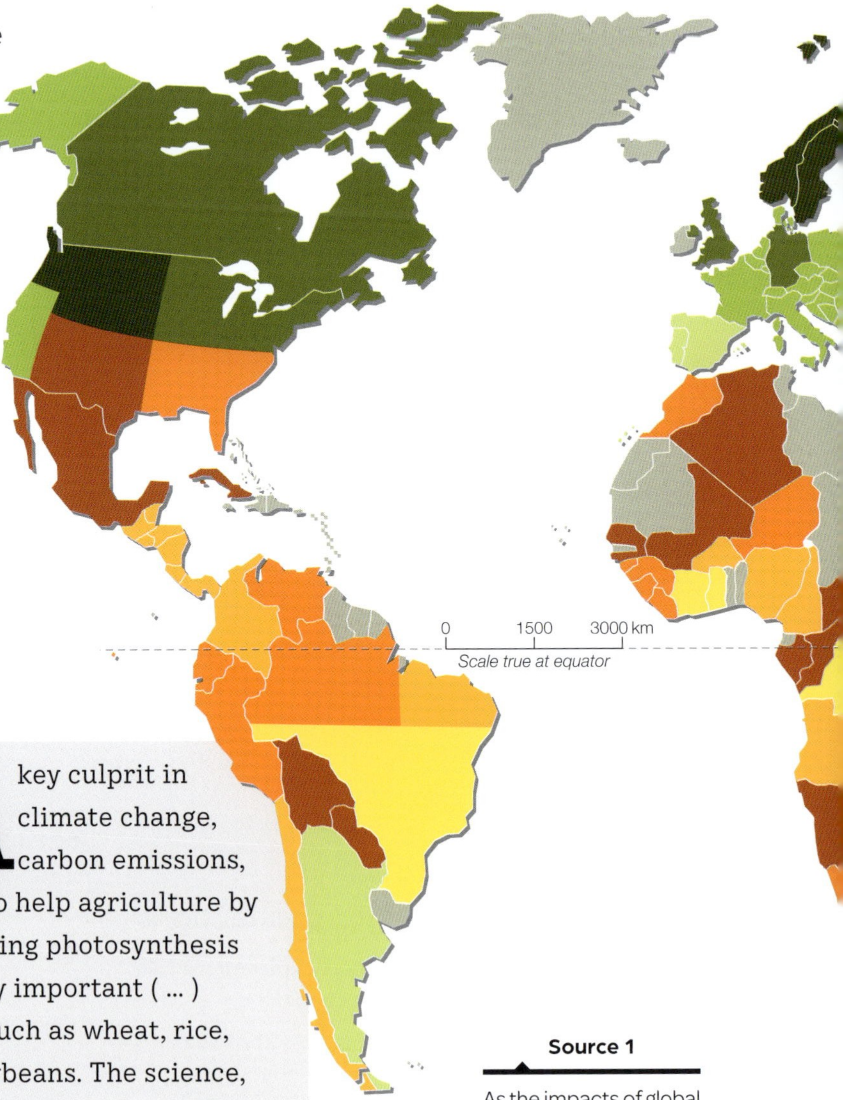

Source 1

As the impacts of global warming increase, agricultural success is expected to change on a global scale.

Source 2

The climate has naturally fluctuated over hundreds of thousands of years.

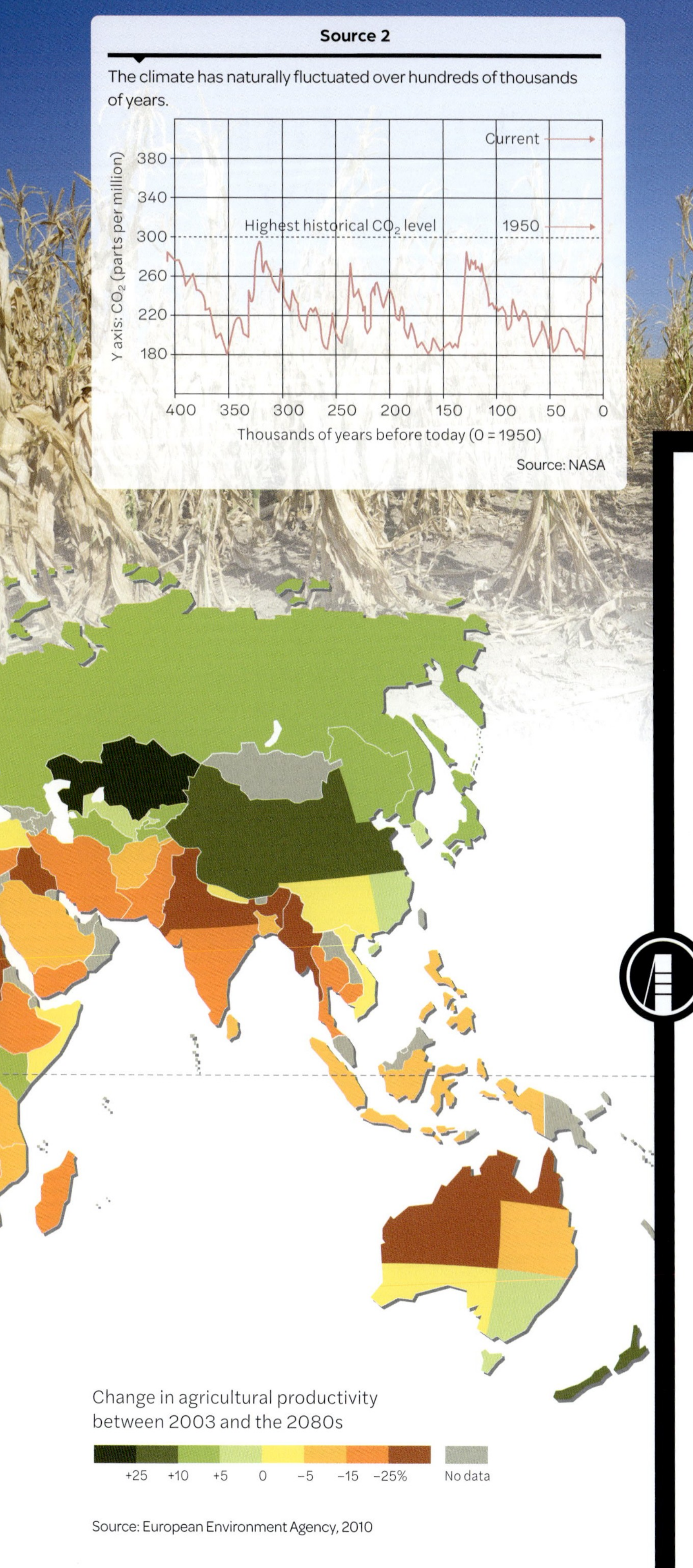

Source 3

Drought and other extreme events may lead to less successful crop yields in already dry environments.

Learning ladder G2.3

Show what you know

1 Outline the difference between the terms 'climate change' and 'global warming'.

2 Why do we use carbon dioxide levels in the atmosphere as a measure of global warming?

3 Source 2: Summarise what is shown in this graph. Outline how it connects to the question, 'How does global warming impact productivity?'

4 Discuss how global warming may further impact fresh water scarcity.

Spatial distributions and patterns

Step 1: I can identify spatial distributions and patterns

5 Source 2: How have carbon dioxide levels in our atmosphere changed over time?

Step 2: I can use data to quantify spatial distributions and patterns

6 Source 2: Identify the predicted change to agricultural productivity rates for South Africa, Canada and Finland.

Step 3: I can describe spatial distributions and patterns

7 Source 1: Outline why some countries do not have data available.

Step 4: I can use data to support exceptions to spatial distributions and patterns

8 Researchers predict a decrease in agricultural productivity in Africa and Australia in the future, while predicting an increase in Canada and Russia. Explain this difference, using references to data from Source 1.

Why do we lack global food security?

As the population of the world grows, so does our need for resources, such as food, water and housing. Since the 1700s, the human population has been increasing rapidly, especially in regions such as Asia and Africa. Improvements in medical knowledge and technology have also increased human life expectancy, so we also require more food per person across a human life span.

What is food security?

According to the World Food Summit (1996), '**Food security** exists when all people, at all times, have physical and economic access to sufficient, safe and nutritious food that meets their dietary needs and food preferences for an active and healthy life'. It is estimated that around one in seven people frequently do not have access to enough food to lead healthy lives. This is known as **food insecurity**.

Food security is not evenly distributed. Regions with higher population densities or rapidly growing populations tend to have lower food security, as there is more competition for available resources. More economically developed nations, such as the USA and Australia, tend to have higher overall food security, but also higher levels of obesity and food wastage. Still, within these richer nations there are marginalised communities that suffer malnutrition, as food security can also vary on a regional and local scale.

Global food security

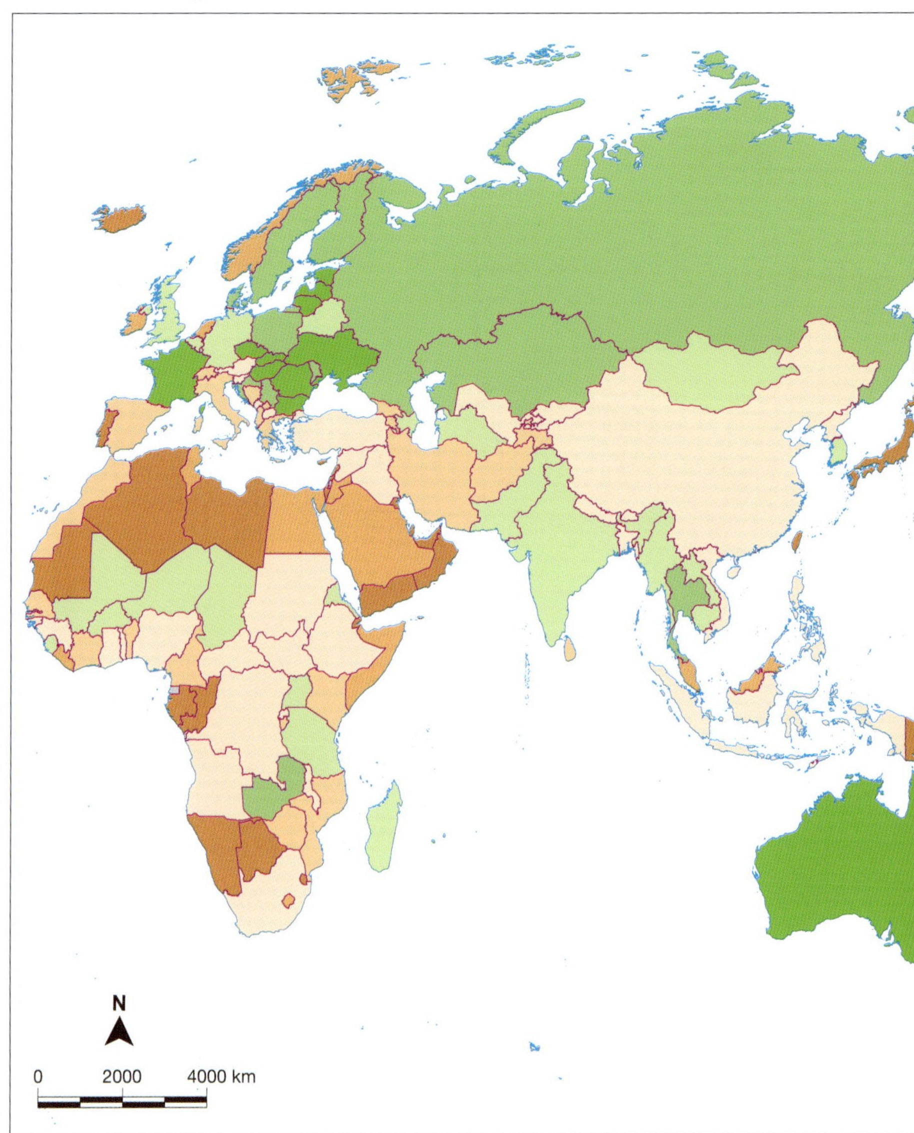

Source 1

Food security is not evenly distributed around the world.

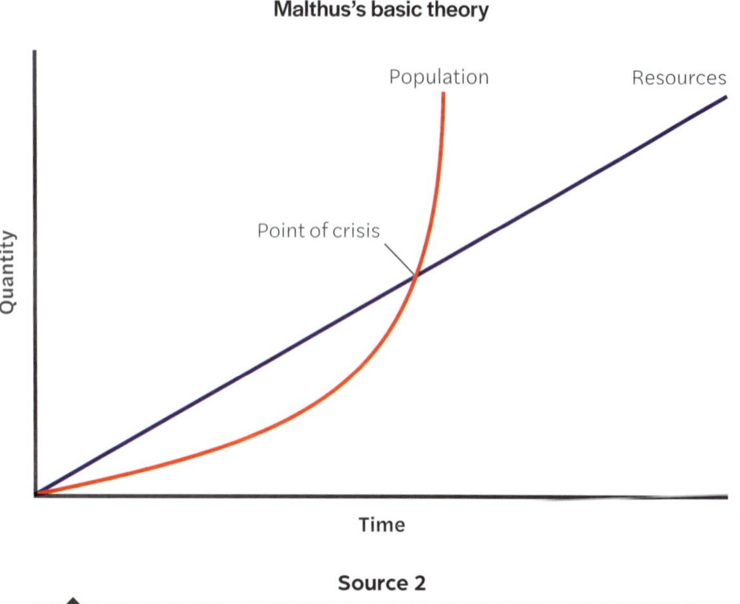

Source 2

Malthus's theory of resource demand and human population growth

Source 3

Malthus developed his 'point of crisis' theory in 1798.

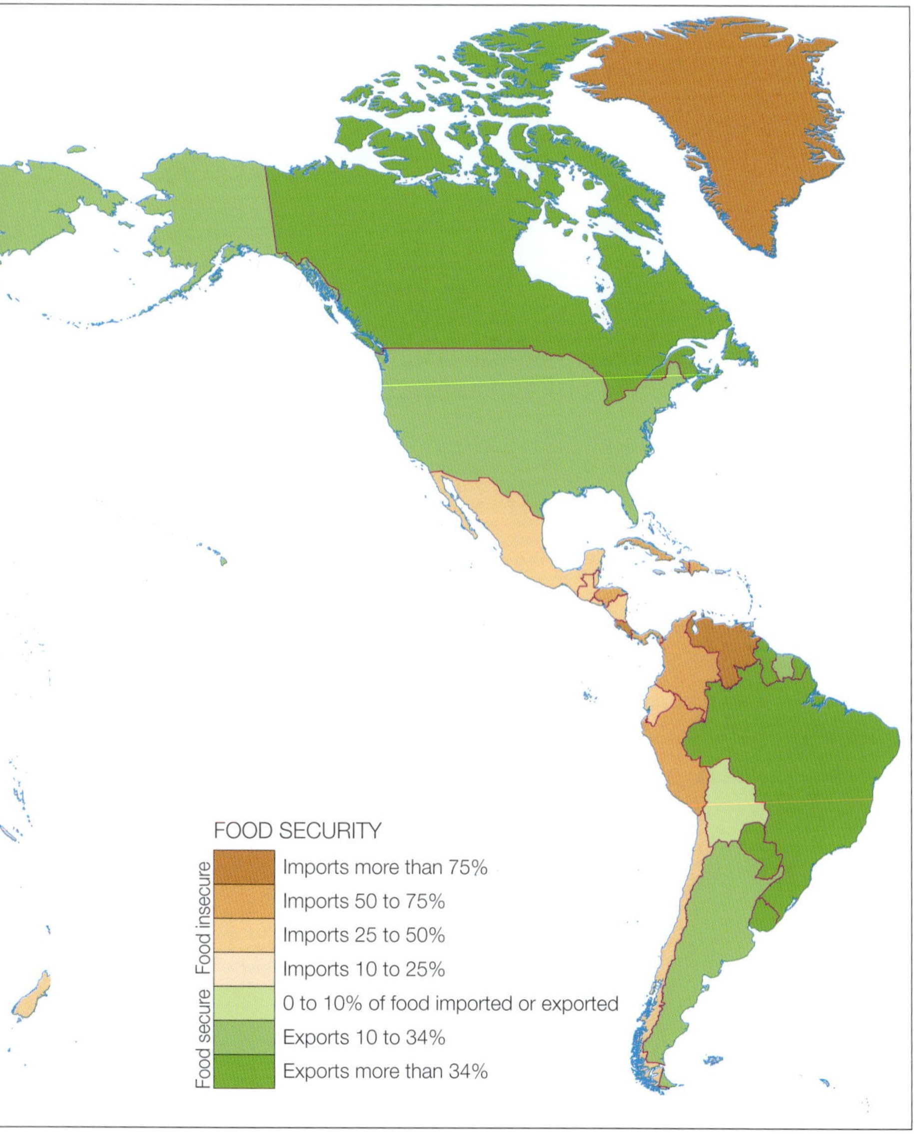

Source: Matilda Education Australia

Who was Thomas Malthus?

Thomas Malthus, a professor of history and political economy, published a book in 1798 called *An Essay on the Principle of Population*. This work outlined his theory that the human population grows **exponentially**, while food production can only increase **arithmetically** (Source 2). In other words, as our population continues to grow, there will be a 'point of crisis' where we no longer have enough food to feed all the people in the world.

Many modern scientists now debate this hypothesis, given advancements in technology and farming techniques. However, because of the uneven distribution of wealth and resources, some communities continue to live in poverty. The United Nations estimates that, tragically, more than 25 000 people die from hunger-related causes every day.

What about the future?

Recent studies suggest that fertility rates (number of children per woman in a population) are decreasing on a global scale. As a result, even though global population numbers keep rising, they are growing at a much slower pace than previous decades. Researchers predict that by 2100, the human population will plateau at around 11 billion people.

Providing food for these additional people is not as simple as just increasing the amount of farming we do. Food security will depend on how populations use, distribute and price food. **Sustainable** production of resources will also be vital, as global warming will alter our crop yields and cause potentially disastrous climate events such as flooding rains or prolonged droughts.

What causes food insecurity?

Save the Children, Press Release 9 July 2019
Fairfield, Connecticut, USA.

Nearly seven million people, or 61 per cent of the population, face acute food insecurity in South Sudan, the highest number of people ever in the country, warns Save the Children. This is an increase of nearly one million people facing acute food insecurity, based on the IPC classification, since the signing of a revitalized peace deal in September 2018.

Unlike its regional counterparts Ethiopia, Kenya and Somalia, which are facing severe food insecurity due to a worsening drought, South Sudan's food crisis is directly linked to the ongoing conflict, which has dramatically disrupted farming activities and livelihoods, and increased displacement.

The latest food crisis and continued violence is a strong indication that the peace process needs greater national and international support.

'Children in conflict don't only die from bullets', Hansraj said. 'Millions of children suffer starvation, brought on through disruption and destruction resulting from the conflict'.

Save the Children has been working in Southern Sudan since 1989 and the organization remains at the forefront in providing lifesaving humanitarian assistance in health, protection, child rights, nutrition and education for children across ten field sites in South Sudan.

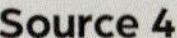

Source 4

This graph shows population growth in each of the world's regions. While the human population continues to increase over time, the growth rate is slowing.

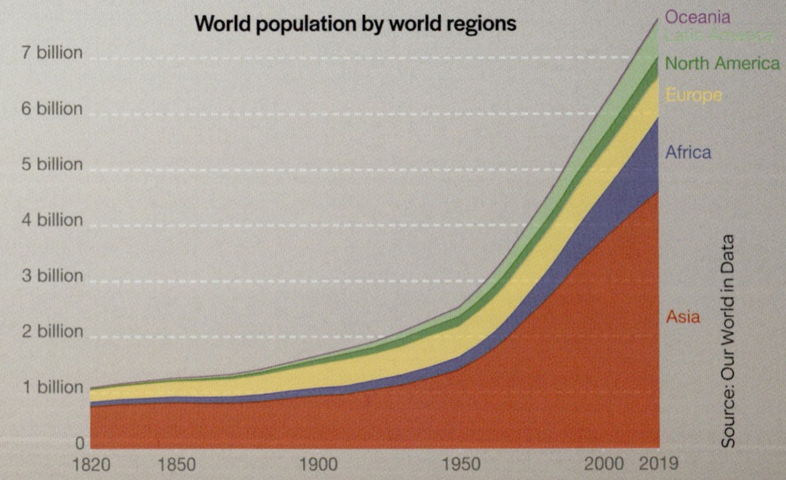

Source 5

Many populations still suffer from severe food insecurity today. This image shows people lining up to register for food distribution in South Sudan, 2016.

Learning ladder G2.4

Show what you know

1 Define 'food security', using examples of a country with high food security and a country with low food security. Discuss why food security is not evenly distributed in today's world.

2 Comment on the link between world population and food/water security over time.

3 As a pair, discuss whether you think global warming may alter food security on a global scale. List some reasons for your ideas.

4 Source **2**: Debate the following statement as a class: 'Malthus' theory is no longer relevant today, given our access to technology and medicines'.

Changes and implications

Step 1: I can identify that changes occur in the characteristics of places over time

5 Visit the Global Food Security Index website at http://mea.digital/GHV9_G2_1. Select one country and record its food security ranking. Why do you think this country is ranked low/high compared to others? (Hint: Consider economic and technological factors.)

Step 2: I can describe how places have changed over time

6 Source 4: Discuss how population has changed over time on a regional scale. Why is that significant when considering food security on a regional scale?

Step 3: I can explain the causes behind the change over time in a place

7 Research a famine event that has occurred over the last 10 years. What were the causes of this event? Could this famine be linked with Malthusian theory; i.e. growing populations?

Step 4: I can make predictions and outline consequences of change over time

8 Refer to http://mea.digital/GHV9_G2_2. Consider the countries that have been ranked highest and lowest for food security (affordability, availability, safety).

a What factors do the top five countries share that increase their food security ranking?

b What factors do the bottom five countries share that reduce their food security ranking?

c Which countries have seen the most change (positive or negative) to their food security ranking over time? Why do you think this is?

d How could population numbers, economic status and access to technology play a role in determining a country's food security ranking? Does this mean Malthusian theory is still relevant today?

thinking globally

Why is food so scarce in the Sahel drylands?

Although a **dryland** appears to be arid and barren like a desert, it is characterised by a fertile period of 1 to 179 days per year. Drylands account for approximately 40 per cent of Earth's available land. The Sahel is a dryland region that covers land between Senegal and Eritrea. Over 41 per cent of people in Africa live in drylands.

What is desertification?

Desertification is a human-induced process in which the quality and fertility of the land is reduced by overcultivation, deforestation and unsustainable farming practices.

Global warming has increased the rate of desertification in the Sahel drylands. From around 1968, the Sahel experienced a number of extreme drought periods that resulted in a severe loss of croplands and livestock, and disastrous famines leading to over 100 000 deaths. However, as the human population of Africa continues to grow at a rate of around 3.1 per cent, so does people's need for food and income, despite these harsh conditions. Therefore, as farmers increase crops to support their communities, they actively increase the rate of desertification in their region.

Despite a number of local and national humanitarian efforts to reduce the impacts of desertification, around 80 per cent of Sahel is affected by land degradation. If this continues, the region may not be able to grow enough food for local communities, leaving them in danger of severe food insecurity.

Source 1

The Sahel region spans northern Africa.

The Sahel region in northern Africa

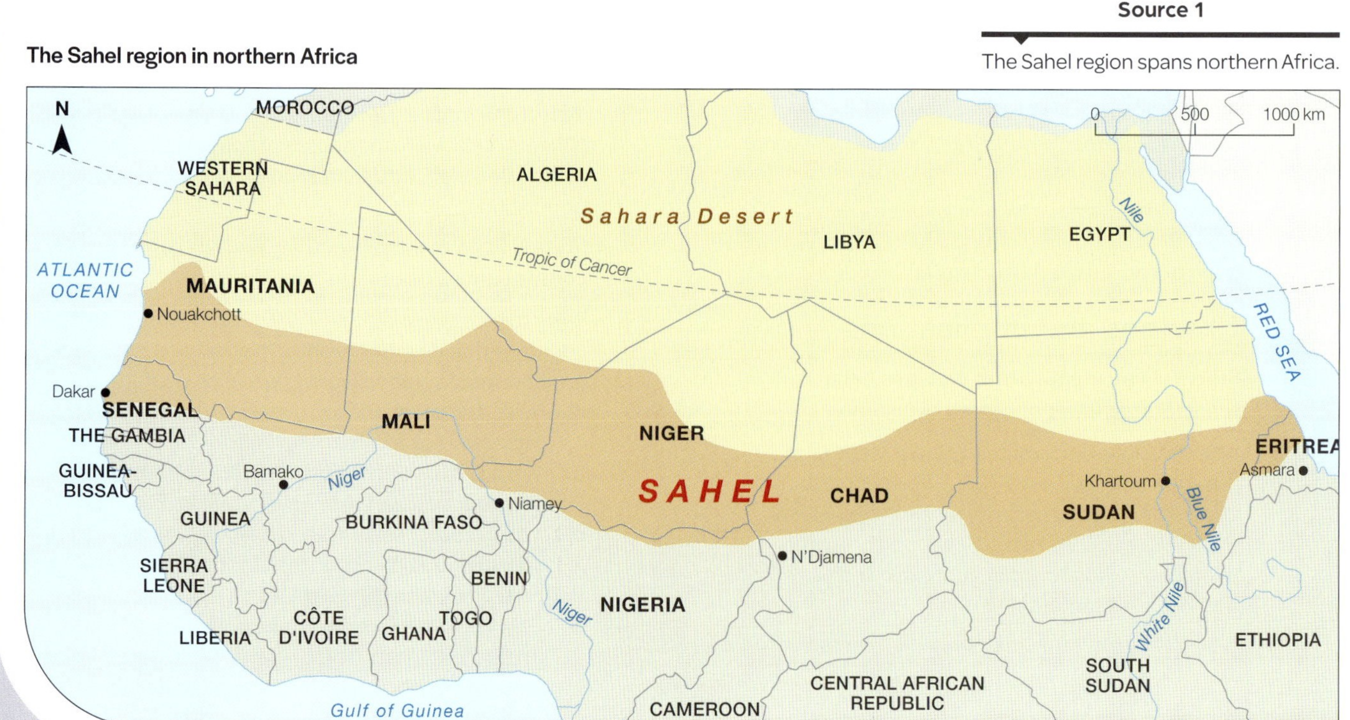

Main causes of soil degradation by region in susceptible drylands and other areas

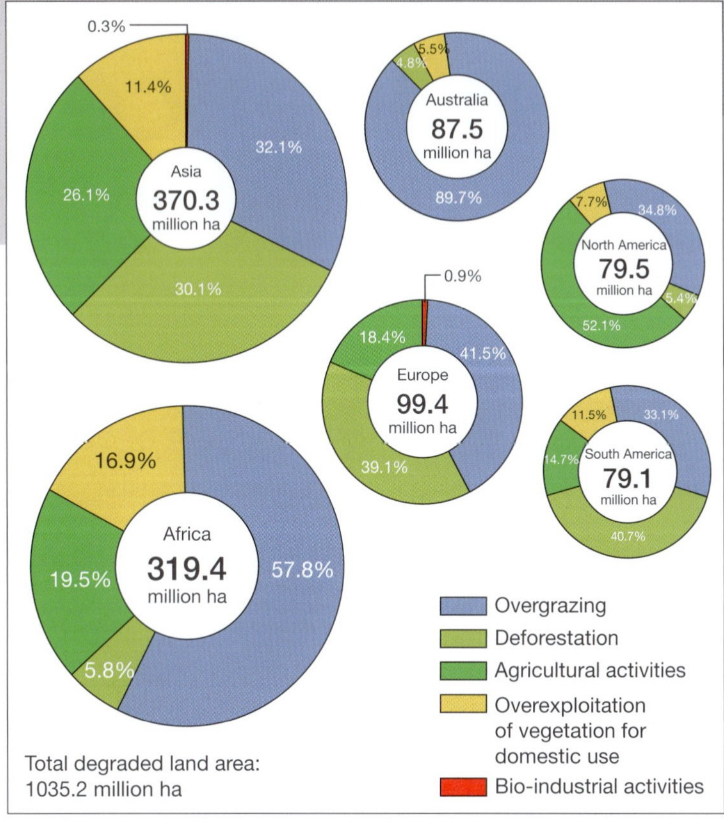

Source 2

There are many contributing factors to desertification in different world regions.

Source: Adapted from World Atlas of Desertification, 2nd Edition, UNEP, 1997

Source 3

A forest guard with tree seedlings he has planted beside a row of anti-soil erosion bricks on his field, Bissiga Village, Yako, Burkina Faso.

Learning ladder G2.5

Show what you know

1 Using Google Maps, describe the relative location of the Sahel region and the absolute location of Niger. Would you describe this place as an LEDC? Why?

2 Define the term 'desertification' and consider how this process is linked with crop yields.

3 Using Source 1 on page 44, describe the location of major food insecurity within Africa.

Patterns and interconnections

Step 1: I can provide short explanations for patterns and interconnections

4 Source 2: Explain why overgrazing is a major contributor to desertification in Africa.

Step 2: I can explain patterns and interconnections

5 Use SHEEPT to summarise why the Sahel region is particularly susceptible to food insecurity.

Step 3: I can use data to support explanations of patterns and interconnections

6 Source 2: Using PQE, describe the main causes of soil degradation. Explain why this may make the Sahel more susceptible to desertification.

Step 4: I can use relevant sources to research further reasons for patterns and interconnections

7 Locate two population pyramids 10 years apart for a country within the Sahel region. Discuss how population change may contribute to desertification.

PQE, page 136
SHEEPT, page 138
Population pyramids, page 146

G2.6

How does the environment influence Australia's productivity?

Australia is a productive nation, with agriculture contributing to 12 per cent of our gross domestic product and providing jobs for at least one in seven individuals.

Over 65 per cent of our produce is exported internationally; however, more than 90 per cent of fresh produce sold in supermarkets is locally grown. In Australia, crops such as grains, sugar and cotton have a gross value of around \$9–15 billion annually. Food waste is a huge issue in Australia, with more than 30 per cent of local produce being thrown out.

Agricultural land use in Australia

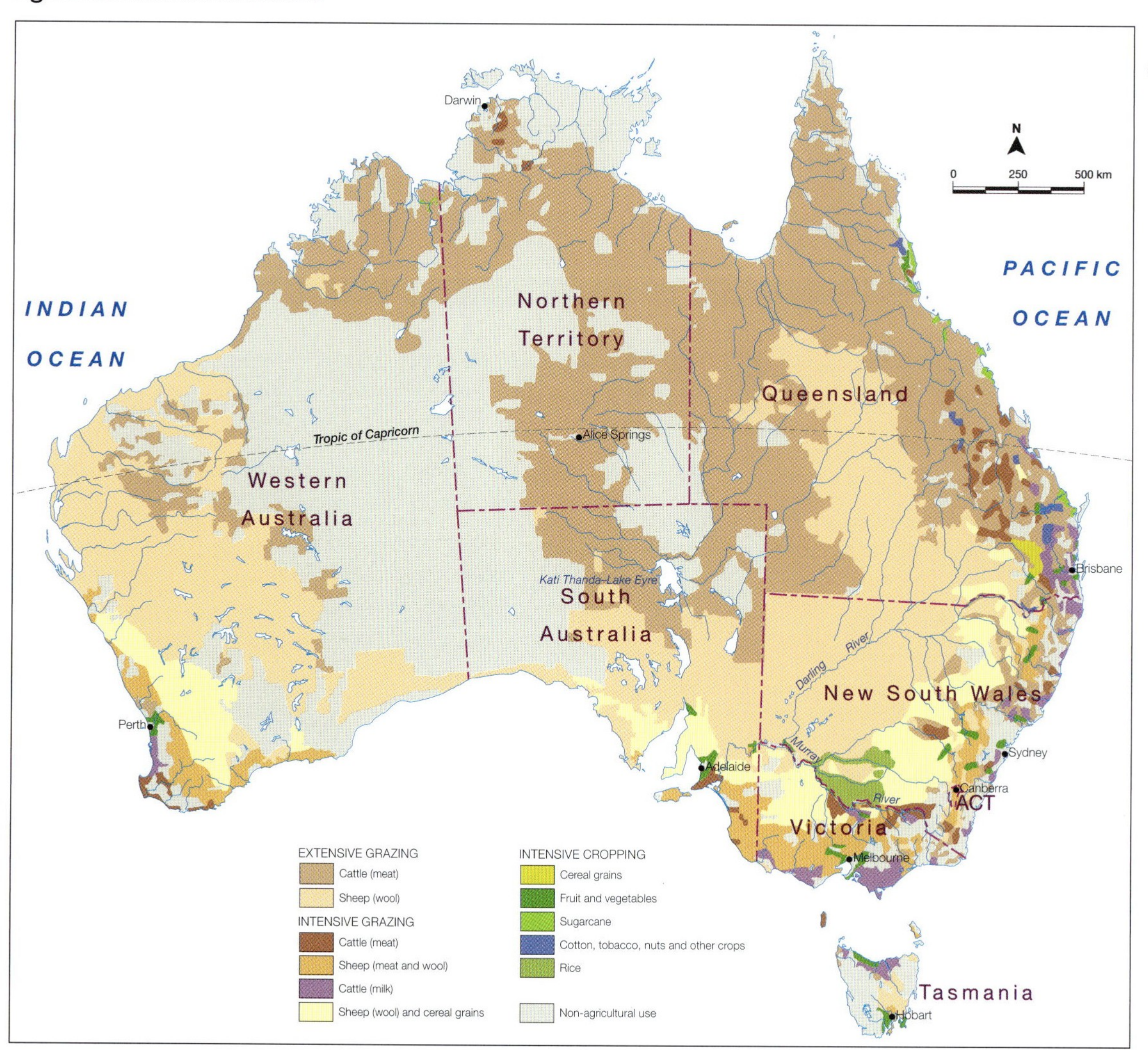

Source 1

Australia is largely used for grazing and livestock agriculture.

Source: Matilda Education Australia

Global warming and food production

As global warming has an impact on our local climate, farmers need to constantly adapt to the changing environment in order to provide for our growing population and sustain our export trade. Farmers have tried to maximise productivity by improving soil health to increase yields from the land, switching to crops that are more drought-friendly and trialling **genetically modified (GM)** crops. Farmers also use spatial technology for surveillance and maintenance of their yield.

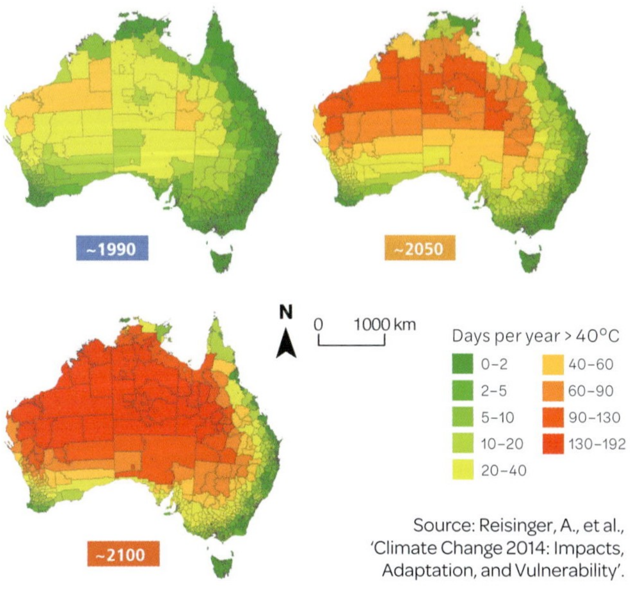

Source 2

Global warming is altering our weather and climate patterns, leading to longer and hotter summers.

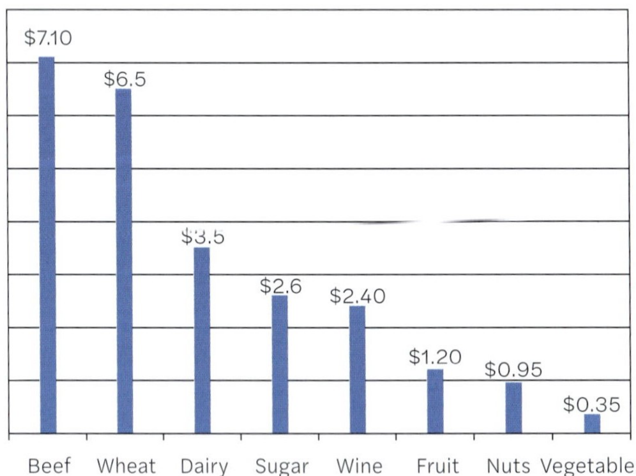

Source 3

Beef is Australia's biggest food export.

Learning ladder G2.6

Show what you know

1 The Murray–Darling Basin is sometimes called 'Australia's food bowl'. What do you think 'food bowl' means in this context?

2 Describe the interconnection between agriculture and changing climate. Note two major connections.

3 Discuss why so much Australian agricultural produce is exported.

Digital and spatial technologies

Step 1: I can interpret different map types using cartographic conventions

4 Source 1: Use PQE to describe the distribution of grazing land in Australia.

Step 2: I can construct paper maps using correct cartographic conventions

5 Create an overlay map of Australia. Have one layer illustrate the distribution of natural biomes and a second highlight the main agricultural uses.

Step 3: I can access and use spatial technology platforms such as GIS

6 Access http://mea.digital/GHV9_G2_3, then:

- select 'Add Data'
- select 'Land Cover and Land Use'
- select 'Land Use and Cover'
- select 'Land Cover'
- select 'Add to the Map' in the top right corner of the data preview map.

Comment on land cover in Australia. List the top two dominant land covers.

Step 4: I can manipulate data using digital and spatial technologies

7 Explore and add other data to the map you accessed in question 6. Comment on how you think humans have acted to alter biomes in Australia over time.

PQE, page 136
Overlay maps, page 148

How did the First Nations Peoples of Australia manage the land?

It was once thought that First Nations Peoples of Australia were hunter-gatherers, travelling to collect food rather than managing the land. However, historians are now challenging this idea and now know that First Nations communities practised **seasonal migration** – moving to particular areas for a set time to ensure they did not exhaust their land of plants and animals.

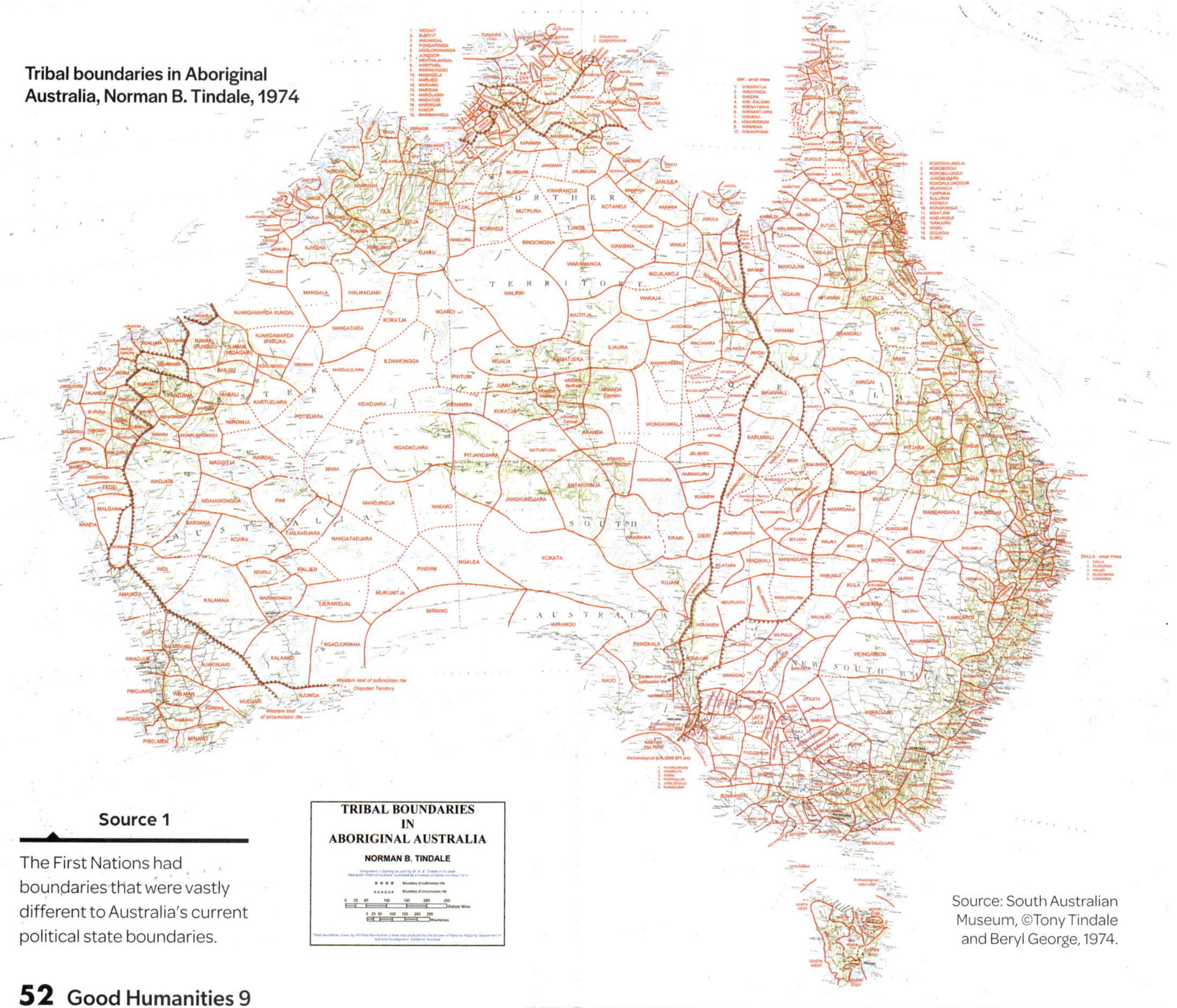

Source 1

The First Nations had boundaries that were vastly different to Australia's current political state boundaries.

Source: South Australian Museum, ©Tony Tindale and Beryl George, 1974.

Source 2

Fire was used to manage the land, hunt and allow for easy harvest.

Australia's first agriculturalists

Using historic journals, drawings and stories, historians have argued that First Nations Peoples engaged in complex agricultural methods that were sustained for thousands of years. There are references to dam building, mass planting and irrigation throughout historical sources, which are in stark contrast to original perceptions. Aboriginal peoples maintained large pastures of murnong (yam daisy) around Melbourne. These crops required deliberate cultivation to ensure the yams were edible. Fire was used to manage the land and to distribute seeds and encourage new growth of vegetation, which in turn attracted grazing animals and allowed easy harvesting. Tools to harvest crops, such as stone knives, have also been discovered.

Food security

The First Nations Peoples of Australia are recorded in historical sources as eating a wide variety of produce, which meant they were able to adapt to changing climates, seasons and weather patterns over time. In contrast, many agricultural farms today are limited to a **monoculture** and rely on the success of a few crop species.

Communities would also move to new locations if drought or other events meant food was scarce in certain locations. Population numbers are said to be have been aligned with food availability, so there were enough resources for everyone.

Learning ladder G2.7

Show what you know

1 Compare First Nations Peoples' approach to agriculture with practices currently used in Australia.

2 Create a photo essay illustrating the comparisons you made in question 1.

Communicate data

Step 1: I can list primary and secondary methods useful for my study

3 Source 1: Is this map a primary or a secondary source of data? Why? How could the data for this map have been collected?

Step 2: I can successfully use data collection methods

4 Collect a series of secondary sources (images, data, written excerpts) online that represent how First Nations Peoples of Australia used, managed and cultivated the land.

Step 3: I can filter collected data

5 Locate secondary data that illustrates First Nations Peoples' agricultural practices. Discuss why historians now conclude that they practised seasonal migration and agriculture.

Step 4: I can organise data collected according to relevance for a research question

6 Imagine you were a geographer assisting historians investigating First Nations' land management. What primary data would be useful for the investigation?

How do bushfires affect what we eat?

Australia is one of the driest continents on Earth; 80 per cent of the country receives less than 600 mm of rain each year. In 2019–20, drier than usual conditions (associated with long-term drought and other climate factors) led to one of the largest bushfire disasters in Australian history.

The 2019–20 bushfires

The fire season began uncharacteristically early, in June 2019, and continued to worsen until January 2020. By February 2020, more than 18 million hectares of land had been burnt, and it was estimated that over one *billion* native animals perished. Many rural communities were devastated, and hundreds of people lost their homes or properties. Tragically, 33 people lost their lives in the fires, including nine fire fighters.

Eastern NSW and eastern Victorian fires and potential livestock impacted

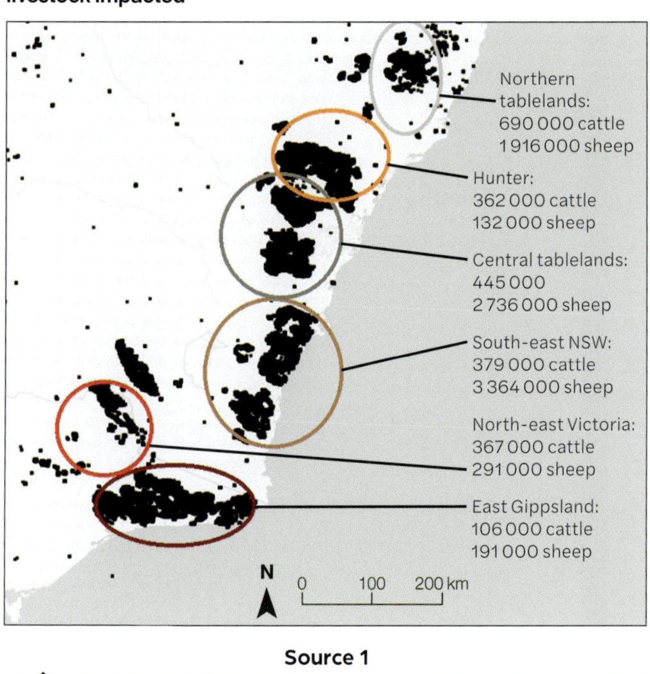

Source 1

The potential impact of the 2019–20 bushfires on livestock production.

Effects on agriculture

The 2019–20 fires had significant impacts on Australia's agricultural sector. More than 14 per cent of the total area burnt in the fires was classified as farmland. Many farms were directly impacted when fires consumed crops and grazing land and it is estimated that 2 per cent of Australia's total cattle and sheep populations were lost. The indirect impacts of the blaze, such as loss of native parklands and the scarcity of fresh water, have had negative effects on other agricultural businesses.

Bushfire damage

New South Wales has suffered the most extensive damage from the ongoing bushfire crisis, with Victoria also seeing widespread impacts (see Source 2).

Recovery efforts

On a local scale, the fires created short-term housing and food security issues. Road closures, loss of infrastructure and smoke haze all made it difficult to distribute food around the country. While some of these issues have been dealt with, others remain.

Nationally, the fires, drought and now the COVID-19 **pandemic** will have long-lasting impacts on our economy. It is more important than ever to 'buy local' to support the industries and communities who, amid the COVID-19 crisis, are still recovering from these fires.

	Conservation	Forests/ plantation	Agriculture	Residential	Total
VIC	428 872	809 689	161 717	16 465	1 416 743
NSW	2 557 544	749 410	558 878	28 581	3 894 412
TAS	18 478	12 505	825	357	32 165
SA	153 010	23 916	98 966	5513	281 404
AUS	3 157 904	1 595 519	820 385	50 915	5 624 723

Source 2

Damage to Australian land from the 2019–20 bushfires (figures in hectares).

Source: Digital Agriculture Services

How Australia's bushfires will affect what you eat

By Bonnie Bayley, ©SBS, 28 Feb 2020

Over in the Snowy Mountains town of Batlow in NSW, Greg Mouat's apple orchard was one of the orchards in the region that were burnt in the fires. 'We have about 14 000 trees, mainly apples and a few cherries, and we lost probably 10–12 per cent of them,' he says.

Oyster farmer Caroline Henry is based at Wonboyn Lake in NSW, an area surrounded by national parks and wilderness. While her oysters weren't directly damaged, 100 per cent of the surrounding catchment was burnt, and the big fear now, ironically, is heavy rain. 'If we get massive rains, it will wash sediment from the fires into the water course, which can lead to algal blooms and issues with the oysters not being able to filter feed,' she says.

Beekeepers are similarly in limbo, including East Gippsland apiarist Ben Murphy. 'We had 450 of our hives in the fire-affected area, and lost a bit over 200 hives, including a lot of our queens,' he says. Murphy's biggest hurdle is that about 75 per cent of the eucalyptus forest sites he leases for his bees were burnt, and could take up to 10 years to regenerate … If bees are disrupted, other industries may be too. 'Almonds, canola, oranges and apples all rely on bees being healthy to pollinate crops,' says Murphy.

The devastation at Ben Murphy's apiary.

Learning ladder G2.8

Show what you know

1 How are climate events such as drought linked to our national productivity?

2 Identify short- and long-term impacts to our overall productivity and crop yields from bushfires.

3 As Earth gets warmer, how might our vulnerability to bushfire and drought events change over time?

Spatial distributions and patterns

Step 1: I can identify spatial distributions and patterns

4 Source 1: What area in Victoria was most badly affected by the 2019–20 bushfires?

Step 2: I can use data to quantify spatial distributions and patterns

5 Source 2: Using appropriate axes, construct a bar graph displaying the amount of agricultural land lost to the fires in four Australian states.

Step 3: I can describe spatial distributions and patterns

6 Source 1: Using PQE, describe the distribution of the livestock harmed by the fires in NSW and Victoria.

Step 4: I can use data to support exceptions to spatial distributions and patterns

7 Source 2: Using data in your response, explain why Tasmania had much less agricultural and residential land damage than the mainland states.

PQE, page 136
Graphing, page 144

economics + business

How do we determine the value of resources?

There are not enough resources in the world to satisfy all our needs and wants, so understanding economics helps to inform our decisions about using scarce or valuable resources.

Source 2

Labour resource: Bakers

Resources in the economy

Whether it is a bakery cooking fresh bread, a publisher producing books or a construction company erecting skyscrapers, all businesses need **resources** as inputs in order to produce the **goods and services** that are their outputs. These resources are the **factors of production** used to make or supply the goods or services the business uses to make a **profit**.

There are four factors of production:

1 **Natural resources**: materials drawn from the natural world that are used to produce goods and services. For example, soil, water and sunlight are natural resources harnessed to grow crops such as wheat and rye, which are used to make flour that is then used to bake bread. Humans can't make natural resources, so they tend to be limited. Natural resources can be **renewable**, such as water and sunlight, or **non-renewable**, such as coal, oil or gas.

Source 1

Natural resource: Wheat

2 **Labour resources**: any human input required for making a good or providing a service. Any work performed by a human being, whether physical or intellectual, to produce profits is referred to as **labour** by economists. The labour resources involved in making fresh bread include not just the bakers but also the counter staff and managers, as well as other staff such as technicians or marketers.

Source 3

Capital resource: Factory and equipment

3 **Capital resources**: human-made resources such as machinery, buildings, equipment and technology used to produce other goods and services. Capital resources used to create and sell bread include flourmills and bakery buildings, ovens and mixers, as well as display counters and cash registers.

Source 5

A farm producing crops such as soya beans, potatoes, peanuts and cane sugar could produce 96 per cent more protein than the same farm producing beef cattle.

Entrepreneurs run the business

4 **Entrepreneurship**: the ability to organise the other factors of production and transform them into a business. Entrepreneurs risk time and money to run a business with the hope of earning a return. Baking entrepreneurs might not make bread themselves but use their business-planning expertise to run a bread-making business.

Scarcity of resources

There are not enough resources to meet all our needs and wants. Economists refer to this situation as **scarcity**, and one part of **economics** is the study of how humans make choices under conditions of scarcity.

Because resources are limited, every time you choose to use them, you must choose to go without other opportunities. Economists use the term **opportunity cost** to describe what is given up in order to obtain something else. For example, a farmer might choose to raise cattle on a farm, but this means they can't grow crops. Forgoing the opportunity to grow crops is the opportunity cost of raising cattle.

Opportunity food loss

Meat production is resource hungry, requiring more land and more water to produce than growing crops. Using scarce agricultural land to grow plant foods, rather than to raise livestock, would create enough food to feed millions of additional people using the same land resources.

Researchers at the Weizmann Institute, in Israel, refer to the opportunity cost of producing meat as 'opportunity food loss'. They compared beef production to producing crops, such as soya beans, potatoes, peanuts and cane sugar, that deliver the same nutrition as meat. They found that using agricultural land for beef production results in an opportunity food loss of 96 per cent per unit of land. In other words, a farm producing 100 grams of protein from plant-based crops would only yield 4 grams from beef.

Learning ladder G2.9

Economics and business

Step 1: I can recognise economic information

1 What different resources are required to make a product or service?

Step 2: I can describe economic issues

2 What is scarcity and why is it important in the study of economics?

Step 3: I can explain issues in economics

3 Use a personal example to explain the concept of opportunity cost.

Step 4: I can integrate different economic topics

4 Why are entrepreneurs important to the production of goods and services?

Step 5: I can evaluate alternatives

5 Explain why all beef producers do not simply change to plant-based protein crops.

thinking locally

How can we plan for a sustainable future?

Over time, the state of Victoria has seen a rapid growth in its population, which increased 1.95 per cent in 2018. It has the second-highest **population density** in Australia.

Inner-city Melbourne is one of the most densely populated areas in Australia, with 17 500 people per square kilometre.

To support our growing population, scientists estimate we will need to produce at least 60 per cent more food in the future. As a result, there will be a significant conflict between the need for agricultural land and the growing **urban sprawl**. Sustainable agricultural practices will be vital in order to provide for Melbourne.

The image in Source 1 was created by researchers at the University of Melbourne as part of their Foodprint Melbourne project. It shows a proposed city foodbowl plan to help Melbourne create a sustainable food future.

Resilient city foodbowl

A vision for Melbourne

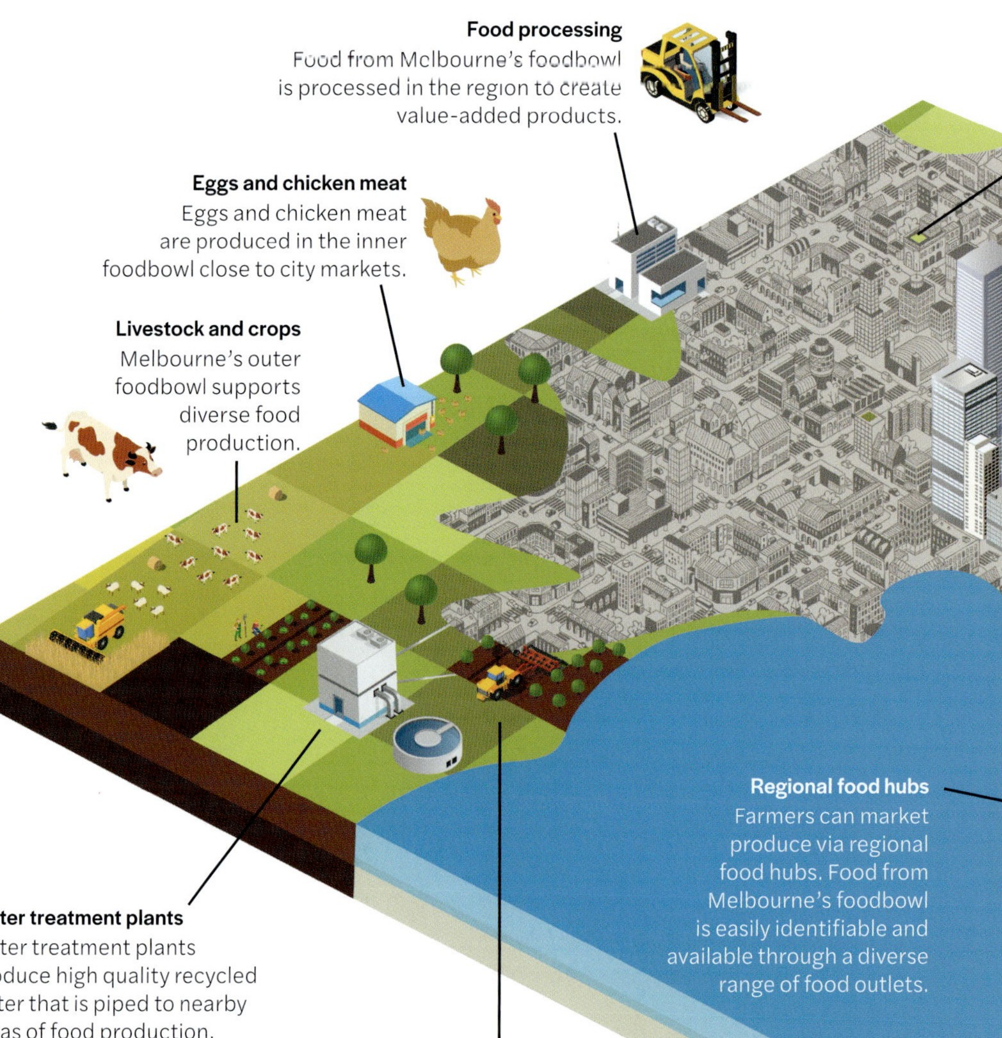

Source 1

Melbourne shown as a resilient city food bowl

Learning ladder G2.10

Show what you know

1 Define 'urban sprawl' and draw a diagram that visually shows this process.

2 How can we link the concept of urban sprawl with human-altered biomes?

3 Discuss how land might need to be used in the future to maintain our demand for both quality housing and food production.

Changes and implications

Step 1: I can identify that changes occur in the characteristics of places over time

4 Outline how the 'Resilient city foodbowl' vision will help change Melbourne and make it more sustainable.

Step 2: I can describe how places have changed over time

5 Research some historical photos of Melbourne's Central Business District (CBD). How has this place changed over time? What factors do you think led to this change? (Hint: Consider SHEEPT factors.)

Step 3: I can explain the causes behind the change over time in a place

6 Consider the suburb of Officer in Victoria. Using satellite images and time-lapse footage you find on the internet, discuss why this region has changed so significantly.

Step 4: I can make predictions and outline consequences of change over time

7 Using your response from question 6, discuss as a class any predictions you have regarding future land use within suburbs such as Officer.

HOW TO

SHEEPT, page 138
Satellite images, page 147

How can we use spatial technology to monitor productivity?

The **Global Positioning System (GPS)**, Google Maps and satellite imagery are all examples of spatial technologies. **Spatial technology** allows geographers and everyday users to visualise patterns on a variety of scales.

Land use and land cover in Victoria, Australia

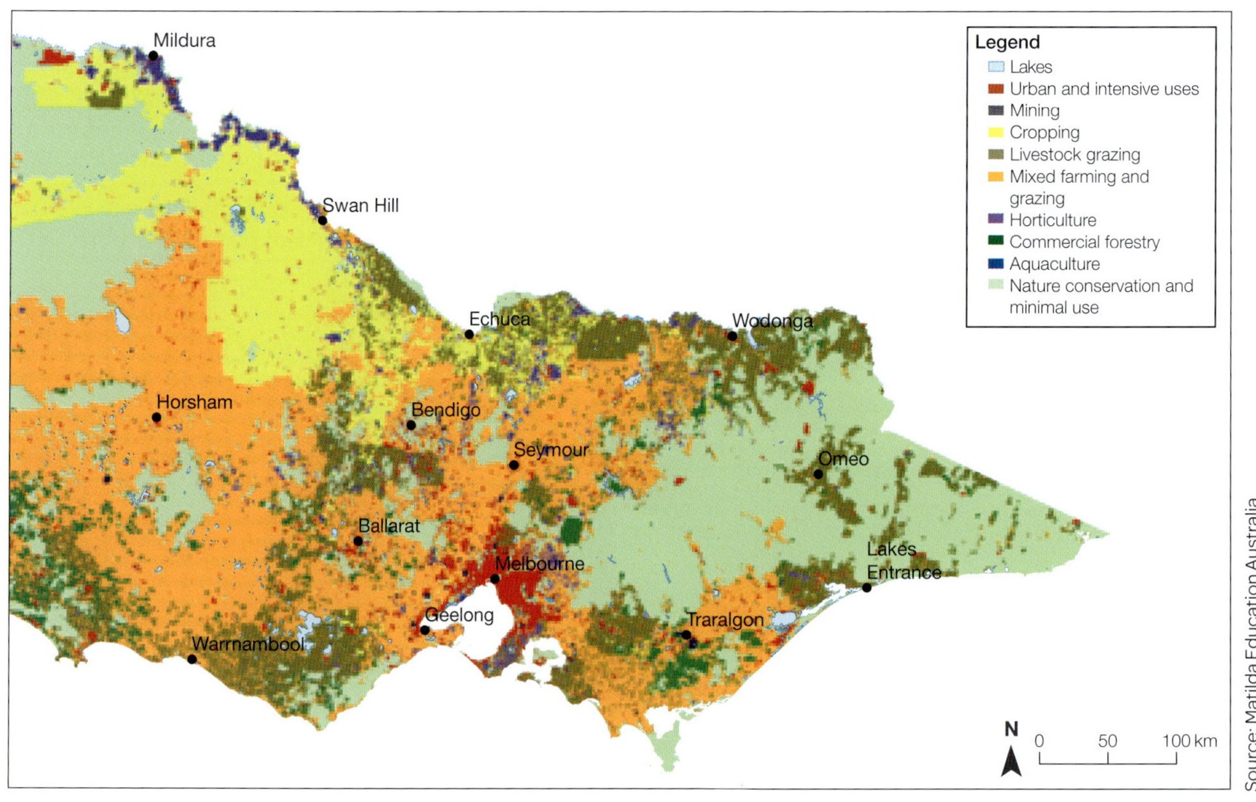

Source 1

Victoria is largely used for cropping and modified pastures.

How can technologies improve food security?

Victoria produces more than $14 billion of agricultural produce and the agriculture sector employs over 87 564 people. As an occupation, farming relies heavily on spatial technology and a detailed understanding of both natural and human geography. To produce successful yields, farmers must understand how to best use their land; be able to adapt to sudden changes in local climate; predict future trade and demand fluctuations; and maintain the fertility, viability and productivity of their land.

Victorian food and fibre export performance 2017–18

Victorian food and fibre exports

$14.1 billion

▲$1.4b (11%)

27% of national food and fibre exports

Meat – Victoria's largest value export

$3.3 billion

▲$910m or 37% on the previous year

USA highest value market for meat exports

$798 million

▲$175m 28% on the previous year

China highest value market for wool exports

$1.5 billion

▲$175m or 13% on the previous year

Japan highest value market for dairy exports

$442 million

▲$91m or 26% on the previous year

Sheepmeat ▲41%

Beef ▲46%

Wine exports to china increased ▲39%

exports to the UK increased ▲36%

Wool value $2.1b ▲18%

Dairy $1.9b ▲9%

Source: State Government of Victoria

Source 2

Victoria's most valuable export is food, particularly meat.

Spatial technology in action

As our population continues to grow, and the impacts of global warming such as extreme drought increase, it is becoming increasingly important for people working in the agricultural industry to ensure that they use the land productively and efficiently.

How spatial technology helps farming

David owns an 8000-acre property north of Griffith in New South Wales. He primarily farms wheat, barley, canola and legumes, such as chickpeas and lupins. David relies solely on natural rainfall to water his crops; therefore, he needs to pay close attention to the local climate and any changes in rainfall and weather patterns. In order to ensure successful yields, David plants crops that he knows are better adapted to drier climates. He also employs GPS and other mapping systems to manage soil quality, chemical use and rotation of crops.

The machinery used in this industry is highly sophisticated. David's tractors operate with a GPS tracker that has a 2-centimetre accuracy steering system to ensure minimal overlap when planting, which reduces fuel and fertiliser waste. Harvesters are also fitted with GPS tracking and yield monitors so maps can later be produced to understand which areas on the property are most productive and which may require soil testing or more fertilisers.

Using technologies such as these can help farmers like David to use the land more sustainably and allow them to monitor and increase the productivity of their farmland over time.

Source 3

Mapping outputs from David's farm. These allow him to assess and monitor the land, and increase its fertility and the viability of his crops.

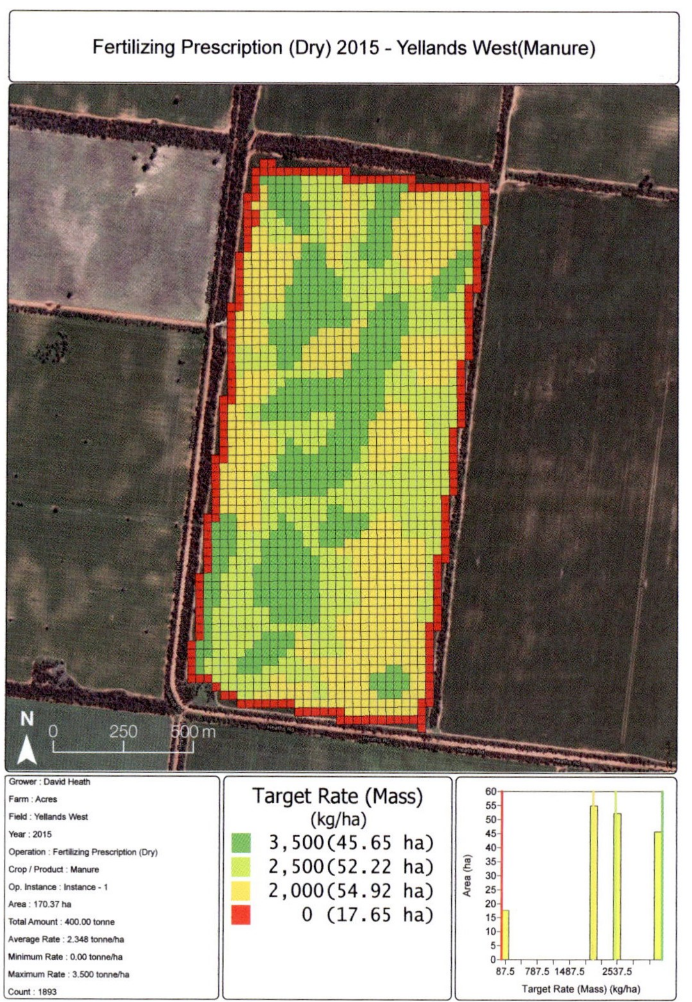

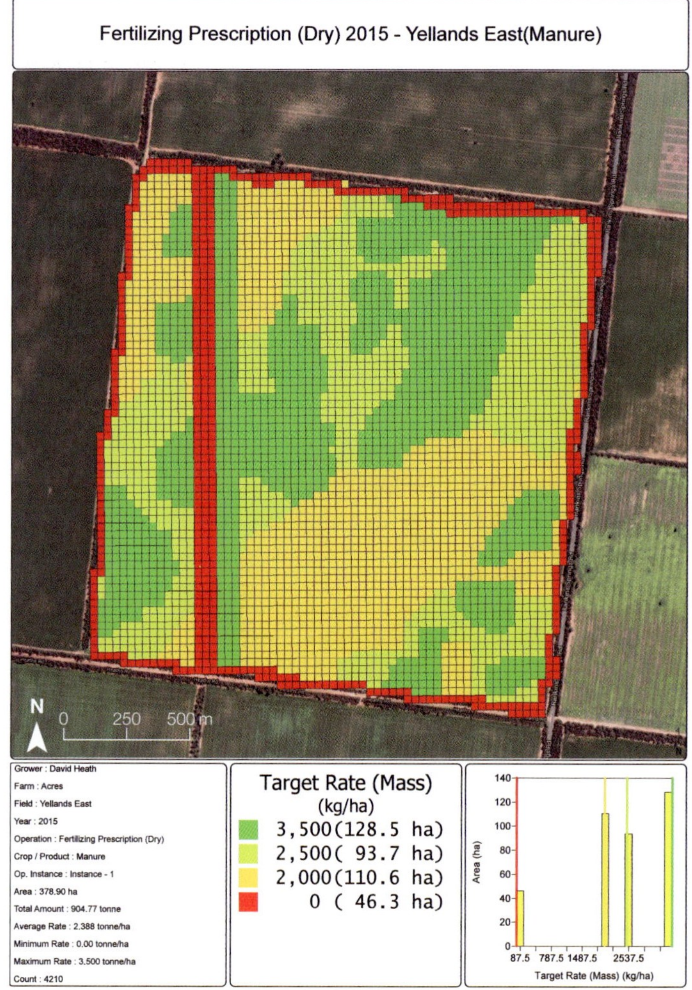

Source 4

Precision farming involves using GPS and other mapping systems to manage farmland and more accurately use resources such as fertilisers.

Learning ladder G2.11

Show what you know

1 Imagine you were teaching a Year 7 class about spatial technology. Write a short summary outlining what it is and how it can be used, and provide two different examples of spatial technologies used today.

2 How can spatial technology be used to monitor food security?

3 Visit the Farming and Food Production page of the CSIRO website at http://mea.digital/GHV9_G2_5. According to CSIRO, how can we better prepare for a more food-secure future?

Digital and spatial technologies

Step 1: I can interpret different map types using cartographic conventions

4 Source 1: State the primary land use of Melbourne.

Step 2: I can construct paper maps using correct cartographic conventions

5 Using correct techniques, draw a sketch map of your local region and highlight any areas that are used for productivity and food production. If there are no areas of food production, consider why this is the case.

Step 3: I can access and use spatial technology platforms such as GIS

6 Describe how spatial technology can be used to map land use around Australia. How is this helpful in determining future food security?

Step 4: I can manipulate data using digital and spatial technologies

7 Source 3: Summarise how maps such as these could be manipulated to assist farmers in maximising their crop production.

BOLTSS, page 132

Masterclass

Learning ladder

Work at the level that is right for you or level-up for a learning challenge!

Global temperature change 2018 compared to the 1951–1980 average

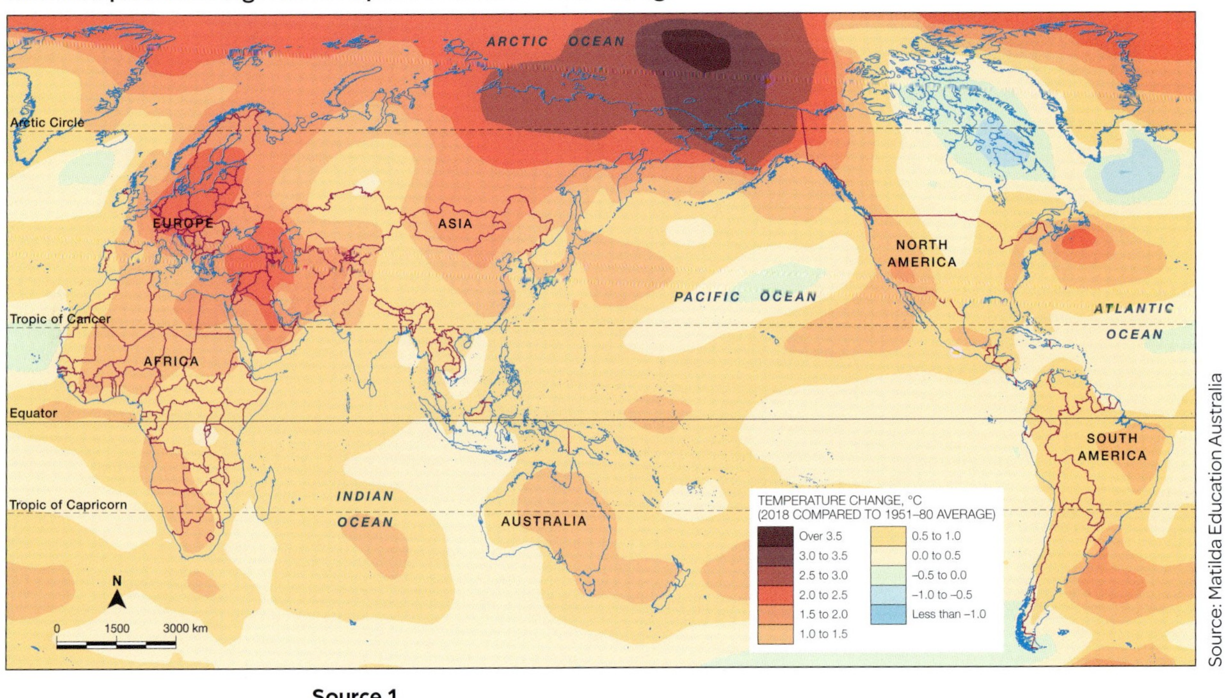

Source 1

Increased greenhouse emissions are leading to global temperature increases.

Global distribution of food security

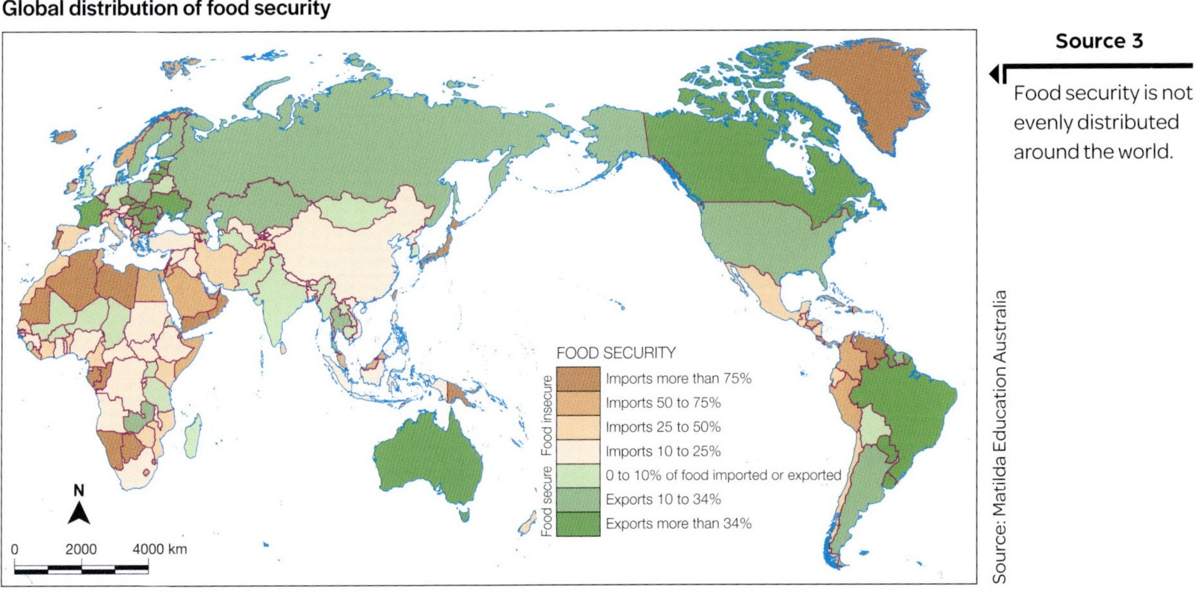

Source 3

Food security is not evenly distributed around the world.

Source 2

While population continues to rise, our overall growth rate is declining.

World population growth, 1750–2100

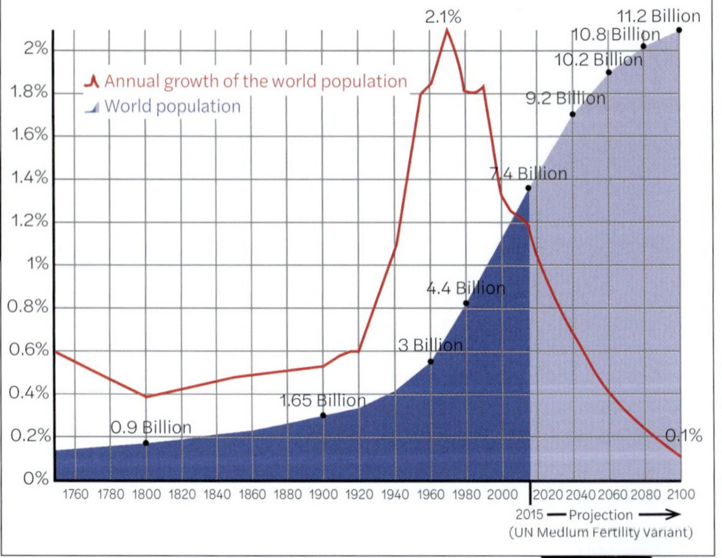

Step 1

a **I can identify spatial distributions and patterns**

Source 1: Identify two countries that are at risk of a 0.15°C temperature change.

b **I can provide short explanations for patterns and interconnections**

Explain why Australia's agricultural industry is dominated by livestock grazing and grain crops.

c **I can identify that changes occur in the characteristics of places over time**

Identify environmental changes that occurred in order to build your school. What effects may this have had on the surrounding region?

d **I can list primary and secondary methods useful for my study**

Source 3: List the primary and secondary data needed to create this map.

e **I can interpret different map types using cartographic conventions**

Describe how topographic maps would be helpful in planning for environmental change.

Step 2

a **I can use data to quantify spatial distributions and patterns**

Source 3: Using data, describe the distribution of food security in Africa compared with North America.

b **I can explain patterns and interconnections**

Source 3: Using SHEEPT, explain the distribution of food insecurity on a global scale.

c **I can describe how places have changed over time**

Describe how your town, suburb or city has changed over time, using examples as evidence in your response.

d **I can successfully use data collection methods**

Use Google Earth to locate your local region. Collect data on the type of biomes in your region and how humans have used or changed them over time.

e **I can construct paper maps using correct cartographic conventions**

Locate population counts for the seven major continents. On a blank world map, illustrate the distribution of the human population in a regional context.

Step 3

a **I can describe spatial distributions and patterns**

Source 1: Describe (PQE) the expected temperature change on a global scale.

b **I can use data to support explanations of patterns and interconnections**

Conduct research to estimate the percentage of Australia that is used for cattle and sheep grazing.

c **I can explain the causes behind the change over time in a place**

Source 1: Provide a brief explanation about why temperature is expected to change on a global scale in the future.

d **I can filter collected data**

'Malthus' theory is still relevant today because of the uneven distribution of food security.' Discuss with reference to data collected from research.

e **I can access and use spatial technology platforms such as GIS**

Comment on how spatial technology can be used to monitor and assess agricultural production.

Masterclass

Step 4

a **I can use data to support exceptions to spatial distributions and patterns**

Source 3: Using examples and data from the chapter and external research, explain why there is uneven food security on a global scale.

b **I can use relevant sources to research further reasons for patterns and interconnections**

Source 2: Using research, explain why world population growth continues to increase while population growth is declining.

c **I can make predictions and outline consequences of change over time**

Source 2: Based on current projections, predict how world population might change in the next century.

d **I can organise data collected according to relevance for a research question**

'Biomes would not change if humans did not alter them.' Discuss whether you agree with this statement, using evidence from the chapter to justify your response.

e **I can manipulate data using digital and spatial technologies**

Visit the National Museum Australia Encounters website to find an interactive map at http://mea.digital/GHV9_G2_6. Explore the different Aboriginal and Torres Strait Islander areas and comment on any links made to agriculture or environmental management.

Step 5

a **I can identify multiple spatial distributions and patterns**

Refer to Sources 1 and 3. Using PQE, identify any interconnections between expected temperature changes and food security patterns on a global scale.

b **I can interpret causes of patterns and interconnections**

Predict how Australia's agricultural land use may change over time due to global warming. Discuss the interconnection between these processes.

c **I can interpret data to quantify predictions based on research**

Access http://mea.digital/GHV9_G2_7.

Select one country. Using the data provided, comment on how affordability, availability and safety affects food security in that place.

d **I can evaluate the success of research methods**

You are asked to make a speech at assembly regarding the impacts of global warming on the world's biomes. You use a basic internet search to gain some data from blogs and forums. Consider the reliability of the data you have collected. Create a list of more reliable sources of information, and then collect three statistics to use in your speech.

e **I can draw conclusions from geographical information in digital and spatial technologies**

Visit the Geoscience Australia website at http://mea.digital/GHV9_G2_8. Navigate to Data & Publications. Click on Interactive Maps. Find the surface hydrology map in the water section. Comment on the distribution of waterways in Australia.

Capstone

How can I understand a human world?

In this chapter, you have learnt a lot about the human world. Now you can put your new knowledge and understanding together for the capstone project to show what you know and what you think.

In the world of building, a capstone is an element that finishes off an arch or tops off a building or wall. That is what the capstone project will offer you, too: a chance to top off and bring together your learning in interesting, critical and creative ways. You can complete this project yourself, or your teacher can make it a class task or a homework task.

mea.digital/GHV9_G2

Scan this QR code to find the capstone project online.

Perceptions and places

G3

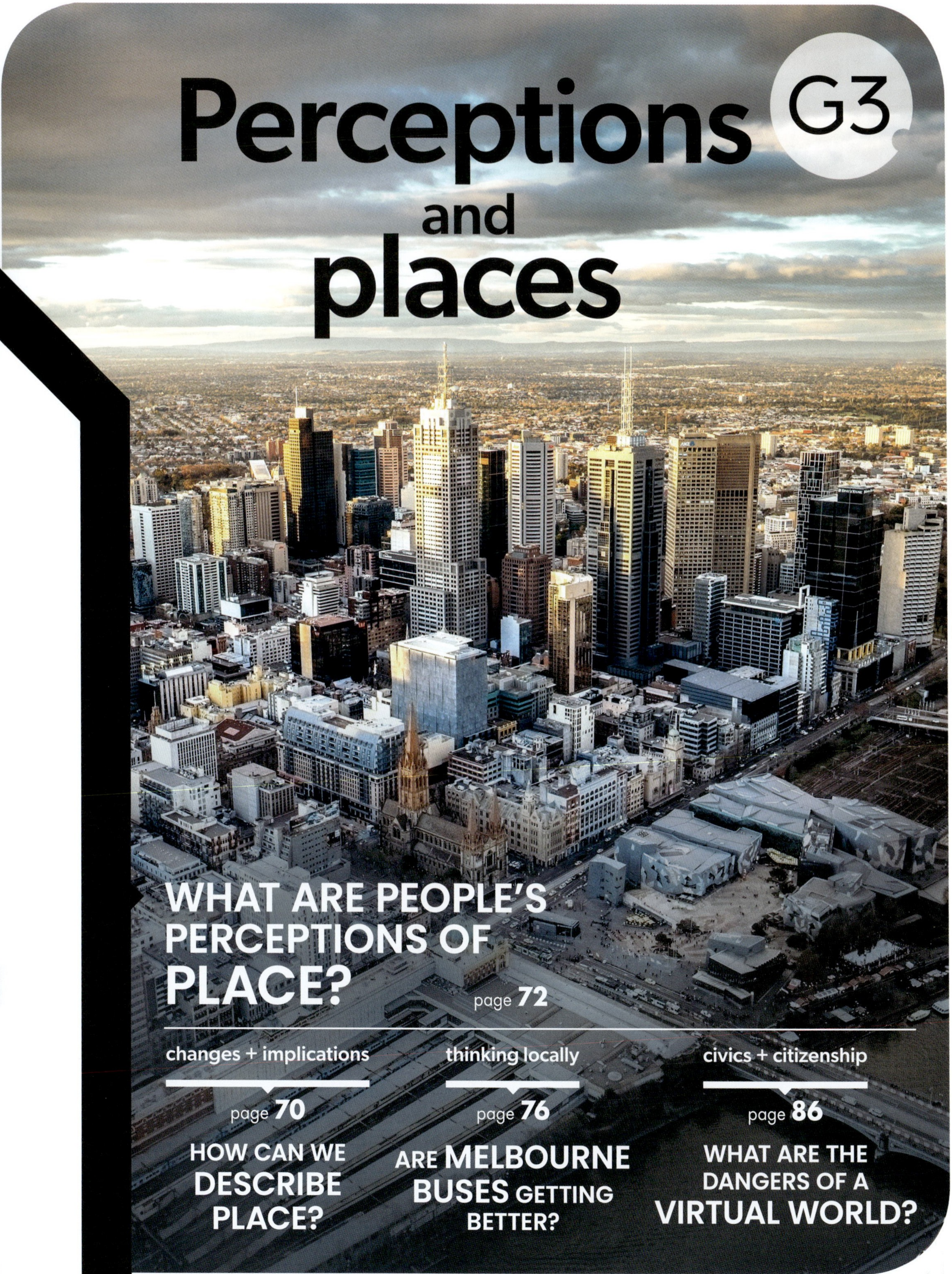

WHAT ARE PEOPLE'S PERCEPTIONS OF PLACE?

page 72

changes + implications

page 70

HOW CAN WE DESCRIBE PLACE?

thinking locally

page 76

ARE MELBOURNE BUSES GETTING BETTER?

civics + citizenship

page 86

WHAT ARE THE DANGERS OF A VIRTUAL WORLD?

How can I understand perceptions and places?

All of us are connected to places. Your home, your school, the place you like to sit at lunch, your favourite holiday location – the concept of 'place' is evident in all of our lives. The *geography of interconnections* focuses on investigating how people are connected to places throughout the world in a wide variety of ways, and how these connections can change places.

learning ladder

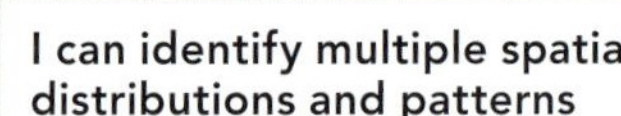

	Spatial distributions and patterns	Patterns and interconnections	Changes and implications
step 5	**I can identify multiple spatial distributions and patterns** I can take my PQE one step further to find links or relationships that exist in places.	**I can interpret causes of patterns and interconnections** I can use multiple sources to find links or relationships that exist in places and can explain 'Why?'.	**I can interpret data to quantify predictions based on research** I can use external data from research as evidence of the positive and negative impacts of a change I have predicted.
step 4	**I can use data to support exceptions to spatial distributions and patterns** I can use data to answer 'Why?' about the exceptions identified in a PQE analysis of places.	**I can use relevant sources to research further reasons for patterns and interconnections** I can use sources other than this textbook to further research patterns I observe in places.	**I can make predictions and outline consequences of change over time** I can use my knowledge of natural processes and world regions to make an educated guess about the positive and negative impacts of change in a place.
step 3	**I can describe spatial distributions and patterns** I can describe patterns, quantify them and point out exceptions (PQE) to describe places.	**I can use data to support explanations of patterns and interconnections** I can use data from a map or graph to explain patterns I observe in places.	**I can explain the causes behind the change over time in a place** I can use my knowledge of natural processes and world regions to explain why changes may occur over time in a place.
step 2	**I can use data to quantify spatial distributions and patterns** I can read data and use it to measure key trends on a map or graph about places.	**I can explain patterns and interconnections** I can identify social, historical, economic, environmental, political and technological (SHEEPT) factors to help me explain perceptions of places.	**I can describe how places have changed over time** I can use specific examples to describe changes over time in places.
step 1	**I can identify spatial distributions and patterns** I can find key trends on a map or graph about places.	**I can provide short explanations for patterns and interconnections** I can write descriptions of patterns and interconnections that I find in places.	**I can identify that changes occur in the characteristics of places over time** I can read information and answer questions about changes over time in places.

Warm up

Source 1

People take risks sitting anywhere they can on an overcrowded train in Dhaka, Bangladesh. They are travelling home for the religious festival of Eid-ul Adha.

Spatial distributions and patterns

1 Compare the community's use of public transport in the place in Source 1 with how people use public transport in Victoria.

Patterns and interconnections

2 Which region of the world the photo in Source 1 was taken in.

Changes and implications

3 Discuss with a partner how the world's growing population may affect how connected people feel to places.

Communicate data

4 Consider your journey to school. What locations along this route do you feel connected to? Why?

Digital and spatial technologies

5 Create a sketch map of the locations outlined in Question 4.

Sketches and annotating, page 140

I can evaluate the success of research methods
On reflection, I can look back and comment on the data collection methods I used and evaluate how successful they were in helping me answer a research question about a place.

I can organise data collected according to relevance for a research question
I can review the data I have collected in the field and display it using graphs, tables, annotations and captions.

I can filter collected data
I can review my collected data and select the most relevant data to answer a research question about a place.

I can successfully use data collection methods
I can use primary and secondary data collection methods in the field and classroom to investigate places.

I can list primary and secondary methods useful for my study
I can create a checklist of methods to investigate a place and categorise them as primary or secondary methods.

Communicate data

I can draw conclusions from geographical information in digital and spatial technologies
I can interpret and analyse patterns by using different layers and features on spatial technology platforms.

I can manipulate data using digital and spatial technologies
I can work with layers and other features on spatial technology platforms to further explore data and interconnections.

I can access and use spatial technology platforms such as GIS
I can use spatial technology platforms to explore data and find patterns.

I can construct paper maps using correct cartographic conventions
I can use a pencil, paper and ruler to construct a map that follows BOLTSS conventions.

I can interpret different map types using cartographic conventions
I understand data found in different types of maps and graphs and use the data to answer questions about a place.

Digital and spatial technologies

How can we describe place?

Place can refer to any area that has defining characteristics and has meaning to people. A meaningful place for you could be as simple as your 'place' at the dinner table, or as complex as an Indigenous Australian person's **connection to Country**. Place can be described by its geographical characteristics, which can be either human or natural.

Source 1

A number plate identifies which place cars have come from as they drive around Australia.

Natural characteristics

Natural characteristics are those that exist largely without human intervention. They may include location, vegetation, animal life, climate, seasonal variations or proximity to landforms.

Human characteristics

The human characteristics of a place may involve social, historical, economic, environmental, political and technological (SHEEPT) factors. For example, a place may have a dominant culture, religion or language (social factors); it may have been affected by war (historical factor); or there may have been an economic boom (economic factor). Places can also be described by their access to technologies and infrastructure, such as schools, hospitals, roads, shops or parklands.

Source 2

First Nations rock art in Kakadu National Park, Northern Territory. This shows an interaction between natural and historical human characteristics.

Location

Places can also be described by their location. A place may be rural or urban, industrial or coastal. A place's location will often determine what we do there – coastal regions are often associated with holidays, recreational activities, fishing or even trade, while city locations are associated with office work, public service, tourism and entertainment.

The natural and human characteristics of places change over time. This may be on a small scale, such as seasonal changes in snowfall at Mount Buller, or large-scale changes in response to population growth and **urban sprawl**. How would you describe your local place? Has it is changed over time?

Source 3

People often describe where they live as either rural or urban.

Learning ladder G3.1

Show what you know

1 Outline the natural and human characteristics of your local region.

2 Compare the terms 'place' and 'space' using examples.

3 Draw a field sketch of your local area, annotating the natural and human characteristics outlined in question 1.

Changes and implications

Step 1: I can identify that changes occur in the characteristics of places over time

4 Create a photo essay of your local region, highlighting changes that have occurred over time.

Step 2: I can describe how places have changed over time

5 Annotate the photo essay from question 4 to show significant impacts to the natural and human characteristics of the area.

Step 3: I can explain the causes behind the change over time in a place

6 Consider the changes you identified in question 4 and 5. Using SHEEPT, explain why these changes may have occurred over time.

Step 4: I can make predictions and outline consequences of change over time

7 Access: http://mea.digital/GHV9_G3_1 and search for your local government area. Using the data displayed, consider how these factors may lead to changes in your region over time. For example, if your area has a growing young population, will more schools be needed?

SHEEPT, page 138
Sketches and annotating, page 140
Photo essays, page 143

What are people's perceptions of place?

A key part of understanding the concept of place is to consider the human connection to a place and the significance of that area. When we have an emotional or spiritual connection to an area, we often say that it has a 'sense of place' for us.

What influences people's perceptions of place?

Each person will have a different opinion about, or attachment to, a place. Your perceptions of a place develop over time and depend on many factors, such as your gender, family history or even tolerance levels. If you don't like loud noises or large groups of people, you may not like major cities. On the other hand, people who grew up in a city might find the busyness, lights and street noise comforting, but struggle with the quiet of rural areas.

Sometimes our perceptions are based on personal experience, such as our connection to home. At other times, our perceptions are influenced by hearing stories from other people. Suburbs or cities can develop negative reputations over time if they are portrayed in the media as areas with high levels of crime and violence.

Why do we connect to different places?

Your connection to a place may make up part of your identity. You may feel most at home in a particular area and have a strong sense of belonging, comfort and cultural connection to that place. Over time, you may develop an attachment to many different places as your needs, wants and desires change.

When communities undergo trauma or disaster, their strong sense of place is often illustrated by their ability to work together to rebuild, support neighbours, provide relief to nearby townships or form groups to aid in the recovery. More than 10 years after the devastating Black Saturday Bushfires of 2009, the people of Kinglake are still working together to support those affected, and to develop strategies to avoid disasters like that in the future.

Source 1

A sign made by locals after the devastating 2009 Black Saturday fires killed 173 people.

Source 2

Religion connects people to place and to other people.

Source 3

Sporting teams and other group activities bond people to a particular community and place.

Learning ladder G3.2

Show what you know

1 Define the term 'sense of place'.

2 Explain how your sense of place can change over time or between regions.

3 Sources 1–3: Create an illustration or mind map that encompasses all of the factors that build your sense of place.

Communicate data

Step 1: I can list primary and secondary methods useful for my study

4 Create a survey to determine which factors are important for creating a sense of place in your class. Include five or more questions.

Step 2: I can successfully use data collection methods

5 Carry out the survey you created in question 4 and discuss your results as a class.

Step 3: I can filter collected data

6 Use the data from your survey to create a summary table or graph that shows the results of your data collection. Is there one factor that your class links more often to 'sense of place' than others? Why do you think this is?

Step 4: I can organise data collected according to relevance for a research question

7 Using your mind map from question 3, and a suitable colour legend, classify the different factors that build your 'sense of place' into the following categories:

- social
- religion
- education
- relationships.

Is there one factor that is dominant when building your sense of place? Compare this with the ideas raised in question 6.

How are different places physically connected?

All places are interconnected. When we discuss how places are connected, we usually think about physical connection such as the roads, transport and trade routes between and within regions. There is an estimated 874 500 kilometres of road in Australia, and around 55 per cent of surface travel (the movement of people or goods by road, train or ship) between places occurs in our capital cities.

Local transport networks

For many people living in cities, public transport is the main way they move between places. As petrol prices and the costs of running private vehicles increase, public transport provides a more sustainable and economic solution for local travel needs. Public transport routes and networks are constantly being upgraded as more advanced technologies emerge and new communities are formed in the outer suburbs of cities.

In Melbourne, the use of trains to get to work has risen by 27 per cent since 2011, which is comparable to an additional 57 trains' worth of passengers.

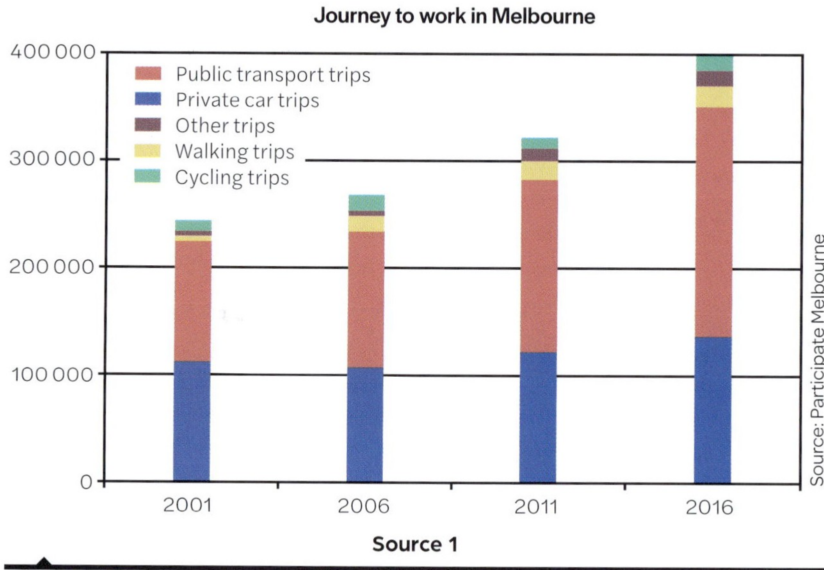

Source 1

The number of people that journey to work in Melbourne using a range of transport options

Source 2

Myki is the ticketing system for public transport in Victoria.

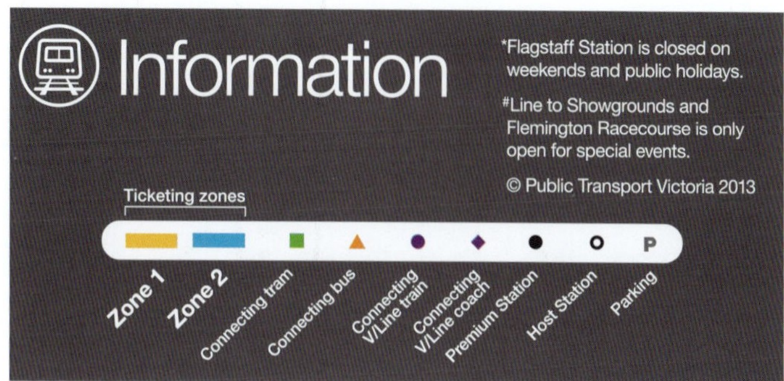

Source 3

Melbourne's train network

Source: © Head, Transport for Victoria 2020

Learning ladder G3.3

Show what you know

1 List five ways that people are physically connected to their surrounding region.

2 Create a photo essay of how your local community is interconnected with the surrounding region.

3 Consider how you are 'physically connected' to your place. Create a list of ways you move around your region – do you ride to school or catch a train to the city?

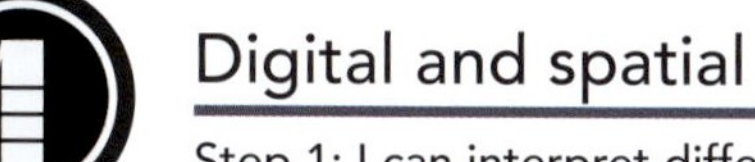

Digital and spatial technologies

Step 1: I can interpret different map types using cartographic conventions

4 Access the map of Australia's roads here: http://mea.digital/GHV9_G3_2. How does transport connect people to regions? Discuss the following ideas:

- a How does trade connect cities and towns?
- b How do public transport routes allow people to move within and between places?
- c Where are most of the transport networks located? Why is this?
- d How could transport networks be improved to allow better connections?

Step 2: I can construct paper maps using correct cartographic conventions

5 Locate a blank map of your local area. Annotate the transport networks that allow connection between you and other places.

Step 3: I can access and use spatial technology platforms such as GIS

6 Access Google Maps. Consider how this technology allows us to feel physically connected to 'place'. How does it help us to more easily move around a 'space'?

Step 4: I can manipulate data using digital and spatial technologies

7 Locate your house on Google Maps. Identify any major roads, highways or freeways nearby. Using the scale, calculate how far they are in kilometres from your house. Discuss how these kinds of roads provide greater physical connection to other places than smaller suburban roads.

Photo essays, page 143

thinking locally

Are Melbourne buses getting better?

Bus lobby pitches to solve Melbourne's transport problems a 'hell of a lot sooner than rail'

By James Oaten and Ben Knight, 25 Oct 2018

Like a lot of Melburnians, Amanda Ralph hates getting the bus to and from work.

It's usually packed, and often late. Sometimes it doesn't even turn up at all. 'We call them ghost buses,' Ms Ralph said. 'They just don't turn up. There's nothing on the PTV app. There's nothing on the Transdev [website] or Twitter account. So, you have no idea.'

But like the two-thirds of Melburnians who don't live near a train or tram stop, Ms Ralph has few alternatives to get from her home in Balwyn North to her job in the city. So, she puts up with the bus – and the long, uncomfortable ride …

Commuters bypass the bus

As Melbourne continues to grow, buses have arguably never been more important to the city's public transport network – servicing new suburbs, and spanning the gaps between the train and tram lines. Yet this year, Infrastructure Victoria found that nearly half of Melbourne's bus network is underperforming. 'Forty per cent of the bus services that we operate aren't getting enough passengers to really justify their existence,' said John Stone, a lecturer in transport planning at Melbourne University.

Melbourne buses compare poorly to Brisbane and Sydney networks. A 2009 study found Sydney buses carry twice as many passengers per kilometre than Melbourne buses. Melbourne buses carry a fraction of the passengers who catch trains and trams, even though the bus network covers more of the city … The main complaint, however, is how infrequently buses run, said Chris De Gruyter from RMIT University's Centre for Urban Research.

'Melbourne has over 300 bus routes. But the majority of these only run every 30 to 40 minutes, and some only run once an hour. We need higher frequency services that people will want to use, rather than have to use because it's their only option.'

Yet for many people in Melbourne, buses are the only option for public transport – especially in the outer growth suburbs, where residents may be waiting decades for a train or tram link. So, it should hardly be surprising that many choose not to spend their time stuck in traffic while standing in the aisle of a bus, and decide to sit in their car instead – adding to congestion …

Bus Association Victoria argues it could fix the frequency problem for $300 million a year – providing services every 15 minutes from 5.30 am to 10.30 pm on weekdays … 'People these days don't want to travel when the timetable says they must,' said Dr Lowe.

'They want to travel when it suits them. And if there's a bus every 15 minutes they don't need a timetable; they just know it's going to show up.'

Melbourne's SmartBus network

Source 1 Melbourne's buses are an important part of the public transport network.

Source: © Head, Transport for Victoria 2020

Learning ladder G3.4

Show what you know

1 What are the main issues with Melbourne's public transport system?

2 How do transport systems need to adapt to meet the needs of the growing population?

3 With a partner, create a list of 3–4 recommendations for the government about how to improve the local interconnection between places through transport.

Patterns and interconnections

Step 1: I can provide short explanations for patterns and interconnections

4 Source 1: Explain the apparent scattered pattern of bus networks in Melbourne. What suburbs do they miss?

Step 2: I can explain patterns and interconnections

5 Using SHEEPT, explain why public transport is important to connect people and places.

Step 3: I can use data to support explanations of patterns and interconnections

6 Consider the situation when more people choose to drive to work than to take public transport. What issues does this create for road networks and the environment? Use at least one point of data from personal research to support your response.

Step 4: I can use relevant sources to research further reasons for patterns and interconnections

7 Access http://mea.digital/GHV9_G3_3, page 3. Review the list of reasons why we rank poorly when it comes to tackling transport emissions. List three things that could help Australia's transport to become more sustainable. You may want to conduct some research on the higher-ranked countries for ideas.

SHEEPT, page 138

How are we virtually connected to place?

Places are not only connected via physical roads, public transport networks and landscapes, but also via virtual pathways, such as social media, internet platforms and other technologies.

Everyday virtual connections

Over time, improvements in technology have altered the way we communicate and connect with places. Social media and photo-sharing applications allow people to travel virtually and develop perceptions about places without physically visiting them. Access to mobile devices, such as phones, tablets and laptops, means that we can connect to any place at any time. These new technologies have allowed for advancements in trade, the launch of online shopping and instant answers to questions about our world.

During the COVID-19 pandemic, international travel became virtually impossible and domestic travel was severely limited. Victorians in particular experienced lockdowns that left us unable to leave our homes.

Suddenly, virtual reality was more important than ever to connect to extended family, friends and places. For many people who live alone, social connection became entirely virtual; for others, live-streamed events such as Phillip Island's Penguin Parade replaced school holiday trips or weekend getaways.

Source 1

Technology has improved our virtual connection to the world.

Source 2

An output from Snap Map – this feature within the application Snapchat allows users to see where their friends are. While you can tag your location on other social media apps, Snap Map turns sharing your location into a visual experience.

Source 3

GPS tracking allows us to watch our food being delivered in real time. It can be used to track our fitness and help us find our way.

Source 4

Social media, such as Instagram, connects us to places we may have never visited.

Learning ladder G3.5

Show what you know

1 How could virtual connection change people's sense of place?

2 Can you be truly connected to a place through technology, or do you need physical access to feel a true connection?

3 Break into two groups and debate the following topic: 'Virtual connection to a place is better than physical connection'.

Communicate data

Step 1: I can list primary and secondary methods useful for my study

4 Consider the following research question: 'Which smartphone app do students use the most to connect to their school environment?' List two primary methods you would use to collect data to answer this question.

Step 2: I can successfully use data collection methods

5 List the apps you use that allow you to connect to other places virtually. As a class, discuss what apps you use and why you use them.

Step 3: I can filter collected data

6 Create a summary table that identifies each of your apps and provide a 1–2 sentence explanation for how each helps you connect to place. (You may wish to record some of the ideas raised in your class discussion from question 5.)

Step 4: I can organise data collected according to relevance for a research question

7 Tally the number of different apps identified in question 5 for everyone in your class. Graph your results.

G3.6

How is technology reducing temporal scale?

As technology has improved, our ability to connect with people and places on the other side of the world has become easier and almost instantaneous. You can now video chat with a family member in Australia when travelling overseas. You can buy the latest fashion online from a shop in the USA and have it land on your doorstep within a week or two. Our ability to connect virtually to place has shortened the **temporal distance** between locations. What once may have taken weeks or even months may now take less than a second.

Temporal scale

In Geography we refer to two main types of scale: spatial scale and temporal scale. Spatial scale refers to the physical size or distribution of a phenomena or process. For example, an urban block is smaller than an agricultural property in spatial scale. Temporal scale is the measurement of time, such as a period of 20 years.

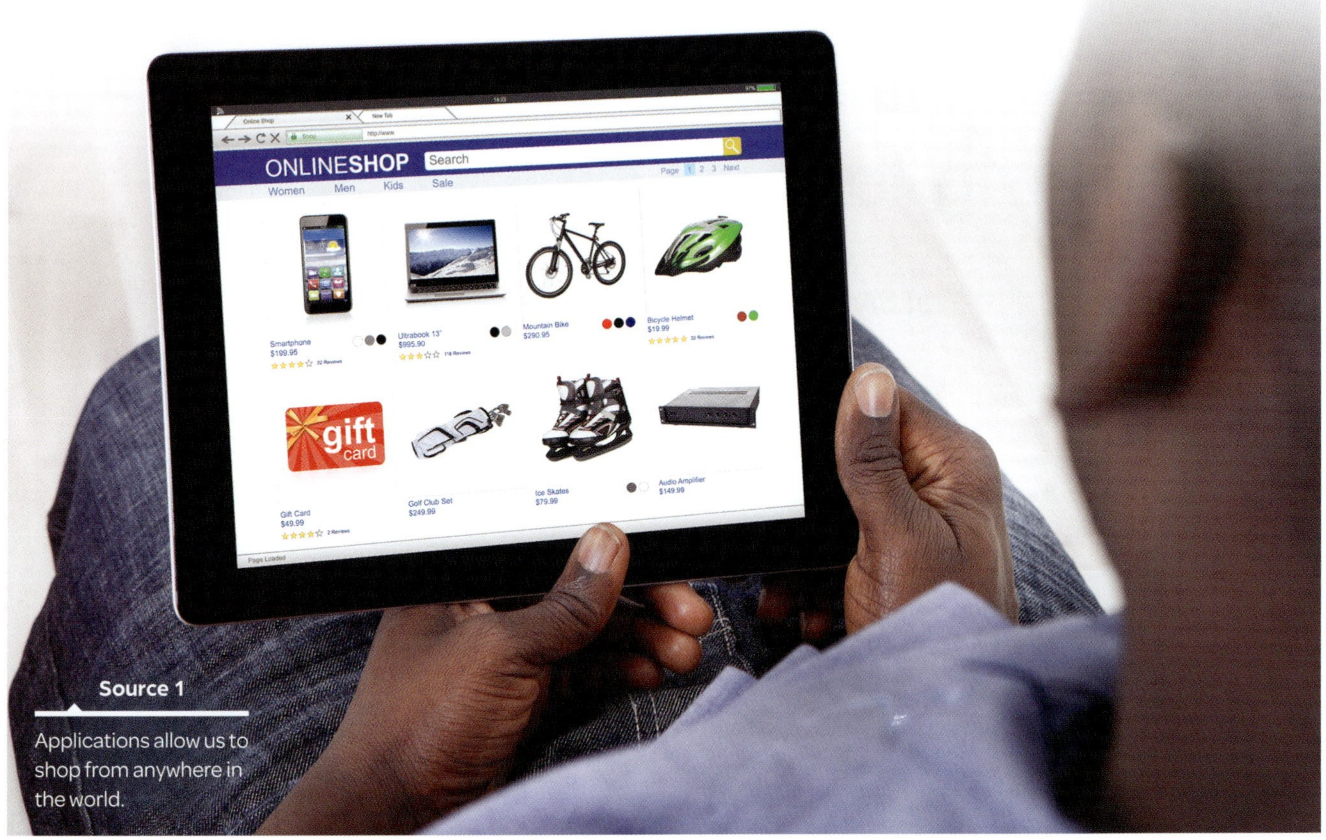

Source 1

Applications allow us to shop from anywhere in the world.

Percentage of adults who use the internet at least occasionally or report owning a smartphone, worldwide 2015

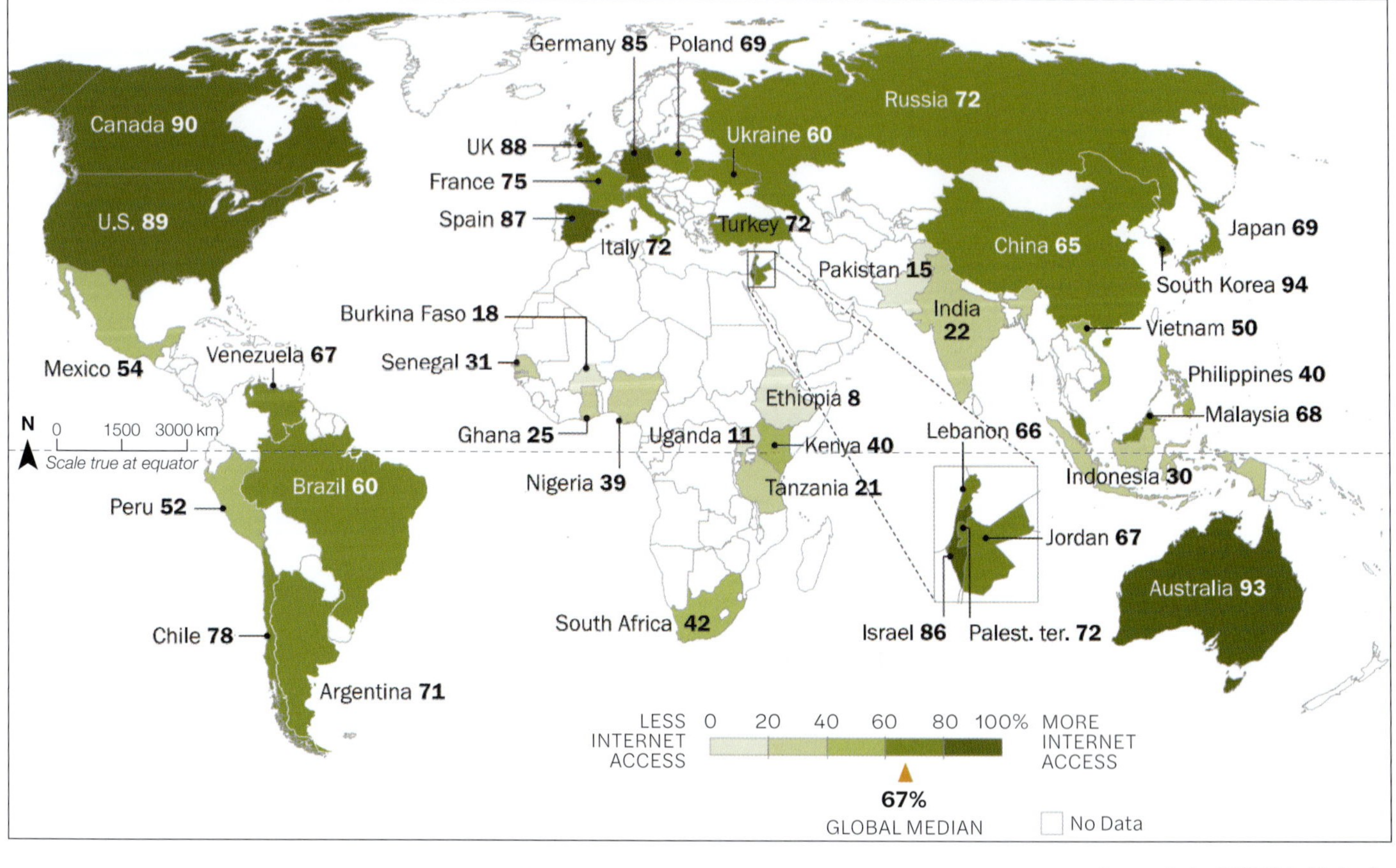

Source: Spring 2015 Global Attitudes Survey, Pew Research Center, 2015

Source 2

This map shows access to the internet on a global scale. In most parts of the world, more than two-thirds of people use the internet. However, fewer peope have access in Africa and South Asia.

Temporal and spatial scale are inherently linked. For example, urban blocks in the suburb of Officer in Victoria have become smaller over the past 10 years to allow for high-density urban development.

Reducing the urban–rural divide

The reduction in temporal distance through technology has meant that people in rural or marginalised communities now have better access to resources enjoyed by residents in more urbanised regions. Historically, rural communities were isolated and often lacked access to higher education opportunities or a range of medical professionals. Thanks to the growth of online education, telehealth and video conferencing, these communities are now able to access these services from their own homes.

Internet access is not evenly distributed

Despite the range of benefits brought about by technology, access to technology is not evenly distributed on a global scale. It is estimated that 94 per cent of youth have access to the internet in **more economically developed countries (MEDC)** compared with 30–67 per cent in **less economically developed countries (LEDC)**. Many countries monitor internet use and restrict access to social media platforms. People in LEDC may not have the financial capacity to purchase devices, or maintain an internet presence, and therefore are less able to access global opportunities to connect to places, people, services and resources. However, while differences are clear, LEDCs are catching up and their access to technology is improving.

Source 3

Some countries block or limit access to the internet, including social media sites.

Levels of internet censorship, 2019

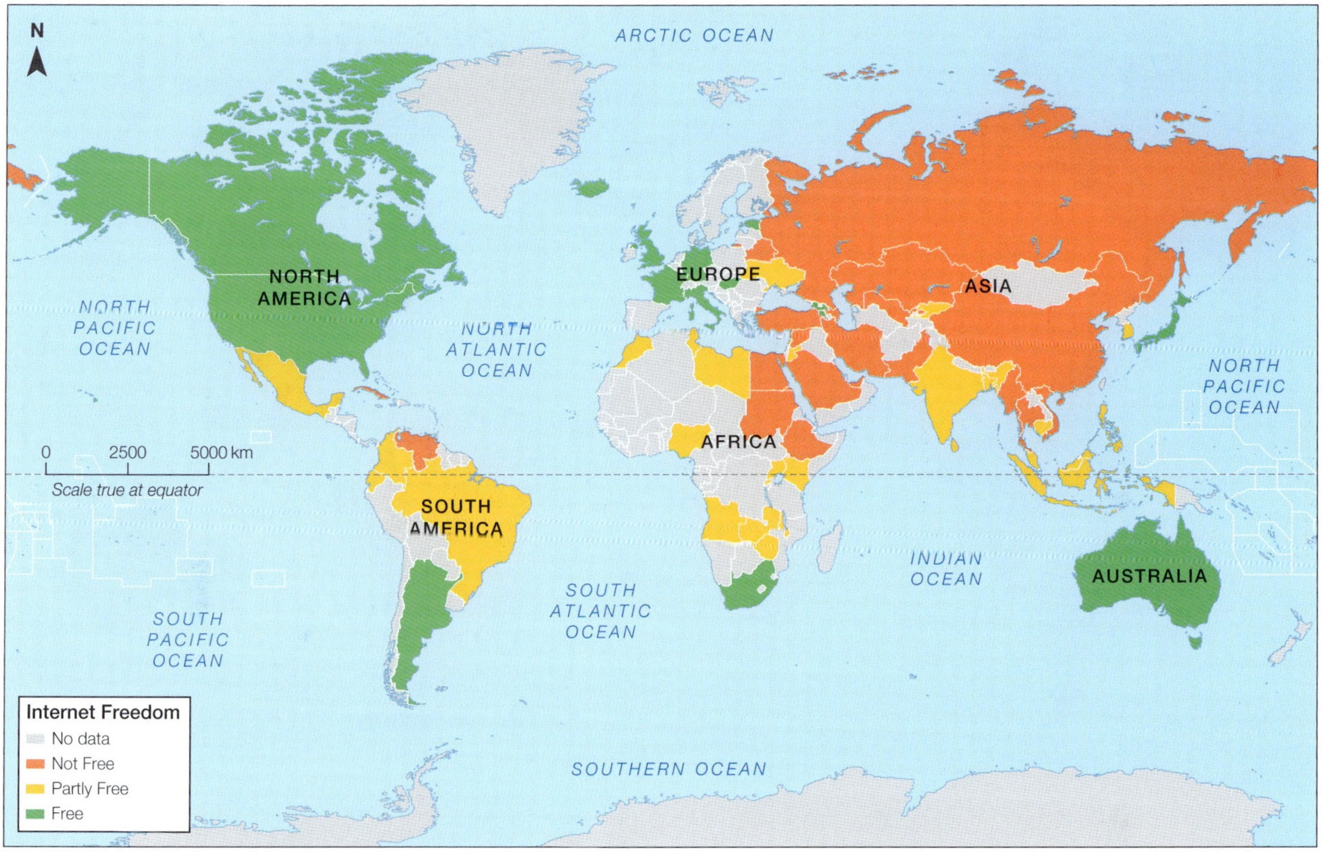

N

ARCTIC OCEAN

NORTH AMERICA

EUROPE

ASIA

NORTH PACIFIC OCEAN

NORTH ATLANTIC OCEAN

NORTH PACIFIC OCEAN

AFRICA

0 2500 5000 km

Scale true at equator

SOUTH AMERICA

INDIAN OCEAN

AUSTRALIA

SOUTH ATLANTIC OCEAN

SOUTH PACIFIC OCEAN

Internet Freedom

- No data
- Not Free
- Partly Free
- Free

SOUTHERN OCEAN

Source: Matilda Education Australia, data from Freedom House

Source 4

Technology has made remote learning accessible to nearly all students.

Source 5

Historically, remote learning meant studying via the School of the Air.

Remote learning

Not only does virtual connection increase our ability to interact with other places and people, it also allows us to access education. Fifty years ago, if you lived in a remote rural community, going to school or university required you either to board at the institution or to travel long distances every day. For some students, the only way they could get their education was via the 'School of the Air'. This remote learning school provided young people with access to teachers and resources by using CB radio connections and broadcasts.

In the present day, remote learning via technology and virtual spaces has developed dramatically, meaning students can complete their VCE online and attend university lectures and seminars virtually.

During the recent COVID-19 crisis, remote learning became vital to help young Australians continue their learning. Victorian school students, in particular, needed to be highly adaptable and to access learning via online platforms for much of the year. Some students reported that online learning fostered independent learning, while allowing them to stay home and learn in their own environment. Others reported that virtual connection for schooling did not allow for the same level of discussion, interaction, class involvement and teacher communication as traditional face-to-face learning. In addition, while remote learning meant that lessons could still take place, school events such as sport days, dances and performances could not, reinforcing that school is more than just what happens in the classroom.

Learning ladder G3.6

Show what you know

1 Compare spatial scale and temporal scale.

2 Explain how virtual connection has narrowed the divide between urban and rural regions.

3 Visit the ICT Development Index 2017 at http://mea.digital/GHV9_G3_4. Using a blank world map, indicate the location of the 10 countries that are the most ICT developed and the 10 countries that are the least ICT developed.

Spatial distributions and patterns

Step 1: I can identify spatial distributions and patterns

4 Source 2: Identify the world region with the highest percentage of adults who use the internet at least occasionally or report owning a smartphone in 2015.

Step 2: I can use data to quantify spatial distributions and patterns

5 Using the map you created in question 3, comment on the distribution of ICT development, using data from the website to support your response.

Step 3: I can describe spatial distributions and patterns

6 Source 3: Locate three countries that censor the internet. Discuss why the governments of these countries may block access for their citizens.

Step 4: I can use data to support exceptions to spatial distributions and patterns

7 Source 2 and 3: Consider why some countries within Africa have greater access to the internet than others. Use data in your response. Consider SHEEPT factors to help expand your ideas.

HOW TO

SHEEPT, page 138

How has COVID–19 affected our reliance on technology?

COVID–19 dominated conversation, news, social media and communities in 2020. National, regional and local restrictions and lockdowns meant that we relied on technology to create a sense of belonging and place like never before.

What is COVID–19?

Caused by a novel (new) coronavirus, 'severe acute respiratory syndrome coronavirus 2' (SARS-CoV-2) is a **viral** infection that causes the disease called 'COVID-19'. COVID-19 affects an infected person's blood vessels and respiratory system, causing symptoms such as fever, lethargy and difficulty breathing, as well as complications such as pneumonia. Researchers described the virus as 'novel' because they had not previously seen it in humans. Medical researchers are still working to understand the disease's full spectrum of symptoms, as well as to create effective treatments or a vaccine.

COVID-19 was first identified in Wuhan, China, with the first known human case recorded on 1 December 2019. From there, the disease was able to spread quickly. On 11 March 2020, the World Health Organisation announced the spread of the virus had reached **pandemic** status.

Responding to the pandemic

To reduce **community transmission**, and ensure the health system could cope with increasing numbers of sick people, Australia (and particularly Victoria) underwent a series of lockdown stages. These ranged from social distancing by standing at least 1.5 metres from others while in the community, to self-quarantining by isolating oneself in the home.

Source 1

Infographic showing ways to prevent transmission.

Technology and our response

As a result of having to physically distance ourselves from others, we needed to 'be together' in new ways. Use of online meeting platforms and social media increased exponentially, and the National Broadband Network (NBN) recorded a 71 per cent increase in user traffic as people tried to work, learn and socialise from home.

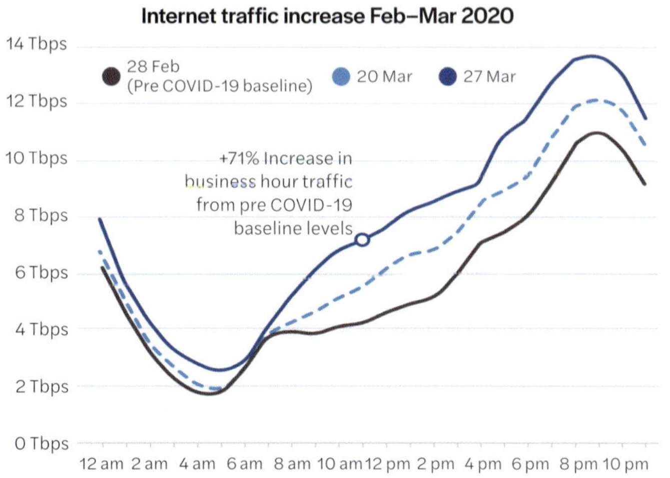

Source 2

Graph released by NBN Co., showing the increase in user traffic during initial lockdowns. For more information, see page 160.

Remote learning became essential to continuing the schooling process, and teachers had to learn new ways to engage with students who were not allowed in the classroom. All over the world, people were quick to download apps that would enable them to stay connected (and entertained!) during the pandemic, as shown in Source 3.

The most downloaded apps in the USA, 26 March–1 April 2020

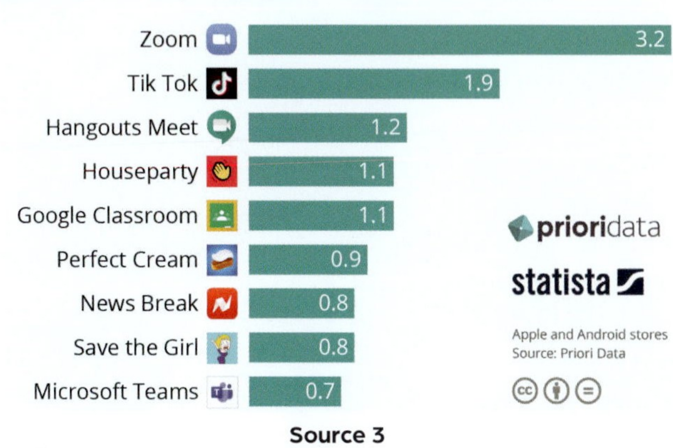

Source 3

The most downloaded apps in the USA from a week early in the pandemic lockdowns.

Learning ladder G3.7

Show what you know

1. Complete a mind map to illustrate your current understanding of COVID-19. Include headings such as symptoms, treatments, prevention methods, responses, news reports and any other categories that you may know of or have heard.
2. Define the term 'pandemic'. Explain why COVID-19 is classified as a global pandemic.
3. Hygiene plays a vital role in reducing the spread of COVID-19. Perfect your handwashing technique by going to http://mea.digital/GHV9_G3_5 to create a new poster for your bathroom using your favourite song.

Changes and implications

Step 1: I can identify that changes occur in the characteristics of places over time

4. Sources 1 and 3: Identify changes to living conditions caused by lockdowns that altered communities and our 'sense of place'.

Step 2: I can describe how places have changed over time

5. Access: http://mea.digital/GHV9_G3_6. Using data from the graphs, comment on the changes to COVID-19 cases over time.

Step 3: I can explain the causes behind the change over time in a place

6. Access http://mea.digital/GHV9_G3_7. Consider why the virus was able to spread so quickly on a:
   a. local scale
   b. regional scale
   c. global scale.

Step 4: I can make predictions and outline consequences of change over time

7. Read through the executive summary for Australia's national COVID-19 response plan at http://mea.digital/GHV9_G3_8. Comment on the government response to the pandemic. Do you think it was effective in the short and long term? Why or why not? Use data and evidence from the report to support your response.

G3.8

civics + citizenship

What are the dangers of a virtual world?

Social media has opened up a virtual world where people can stay in contact anywhere and at any time. The ease of connecting to people has opened up new opportunities for cyber criminals to undertake fraud, theft or harassment. Lawmakers and enforcers are working hard to keep up with these new threats.

Law enforcement in a virtual world

Our ability to connect virtually to place has greatly shortened the temporal distance (see pages 80–81) between locations. Social media applications allow people to travel virtually and access to mobile devices, such as phones, tablets and laptops, means that we can connect to any place at any time. These changes in technology have altered the way we communicate and connect with places. It has also had a major impact on law enforcement.

The development of a virtual world has created new classes of **cyber crime** and allowed existing offences to be committed in new ways, such as:

- crimes committed to disrupt, damage or infect computers and computer systems
- **computer intrusion** – unauthorised access to a digital device or network
- using technology to commit traditional crimes such as fraud, theft or harrassment.

Cyber crimes can be difficult to police because of the lack of geographic boundaries in the virtual online world. Australian law enforcement authorities can usually only take action when the perpetrator or computer server resides in Australia. Authorities in Australia have increasingly joined global law enforcement networks to fight the rising tide of cyber crime.

Law enforcement is also using the capabilities of new online technologies to fight back against cyber criminals. The Cyber Security Operations Centre coordinates cyber security capabilities across the Australian government and shares information with the Australian Federal Police, the Australian Security Intelligence Organisation and the Australian Signals Directorate. It uses its technological capabilities to disrupt and deter offshore cybercriminals.

Source 1

The Australian Cyber Security Operations Centre opened in 2010.

Source 2

Victims should report acts of cyberbullying to a trusted adult and block the bully (where possible). All technology users should avoid sharing personal information and should protect their passwords.

Cyberbullying

Cyberbullying is the use of technology to harass, threaten or embarrass another person with the intent to hurt them. This form of bullying can be particularly harmful because the victim can be publicly tormented wherever they are, while the bully can be anonymous.

Cyberbullying may occur through:

- abusive or aggressive messages or posts
- humiliating messages, photographs or videos
- pretending to be someone else online
- spreading upsetting online gossip.

If cyberbullying is happening to you, or you are a witness to it, tell an adult you trust. This can be difficult to do as you may not know the identity of the bully or may feel embarrassed. However, in severe cases the police may be able to identify an anonymous cyberbully, so it is important to report it.

Legal protection against cyberbullying

Lawmakers in parliament and the court system, along with law enforcement agencies, are working to keep up with the use of new technologies. New laws, penalties and enforcement strategies can help children and young people stay safe in the virtual world. These include:

- Section 474.17 of the *Criminal Code Act 1995* (Cwlth) makes it illegal to use a phone or online platform to threaten, harass or seriously offend a person. The maximum penalty for this offence is three years imprisonment.
- Section 21A(2)(d) *Crimes Amendment (Bullying)* Act 2011 (Vic) amended the *Crimes Act 1958* (Vic) to include making threats, using offensive and abusive words and acting with the intention of causing the victim to fear physical or mental harm. This offence attracts a maximum penalty of 10 years imprisonment.
- *Sex Discrimination Act 1984* (Cwlth) criminalises the sexual harassment of people. In an online environment this could be sending unwanted sexual messages or images, making inappropriate and upsetting advances on social media or posting sexually explicit images.

Learning ladder G3.8

Civics and citizenship

Step 1: I can identify topics about society

1 What are two positive aspects and two negative aspects of social media?

Step 2: I can describe societal issues

2 What is cyber crime and how are law enforcement authorities fighting it?

Step 3: I can explain issues in society

3 What is cyberbullying and why might it pose a particular danger for young people?

Step 4: I can explain different points of view

4 Why is it difficult for police and other law enforcement agencies to detect and prosecute cybercriminals?

Step 5: I can analyse issues in society

5 There is no specific cyberbullying law in Australia. Do you think that existing laws adequately protect victims of cyberbullying?

How does trade connect places and people?

Trade has been occurring between places for thousands of years. Historians believe that the first long-distance trade was the exchange of expensive goods, such as textiles and precious metals, around 3000 BCE in Pakistan.

Global trade is the term now used for the import and export of goods and services that occurs within and between countries.

The benefits of trade

Today, 25 per cent of all products produced globally are exported. Trade is good for a country's **gross domestic product (GDP)**. GDP is the amount of money that a country earns from all its goods and services, and is often measured over a year. When a country exports more goods than it imports, the value of its GDP increases.

The cost of exporting goods has decreased over time due to technological advancements, which have reduced transaction and movement costs. This, along with other factors, has resulted in an increase in global exports from 17 per cent in 1979 to 24 per cent in 2017.

World shipping routes

Source 1

Trade routes on a global scale

Source 2

Shipping containers full of goods for export waiting to be loaded on a dock

Source: Matilda Education Australia

Learning ladder G3.9

Show what you know

1. Compare the terms 'goods' and 'services'.
2. Source 1: Comment on the interconnection between trade, technology and temporal scale.
3. Create a timeline showing how trade has changed over time, using dates and data to support your summary.

Spatial distributions and patterns

Step 1: I can identify spatial distributions and patterns

4. Access http://mea.digital/GHV9_G3_9. Identify what this interactive map is showing.

Step 2: I can use data to quantify spatial distributions and patterns

5. Access http://mea.digital/GHV9_G3_10. Select one country on the globe and identify what goods are imported or exported to that place. Research two pieces of data to support your observations.

Step 3: I can describe spatial distributions and patterns

6. Visit http://mea.digital/GHV9_G3_11. Comment on the movement of goods during 2012. Alter the filters and colours and explore the changes it makes to the map's appearance.

Step 4: I can use data to support exceptions to spatial distributions and patterns

7. Visit http://mea.digital/GHV9_G3_12. Are there any locations where shipping appears not to occur? Consider why this is the case. Research how these places may receive goods and services other than via shipping.

Australia's biggest trade partners

Trade is one of Australia's biggest economic contributors – the country earned $387 billion in 2017 from trade alone. As demand for goods increases globally, especially in Asia, the value of Australia's goods grows. In 1990 trade contributed 32 per cent to Australia's GDP; by 2017, it accounted for 42 per cent of GDP.

Trade partners

In Australia, our top five trading partners are China, Japan, the United States, South Korea and India. Australia is China's seventh largest trading partner. In 2016–2017, our largest import was tourism and travel; Australians spent more than $28.6 billion overseas. Iron ore was our largest export and earned the country $66 billion.

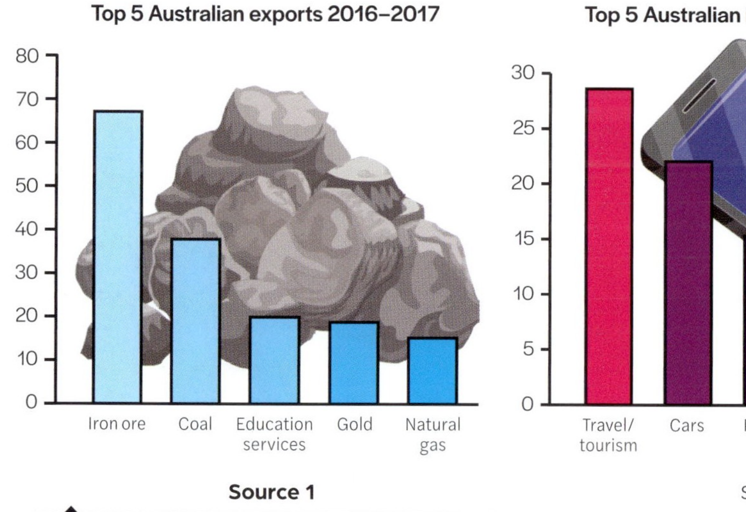

Source 1

Australia's top import and export goods and services

Source: data from SBS.com

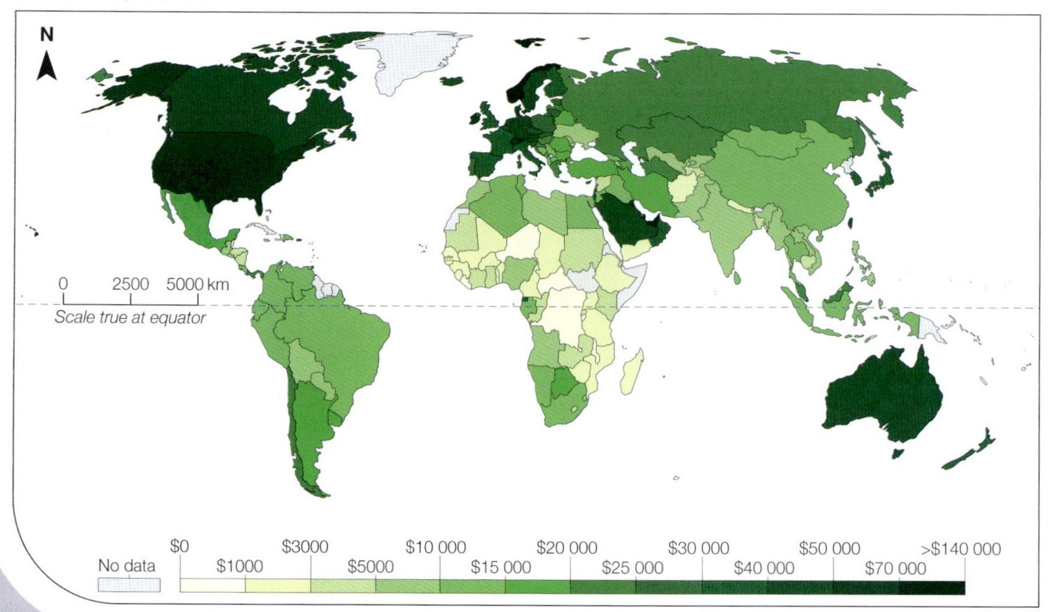

Source 2

How countries share the world's wealth

Source: data from Worldbank

Source 3

An enormous bucket wheel excavator operating at one of Australia's many open cut coal mines. Coal is currently Australia's most valuable export and Australia is the second largest exporter of thermal coal in the world, earning $26 billion in 2018. Around 20% of exported coal went to China.

The impact of COVID-19 on trade

The COVID-19 pandemic has caused much uncertainty when it comes to trade, especially with our major partner China. Reports on changes to global goods values, import/export quality checks and suspensions show that many industries are concerned about a 'global economic shutdown'.

Currently, most governments are primarily focused on health and maintaining health services, which is very different to other recessions, such as the **Great Depression** of the 1930s. As a result, lockdowns and restrictions have meant major job losses, financial debt and alterations to governmental budgets. These budget changes aim to support both communities and industries, so that after the pandemic, the countries and their trade capacities are able to recover.

Exports from China from Oct 2019 to Jun 2020

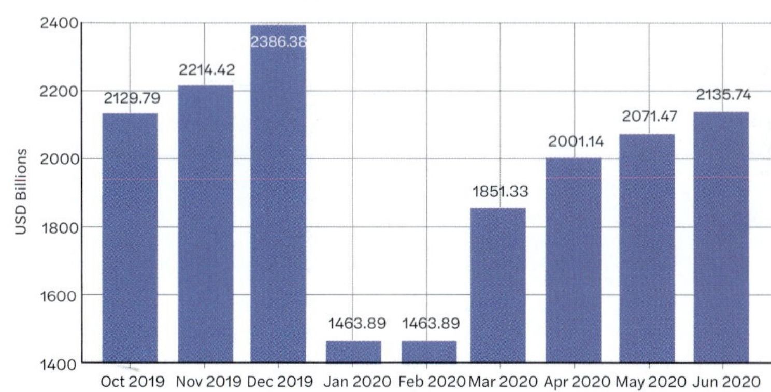

Source 4

This graph shows the sudden, massive drop in exports from China in January and February 2020, as well as the gradual increase from that point.

Learning ladder G3.10

Show what you know

1 Compare the terms 'imports' and 'exports'.

2 Why is trade important to Australia's economy and connection with surrounding regions?

3 Consider Source 2 and rank the countries listed according to their share in the world's Gross Domestic Product (GDP).

Patterns and interconnections

Step 1: I can provide short explanations for patterns and interconnections

4 Discuss why China is one of our most important trade partners.

Step 2: I can explain patterns and interconnections

5 Source 4: Explain why COVID-19 would create such uncertainty in global trade.

Step 3: I can use data to support explanations of patterns and interconnections

6 Source 2: How would having only a few countries holding most of the world's wealth affect economies of LEDCs?

Step 4: I can use relevant sources to research further reasons for patterns and interconnections

7 Because of COVID-19, 2020's trade and stock data does not match what was predicted prior to the crisis. Conduct some research and report on any changes that have occurred in Australia's major trade industries as a result of the pandemic.

Masterclass

Learning ladder

Work at the level that is right for you or level-up for a learning challenge!

EU trade partners and susceptiblity to climate change

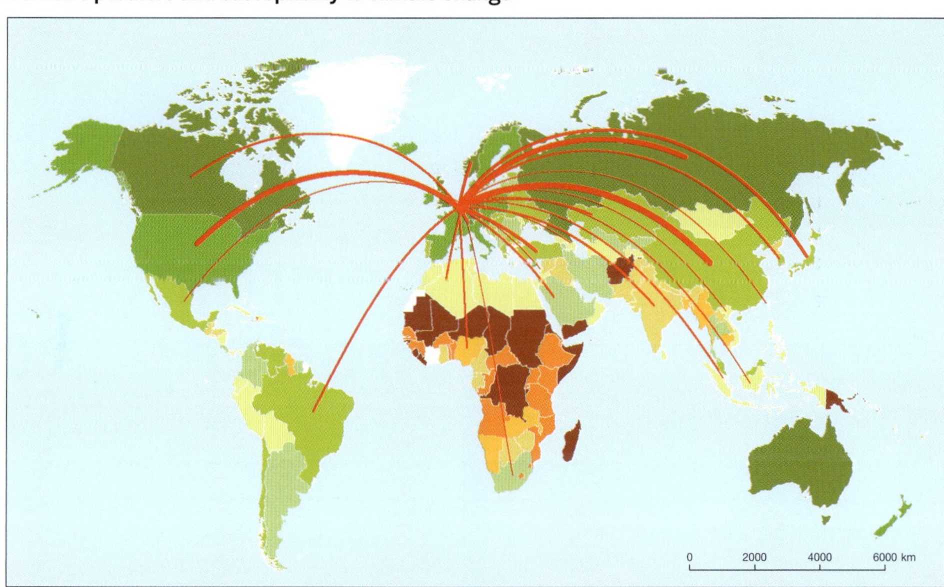

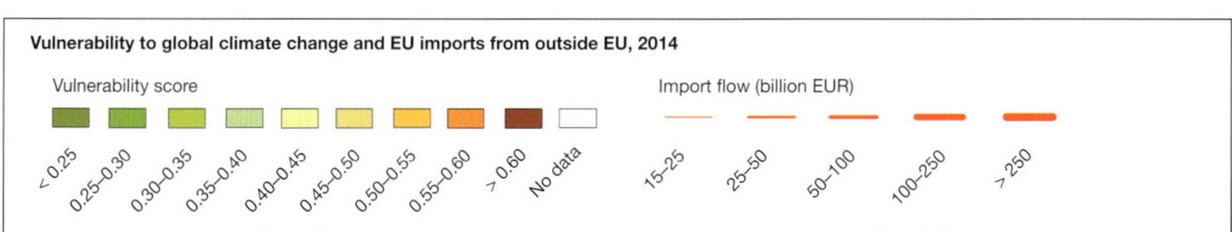

Source: Matilda Education Australia, European Environment Agency, 2014

Source 1

European Union (EU) trade import countries and their susceptibility to global climate change.

Global trade as a share of GDP, 2016

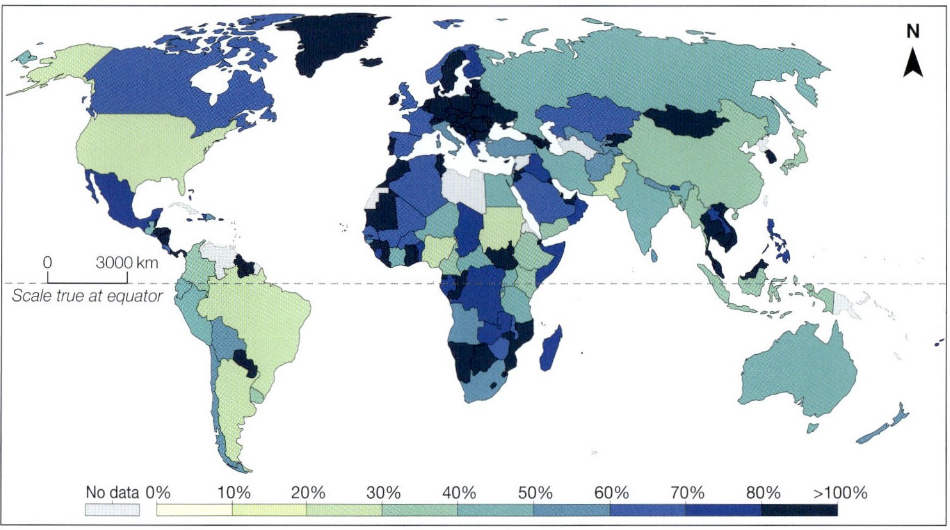

Source: Matilda Education Australia, World Bank, WDI, 2016

Source 2

Trade (exports plus imports) as a share of GDP, 2016. Figures correspond to the 'trade openness index': the sum of exports and imports of goods and services, divided by GDP.

Step 1

a I can identify spatial distributions and patterns

Source 1: List the countries that the EU imports goods from.

b I can provide short explanations for patterns and interconnections

Source 1: Provide a brief explanation as to why regions in Africa may be more vulnerable to the impacts of climate change than Australia.

c I can identify that changes occur in the characteristics of places over time

Consider your time at high school so far. How have your experiences helped develop your sense of place as a student?

d I can list primary and secondary methods useful for my study

Create a list of primary and secondary methods that would be helpful in investigating people's sense of place in your local region.

e I can interpret different map types using cartographic conventions

Visit http://mea.digital/GHV9_G3_13. Describe the distribution of billionaires in the world.

Step 2

a I can use data to quantify spatial distributions and patterns

Source 1: Rank the countries that the EU imports goods from according to potential risk of impact from climate change.

b I can explain patterns and interconnections

Source 1: Using SHEEPT, explain how global climate change may affect EU imports.

c I can describe how places have changed over time

Refer to Source 2. Describe how the outbreak of COVID-19 may impact exports and imports on a global scale.

d I can successfully use data collection methods

Locate secondary sources about Australia's imports and exports for this year. Create a table highlighting Australia's key imports and exports for this year.

e I can construct paper maps using correct cartographic conventions

On a blank world map, display the top five countries for GDP growth in 2017 from the data in Source 3.

Country	GDP per capita growth (%) 2018	GDP per capita ($US) 2018
Afghanistan	−0.6	524
Australia	1.4	57 396
Sierra Leone	1.3	534
Egypt	3.2	2549
New Zealand	2.8	42 950
Germany	1.2	47 639
Sweden	0.7	54 589
Japan	0.5	39 159
USA	2.4	62 840
Chile	2.5	15 924

Source: World Bank, WDI

Source 3

Trade data for selected countries, 2018

Step 3

a I can describe spatial distributions and patterns

Conduct a PQE analysis on Source 2, describing the patterns of countries whose GDP rely on trade.

b I can use data to support explanations of patterns and interconnections

Visit http://mea.digital/GHV9_G3_14. Compare the rail networks for the different regions highlighted in the link above. Consider how these differences may impact local people in terms of connection to other people and places.

Masterclass

c I can explain the causes behind the change over time in a place

Given people's strong connection to place, discuss why town planners would choose to alter a place over time.

d I can filter collected data

Source 3: Which column of data would be most helpful in creating a map like the one shown in Source 2?

e I can access and use spatial technology platforms such as GIS

Visit http://mea.digital/GHV9_G3_14 again. How does this interactive transport map assist locals and tourists to navigate Europe?

Step 4

a I can use data to support exceptions to spatial distributions and patterns

Visit http://mea.digital/GHV9_G3_15. Using data, describe the distribution of income on a global scale.

b I can use relevant sources to research further reasons for patterns and interconnections

Source 1: Identify why some countries may be more susceptible to climate change than others.

c I can make predictions and outline consequences of change over time

Visit http://mea.digital/GHV9_G3_16. Using different layers from the Content tab, discuss how the cost of products varies between places. How may this have an impact on a person's sense of place?

d I can organise data collected according to relevance for a research question

Using data from Source 3, graph the GDP per capita for at least five countries.

e I can manipulate data using digital and spatial technologies

Visit http://mea.digital/GHV9_G3_17. Use the legend and zoom button to explore the locations of Nike workers on a global scale. Note their locations and consider why they might be in these areas.

Step 5

a I can identify multiple spatial distributions and patterns

Suggest how the outbreak of COVID-19 may alter human landscapes and people's sense of place.

b I can interpret causes of patterns and interconnections

Refer to Sources 1 and 2. How will increased climate change have an impact on imports and exports worldwide?

c I can interpret data to quantify predictions based on research

Research predictions of how technology may change in the future, and then suggest how such changes may influence our connections to people and places.

d I can evaluate the success of research methods

Now you have explored some of the weblinks listed in this Masterclass, write one or two paragraphs evaluating how successful spatial technologies are in understanding interconnections between people and places.

e I can draw conclusions from geographical information in digital and spatial technologies

Visit the Center for International Development at the Harvard University website at http://mea.digital/GHV9_G3_18. Comment on the complexity of world trade and explain why it is such an important way of connecting with other places.

Capstone

How can I understand perceptions and places?

In this chapter, you have learnt a lot about perceptions and places. Now you can put your new knowledge and understanding together for the capstone project to show what you know and what you think.

In the world of building, a capstone is an element that finishes off an arch or tops off a building or wall. That is what the capstone project will offer you, too: a chance to top off and bring together your learning in interesting, critical and creative ways. You can complete this project yourself, or your teacher can make it a class task or a homework task.

mea.digital/GHV9_G3

Scan this QR code to find the capstone project online.

Choices and changes

G4

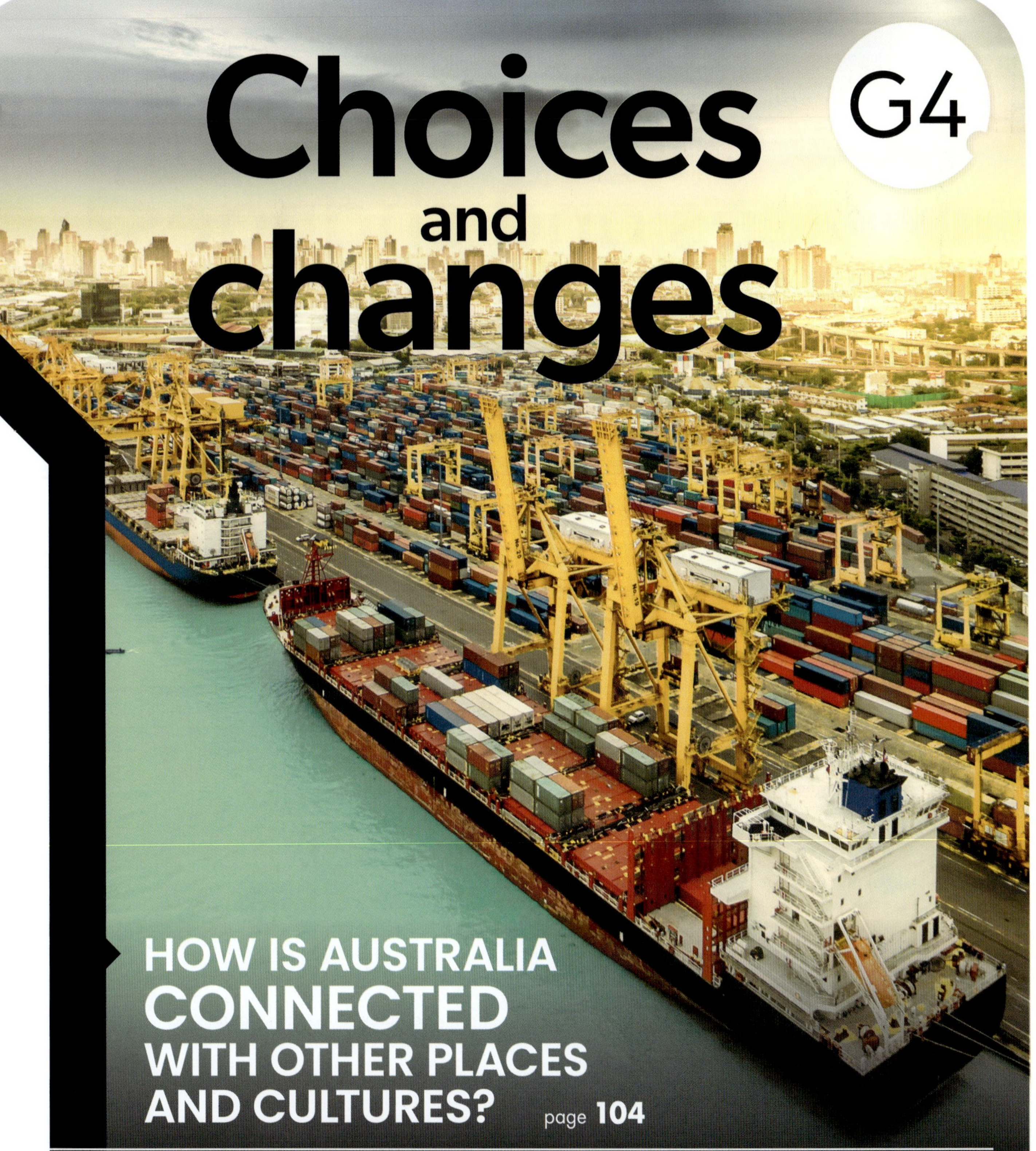

HOW IS AUSTRALIA CONNECTED WITH OTHER PLACES AND CULTURES?

page 104

patterns + interconnections

page 100

HOW DOES CONSUMERISM IMPACT PEOPLE?

economics + business

page 102

HOW DOES AUSTRALIA EXPORT EDUCATION?

communicate data

page 114

CAN TOURISM BE SUSTAINABLE?

How can I understand choices and changes?

The choices we make about what we eat, what technology we use and where we travel have huge implications for the environment. We need to ensure that we are making sustainable choices, so that future generations can enjoy the same experiences we do today.

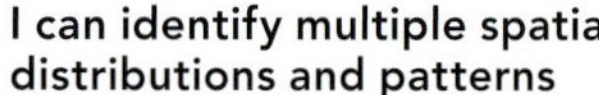

Step	Spatial distributions and patterns	Patterns and interconnections	Changes and implications
step 5	**I can identify multiple spatial distributions and patterns** I can take my PQE one step further to find links or relationships that exist in places in relation to choices and changes.	**I can interpret causes of patterns and interconnections** I can use multiple sources to find links or relationships that exist in relation to choices and changes and can explain 'Why?'.	**I can interpret data to quantify predictions based on research** I can use external data from research as evidence of the positive and negative impacts of a change I have predicted.
step 4	**I can use data to support exceptions to spatial distributions and patterns** I can use data to answer 'Why?' about the exceptions identified in a PQE analysis of places in relation to choices and changes.	**I can use relevant sources to research further reasons for patterns and interconnections** I can use sources other than this textbook to further research patterns I observe in places in relation to choices and changes.	**I can make predictions and outline consequences of change over time** I can use my knowledge of natural processes and world regions to make an educated guess about the positive and negative impacts of change.
step 3	**I can describe spatial distributions and patterns** I can describe patterns, quantify them and point out exceptions (PQE) to describe places in relation to choices and changes.	**I can use data to support explanations of patterns and interconnections** I can use data from a map or graph to explain patterns I observe in places in relation to choices and changes.	**I can explain the causes behind the change over time in a place** I can use my knowledge of natural processes and world regions to explain why changes may occur over time.
step 2	**I can use data to quantify spatial distributions and patterns** I can read data and use it to measure key trends on a map or graph about places in relation to choices and changes.	**I can explain patterns and interconnections** I can identify SHEEPT factors to help me explain places in relation to choices and changes.	**I can describe how places have changed over time** I can use specific examples to describe changes over time.
step 1	**I can identify spatial distributions and patterns** I can find key trends on a map or graph about places in relation to choices and changes.	**I can provide short explanations for patterns and interconnections** I can write descriptions of patterns and interconnections that I find in places in relation to choices and changes.	**I can identify that changes occur in the characteristics of places over time** I can read information and answer questions about changes over time.

Source 1

The island of Koh Phi Phi, before and after its tourism boom

Spatial distributions and patterns

1 Looking at Source 1, describe how the distribution of vegetation has changed over time in Koh Phi Phi.

Patterns and interconnections

2 Describe the interconnection between tourism and environmental degradation in Koh Phi Phi.

Changes and implications

3 How might tourism alter Koh Phi Phi in the future?

Communicate data

4 Research how tourist numbers to Koh Phi Phi have changed over time. Are they predicted to increase in the future?

Digital and spatial technologies

5 How could spatial technology be used to assess the changes caused by tourists over time?

I can evaluate the success of research methods
On reflection, I can look back and comment on the data collection methods I used and evaluate how successful they were in helping me answer a research question in relation to choices and changes.

I can organise data collected according to relevance for a research question
I can review the data I have collected in the field and display it using graphs, tables, annotations and captions.

I can filter collected data
I can review my collected data and select the most relevant data to answer a research question in relation to choices and changes.

I can successfully use data collection methods
I can use primary and secondary data collection methods in the field and classroom to investigate choices and changes.

I can list primary and secondary methods useful for my study
I can create a checklist of methods to investigate a choice or a change and categorise them as primary or secondary methods.

Communicate data

I can draw conclusions from geographical information in digital and spatial technologies
I can interpret and analyse patterns by using different layers and features on spatial technology platforms.

I can manipulate data using digital and spatial technologies
I can work with layers and other features on spatial technology platforms to further explore data and interconnections.

I can access and use spatial technology platforms such as GIS
I can use spatial technology platforms to explore data and find patterns.

I can construct paper maps using correct cartographic conventions
I can use a pencil, paper and ruler to construct a map that follows BOLTSS conventions.

I can interpret different map types using cartographic conventions
I understand data found in different types of maps and graphs and use the data to answer questions in relation to choices and changes.

Digital and spatial technologies

How does consumerism impact places?

As our world changes, there is an increasing desire to keep 'up to date' and own the most recent technology, cars, clothes and gadgets. **Consumerism** can be defined as our desire to own products that exceed our basic human needs.

Consumerism can be a positive **phenomenon** because our increasing desire for goods and services means that industries and economies expand, and employment increases. However, consumerism can also have negative effects on social and environmental conditions.

In 2005, around 59 per cent of the world's resources were bought and owned by 10 per cent of the population, creating huge disparities between the haves and have-nots. According to a 2017 report by international bank Credit Suisse, the richest 1 per cent of the world's population owns more than 50 per cent of the world's wealth.

Mobile phone subscriptions, per 100 people

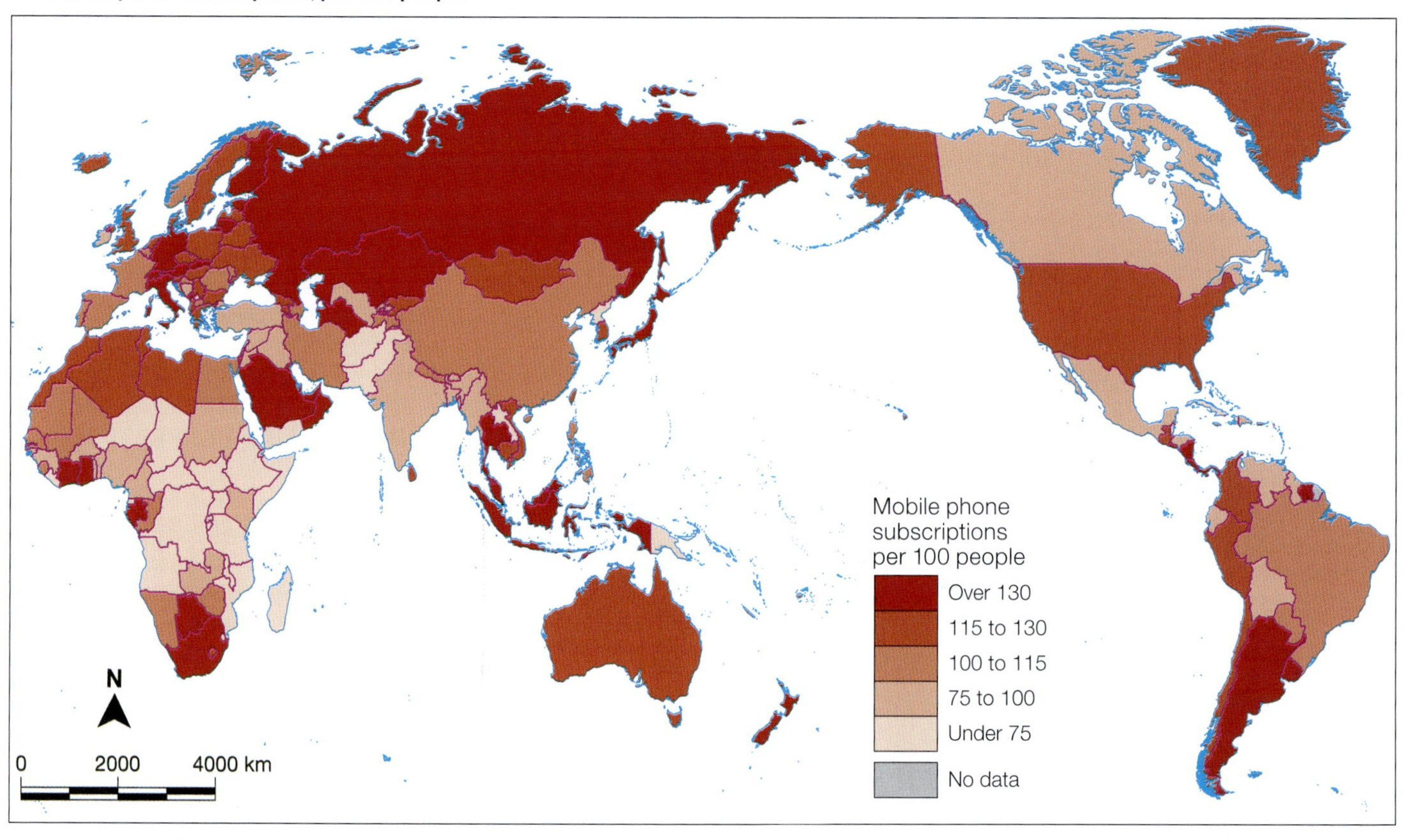

Source 1

Mobile phone usage on a global scale

Source: Matilda Education Australia

BUT YOUR SMART CHOICES CAN MAKE A DIFFERENCE!

HOW TO BE SMART ABOUT YOUR SMART PHONE

Americans get rid of millions of cell phones every year. Minimize your environmental impact by maximizing the life of your phone!

CHOOSE YOUR SERVICE PROVIDER CAREFULLY. Many companies sell their own phones, which are not always transferable between companies.

SELECT A PHONE WITH FEATURES YOU NEED AND A STYLE YOU LIKE, so you'll keep it longer. Bonus: you'll save money!

TAKE CARE OF YOUR PHONE.

- Protect it with a case
- Keep it out of extreme temperatures and away from liquids
- Charge the battery according to manufacturer's directions

GIVE YOUR OLD PHONE A NEW LIFE!

You can reduce the amount of raw materials mined and energy used to produce more phones, as well as the packaging used to transport them—and make a difference in our environment!

GIVE OR DONATE
Help someone in need.

- **Give** your phone to a family member or friend
- **Donate** it to charity

RESELL
Make some extra money and help someone out.

- **Sell** your phone using a reputable website

Take advantage of a service provider's **buy-back program.**

RECYCLE
Drop your old phone off at an **electronics recycling** location or a take-back program. Phone parts and materials can be used in a new phone or product. For example:

- **Metals** can be used in automotive parts and jewelry
- **Plastics** can be used in garden furniture
- **Batteries** can be used in more batteries

United States Environmental Protection Agency
www.epa.gov/recycle

Source 2

These two infographic panels from the US Environmental Protection Agency address smartphone technology, consumerism and the environment.

Environmental impact of smartphones

Our attachment to smartphones is also detrimental to the environment. Although operating them does not use a lot of energy, researchers estimate that 85 per cent of a smartphone's emissions impact occurs during manufacturing. By 2040 it is predicted that technology will contribute more than 14 per cent of the world's total carbon footprint. Part of the problem is that people replace their technology too quickly, because they want the latest model, creating unnecessary **e-waste**.

	2013	2014	2015	2016
United States	20.5	20.9	21.6	22.7
China	18.6	21.8	19.5	20.2
European Union	18.3	19.5	20.4	21.6

Source: Kantar Worldpanel

Source 3

Smartphone life cycles (in months)

Learning ladder G4.1

Show what you know

1 Describe one positive aspect of consumerism and one negative aspect.

2 Explain why smartphones are a good example of the detrimental impacts of modern consumerism.

3 Use Source 3 to create a graph that illustrates the life cycle of smartphones in the United States, China and the European Union.

Spatial distributions and patterns

Step 1: I can identify spatial distributions and patterns

4 Source 1: Identify five countries that have high mobile phone usage.

Step 2: I can use data to quantify spatial distributions and patterns

5 Using data, identify how mobile phone usage varies within Asia.

Step 3: I can describe spatial distributions and patterns

6 Source 1: Use PQE to describe the distribution of mobile phone subscriptions on a global scale.

Step 4: I can use data to support exceptions to spatial distributions and patterns

7 Do any of the countries that have a high number of mobile phone subscriptions per 100 people surprise you? Why is that the case? Use data in your response.

PQE, page 136
Graphing, page 144

How does consumerism impact people?

Global trade brings many benefits to people around the world. Consumers are able to purchase the products they want while the manufacturing industry, through employment, can contribute to alleviating poverty in less economically developed countries (LEDCs). Manufacturing in LEDCs can, however, have significant negative impacts at the individual level. Many people work in unsafe conditions and do not earn enough income to lift themselves out of poverty.

The human cost

Profit is the major driver of trade; most businesses want to produce cheap products quickly and sell them at a global scale for a higher price than they cost to make. In order to reduce production costs, companies often outsource the manufacturing or development of their products to other countries where labour is cheaper.

As a result, many people who work in manufacturing for large companies live in LEDCs and earn below the minimum wage. In some cases, even children are forced to work. According to World Vision, there are 73 million child labourers around the world, aged between 5 and 11 years old, and more than 50 per cent of them work in agriculture.

Fairtrade

You may have noticed the Fairtrade mark (Source 1) on some products you buy, such as coffee or chocolate. This mark means that the products have been made by small-scale companies that meet Fairtrade Australia and New Zealand's standards for workers' rights, environmental standards and pay.

The Fairtrade program is an international initiative that promotes education and awareness about fair trade in the broader community. In Australia and New Zealand, the foundation also helps marginalised and rural communities, especially those in the Pacific Islands, to earn a sustainable income and move out of the poverty cycle.

Distribution of Fairtrade (FT) producers and purchasers on a global scale

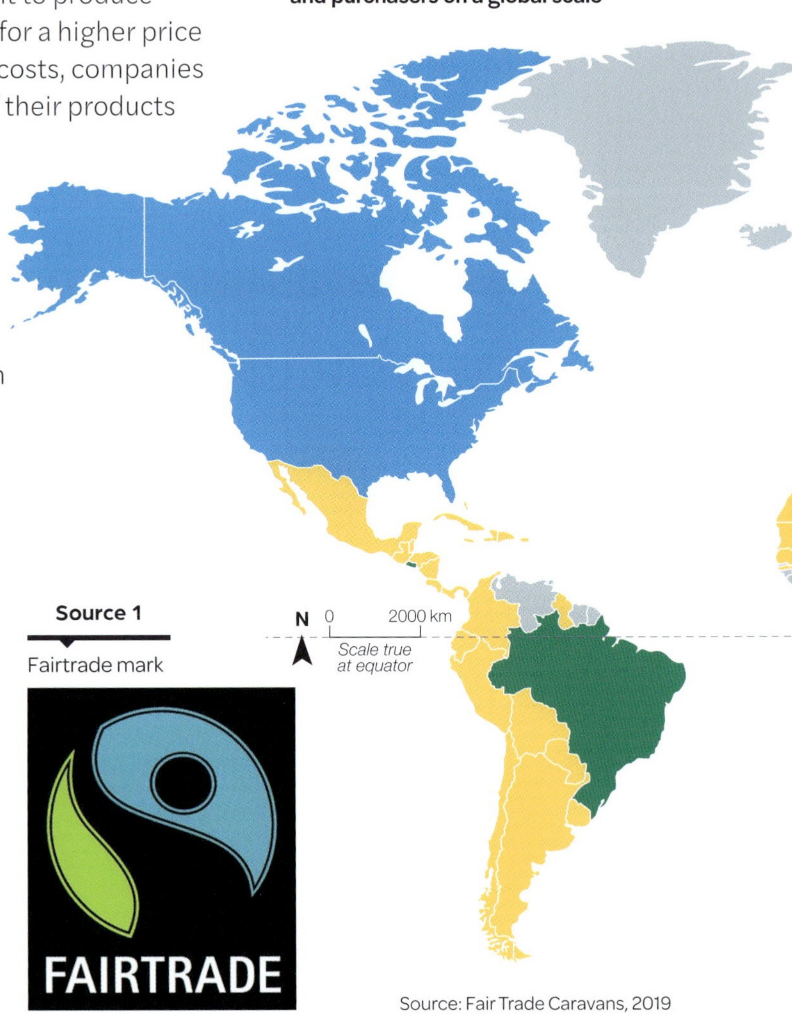

Source 1

Fairtrade mark

Source: Fair Trade Caravans, 2019

Source 2

Fairtrade aims to assist people in LEDCs.

Source 3

Countries that produce or buy Fairtrade (FT) products

Learning ladder G4.2

Show what you know

1. Explain the economic benefit for global companies of outsourcing their manufacturing to LEDCs.
2. Outline how the Fairtrade program aims to reduce the impact of consumerism on LEDCs.
3. List at least three ways you could reduce the impact of your consumerism on LEDCs.

Patterns and interconnections

Step 1: I can provide short explanations for patterns and interconnections

4. Locate a list of Fairtrade products that are sold in Australia. Explain why some products are not yet sold with the 'fair trade' philosophy.

Step 2: I can explain patterns and interconnections

5. Source 3: Using SHEEPT, explain the distribution of countries that produce Fairtrade products.

Step 3: I can use data to support explanations of patterns and interconnections

6. 'Countries that buy Fairtrade products tend to be more economically developed countries (MEDCs).' Justify this statement using data from Source 3.

Step 4: I can use relevant sources to research further reasons for patterns and interconnections

7. Research the Fairtrade program and its impact on LEDCs. Create an infographic that contains at least five facts and two images showing this impact.

HOW TO

SHEEPT, page 138

How does Australia export education?

While coal, minerals and other natural resources make up most of our exports, Australia also exports education services to other nations. We are the world's third-largest provider of international education services, and education is currently our fourth-largest export industry. However, the COVID-19 pandemic has severely affected the education industry, and it may take a long time to recover.

Tertiary education as an export

International students travel from their home country to another nation in order to get the education they desire. Universities around the world seek to attract students from other countries. Because these students bring money into the country, education is considered an export, even though no materials or services are sent overseas.

A variety of factors, including changes to **immigration** laws in other nations and a deliberate attempt by the government to market Australia to international students, led to a huge increase in Australia's income from international education over the last decade.

By 2019, international education contributed more than $37 billion to Australia's economy, with more than 300 000 overseas students studying in this country. This income includes both the fees paid by students for their education and related spending such as the rent, taxes and visa fees the students pay while living in Australia.

More than half of Australia's international students come from five other nations: China (30%), India (11%), Nepal (5%), Malaysia (4%) and Brazil (4%). Universities, education providers and governments have focused on China and India in particular to attract students.

The impact of COVID-19

The COVID-19 pandemic affected the education industry in a number of ways – all of them negative. In the first stages of the pandemic, many students from China were barred from entering Australia, leading to a drop in enrolments. In the first quarter of 2020, international student arrivals were 21 per cent lower than in the first quarter of 2019; arrivals from China dropped by almost 50 per cent.

Source 1

Australia's income from international education sources in the 2018–19 financial year.

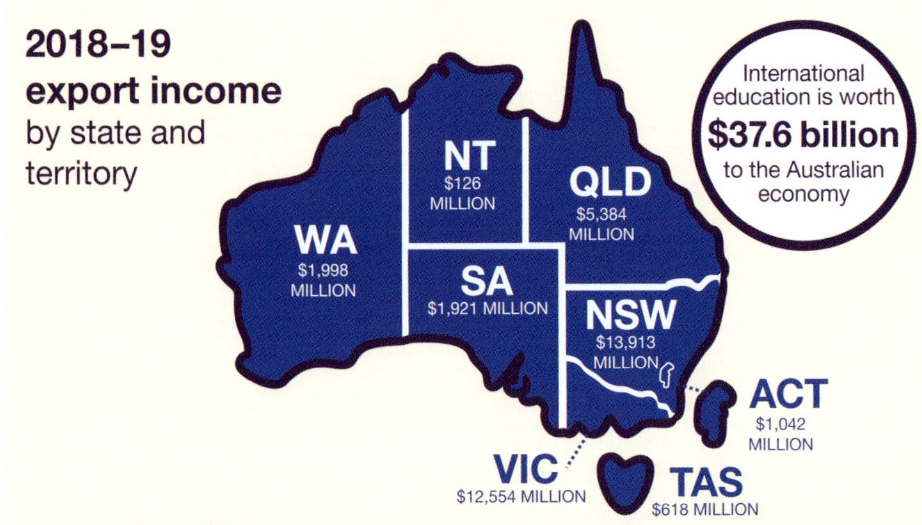

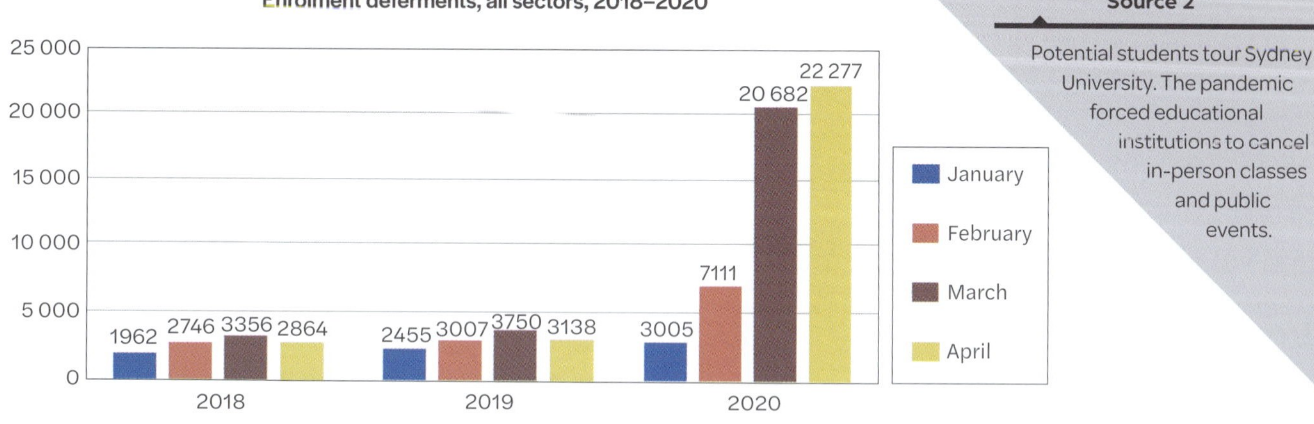

Source 2

Potential students tour Sydney University. The pandemic forced educational institutions to cancel in-person classes and public events.

Source 3

Deferment numbers for international students in Australia rose sharply in March and April 2020.

The situation escalated in March 2020, when Australia closed its borders to *all* international travel. Universities closed their campuses, cancelling classes or moving them online, and many international students deferred their studies for a year or more while attempting to find jobs or protect their health.

The university sector has been severely damaged by the drop in numbers, as most Australian universities used the income from international students as a big part of their revenue. It is estimated that education export revenue will fall by at least $3 billion across the country, leading to the loss of more than 20 000 jobs in Australia's 38 universities. It remains to be seen whether all of those universities will continue operating after the pandemic ends.

Learning ladder G4.3

Economics and business

Step 1: I can recognise economic information

1 Consider Source 1. Identify how much income Victoria received from international education exports in 2018–19.

Step 2: I can describe economic issues

2 Consider Source 3. Describe how this graph reflects how international students responded to the COVID-19 pandemic in the first quarter of 2020.

Step 3: I can explain issues in economics

3 Explain why the COVID-19 pandemic had such a significant impact on Australia's education export industries.

Step 4: I can integrate different economic topics

4 Explain why a drop in international student numbers would affect other industries and economies, not just the education sector.

Step 5: I can evaluate alternatives

5 Choose two Australian universities. Research the financial effects that the pandemic had upon these universities, and the actions they took to cope with those effects. Which university had the most effective response? Justify your answer using data.

How is Australia connected with other places and cultures?

As Australians, we are connected to people from around the globe through personal and virtual connections and through the pursuits we share: cultural, social, religious, culinary and so on. Trade, migration and tourism are three major pathways of connection.

Trade

Trade connects us with different regions of the world. Our major trading partners include East Asia and Europe. In 2017–18, our exports to East Asia and Europe earned Australia $231 002 000 and $21 712 000 respectively. China is a major market for imports and exports. Our main exports are primary products, including food, minerals and fuel.

Migration

Migration is the movement of people into (immigration) and out of (emigration) a country. Australia's migrant intake is adjusted each year and includes places for skilled workers (designed to fill gaps in Australia's labour market), family members of existing residents or citizens, and people seeking asylum.

Source 1

International arrivals to and departures from Australia

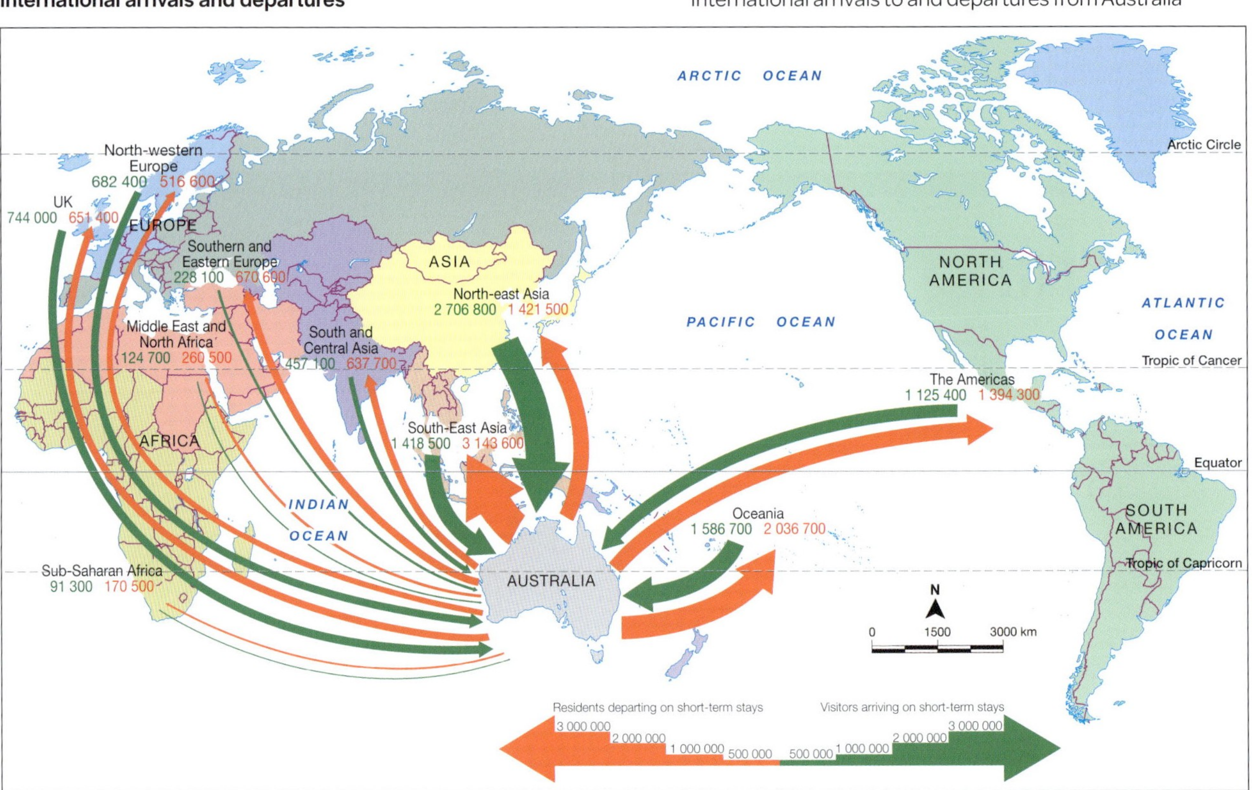

Source: Matilda Education Australia

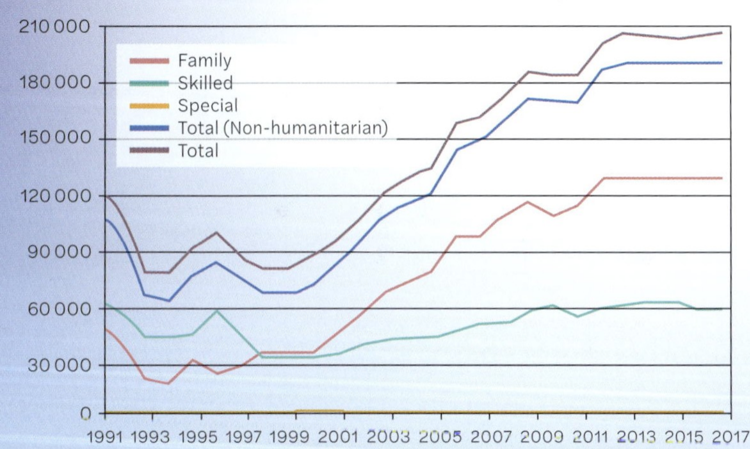

Source 2

Australia's migration program since the 1990s

Source: Department of Immigration and MacroBusiness

Source 3

Advertisements like this one promote tourism in Australia

Tourism

Tourism is defined as a person, or group of people, visiting a place other than where they live or work, for more than 24 hours and less than 12 months. In 2018, the Australian tourism industry employed over 924 000 people, and it contributed around $55.3 billion to the nation's GDP. Tourists visiting Australia spent $135.4 billion.

Learning ladder G4.4

Show what you know

1 Source 1: Comment on how Australia is interconnected with other regions in the world.

2 How do migration and tourism play a major role in connecting us with other people, cultures and places?

3 Source 1: Create a graph showing the origins of people who travel to Australia.

Digital and spatial technologies

Step 1: I can interpret different map types using cartographic conventions

4 Locate a map that shows where migrants to Australia have come from.

Step 2: I can construct paper maps using correct cartographic conventions

5 Using a blank map of Australia, illustrate the rate of migrant settlement in each state.

Step 3: I can access and use spatial technology platforms such as GIS

6 Access http://mea.digital/GHV9_G4_1 and locate your town or local area. Identify the percentage of migrants that live there and their origins.

Step 4: I can manipulate data using digital and spatial technologies

7 Access http://mea.digital/GHV9_G4_2 and compare ABS migration data for both Victoria and NSW. Are there any key similarities or differences? Using one SHEEPT factor, write one reason for these similarities or differences in trends.

Mapping skills, page 134
SHEEPT, page 138

What is global tourism?

Another outcome of improving technologies is the increase in global tourism. Advancements in air transport mean that people can travel internationally within the space of a few hours – the **temporal distance** between countries is getting smaller.

Recently, Qantas began flying passengers to the UK from Australia without stopping. Improvements such as this will further reduce temporal distance and encourage the mass movement of people between places.

In 2017, France was the most visited country in the world, with over 87 million visitors, while Kuwait and Bhutan were two of the least visited countries. As the cost of travel reduces over time, it is becoming more accessible to young people. The World Tourism Organization (UNWTO) estimated that young people would have spent US$400 billion globally on tourism in 2020 – if they had been able to travel.

The impact of COVID-19

The health crisis created by the outbreak of COVID-19 changed travel and migration drastically. Suddenly tourism – a huge contributor to economies and livelihoods and a way to connect with other people, places and cultures – was viewed as a high transmission risk for COVID-19.

Suddenly, most countries disconnected from the rest of the world. Borders were closed or restricted, quarantining made allowing new people into regions expensive, and social distancing and lockdowns hindered internal travel. Some Australian travellers were trapped overseas, having to extend their hotel stays for weeks or even months.

Source 1

An advertisement for Contiki tours, created before the impact of COVID-19 was realised.

The long-term impact of the pandemic on tourism is not yet known. International flight numbers have plummeted, which has had a devastating impact upon our island nation. It may be a long time before the Australian tourism industry recovers, or before Australian tourists are able to travel overseas again.

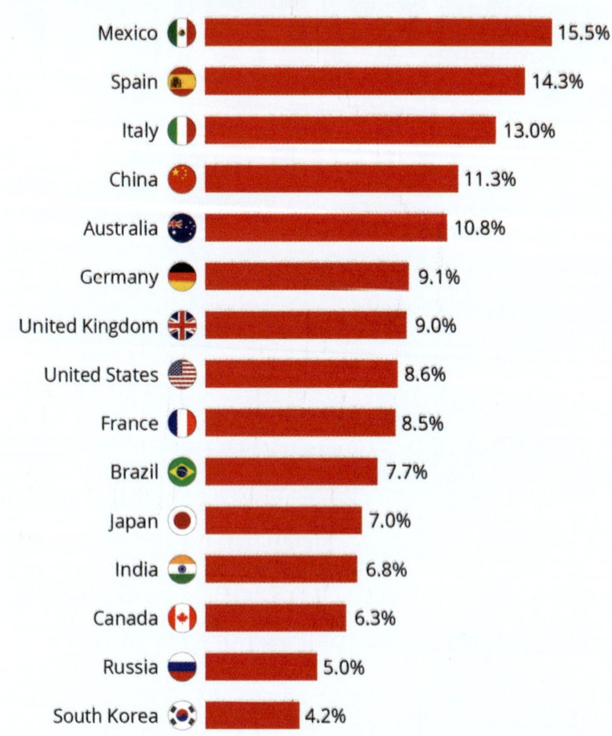

Source 2

Numbers of national and international flights have dropped sharply during the pandemic.

Source 3

The countries most vulnerable to the pandemic's impact on tourism; Australia is fifth on the list

Learning ladder G4.5

Show what you know

1 Explain the interconnection between tourism and the reduction in temporal scale.

2 Describe how global tourism has changed over time

Communicate data

Step 1: I can list primary and secondary methods useful for my study

3 Create a survey to carry out in your class to answer the research question: 'Where have people travelled in the world?'

Step 2: I can successfully use data collection methods

4 Carry out the survey created in question 4. Graph the outcome and discuss your results as a class.

Step 3: I can filter collected data

5 Source 3: Consider how COVID-19 impacted travel and tourism in 2020. Which countries were affected most by the travel restrictions? Do they have any similarities? Why do you think these countries were most affected?

Step 4: I can organise data collected according to relevance for a research question

6 Consider the impact that the lack of travel has had on local and national economies. Choose a focal country and investigate the economic impact that tourism has had on that location before and after the outbreak of COVID-19.

Graphing, page 144

G4.6

What determines where people travel?

Imagine you are planning your dream trip. There are no limits – you can go anywhere you want, experience new countries and cultures, explore beautiful natural locations or bustling cities. How do you start to plan? What factors will determine where, when and for how long you travel?

Source 1

The COVID-19 pandemic led to mass cancellations of international flights and airport shutdowns.

While financial restrictions may be at the forefront of your mind, you might also consider the cultural experience you want to have, the language spoken in the host country, how safe the host country is, the social or religious rules of the host country, and how you would 'fit in' with the local people. Proximity and cultural ties are other factors that influence where people choose to travel.

In recent history, recreational travel was largely based on choice and the interests of the individual tourist. However, during the peak of the global COVID-19 pandemic, border closures and strict travelling regulations both within and between countries significantly altered our ability to make these choices. Many people had to cancel their travel plans or became stranded in their host country. Travel suddenly became a necessity to return home to be with family and access healthcare. It is yet to be seen how the pandemic will change tourism and our travel choices in the future.

States of peace on a global scale, 2018

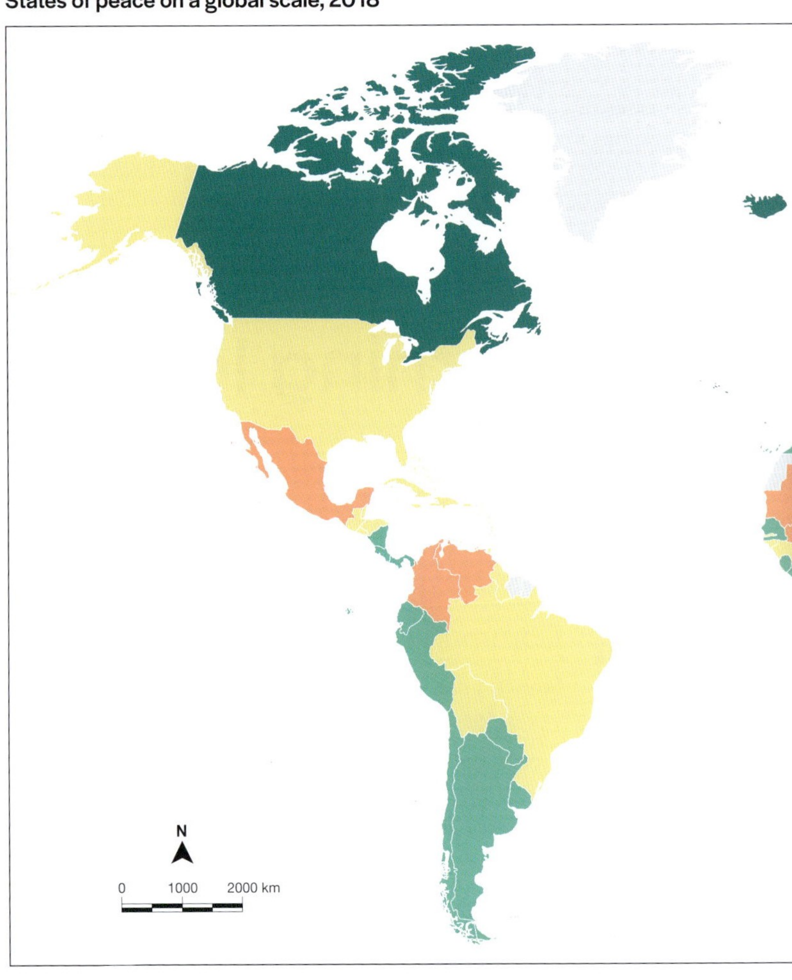

International travel controls during the COVID-19 pandemic.

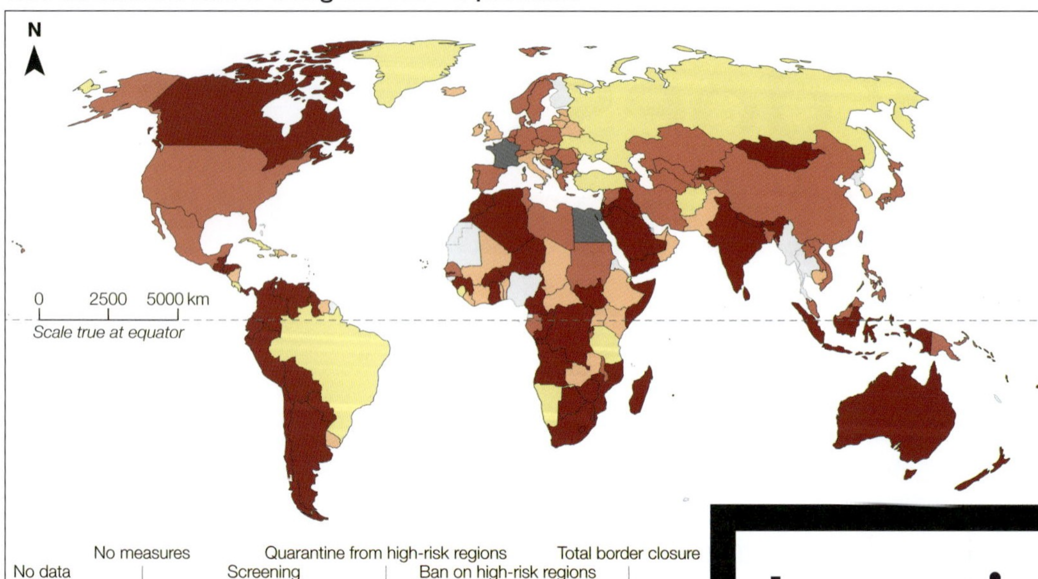

Source 2

International travel controls during the COVID-19 pandemic, 12 August 2020.

Source 3

This map comes from the Institute for Economics and Peace. Their Global Peace Index ranks 163 independent states and territories according to their level of peacefulness.

The state of peace

Very high | High | Medium | Low | Very low | Not included

Source: Institute for Economics and Peace, Global Peace Index Report 2019

Learning ladder G4.6

Show what you know

1 Define the term 'tourism'.

2 Create a table with the column headings 'Factors that encourage tourism' and 'Factors that restrict tourism'. Complete the table, considering the factors that would determine where, when and why you travelled to different places.

Spatial distributions and patterns

Step 1: I can identify spatial distributions and patterns

3 Identify the countries in Source 2 that had no measures in place for international travel during August 2020.

Step 2: I can use data to quantify spatial distributions and patterns

4 Source 3: Describe, using data, the distribution of the safest countries to visit on a global scale.

Step 3: I can describe spatial distributions and patterns

5 Consider what is meant by 'peace' in Source 3. How does this compare with the countries that have high-level travel restrictions during COVID-19? Outline any similarities and differences between these two maps.

Step 4: I can use data to support exceptions to spatial distributions and patterns

6 Use PQE to describe the distribution of countries that have high-level travel restrictions. Provide data to support both your pattern and exception.

HOW TO

PQE, page 136

How does travel impact the culture of a region?

Tourism can have positive and negative effects on place. One major benefit is the huge economic contribution that the industry makes to local economies, creating employment opportunities for the host country. For example, tourism accounts for over 76.2 percent of the GDP of Palau, a small Micronesian nation in the Pacific. One major negative effect of tourism is when a local culture is **commodified**. Important practices can be cheapened if they are marketed and sold to tourists.

Source 1

Some people visit and take photos at cultural sites for Instagram likes

The commodification of culture

When travelling, many tourists want to experience different cultures, and so will pay to attend shows, festivals or rituals. While this may initially have a positive effect by renewing interest in traditional practices, it may become difficult for locals to separate sacred events from spectator events.

When an event, ritual or practice is packaged and sold for monetary gain it is said to have become commodified. By putting a price on culture, its value and meaning for communities may be diminished. Consider someone buying a boomerang as a toy, rather than appreciating its significance as part of Indigenous Australian culture, or joining a tour group to have a private 'cultural experience' when overseas? These are both examples of cultural commodification.

Source 2

Almost no destination in the world is free from the commodification of local culture and customs.

Source 3

Tourism industries often promote traditional weapons and tools as designer products rather than cultural artefacts.

Learning ladder G4.7

Show what you know

1. Define the term 'commodification' in relation to tourism.
2. What are some positive outcomes of commodification?
3. As a class, create a list of items or souvenirs you have bought that might have supported cultural commodification.

Changes and implications

Step 1: I can identify that changes occur in the characteristics of places over time

4. Explain how tourists alter local culture and traditions in a host country.

Step 2: I can describe how places have changed over time

5. Consider one country you have visited or would like to visit. Research how this country has changed over time due to tourism and commodification.

Step 3: I can explain the causes behind the change over time in a place

6. Comment on how commodification could assist a country's economy, especially in regards to local jobs and industry.

Step 4: I can make predictions and outline consequences of change over time

7. Commodification is one example of how culture is impacted by travel. Using examples, suggest other ways tourists could bring change to host countries. Consider both positive and negative changes that could occur in these places.
8. In an attempt to reduce the spread of COVID-19 in 2020, many countries closed or put hard restrictions on their international borders. How could the resulting decline in international tourism reduce commodification of cultures in some places?

G4.8

How do tourist destinations change and adapt over time?

In 1980, Geography researcher Richard W. Butler proposed a six-stage model to describe the impact tourism can have on a location. His model shows that tourist destinations initially benefit from an increase in tourism, but must carefully manage ongoing tourism to avoid stagnation while holding on to what makes a place special for both tourists and residents.

Butler's model

As shown in Source 2, once tourists 'discover' a place (1), the news of the location usually spreads and leads to an increase in tourists arriving each year (2). With this increase, infrastructure such as restaurants and hotels is constructed to support this new temporary population. When the place is adequately developed, tourism spending usually increases (3), which can lead to further improvements to the location (4).

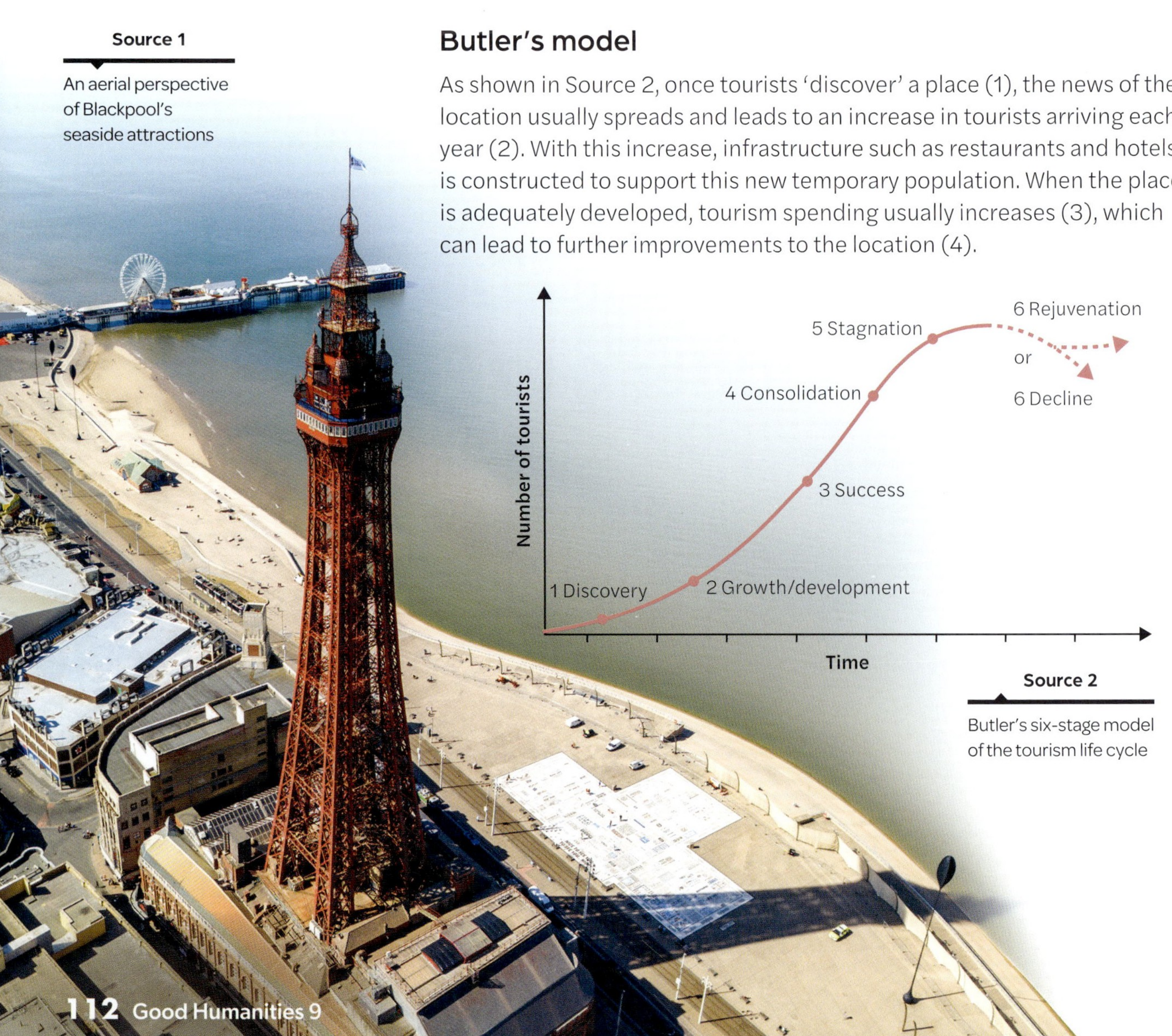

Source 1

An aerial perspective of Blackpool's seaside attractions

Source 2

Butler's six-stage model of the tourism life cycle

Source 3

Blackpool at night

However, infrastructure ages and mass tourism can contribute to the degradation of a space over time through pollution, environmental damage or building decline (5). Once a location reaches this stage, it may undergo rejuvenation through new government or business funding (6) or continue to decline to the point that tourists no longer wish to visit (6).

The Blackpool example

The development of Blackpool, a seaside town in the UK, is a good example of the six-stage model. In the 1800s, wealthy tourists travelled to Blackpool as a coastal destination. As the town's popularity increased and transport improved, infrastructure was developed and fairgrounds, light shows and hotels were built. Tourism numbers peaked in 1939; as overseas holidays become more affordable, visitor numbers declined and Blackpool went into a state of disrepair.

In more recent years, Blackpool's Town Council has spent over £100 million to rejuvenate the main street, build a modern conference centre, extend public transport systems and improve existing roads and networks, all to encourage tourists back to the seaside township.

Learning ladder G4.8

Show what you know

1 Describe each stage of Butler's model using examples from Blackpool.

2 Create a photo essay on Blackpool, with at least six images representing each stage of Butler's model.

Changes and implications

Step 1: I can identify that changes occur in the characteristics of places over time

3 Identify a tourism hotspot in Victoria. How has this region changed over time as a result of tourism? Outline these changes and discuss them with a partner.

Step 2: I can describe how places have changed over time

4 Source 2: How might the growth/developmental stage of Butler's model have positive impacts on a place? (Hint: Consider the SHEEPT factors in your response.)

Step 3: I can explain the causes behind the change over time in a place

5 Using examples, explain why Blackpool changed so significantly over time. What were the main causes of this change?

Step 4: I can make predictions and outline consequences of change over time

6 Do some further research on Blackpool today. Predict whether Blackpool will become popular again or continue to decline.

SHEEPT, page 138
Photo essays, page 143

Can tourism be sustainable?

Although tourism can benefit economies and employment in different regions, it also contributes to more than 10 per cent of global carbon emissions per year. Consider what you use when you travel. You may travel by plane or car to your destination, eat packaged food, take longer showers, have fresh towels sent to your hotel room each day and use the air conditioner for hours on end (depending on the climate). All of these actions have an impact on the environment.

Source 1

Ecotourism is not just about exploring the natural environment of a location. It is also about travelling sustainably, being educated about local practices, environments and traditions, and ensuring waste is minimised.

Source 2

Do you agree with this cartoon?

Source 3

Ecotourism does not require you to camp or 'rough it'. In fact, some luxury resorts have been certified as ecotourism destinations in Australia. The way they operate, educate and market are seen to meet the requirements for ECO certification.

Ecotourism

The primary aim of **ecotourism** is to experience the natural world and learn how to protect, sustain and conserve the environment for future generations. In order to be considered an ecotourism destination, tours or resorts must meet the following criteria:

- protection of the environment through habitat, species and traditional culture conservation and the minimisation of waste
- presentation of information that educates tourists about local values, heritage and culture
- honesty in marketing, reporting of pollution, managing protected areas and supporting local Indigenous communities.

ECO certification in Australia

To become a certified ecotourism destination in Australia, tours, resorts and other operators need to apply through Ecotourism Australia, a non-profit organisation established to promote ecotourism in our region. The operator's application is reviewed to ensure it meets the above criteria.

As travellers become increasingly aware of how their carbon footprint contributes to our changing climate, demand for ecotourism destinations is increasing. By supporting ecotourism, tourists can travel, learn about different environments, minimise their local and regional impact and help local communities to be more sustainable.

Learning ladder G4.9

Show what you know

1 Outline the criteria that must be met for a location to promote itself as an ecotourism destination.

2 Explain how ecotourism is different to other forms of tourism.

3 Given the carbon emissions caused by travelling, exploring and staying in a new location, can tourism even truly be 'sustainable'?

Communicate data

Step 1: I can list primary and secondary methods useful for my study

4 Imagine you are assessing a new resort to determine whether it could claim to be an ecotourism destination. What primary data collection methods would you require to gain evidence on their eligibility?

Step 2: I can successfully use data collection methods

5 Select a popular travel destination and investigate how many criteria it meets for ecotourism. What simple things could it change to become more sustainable for travellers?

Step 3: I can filter collected data

6 Research ecotourism destinations in Australia. Critically investigate the activities offered at these places and discuss as a class whether they are true ecotourism destinations.

Step 4: I can organise data collected according to relevance for a research question

7 Choose one of the locations you investigated in question 6 that meets all the criteria. Create a sketch map of the location and annotate your map with examples of how it meets the criteria.

thinking locally

How has tourism changed Phillip Island?

Victoria's Phillip Island has long been a holiday destination for domestic and international tourists. The island is widely recognised for its beaches, natural landscapes and wildlife.

Phillip Island has been marketed as a natural tourist site since around 1840, when wealthy families would sail to the island from the mainland. These visitors introduced invasive species such as rabbits, pheasants and foxes in order to hunt them for sport.

In 1940, a bridge connecting Phillip Island to the mainland was built, which led to an ongoing influx of tourists. Currently, Phillip Island hosts more than 4 million visitors annually. Natural tourism attractions are highly developed, with advanced viewing platforms and an education centre.

Phillip Island has a small community of 7000 permanent residents. The local council has forecasted that the population will be stable over the next few years, and increase slowly from 2028 (Source 1). However, the island has four major supermarkets and many restaurants open year-round. This imbalance between the local population's needs and infrastructure for visitors is another example of how seasonal tourism can have a large impact on how a place changes over time.

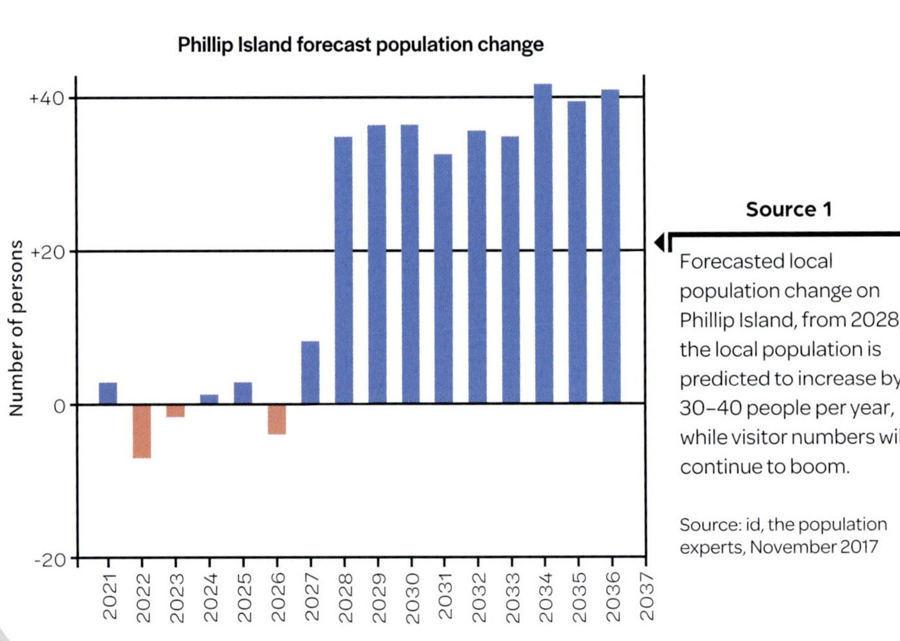

Source 1

Forecasted local population change on Phillip Island, from 2028, the local population is predicted to increase by 30–40 people per year, while visitor numbers will continue to boom.

Source: id, the population experts, November 2017

Source 2

Overlooking the penguin reserve at Summerland Beach, Phillip Island, c. 1920–1954.

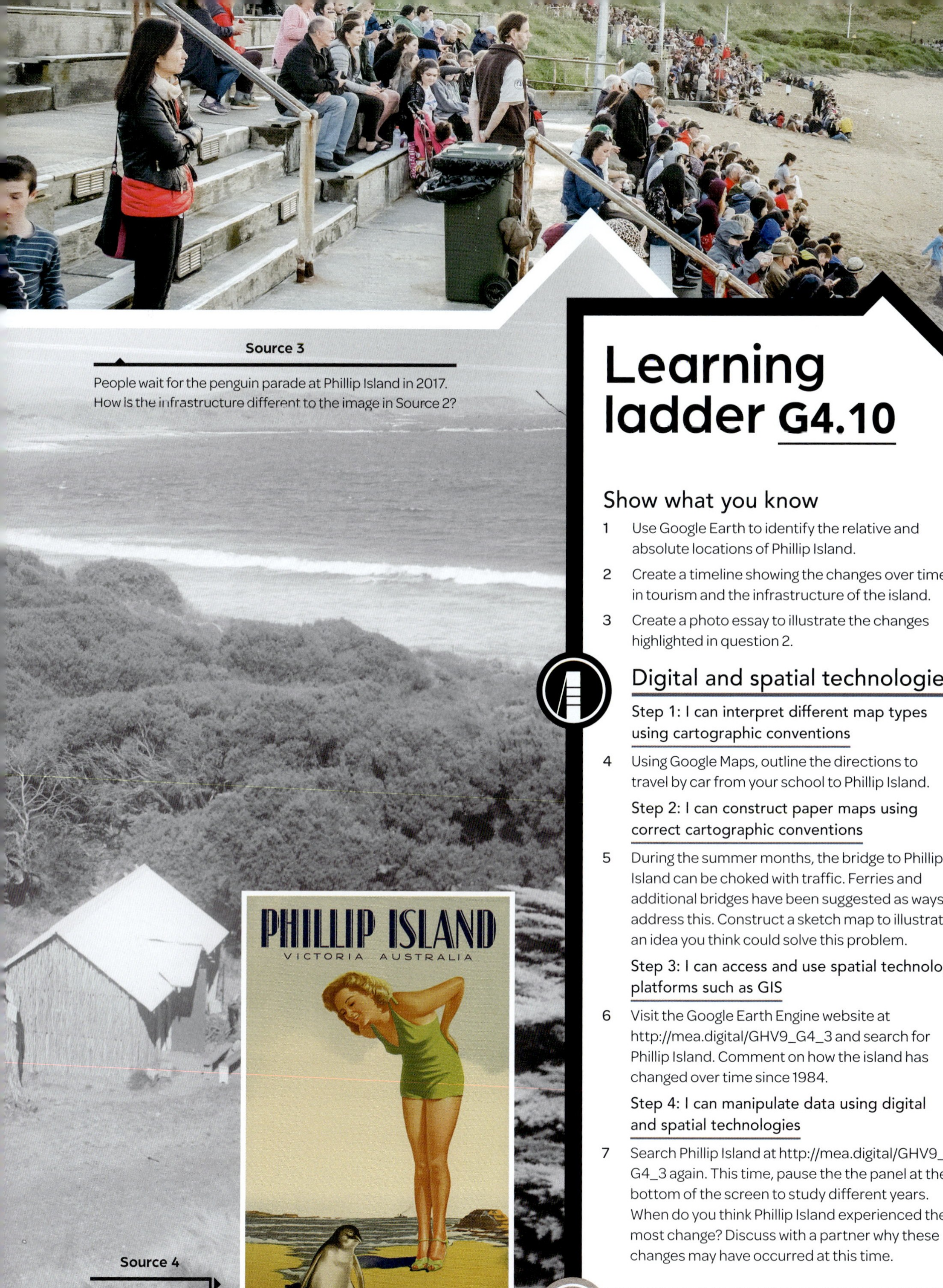

Source 3

People wait for the penguin parade at Phillip Island in 2017. How is the infrastructure different to the image in Source 2?

Source 4

A travel poster from the 1930s for Victorian Railways.

Learning ladder G4.10

Show what you know

1 Use Google Earth to identify the relative and absolute locations of Phillip Island.

2 Create a timeline showing the changes over time in tourism and the infrastructure of the island.

3 Create a photo essay to illustrate the changes highlighted in question 2.

Digital and spatial technologies

Step 1: I can interpret different map types using cartographic conventions

4 Using Google Maps, outline the directions to travel by car from your school to Phillip Island.

Step 2: I can construct paper maps using correct cartographic conventions

5 During the summer months, the bridge to Phillip Island can be choked with traffic. Ferries and additional bridges have been suggested as ways to address this. Construct a sketch map to illustrate an idea you think could solve this problem.

Step 3: I can access and use spatial technology platforms such as GIS

6 Visit the Google Earth Engine website at http://mea.digital/GHV9_G4_3 and search for Phillip Island. Comment on how the island has changed over time since 1984.

Step 4: I can manipulate data using digital and spatial technologies

7 Search Phillip Island at http://mea.digital/GHV9_G4_3 again. This time, pause the the panel at the bottom of the screen to study different years. When do you think Phillip Island experienced the most change? Discuss with a partner why these changes may have occurred at this time.

HOW TO

Sketches and annotating, page 140
Photo essays, page 143

Masterclass

Learning ladder

Work at the level that is right for you or level-up for a learning challenge!

International tourism: Number of arrivals, 2016

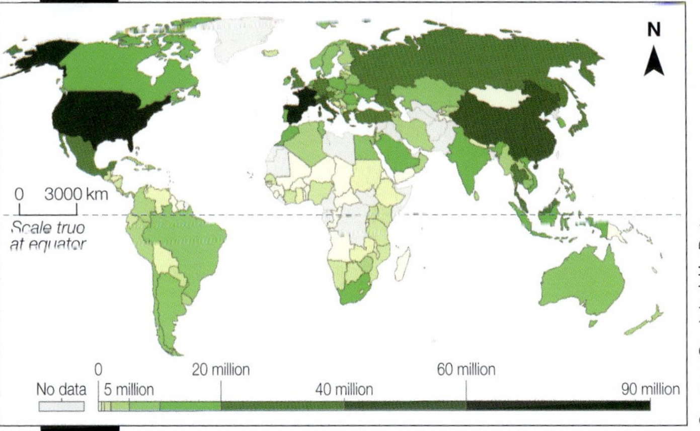

Source 1

For this map, tourists are described as overnight visitors who travel to a country and whose main purpose in visiting is not commercial.

Tourist visits globally, 2017

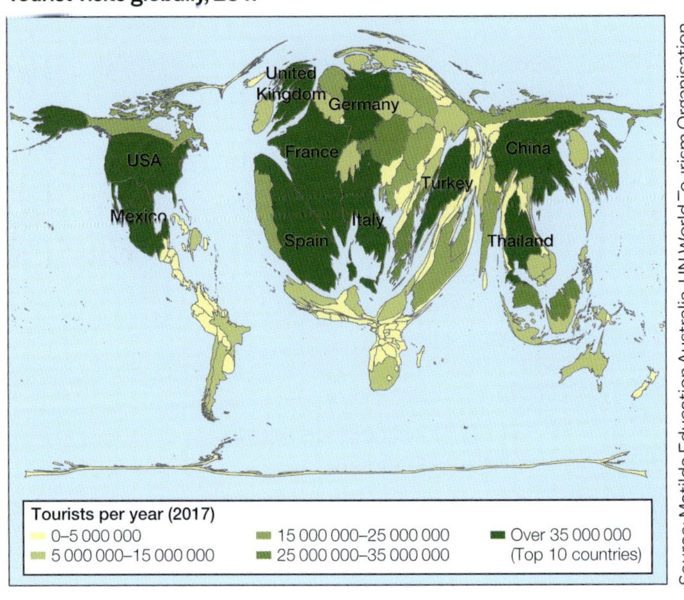

Source 2

This cartogram represents the most popular tourist destinations as the biggest countries on the map

Source 3

This map shows how patterns of migration are changing. Some areas have seen increased migration in recent years.

Traditional and emerging centres of migration growth, 2017

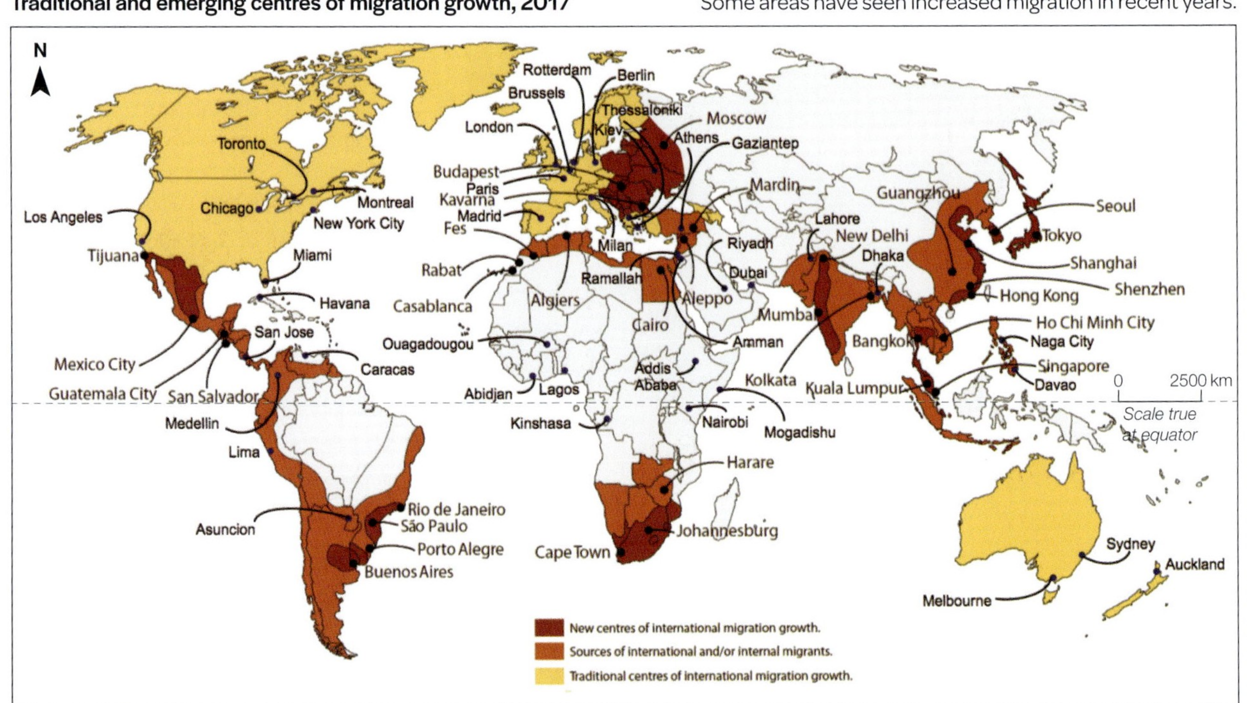

Step 1

a **I can identify spatial distributions and patterns**

Source 2: Identify three countries that have the highest number of foreign visitors.

b **I can provide short explanations for patterns and interconnections**

Explain why some countries attract more migrants than others. (Hint: consider the SHEEPT factors.)

c **I can identify that changes occur in the characteristics of places over time**

Consider the example of Phillip Island on pages 116–17. Outline the major changes that have occurred in this place over time as a result of tourism popularity.

d **I can list primary and secondary methods useful for my study**

Imagine you were carrying out a research project investigating how a local tourist site fits in with Butler's model (see pages 112–13). Create a list of primary and secondary methods that would help you collect relevant data for this study.

e **I can interpret different map types using cartographic conventions**

Interpret the key trends highlighted in Source 1.

Step 2

a **I can use data to quantify spatial distributions and patterns**

Source 2: Using data, rank the continents according to tourism popularity.

b **I can explain patterns and interconnections**

Using SHEEPT, explain the movement of people around the world in Source 3.

c **I can describe how places have changed over time**

Describe how commodification may influence change in tourism destinations.

d **I can successfully use data collection methods**

Visit a series of reputable websites and create a summary table of data to answer the following research question: From where do most tourists who travel to Australia originate?

e **I can construct paper maps using correct cartographic conventions**

Visit the National Geographic Migration Data Table at http://mea.digital/GHV9_G4_4. Create a map that shows the origin of migrants who move to Australia.

Step 3

a **I can describe spatial distributions and patterns**

Source 2: Using PQE, describe the distribution of tourism popularity on a global scale.

b **I can use data to support explanations of patterns and interconnections**

Using data, explain how mobile phones and PCs have changed in popularity over time.

c **I can explain the causes behind the change over time in a place**

Suggest how consumerism causes further segregation between LEDCs and MEDCs on a global scale.

d **I can filter collected data**

Create a photo essay using secondary sources that shows the impact of commodification on regional tourist destinations.

e **I can access and use spatial technology platforms such as GIS**

Visit the Cool Maps website at http://mea.digital/GHV9_G4_5. Comment on the distribution of shipping container traffic worldwide and how it has changed over time.

Masterclass

Step 4

a I can use data to support exceptions to spatial distributions and patterns

Source 3: Using data to support both your pattern and exception, describe global migration patterns.

b I can use relevant sources to research further reasons for patterns and interconnections

Using the internet, locate satellite images that represent change over time in a location near you. Discuss how these changes may have been caused by trade, migration or tourism in the area.

c I can make predictions and outline consequences of change over time

Predict how Phillip Island will continue to change in the future based on assumed increased resident and tourism numbers.

d I can organise data collected according to relevance for research question

Visit the World Bank Data website at http://mea.digital/GHV9_G4_7. Create a summary table showing countries that have experienced an increase in tourism and a decrease in tourism over time.

e I can manipulate data using digital and spatial technologies

Visit the GIS for Schools website at http://mea.digital/GHV9_G4_8. Using the layers in the 'content' tab, explore migration within Australia and discuss any major patterns or findings as a class.

Step 5

a I can identify multiple spatial distributions and patterns

Conduct a PQE on Sources 2 and 3. Comment on the interconnection between tourism popularity and the countries that people choose to migrate to.

b I can interpret causes of patterns and interconnections

Refer to Sources 1 and 2. Discuss the data presented in both maps and how the representations interconnect.

c I can interpret data to quantify predictions based on research

Using research, predict how future population growth may positively and negatively impact people and places. Consider tourism, migration, consumerism and need for resources.

d I can evaluate the success of research methods

Visit the Cool Maps website at http://mea.digital/GHV9_G4_6 again. Evaluate the success of this map as a resource showing how trade creates interconnections between people and places.

e I can draw conclusions from geographical information in digital and spatial technologies

Visit the Cool Maps website at http://mea.digital/GHV9_G4_9. 'Global migration is evenly distributed on a global scale over time.' Discuss with reference to the source provided.

Capstone

How can I understand choices and changes?

In this chapter, you have learnt a lot about choices and changes. Now you can put your new knowledge and understanding together for the capstone project to show what you know and what you think.

In the world of building, a capstone is an element that finishes off an arch or tops off a building or wall. That is what the capstone project will offer you, too: a chance to top off and bring together your learning in interesting, critical and creative ways. You can complete this project yourself, or your teacher can make it a class task or a homework task.

mea.digital/GHV9_G4

Scan this QR code to find the capstone project online.

G5 Fieldwork

HOW DO I CREATE AND ANSWER A RESEARCH QUESTION?

page 122

fieldwork task 1

page 126

HOW CAN WE EXPLORE BIOMES VIRTUALLY?

fieldwork task 2

page 128

HOW CAN WE EXPLORE OUR INTERCONNECTION WITH *PLACE*?

fieldwork skills

page 130

HOW DO I WRITE A FIELDWORK INTRODUCTION

How do I create and answer a research question?

Fieldwork is an important part of Geography. It is how we discover new things, learn about patterns and relationships and explore the world around us. In the previous Geography chapters, you explored biomes, food security and interconnections between people and places. Now you will conduct your own research within your local area to answer a research question and write a fieldwork report.

Creating a research question

A **research question** is an overarching idea you want to investigate. Once you develop a research question you will need to write a hypothesis. A **hypothesis** is an 'educated guess' about what the answer to the research question might be. In this chapter, you will be guided to create your own research question and hypothesis and then design fieldwork to answer your research question.

Choosing a fieldwork location

Choosing the right location to conduct fieldwork is very important. The place you choose will depend on your topic, research question and hypothesis. Make sure you confirm the location of your fieldwork before you begin to conduct research and collect data.

Pre-fieldwork research

Before you start your fieldwork, do some background research so that you understand the field site, its characteristics and other relevant information.

Researching via the web is the most obvious and accessible source of information; however, you should only use reliable websites. Government, educational and other approved sites are most useful and contain data that is reliable. Blogs, social media and other public-based sites may be **biased** and misleading. Books, handouts and other paper resources are also useful when conducting research.

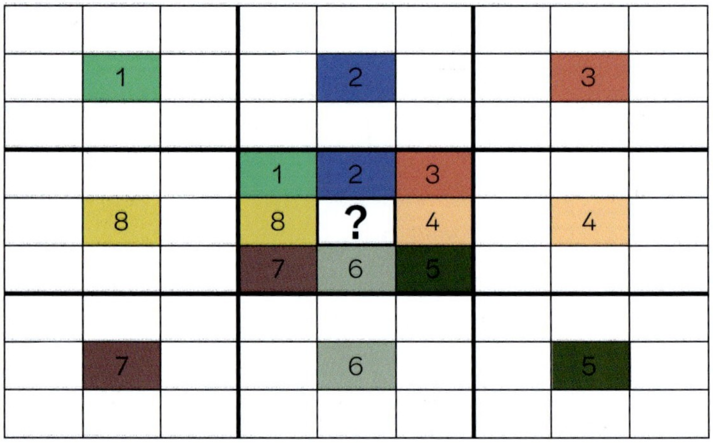

	1			2			3	
			1	2	3			
	8		8	?	4		4	
			7	6	5			
	7			6			5	

Source 1

A lotus diagram is a tool to help you develop key ideas and themes regarding a research question. To use the diagram:

1 Write your research question into the central rectangle.

2 Come up with eight related ideas you wish to investigate. These ideas might include the SHEEPT factors for your chosen location (see page 138). Write each idea in the boxes marked 1–8.

3 Conduct some research for each idea, and fill out the remaining squares with relevant notes.

Method	Primary or secondary?	Description	This data will help me answer my research question because …	How will I gather the data?
Photos	Primary	I will take five photos of infrastructure, natural environments and recent developments in the field site using a digital camera.	Taking photos provides a visual representation of the human and natural environments present at the research site.	Using a digital camera, which can then be uploaded to my report

Source 2

Record your research methods clearly and consistently so that you can repeat them.

Collecting data

Conducting fieldwork involves collecting data. There are two main kinds of data: **quantitative** and **qualitative**. Quantitative data tends to be recorded in numbers. The number of people in a population and the flow rate of a river are examples of quantitative data. Qualitative data tends to be more observational: descriptions of a place, field sketches and photos all provide qualitative evidence to answer your research question. Both aspects of data collection are important. Collect both quantitative and qualitative data in your fieldwork.

Primary and secondary methods of data collection

As a class, you need to develop a list of **primary** and **secondary methods** of data collection. Primary methods are tasks that you will do in the field to find data to answer your research question, and prove or disprove your hypothesis. Secondary methods are ways you can collect data back in the classroom to help you answer the questions, such as research using websites, books or other publications.

Some examples of primary methodologies:

- observations
- photographs
- note taking
- surveys (questionnaire or observational)
- field sketches
- mapping.

Using the table in Source 2 as a template, discuss and record some primary and secondary methods that may help you answer your research question.

Source 3

Note taking is an important part of fieldwork.

Preparation for fieldwork

Before you go out in the field, create a checklist of items you need to bring and data you need to collect. For example:

- clipboard
- paper (blank and lined)
- pencils
- camera
- number of cars that pass through an intersection in 30 minutes
- survey questions.

Source 4

Share your research data with classmates to assemble a larger and more useful bank of information.

Communicating your findings

Once you have collected data in the field using primary and secondary methods, you need to communicate your findings. You could communicate your research via a report or display it as a poster or presentation. You can also present your photos and sketches in a slideshow presentation, or as a hardcopy folio of evidence.

When you communicate findings, ensure that you highlight any patterns, interconnections or significant data that answers the research question and proves or disproves the hypothesis. Presenting data clearly is vital so that your reader can understand your analysis and conclusions.

Writing research reports

To prepare for your future studies, begin practising communicating fieldwork results and analysis as a formal report. Using subheadings, referencing and acknowledging sources are important aspects of formal reports because they show the depth of your research and understanding, and give your reader a clear structure to follow.

Your subheadings might be:

- Introduction (background research, question and hypothesis)
- Primary methods
- Secondary methods
- Presentation of data
- Data analysis
- Conclusions
- Evaluation
- References (bibliography).

Evaluating the success of your methods

Before you go out in the field, you need to decide on the methods you are going to use. Unfortunately, sometimes the methods selected are not successful, and need to be improved before they can help collect more valuable data. The evaluation section of your report requires critical reflection on your use of data collection methods. You do not need to reflect on your time management or how much you enjoyed the trip. Your evaluation should focus on the following questions.

- What was the most successful primary method and why?
- What was the least effective primary method and why?
- What was the most successful secondary method and why?
- What was the least effective secondary method and why?
- If you were to repeat this study, what would you do differently next time?

Source 5

You need to be organised when planning a fieldtrip and writing your report.

G5.2

fieldwork task 1

How can we explore biomes virtually?

Background to fieldwork approach

Biomes vary on a number of scales. Because they are so diverse, the way we use them to produce resources, such as food, also varies. Dry, desert like areas produce less than regions that receive high levels of rainfall and sunlight. Studying biomes can help us understand our world better.

Our choice of fieldwork location is usually limited to accessible places near school. However, we can use virtual fieldwork as a viable alternative to investigate research questions. In this section, you will conduct a virtual fieldwork inquiry using Google Earth to answer a research question you design as a class.

Research question

The research question you create needs to be focused on how the world's biomes have changed over time because of human actions. As a class, brainstorm these terms and craft a question that you will be able to answer via virtual fieldwork. Some questions you may consider:

- How have humans changed Earth over time?
- How do human actions affect our food security?
- How does increasing human population lead to loss of agricultural land?

Once you have a question, you need to create a hypothesis. You may all have a slightly different hypotheses, but that will create a more interesting analysis.

Pre-fieldwork research

Once you have decided on a research question, you are ready to complete some pre-fieldwork research. Explore the internet, books and other resources for information about key concepts or processes that readers must know in order to understand your report.

Source 6

This diagram uses 11 categories of biome.

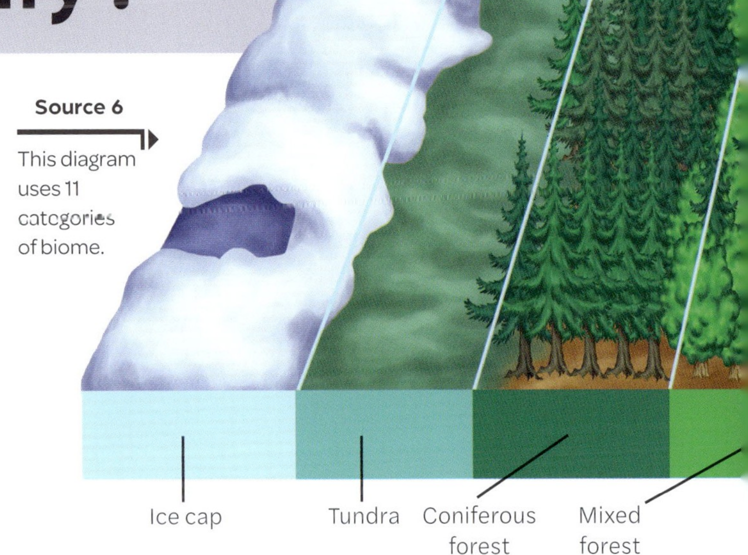

Introduction and method

1 State your research question and hypothesis.
2 Provide some background information so that readers will understand all key terms and concepts in your report.
3 State two primary fieldwork techniques you used to complete this fieldwork and describe how they were helpful in answering the research question.
4 State two secondary resources you used to prepare for your fieldwork and describe how they were helpful in answering the research question.

Collecting your data

As this is a virtual fieldtrip, you need to decide as a class which area of the world you will focus on. You might instead choose to break into research groups and look at different parts of the world.

These weblinks may help when exploring your chosen question.

- Visit the Google Earth Engine website at http://mea.digital/GHV9_G5_1.

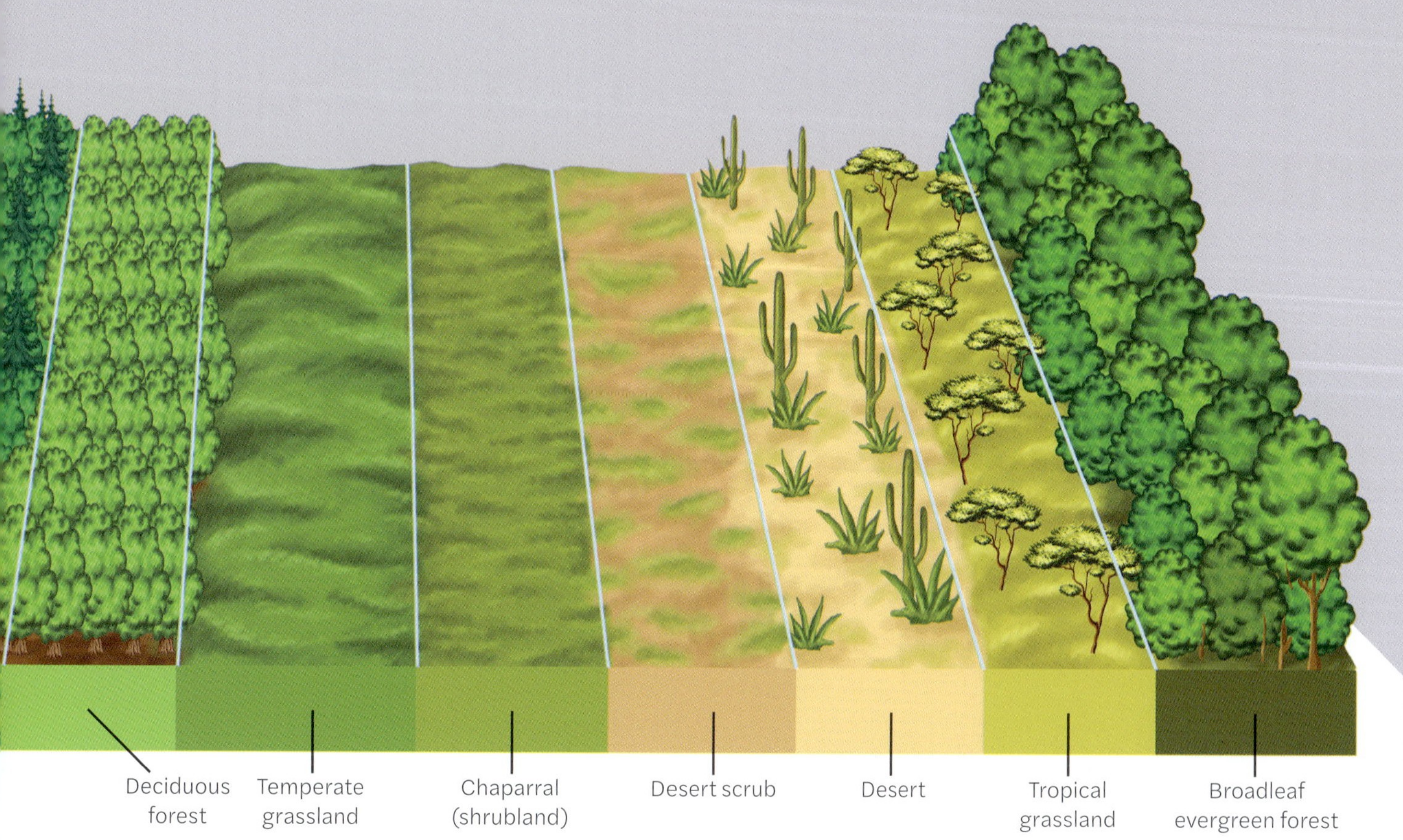

- Read an article about time lapse data at http://mea.digital/GHV9_G5_2.
- Go on a virtual nature tour from Google Earth at http://mea.digital/GHV9_G5_3.
- Visit the Biome viewer website at http://mea.digital/GHV9_G5_4.
- Discover world human population change over time at the World Population History website at http://mea.digital/GHV9_G5_5.

Presenting your data

Present the data you collected using tables, graphs, annotated sketches and photographs.

Analysing your data and drawing conclusions

Using the data you collected while on your virtual fieldtrip, consider how this information helps you to answer the research question. Does it support your hypothesis? Or are you beginning to change your mind? Complete the following tasks.

1. Summarise the key findings of your research.
2. Was your hypothesis correct (supported)? Why/why not?
3. Record five key pieces of evidence (data, photos, sketches or surveys) that prove or disprove your hypothesis.
4. Which methods were the most successful for collecting data to answer the research question? Why?
5. Which methods were the least successful for collecting data to answer the research question? Why?
6. What changes to your fieldwork approach would you suggest to future Year 9 students completing the same task?

Handing in your report

Once you have completed all the steps of your fieldwork report, you are ready to collate your work into a folio and submit it to your teacher.

fieldwork task 2

How can we explore our interconnection with *place*?

Background to fieldwork approach

Exploring interconnections between people and places has been a major theme throughout the Geography chapters of this textbook. People can be physically connected to places through transport networks. They can also be connected to places virtually through technology. This is particularly important for rural communities who may otherwise not have access to the same services and resources as people who live in urban areas.

In this fieldwork project, you will investigate how people are connected both physically and virtually to your local place. You will explore other people's sense of place and create a summary of the services that attract people to the region.

Research question

You may choose to create your own research question for this investigation or use one of the suggestions below. Either way, your question should encompass place, people and interconnection. As a class, brainstorm these terms and formulate a question that you will be able to answer through doing fieldwork in your local area. Some questions you may consider:

- What facilities or services in your local area increase the community's sense of place?
- How does sense of place differ between people within your local community?
- How does virtual connection to place play a role in your local community?
- Is virtual or physical connection to place more important in your community?

Once you have decided on a question, you need to create a hypothesis. You may all have slightly different hypotheses, but that will create a more interesting analysis.

Pre-fieldwork research

Once you have decided on a research question you are ready to complete some pre-fieldwork research. Explore the internet, books and other resources for information about key concepts or processes that the reader will need to know in order to understand your report.

Introduction and method

1. State your research question and hypothesis.
2. Provide some background information so the reader will understand all key terms and concepts used in your report.
3. State two primary fieldwork techniques used to complete this fieldwork and describe how they helped to answer the research question.
4. State two secondary resources you used to prepare for your fieldwork and describe how they were helpful in answering the research question.

Collecting your data

The best way to collect data on people's opinions is through surveys. As a class, design a survey that asks locals about their sense of place, the services they feel connected to and the facilities they think are required to increase the sense of connection to their local area.

Once you have created the surveys, go into the community and gather data. If you all do a couple of surveys each and collate the class data, your sample size will be larger and your results will be more reliable. Remember, your class is your research group. Work together to share photos, survey results, data and other secondary sources.

Presenting your data

Present the data you collected using tables, graphs, annotated sketches and photographs.

Source 7

Community gardens, such as this one in central Melbourne, are a great source of connection.

Source 8

Technology can both connect and divide us

Analysing your data and drawing conclusions

Using the data you collected while on your fieldtrip, analyse how this information helps you to answer the research question. Does it support your hypothesis? Or are you beginning to change your mind? Complete the following tasks.

1 Summarise the key findings of your research.
2 Was your hypothesis correct (supported)? Why/why not?
3 Record five key pieces of evidence (either data, photos, sketches or surveys) that prove or disprove your hypothesis.
4 Which methods were the most successful for collecting data to answer the research question? Why?
5 Which methods were the least successful for collecting data to answer the research question? Why?
6 What changes to your fieldwork approach would you suggest to future Year 9 students completing the same task?

Handing in your report

Once you have completed all the steps of your fieldwork report, you are ready to collate your work into a folio and submit it to your teacher.

G5.4

How do I write a fieldwork introduction?

Where do you start when writing a fieldwork report? Your introduction provides a background on what you are investigating and why it is important. In other words, it provides context for your research.

- When writing an introduction, start with your broadest ideas, and finish with your research question and hypothesis. Your opening lines may need to define some key terms or provide background information about your research topic.
- A formal fieldwork report does not use pronouns such as 'me' or 'I'.
- Cite research to show where you sourced information and to demonstrate you have used reliable sources.

Use the SPICESS concepts as prompts to help you explore your research question and create a detailed introduction. These are annotated on the sample shown.

Example research question: *What influences young people's sense of place in my region?*

Example hypothesis:

Sense of place for young people is developed through local community engagement.

In our constantly changing world, it is more important than ever to feel engaged and that we belong. Individuals develop a 'sense of place' by forming emotional or spiritual connections to their local area. Each person may have a different opinion about, or attachment to a place. These perceptions develop over time and depend on factors such as family history and economic status and are influenced by tolerance to negative factors such as traffic or noise. People can develop a 'sense of place' through involvement with community, such as with sporting teams or other local groups, and through sharing common interests.

Officer is a small suburb located in southeast Victoria (38.0611° S, 145.4151° E), 48km from the CBD (Google Maps, 2020). Officer is part of the southeast growth corridor and has a rapidly growing population reaching just over 7000 residents (ABS, 2016). Before urbanisation, Officer was a small rural town, characterised by open farmland, low-density housing and small-scale industry. As the demand for residential spaces close to the CBD increased, Officer began to expand with a range of high-density housing estates and residential streets now dominating the landscape.

Because of its rapid population growth, Officer township and other necessary infrastructure are yet to be constructed and so locals need to travel to surrounding townships for resources. Given the ongoing changes, it may be difficult for new or existing locals to develop a sense of community and place. In particular, it is important to understand the needs of the youth in this growing community in order to provide appropriate groups and other resources to help create a sense of belonging for young people. Therefore, the question was asked, 'What influences young people's sense of place in my region?' and it was hypothesised that young people could develop a sense of place through access to more local groups for community engagement.

Interconnection and change: How does this place connect to surrounding regions? How has this place changed over time? Why is this a good place to conduct your research?

Environment and sustainability: Describe the human and natural characteristics of this place. Are there any other environmental factors your reader needs to understand?

Space and place: Describe the location you are using for your study. What is the absolute and relative location? Why is this a good place to conduct your research?

Scale: Think about the scale of your research. In what areas will you be collecting data and for how long? How big is the space you are collecting data in? Why is this a good place to conduct your research?

G6

Geo How-To

Knowing the locations of countries, their capitals and even their flags can be useful geographic knowledge. However, the key to success in Geography is understanding key skills and being able to apply them in different situations. This chapter will walk you through some of the key skills for Year 9 and provide examples of how to use them.

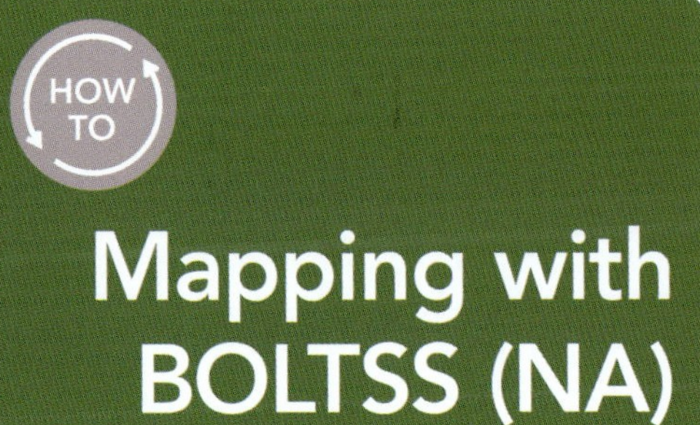

Mapping with BOLTSS (NA)

Every map requires BOLTSS in order to be understood by the geographic audience, and some people add (NA) to the acronym to remind them to be neat and accurate. When you construct maps, use BOLTSS like a checklist to ensure you complete your mapping tasks correctly.

B Border

A border is important to show the edges of the mapping field. It provides a clear area inside which to construct your map and makes it appear clear and neat to readers.

O Orientation

An orientation, or compass, helps us understand direction when reading the map. Orientations can be drawn as a 4-point, 8-point or 16-point compass.

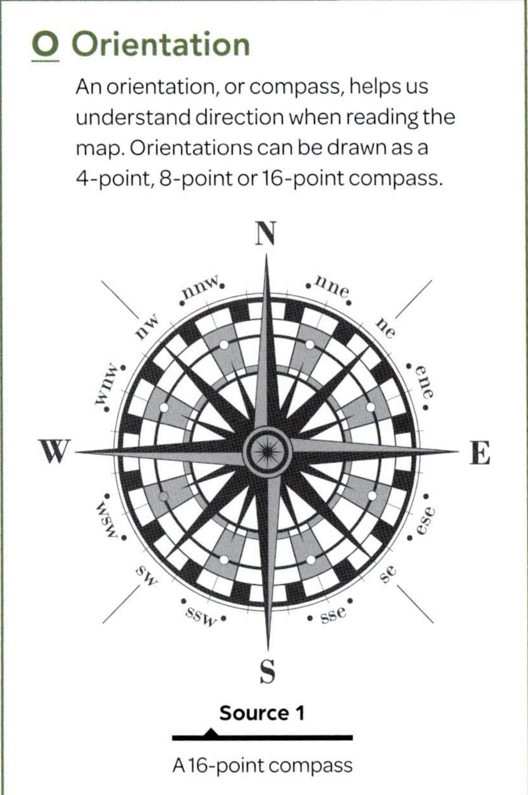

Source 1

A 16-point compass

Map of central Portland, Victoria

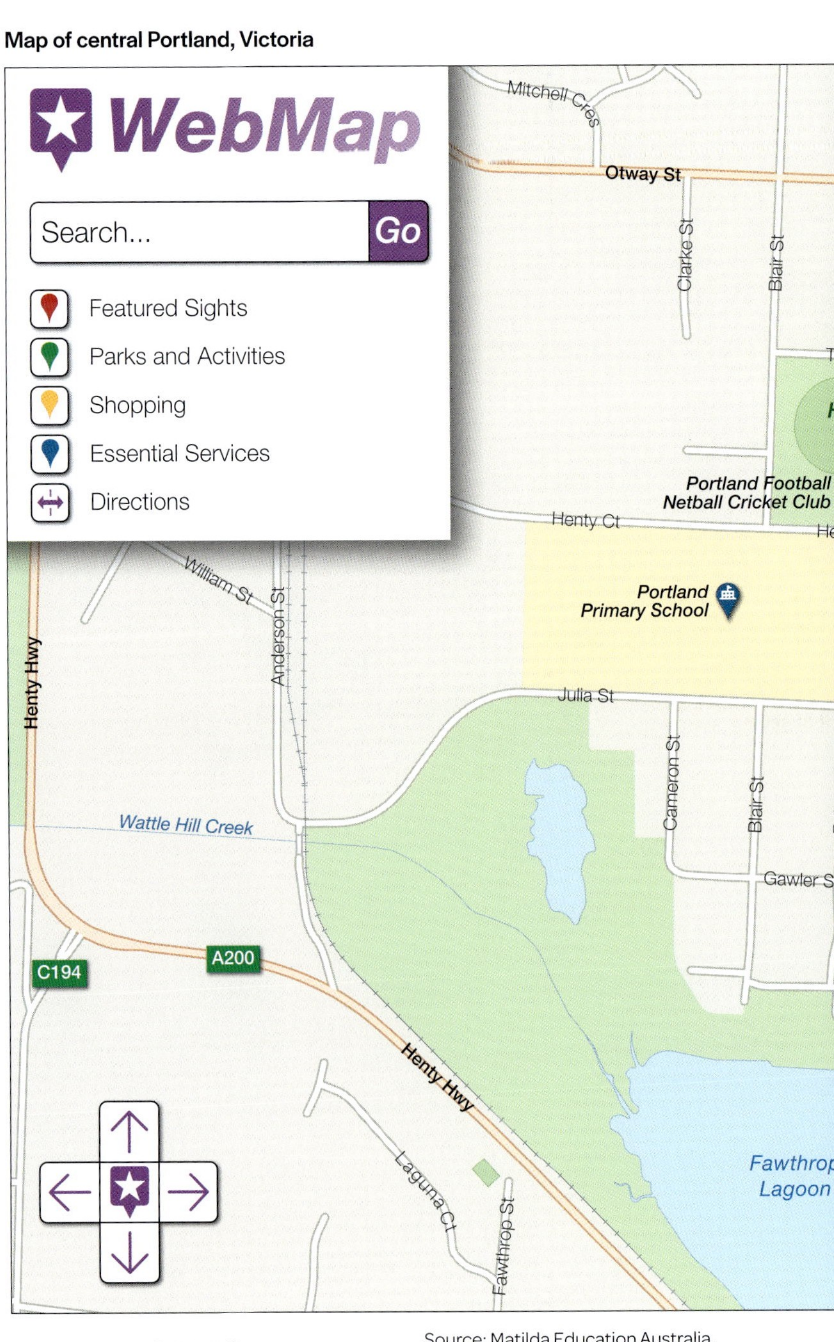

Source 2

This map features all aspects of BOLTSS

Source: Matilda Education Australia, Open Street Maps

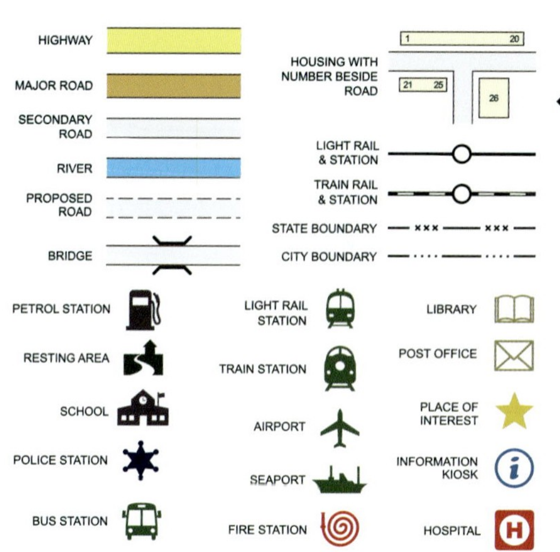

Source 3

A sample legend

L Legend

A legend, or key, is vital to understanding the map. Without a legend, colours and symbols would not make sense and we would not be able to interpret patterns or distributions.

T Title

A title gives us an understanding of what the map is showing. If you are drawing a sketch map, you should also provide a date and time. This allows you to monitor change in a location over time.

N
0.25 km
Portland District Hospital
Otway St
Percy St
Otway Ct
Bentinck St
Nuns Beach
Portland CFA
Market Ct
St Johns Lutheran Church
Lee Breakwater
Portland Presbyterian Church
Tyers St
Supermarket
All Saints Catholic Church
Portland Maritime Discovery Centre
PORTLAND BAY
Henty St
Bank
Richmond St
Maritime Discovery Centre
Portland
St Stephen's
Julia St
Lee Breakwater Rd
Portland Yacht Club
Percy St
Supermarket
Bentinck St
Gawler St
Gawler St
Cliff St
Charles St
R.B Anderson Rd
Barton Pl
Portland Police Station
Powerhouse Car Museum
Glenelg St
Glenelg St
Portland Leisure & Aquatic Centre
Botanical Gardens
Powerhouse Museum

S Scale

A scale provides us with information on how big something is in real life. While a house on a map may be 2 centimetres across, it may be representing a 15-metre wide three-bedroom home. Scales can come in many forms: linear, ratio or fraction.

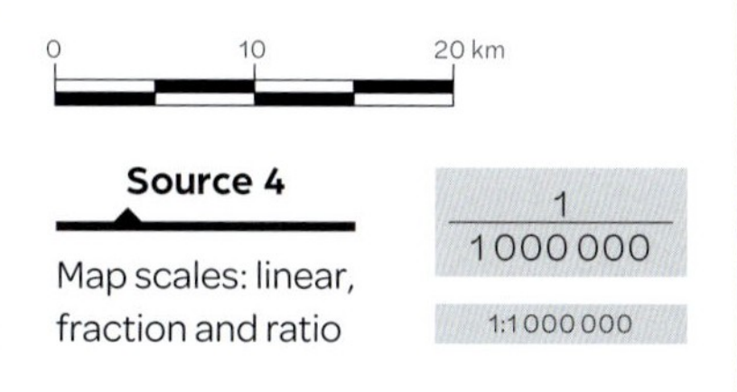

Source 4

Map scales: linear, fraction and ratio

S Source

Always acknowledge the source of the information that you illustrate on your map. The source can also indicate whether the map is reputable or not.

(NA) Neatness and Accuracy

Some people add the final letters 'NA'. When we read a map, we rely on it being neat and legible and we expect that its data is displayed accurately. Therefore, when you construct a map it is important that you take the time to correctly illustrate the patterns and distributions you see in the data.

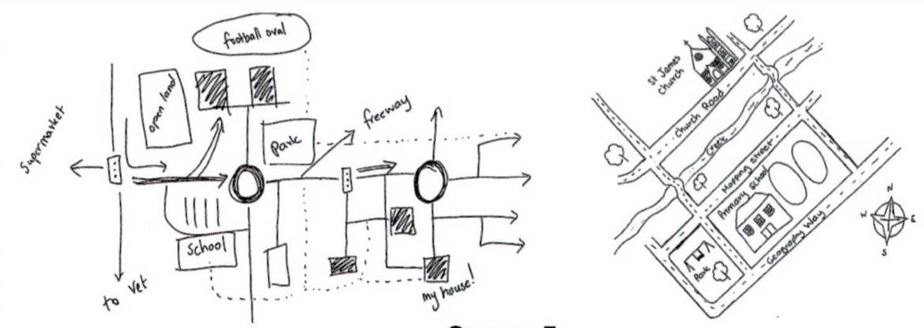

football oval
open land
supermarket
Park
freeway
School
to Vet
my house!
St James Church
Church Road

Source 5

Which sketch map is easier to read?

HOW TO

Mapping skills

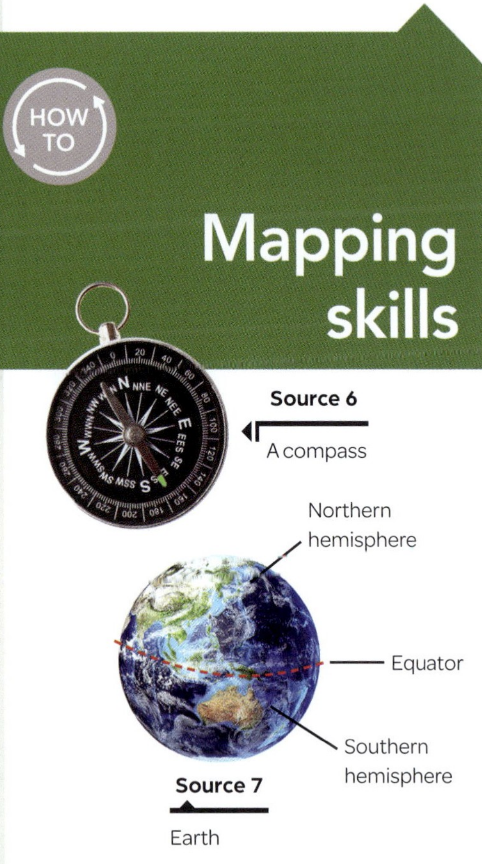

Source 6

A compass

Source 7

Earth

Direction

In day-to-day life, we tend to use directions such as 'left' and 'right', 'above' or 'below'. While these words are helpful, in Geography we also need to use compass points: north, south, east and west.

Consider the world globe in Source 7. Around the centre there is a line of latitude called the **equator**. North of the equator is the **northern hemisphere**, which includes the continents of North America and Europe. South of the equator is the **southern hemisphere**; this is where we live!

It's a common mistake to use the word 'above' instead of 'north' or 'below' instead of 'south'. If you say something is 'above' the equator, you are actually saying it is floating in the air over the top of it! If you say something is 'below' the equator, you are describing something buried beneath it! Use directional terms carefully to ensure you are sending people in the right direction.

Latitude and longitude

How do you identify where you live? Most people have a street address that gives the exact location of their house. However, how do we identify the location of a place that doesn't have an address? Geographers use grid references and latitude and longitude to help identify these locations.

Source 8

A world map with lines of latitude and longitude shown

Latitudinal lines are shown running around Earth like a series of belts. We use the equator as a reference point, so it represents 0° **latitude**. Lines north of the equator are numbered 1, 2, 3 etc. degrees north (°N), all the way to 90°N at the North Pole. Lines south of the equator are 1, 2, 3 etc. degrees south (°S), all the way to 90°S at the South Pole.

Longitudinal lines are like jail bars wrapping around the globe from east to west. These lines are measured starting from the prime meridian (the line that passes through Greenwich, in the UK), which is 0° **longitude**. Lines of longitude east of the prime meridian are 1, 2, 3 etc. degrees east (°E) and lines of longitude west of the prime meridian are 1, 2, 3 etc. degrees west (°W). On the map in Source 8, Lima can be described as having an absolute location of 12°S and 77°W, as it is south of the equator near the 15° line of latitude and just over 75° west of the prime meridian.

Scale

Look at Source 7 again. Logically, we know that this image of Earth is not to scale. In reality, Earth is over 12 700 km in diameter!

Scale is important in Geography as it allows us to shrink maps and other images down to a size where we can see patterns, distributions and changes. Typically, scale is displayed using a line that acts like a ruler, showing you how many centimetres on the map represent the real distance.

Scale can also be used to describe changes that occur over time. A 'large-scale change' indicates that something has caused major alterations to a region. A 'small-scale change' indicates that little has been altered. Review the images in Source 10. What is the scale of the city's change?

Source 9

Obviously, this map is not to scale. It is a large area that has been shrunk down to fit on this page. The scale on this map helps us determine how big things are in real life.

Source 10

Change over time to Las Vegas, 1967 to 2014

In Geography we use maps and graphs to understand what is happening around us. In many cases, these resources provide us with information on patterns and spatial distributions of phenomena.

The formula PQE helps us describe these patterns and distributions. P stands for pattern, Q stands for quantify and E stands for exception.

P Pattern

A pattern is a trend in the data. When looking for a pattern, read the legend and interpret what the colours or symbols mean. On a graph or map, you may notice that all the data points tend to be clustered in one spot or that there is an uneven distribution of data points. You may need to use the names of places or even your compass points to describe where these clusters appear on the map. For example, when observing Source 11 to the right, we notice that people in countries in the northern hemisphere tend to consume more eggs than those in the southern hemisphere.

Descriptive words that may help you describe patterns are: *clustered, even, uneven, highly distributed, north, south, east, west, increase, decrease* and *fluctuate.*

Q Quantify

When we quantify our pattern, we use numerical data to provide evidence of what we see. You can gather data by using the legend, measuring with the scale or doing a count. Ensure your data relates directly to the pattern you recorded earlier. For example, we noticed that people from countries in the northern hemisphere tend to consume more eggs than those in the southern hemisphere. To quantify, use the legend on the map. It shows that many countries in the northern hemisphere consume more than 12 kilograms of eggs per person per year, while in most southern hemisphere countries people consume less than 10 kilograms per person, per year.

E Exception

An exception is a trend on the map or graph that doesn't 'fit in' with our original pattern statement. When we observe an exception, it is also good to quantify it to provide a comparison to our original statement. For example, we noticed in Source 11 that people in countries in the northern hemisphere tend to consume more eggs than those in the southern hemisphere. However, people in Argentina, which is in the southern hemisphere, still consume quite a lot of eggs: more than 16 kilograms of eggs per person per year.

What is the difference between qualifying data and quantifying data?

PQE helps us describe patterns and distributions. When we describe, we 'say what we see'. In a PQE analysis, we do not explain or give a reason why we see patterns; we do this in a SHEEPT analysis (pages 138–39). To **quantify** means to use percentages, counts, ratios or data to provide details about the patterns you are describing. To **qualify** a statement means to use general describing words such as 'large', 'many', 'broad' or 'small' to describe a pattern or change.

By using quantifiable data, we can more easily see key differences between locations or monitor change over time. If your PQE analysis states: 'Egg consumption varies worldwide', does this describe things clearly? Or does this quantified statement provide more detail: 'In the southern hemisphere, 29 out of 32 countries consume less than 10 kilograms of eggs per person, per year.'?

How do I start my PQE sentences?

When writing a PQE analysis, start sentences with the following key terms:

Pattern:

Overall…

For example:

Overall, countries in the northern hemisphere tend to consume more eggs than those in the southern hemisphere.

Quantify:

To quantify…

For example:

To quantify, according to the legend, people in 29 out of 32 countries of the southern hemisphere consume less than 10 kilograms of eggs per person, per year.

Exception:

However…

For example:

However, Argentina is in the southern hemisphere and they consume over 16 kilograms of eggs per person per year.

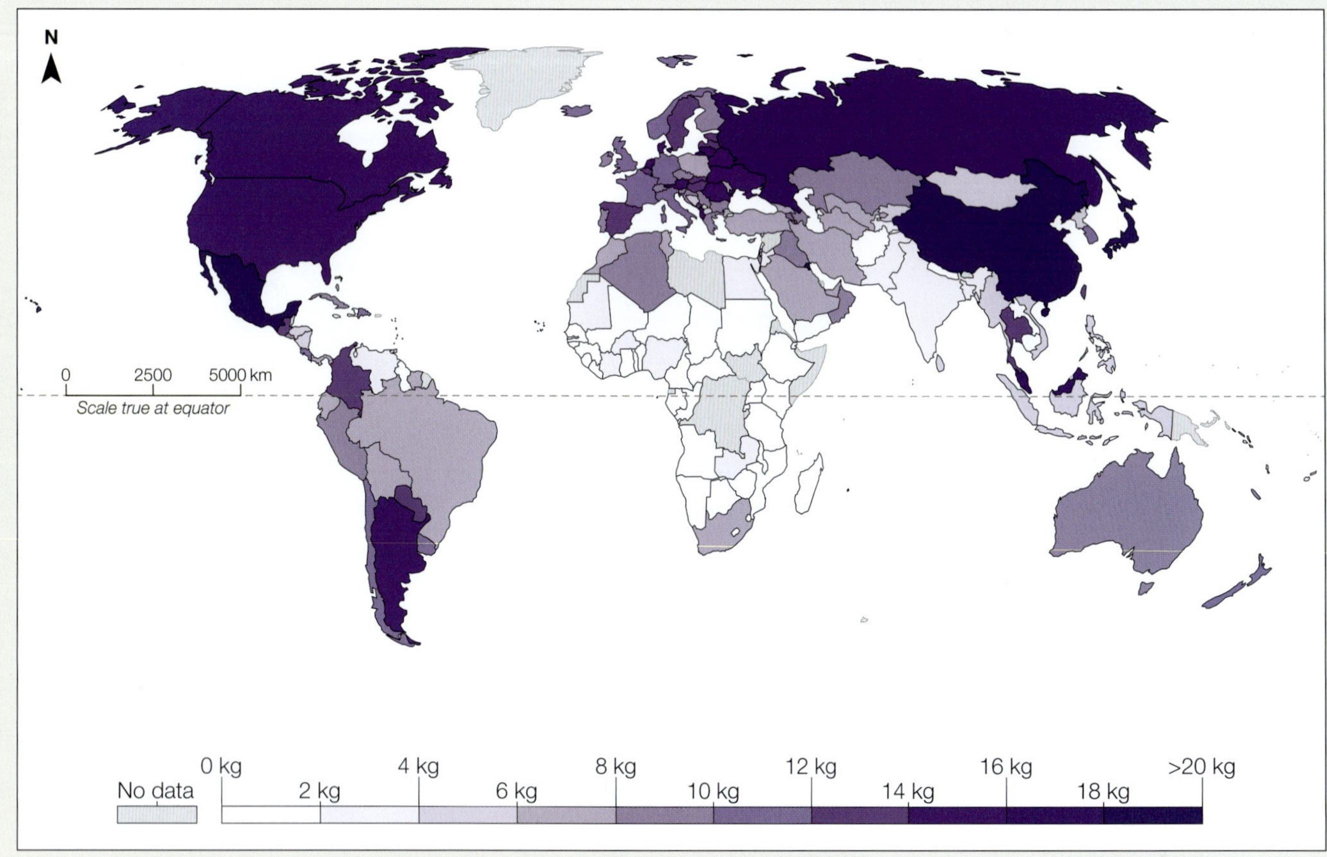

Source 11

Egg consumption throughout the world in 2017

Source: Our World In Data

SHEEPT

SHEEPT is an acronym that helps you remember the reasons why a spatial pattern occurs. It stands for: Social, Historical, Economic, Environmental, Political and Technological.

Source 12

Sydney, NSW

S Social

Social factors are anything to do with people. Social factors include population, culture, language and religion.

H Historical

Historical factors are anything to do with our past. Historical events, buildings, people and changes to climate all influence what we see in our world today. Sydney's urban landscape has changed significantly over time.

E Economic

Economic factors are those relating to money. In Geography, income, costs of things and how much money is spent can provide us with information on a place.

How do I write a SHEEPT analysis?

SHEEPT is usually used to explain why patterns or distributions may occur in a particular region. It can also be used to expand our thinking when annotating images or considering new geographical content. The text on the right is an example of how to write a SHEEPT analysis for an image. The highlighted terms indicate the use of a SHEEPT term. Can you identify all of them? (The analysis does not need to include every term.)

Source 12 is an image of the Sydney CBD and its famous landmarks. The Sydney Harbour Bridge is a historical site that took eight years to construct and was opened by Premier Jack Lang in 1932. Today, the bridge is a major tollway and thoroughfare, with eight car lanes, train tracks and paths for bikes and foot traffic. The bridge is also a tourism site, with tourists paying to climb the structure and see the speculator views of the city, harbor and surrounding environment from the top.

E Environmental

Environmental factors are those relating to the natural or human environments on Earth. Humans can manipulate the environment to suit their needs. Here we see how parklands have been urbanised to support a growing population.

P Political

Political factors are those to do with the government or leading groups. When we consider political factors, we refer to laws and policies.

T Technological

Technological factors relate to the different kinds of technology that we have access to. This could be in the form of gadgets, **spatial technology**, medical technology or even roads, transport or basic machines used in the home.

Sketches and annotating

Field sketches are an excellent way of recording data when you are investigating a research question.

Sketches allow you to annotate movement, patterns or any interconnections you see. Field sketching is not a test of your artistic skills – the idea is to record a simplified version of what you can see.

T Title (1)

Create a heading for your sketch that is clear and provides information on the location. You may wish to record both the **absolute** and **relative locations**.

O Orientation 

An orientation shows the direction that you were facing when you conducted your sketch. In order to record a correct orientation, you need to use a compass.

A Annotations 

Annotations are the most important thing to complete when drawing a field sketch. Annotations allow you to record details about what you see and explain how elements of your drawing relate back to the research question. Ensure that lines pointing to your annotations are completed with a ruler and do not overlap.

THE GREAT ALPINE ROAD (1)
Bright, Victoria:
9.8.2020 9.30 am (5)
(6)
Green trees and lush mountain landscape indicates it is spring or summer and so not peak tourist seaon in this region

TOASTIE will help you remember the key skills when making a field sketch. It stands for Title, Orientation, Annotations, Scale, Time, Information and Edge.

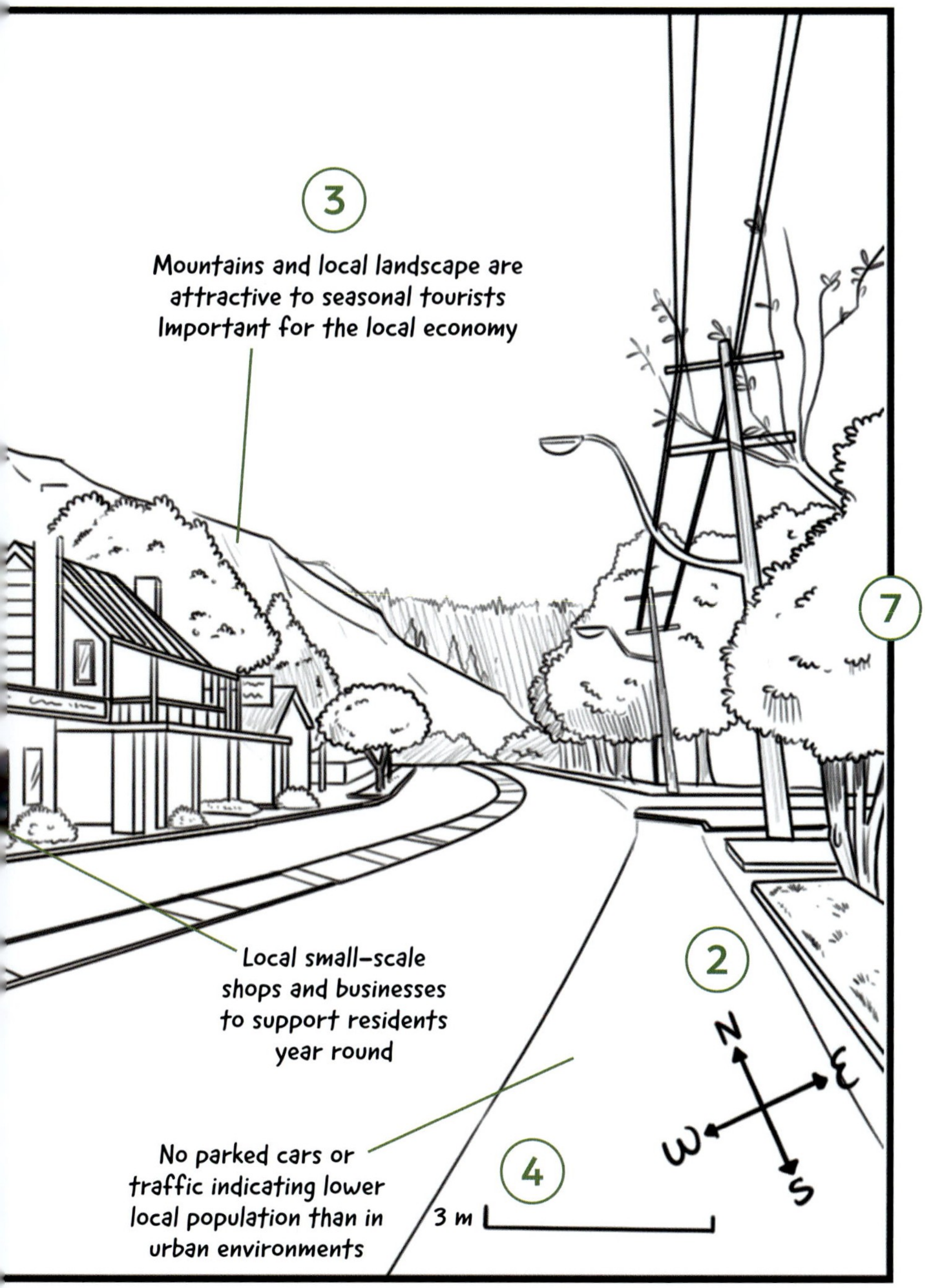

3

Mountains and local landscape are attractive to seasonal tourists Important for the local economy

7

2

Local small-scale shops and businesses to support residents year round

N
E
W
S

No parked cars or traffic indicating lower local population than in urban environments

4

3 m

Source 13

An annotated field sketch with accompanying photo.

S Scale 4

Most sketches are not to scale. However, all geographical maps and sketches require a scale to give the reader some indication of size. To estimate a scale, use a metre-rule or pace out an area that you have sketched. Then, using a small ruler, identify how large the same area is on your drawing. For example, you may estimate that the path you are looking at is 1 metre wide, and when you measure your drawing of the path it is 1 centimetre wide. Therefore, your rough estimated scale is 1 m = 1 cm.

T Time

By recording the time your sketch was completed, you can analyse how the environment changes over the course of a day, a month or even years!

I Information

Provide more than just one-word annotations on your sketch. Annotations should be at least one sentence and help your reader to identify any patterns.

E Edge

Draw a border so it is clear where your sketch starts and ends.

Visual communication

Flow diagrams

Flow diagrams help show processes or how different parts of the environment are *interconnected*. The arrows represent the movement between stages or the connection between things.

Source 14

A flow diagram

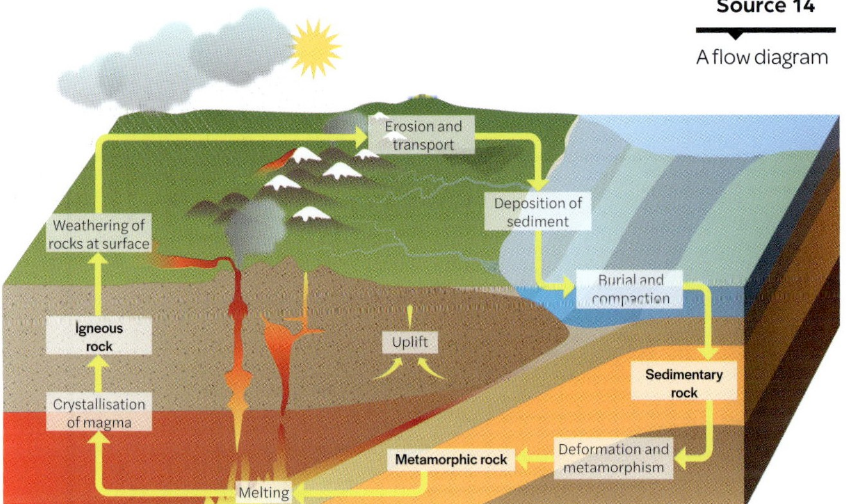

Block diagrams

Block diagrams are a useful way of drawing landscapes to show what is happening both above and below ground. For example, when we consider the movement of water, block diagrams help us visualise how water can move both on the surface of Earth through rivers, oceans and streams, as well as under the ground through water tables and pipes.

When you create a block diagram, you need to ensure that you draw it three dimensionally, with equal amounts of drawing room above and below the surface of Earth. Annotations are very important when drawing a block diagram, as this is how your audience can interpret your illustration.

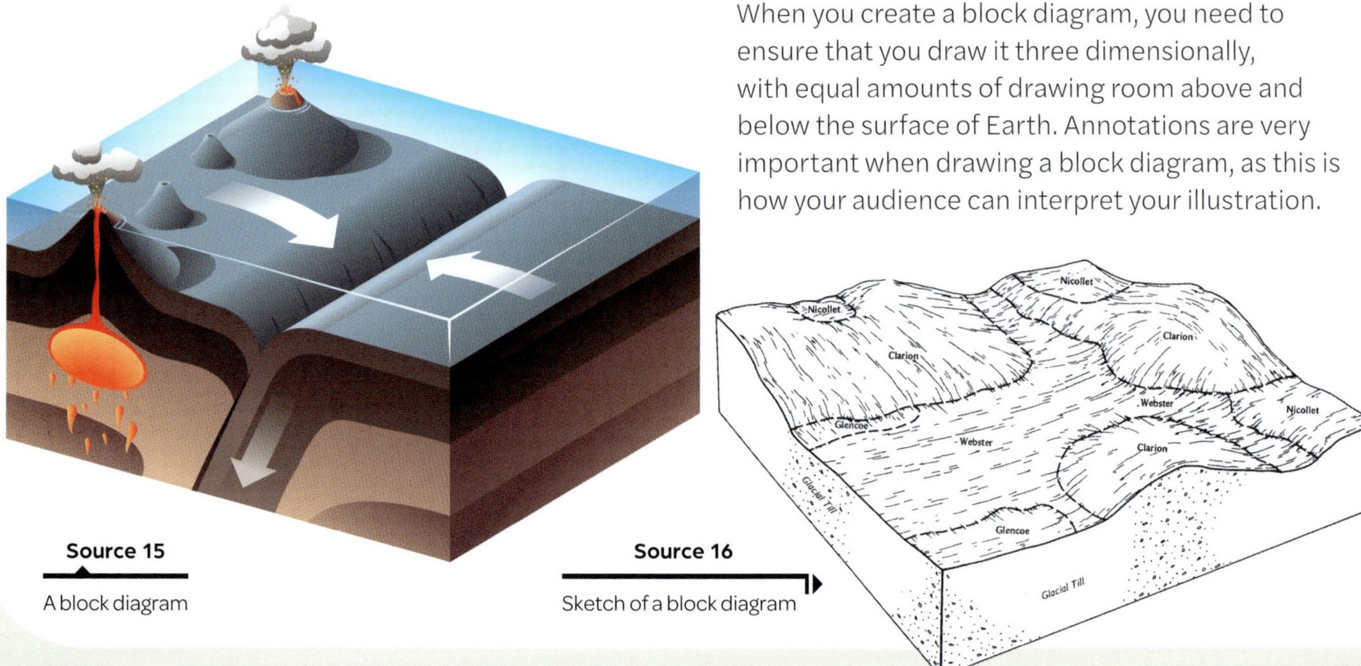

Source 15

A block diagram

Source 16

Sketch of a block diagram

Photo essays

A **photo essay** is a way of presenting information visually to show characteristics of a place or process. A photo essay usually includes a series of photos with specific annotations or captions that provide a brief background into the key features of the image or the meaning behind the image choice.

Source 17

A photo essay

An open field that has been altered by humans for agriculture

Evidence of vehicles and human interaction

Cleared and maintained for agriculture and farming

This is a mountainous biome. Due to the height of the mountains, snow may fall at their peaks.

Vegetation grows around the base and in the valleys.

Meltwater from the mountains runs off and forms a lake at the base in low-lying landscapes.

A coastal biome that contains a series of islands and rocky formations

Evidence of human interaction with this biome for recreation or fishing

These formations may have been formed by weathering and erosion over many years.

A riverine biome bordered by lush vegetation, which creates a range of suitable habitats for animal life

Vegetation uses the river for water to grow and photosynthesise.

The river provides water and habitat to animals living in this biome.

Graphing

Choosing a suitable graph

Graphing is an important way to display geographic information. By using the appropriate graph type we can clearly show patterns and changes over time.

Bar graphs are most suitable for comparing small changes over time that are harder to interpret on a line graph. For example, in Source 19, precipitation is represented in bars as it only ranges from 45–70 mm over the year.

Line graphs are more suitable to show and compare larger changes that occur over longer time periods. For example, in Source 18, population is shown to change significantly over many years.

The two fluctuating lines represent Australia's total population change and its immigrant population change. The steadier line shows the natural increase which has not changed as much over time.

Pie charts are best used when you want to show proportions. These charts are not suitable for showing change over time.

When creating a graph, use the acronym **SALTS** to guide you: scale, axis, legend, title and source.

Scale: The scale of your graph will depend on the data you are trying to visualise. To choose your axis scale, identify the lowest and highest value (range), then fill in the numbers in between to mark your data points.

Axis: Each graph has an *x*-axis and *y*-axis. The *x*-axis is the horizontal axis and the *y*-axis is the vertical. Make sure you label each axis!

Legend: A graph often uses colours to represent data. A legend indicates to the audience what these colours mean and how to read the data.

Title: A title lets your audience know what your graph is showing.

Source: When you graph information, you must acknowledge where you obtained your data. By stating the source of your information, the reader knows how reliable it is.

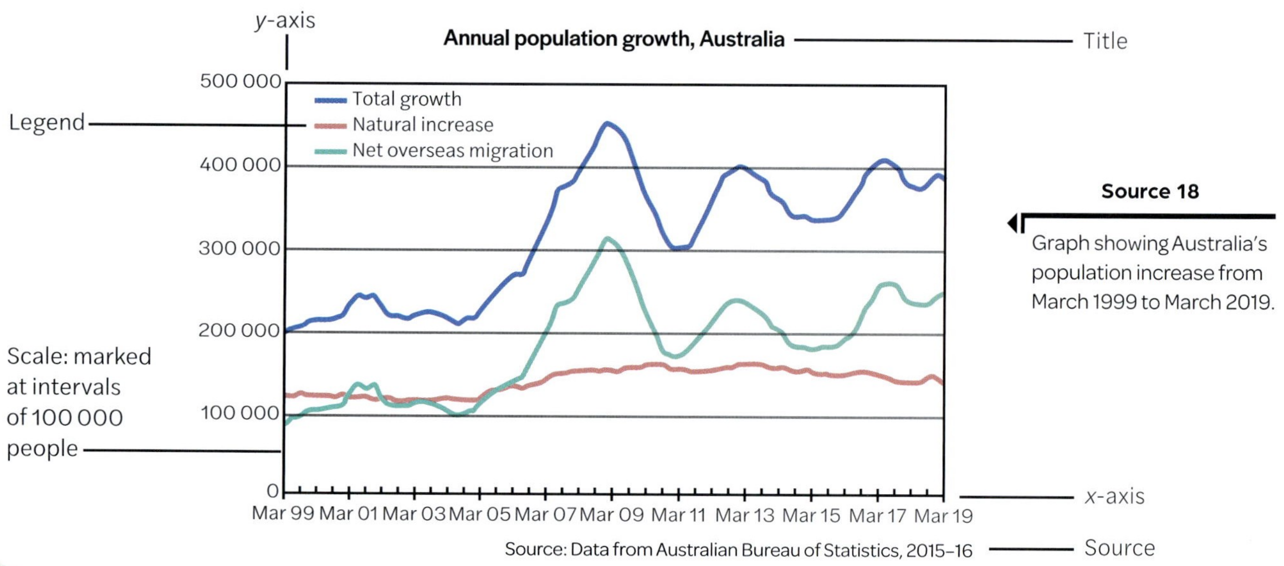

Source 18

Graph showing Australia's population increase from March 1999 to March 2019.

Climate graphs

A climate graph is simply two graphs in one.

The graph below shows the temperature (degrees Celsius) on the left *y*-axis, which relates to the red average temperature line; and the amount of precipitation (mm) on the right *y*-axis, which relates to the blue columns. The horizontal axis or *x*-axis, shows us the months of the year.

We notice that the red line, or average temperature, fluctuates throughout the year, peaking in December to March and decreasing from April to July. Precipitation fluctuates in a similar way, but tends to be highest in the spring from August to November, with the exception of another peak in May. Using two PQE analyses (one for temperature and one for precipitation) is the best way to describe the patterns we observe in a climate graph.

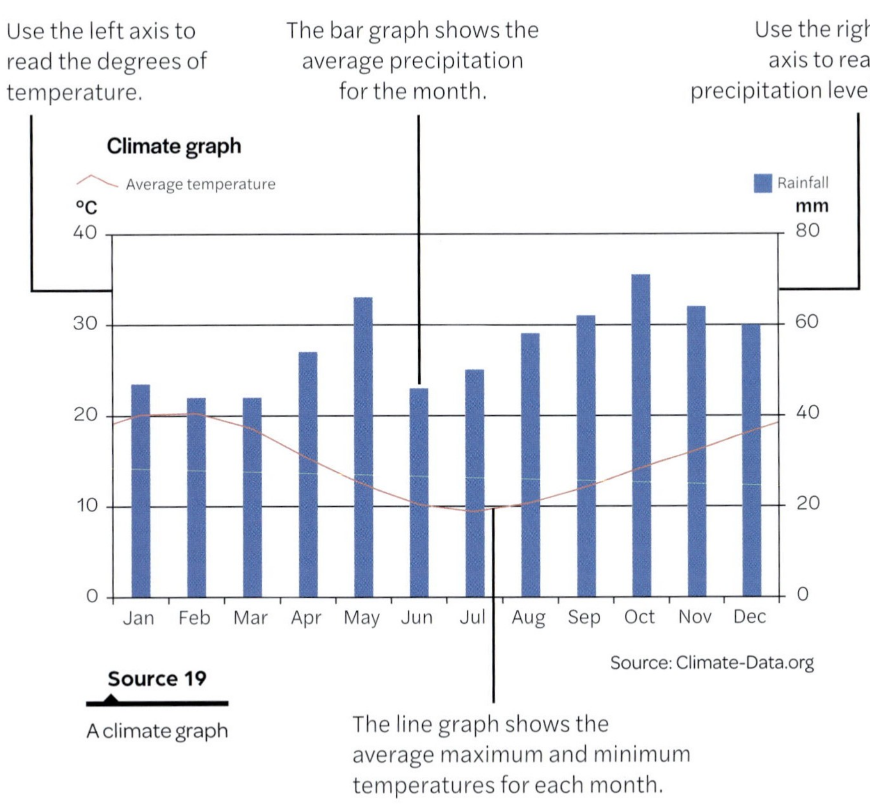

Source 19

A climate graph

Here we need to apply SALTS! Label the axes, provide a title and a source. The legend is part of the axis labels: red line denotes average temperature in degrees Celsius; blue columns denote rainfall in mm.

What is the difference between weather and climate?

Look outside your window and describe today's temperature and the amount of rainfall. What you have just described is the weather. Weather changes daily but we can usually predict it up to 10 days in advance.

Now close your eyes and describe the 'climate' of Australia. Do you imagine Australia as mostly hot and dry? Just because we can describe Australia as hot and dry, does not mean it is like this everywhere, all year round. Unlike the weather, climate helps us describe the yearly (annual) average temperature and level of precipitation (rainfall, snow etc.) in a region or country. Climate graphs help us visualise the climate of a region or country. Climate change describes how the average temperature and precipitation levels of a location change over time.

Population pyramids

Population pyramids are graphs that show the number of females and males in particular age groups in a population; they are like bar graphs turned on one side.

Population pyramids can be made on a local, national or global scale. On a population pyramid, female data is normally shown on the right side and male data on the left. The length of each bar represents the number of males or females within that age group.

The shape of the pyramid tells us about the population:

- Triangular means the population is growing, because there are more young people than old
- Box-like means growth is slow or stable, as there are roughly equal numbers of old and young people.
- If the shape becomes wider towards the top, like a reduced pentagon, it represents an ageing or declining population, as there are more middle-aged people than young people.

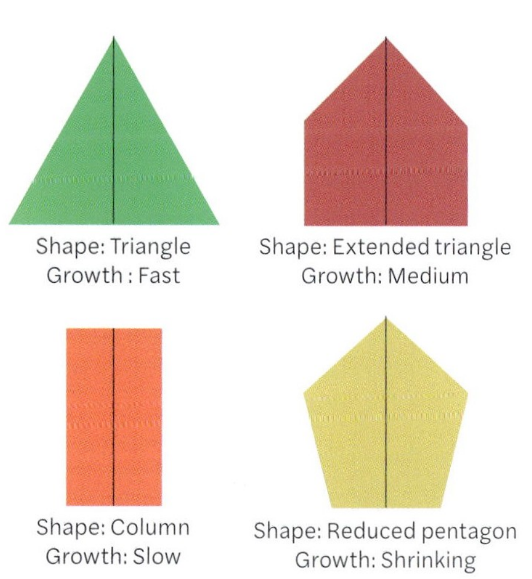

Source 20

Various population pyramids. The shape of the pyramid tells you about the population.

How do I interpret a population pyramid?

Source 21 shows population pyramids for Australia in 1955 and 2018. In 1955, there were many young people in the age range 0–14. Approximately 5.5 per cent of the population were 0–4-year-old males and 5.3 per cent were 0–4-year-old females. We also had many working-age people. For example, 4.1 per cent of the population were 30–34-year-old males and 3.8 per cent were 30–34-year-old females. In 2018, there are significantly fewer young people and more old people in our population. Our population also grew to 25 million people.

Source 21

Population pyramids showing Australia's population in 1955 (left) and 2018 (right)

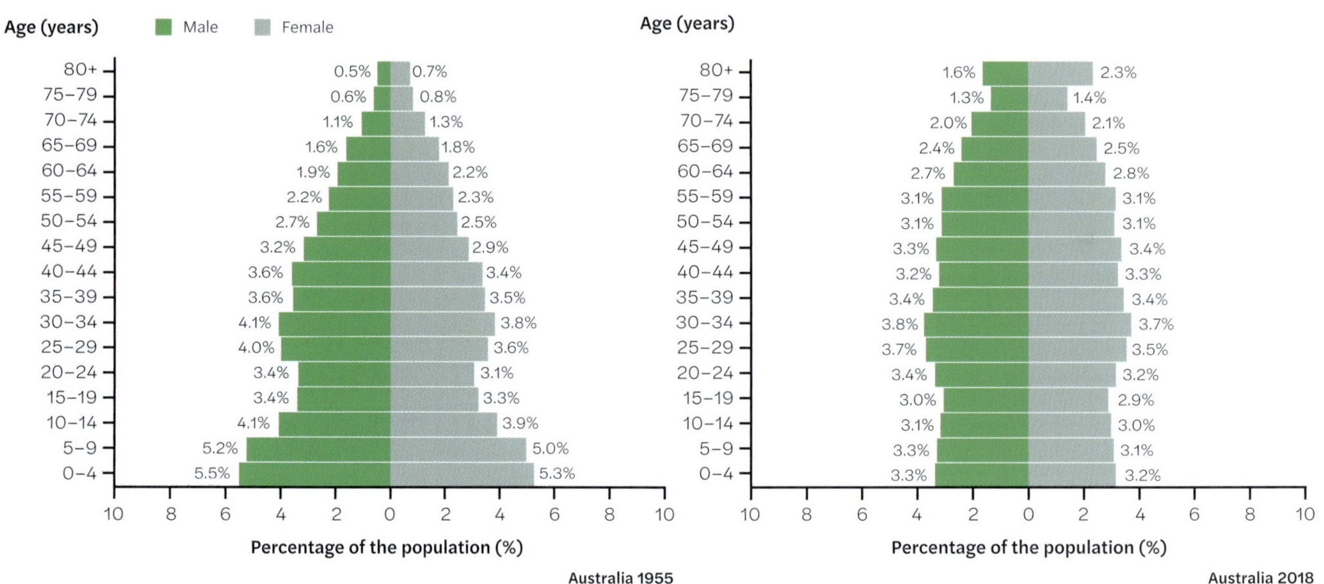

Source: PopulationPyramid

HOW TO

Satellite images

Satellite images are pictures of Earth taken from space. Satellite images give us the most information if taken during the day. Satellites orbit Earth and constantly take images, and we can use this data to see a change over time in land cover or other spatial patterns.

Source 22

Satellite images taken at night provide different information than those taken during the day. This photo of the US shows the extent of its urban areas.

Autumn 17 Oct 2012

N 0 40 80 km

Winter 7 Jan 2013

N 0 40 80 km

Spring 20 April 2013

N 0 40 80 km

Summer 14 June 2013

N 0 40 80 km

Source 23

These satellite images show the forests covering the Great Smoky Mountains of the southeastern US over four seasons. Satellite images are useful for seeing change over time.

When trying to interpret a satellite image, look for colour and shapes. Colour can give you an idea of land cover. For example, green usually indicates vegetation, blue is the colour of water and brown is desert or barren land. White can indicate either snow or clouds, so we use shapes to tell them apart. Clouds tend to look fluffy and can sometimes (depending when the images were taken) cast a shadow. Snow usually appears on the tops of mountains and you can observe where it is melting or following the slopes. Looking for shapes is also helpful when identifying rivers or reservoirs, which can be seen as meandering blue lines.

Seasonal variations can sometimes be very clear in satellite images – in spring and summer, green vegetation flourishes, while during winter, white snow may fall or vegetation may die, leaving brown bare ground.

Source 24

This satellite image shows the various geographical features of central Chile and Argentina. We can use this information to answer geographical questions and understand spatial patterns.

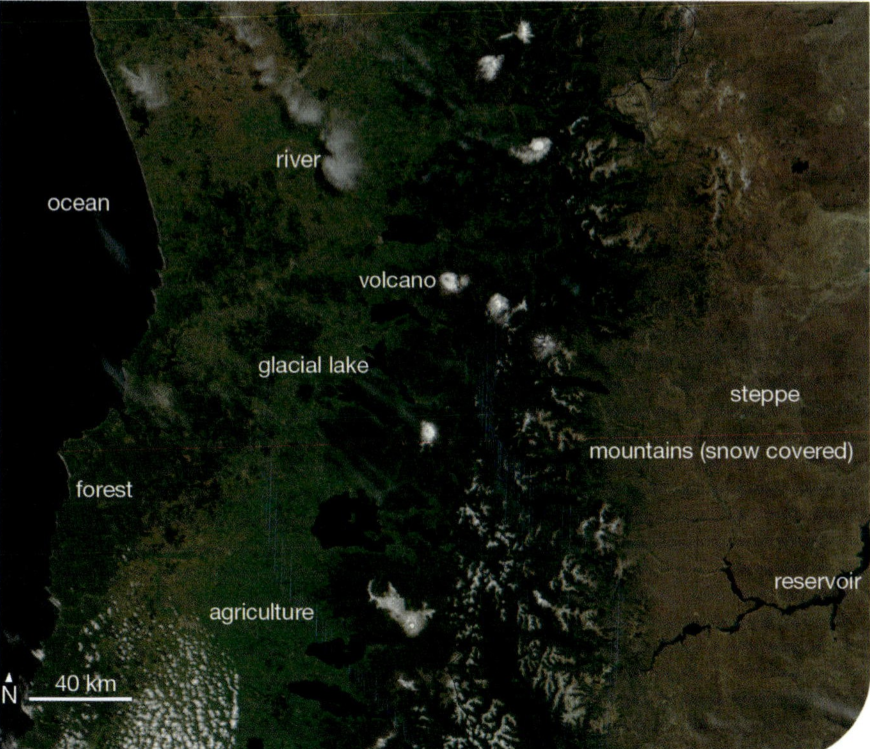

Overlay maps

Overlay maps are a way of creating layers on a paper map. Usually layers are presented digitally on software such as Geographic Information Systems (GIS). Layers are put over the top of a base map, which might be an outline map of the world or place, or even a terrain map. Each layer contains a different set of information or data that provides an insight into a process or spatial distribution occurring on that scale. By viewing multiple layers at the same time, you can observe interconnections between spatial patterns.

To create a layered paper map, you require the correct base map and some tracing or baking paper. Your base map should contain all elements of BOLTSS (see pages 132–33) except for a legend. Place a piece of paper over the top of the base map and colour in your first distribution. You do not need to trace the outlines of the countries, as this is what the base map is for.

Each layer requires a border (so you can line it up correctly with the base map) and a legend so it is clear what the layer is showing. You can create multiple layers and then overlay them all to observe patterns and trends.

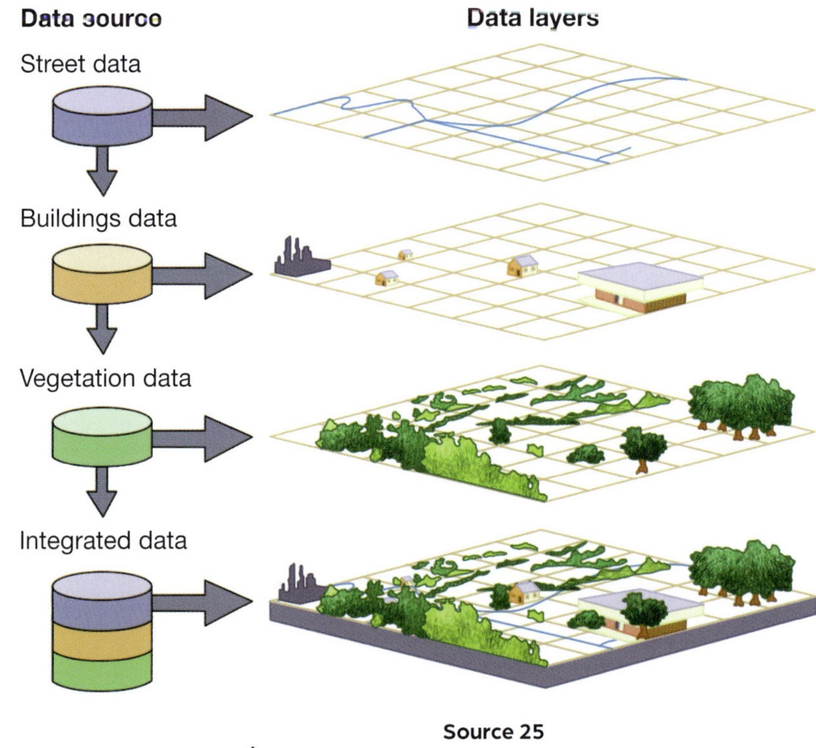

Source 25

Digital layers created to observe interconnections and trends

Spatial technology

Imagine you were given an Excel spreadsheet with over one million data points, all stating the latitude and longitude of McDonald's locations around the world. How would you analyse this data and explain or describe patterns or trends? Geographers use spatial technology such as GIS, a digital mapping platform that helps us visualise data.

Spatial technology is used every day. You probably are not even aware that you are using it! Google Maps (GPS) is a prime example, where you can visualise your location and destination and be guided along the shortest route. Uber Eats, Snap Maps, Instagram and other location services all use spatial technology to give their apps location data.

Transects

Transects are straight lines created on a surface (often with measurements marked on them like rulers) that help us observe a sample of the surface environment. They are useful tools for collecting information about species richness and investigating the characteristics of places.

Transects can be created both on land and underwater. Researchers and surveyors often use transects as a way to measure change over time in reefs or monitor species richness, as shown in Source 26. As a student, you will most likely undertake transects on land.

To complete a transect, first select a suitable location. For example, if you are investigating how urbanisation has affected the natural environment in your local area, you might lay down a 1-metre ruler in a field, then count the different species of plant you identify along the ruler transect. You would then repeat this transect technique in other random places within your study site, to create a more accurate picture of the impact of urbanisation.

You can also use transects to analyse photographs and consider how a place changes over a set distance, as was done for the Aral Sea, Source 1 on page 40. These transects are particularly helpful when identifying land use or environmental types.

Source 26

Amphibian researchers conduct a transect in a gallery forest in the savanna to identify amphibian biodiversity.

Cross-sections

A cross-section shows the shape of a geographical landscape or landform from the side, as if it had been sliced by a knife. This type of drawing can be helpful in visualising landforms from different perspectives. When hand drawing a cross-section, we finish with a 2D side-on view of the terrain; however, computer programs can create 3D representations to enhance our understanding.

What you see on your map

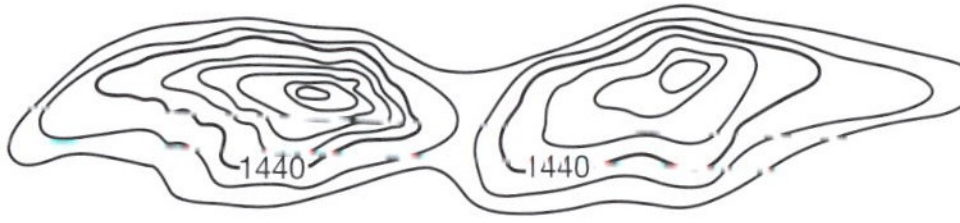

3D view of landmark

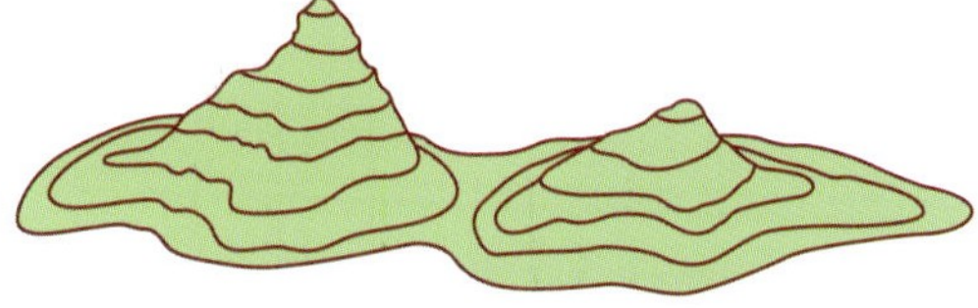

Source 27

The difference between what you see on a cross-section versus the 3D view of a landmark

Source 28

An aerial photo of Uluru with contour lines overlaid.

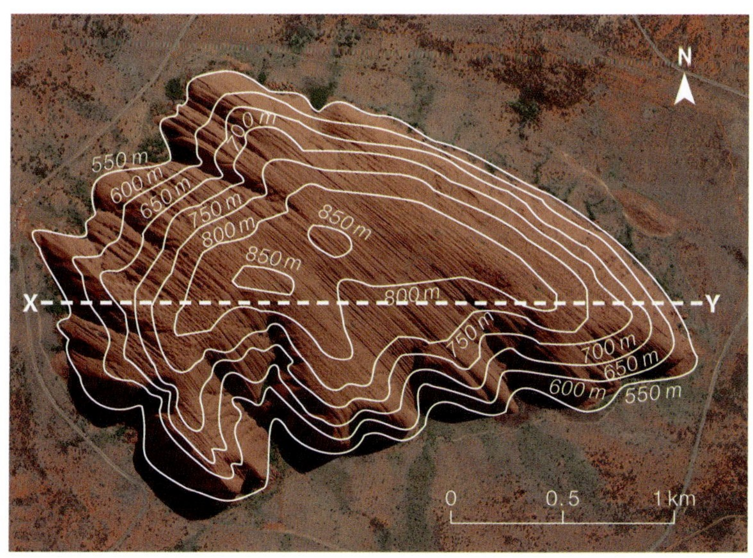

Source: Contour map from Matilda Education Australia, Geoscience Australia

Contour lines are used to construct a cross-section. The overlay on the aerial photograph of Uluru in Source 28 is a contour map. The contour lines are the numbered lines that join places of equal height. Close contour lines indicate a steep slope and widely spaced contours mean a gentle slope.

Source 29

A simple cross-section of Uluru, x and y are matched to the transect line in Source 28.

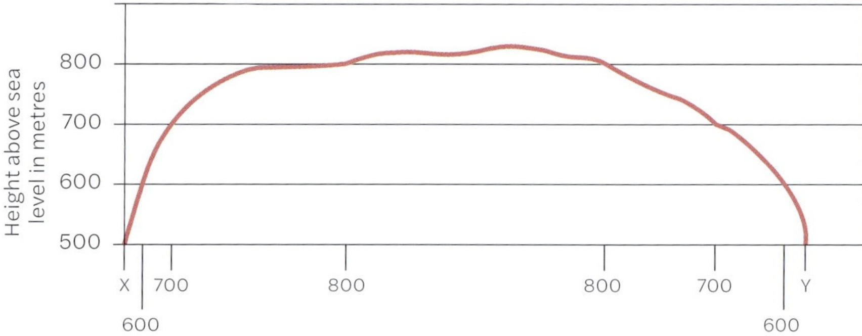

How to draw a cross-section

a Place a straight edge along the line between points A and B. Mark each contour line at the point it touches the edge of the paper and record the height.

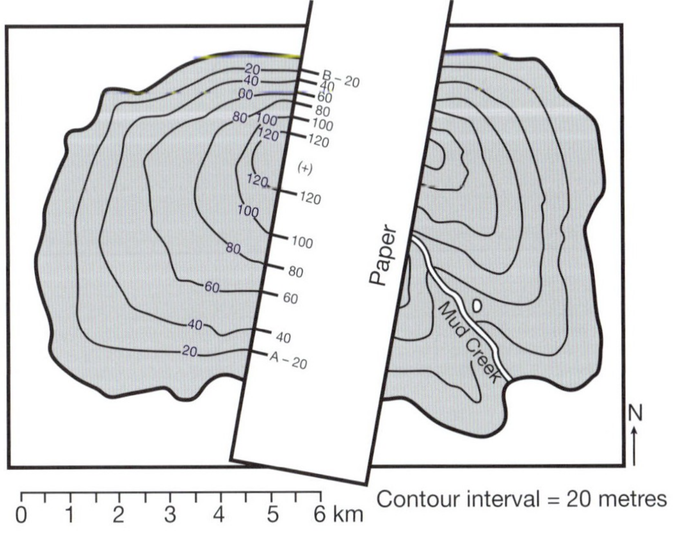

b Draw a vertical scale like the one below. You need to decide on a scale depending on the range of elevations presented on the contour map. For example, for the Uluru cross-section in Source 29, 1 centimetre represents 100 metres.

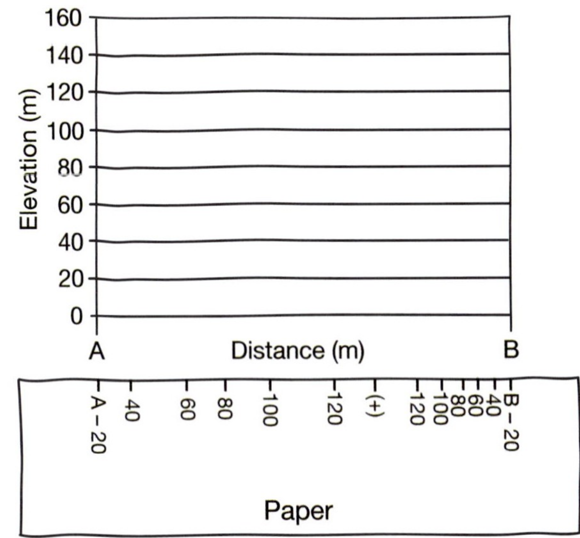

c Place the marked edge of the paper underneath the vertical scale and mark the appropriate height directly above with a dot.

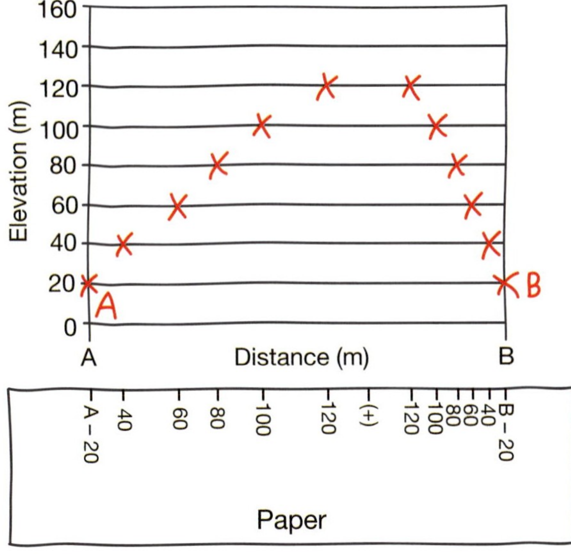

d Join the dots with a smooth line and add labels.

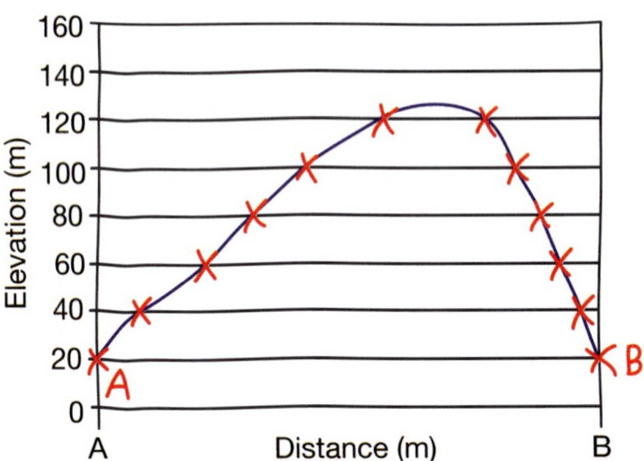

Glossary

absolute location a precise description of a place's location; for example, an address or geographical coordinates

arithmetically constant growth over a set period of time; for example, 1, 2, 3, 4 ...

atmosphere all the gases that surround Earth

awareness campaigns movements that promote understanding about an issue in the local or global community

bias to be for or against an idea, a person or a group, especially in a way that could be thought of as unfair

biodiversity the measure of variation in a community of animals or plants

biome a way of classifying environments based on their defining characteristics such as climate, soil type or vegetation

biosphere all of the living things on Earth

block diagram a three-dimensional diagram that shows how a system or feature works

BOLTSS(NA) the acronym for Border, Orientation, Legend, Title, Scale and Source (Neatness and Accuracy); used when constructing maps

capital resources human-made resources such as machinery, buildings, equipment and technology used to produce other goods and services

climate average temperature and precipitation in a certain place, over a year

climate change the natural fluctuation in temperature and precipitation over a long time period

climate graph a graph showing average rainfall and temperature for an area over a period of time

commodification the process of selling cultural traditions or artefacts for profit to tourists

community a group of people or animals connected in some way, by geography or by common interest

community transmission transmission of a virus within a community where there is no identifiable link to a source of the exposure (i.e. in COVID-19, transmission in the community that cannot be explained by an infected person having recently returned from overseas or being in contact with someone who has)

computer intrusion unauthorised access to a digital device or network

connection to Country the deep spiritual, physical, social and cultural relationship between Indigenous Australians and the land

consumerism our desire to own products that exceed our basic human needs

contour lines lines on a map that join points of the same height above or below sea level

cross-section an image that shows the shape of a geographical landscape or landform from the side; it usually creates a 2D side-on view of the terrain

cyber crime a crime committed via technology, such as the internet

cyberbullying the use of technology to harass, threaten or embarrass another person with the intent to hurt them

deforestation clearing of a forest to make room for a different land use such as farming

desertification occurs when land along the edge of a desert, or land previously not classified as desert, becomes damaged by drought and through overuse by humans and is replaced by desert

dryland an arid region; devoid of water

e-waste the amount of rubbish produced by the technology industry, such as old mobile phones

economics the study of how we produce and consume products and how wealth is established and distributed

ecosystems a community of animals and plants that are interconnected in space and time

ecotourism experiencing the natural world with the sole aim to learn how to protect, sustain and conserve the environment for future generations; ecotourism destinations need to prove they are protecting the environment, educating tourists and supporting local communities

entrepreneurship the ability to organise the factors of production and transform them into a business

equator the invisible line drawn around the centre of Earth to help describe phenomena, locations and hemispheres

erosion the process whereby rock particles are moved by flowing air (wind), water (rivers, sea and rain) or ice (glaciers)

exponentially refers to an increase that becomes more and more rapid; for example, 1, 2, 4, 8, 16, 32 ...

factors of production the resources that are used to create goods and services

flow diagram a diagram showing how one thing flows to another; arrows show the movement between stages, or show how different items are connected

food insecurity limited access to or unreliable access to safe and quality food or resources

food security consistent access to safe and quality food and resources in order to grow and thrive

genetically modified (GM) refers to organisms, often food crops, that are altered on a genetic level to enhance their growing capacity

geographic characteristic a naturally occurring feature of a place, such as its landforms and ecosystems

global citizenship a description of the way an individual acts and interconnects with the diversity of people and communities around them and across the world

global warming the unnatural rise in the Earth's temperature linked to the increase of fossil fuel use since the Industrial Revolution

goods and services resources that are produced or delivered by people in a population

Global Positioning System (GPS) a satellite navigation system used to determine the position of an object or person on the ground or at sea

Great Depression a severe economic downturn that began in 1929 and lasted for approximately 10 years

Gross Domestic Product (GDP) The value of the goods and services provided or produced by a country over a given time period (usually a year)

Hadley cell the atmospheric circulation of air moving upwards at the equator and downwards 30 degrees north and south

hydrosphere all the water on Earth

hypothesis a proposed explanation used as the starting point for further investigation

ice ages periods of time defined by wide expanses of glaciers across the globe

immigration the movement of people into a country to live or to work

irrigation the watering of crops and pastures from a water source other than precipitation; for example, with water from a river or lake

labour the physical work of people in industry; any work performed by a human being, whether physical or intellectual, to produce profits

labour resources any human input required for making a good or providing a service

latitude a coordinate that describes the location of a place north or south of the equator on the Earth's surface (0–90 degrees)

less economically developed country (LEDC) a low-income country experiencing severe barriers to development; also referred to as a developing country

leaching the loss of water-soluble nutrients from soil through rainfall or irrigation

lithosphere Earth's rigid outer layers; made up of the crust, upper mantle and tectonic plates; all the soil, rock, mountains etc. on the surface of Earth

longitude a coordinate that describes the location of a place east or west of the prime meridian on Earth's surface (0–90 degrees)

migration the movement of people into (immigration) and out of (emigration) a country

monoculture the practice of cultivation of a single crop or animal species in one area at a time

more economically developed country (MEDC) a country with a strong economy, in which most people have access to good education, health care and employment opportunities

natural resources assets that occur naturally as part of the environment

net primary productivity a quantitative measure of how efficient a biome is

non-renewable resource a resource that cannot be regenerated or replenished within a human timescale once it is used up (for example, coal, oil)

northern hemisphere the half of Earth north of the equator

opportunity cost what is given up in order to obtain something else; e.g. forgoing the opportunity to grow crops is the opportunity cost of raising cattle

overlay map when layers containing different sets of information or data are created on top of a map

pandemic a disease that spreads in multiple countries around the world at the same time, usually affecting a large number of people

phenomenon something that is observed to exist or happen

photo essay a series of photographs used to present information visually to show characteristics of a place or process

photosynthesis the natural process that primary producers undertake to change carbon dioxide, water and sunlight into energy (glucose)

place a location on Earth's surface that has defining characteristics and has meaning to people

population density a measure of how many people live in a particular region or area

population pyramid a graph showing the number of males and females living in age groups in a population

PQE the acronym for Pattern, Quantify and Exceptions, used to describe spatial patterns or graphs

precipitation water that falls from the atmosphere to the ground as rain, snow, hail or sleet

primary method a data-gathering activity undertaken in the field, such as field sketches

primary producer a plant or animal that creates its own glucose (energy) through photosynthesis

profit money that is received after expenses are taken out

producer an organism that uses sunlight, water and carbon dioxide to photosynthesise and produce its own energy

productivity the energy produced by the primary producers in an environment

quantify to use percentages, counts, ratios or data to provide details about the patterns you are describing

qualify to use general describing words such as 'large', 'many', 'broad' or 'small' to describe a pattern or change

qualitative data non-numerical data based on qualities or characteristics

quantitative data numerical data based on measurements or counts

relative location a description of where a place is located in the world; for example, describing where your house is in relation to local landmarks

renewable resource a natural resource that can regenerate or replenish within a human timescale (for example, trees)

research question an idea to be investigated or a problem to be solved through research

resource a source of energy or material; a resource can be natural or man-made, or renewable or non-renewable

saline containing or resembling salt

SALTS the acronym for Scale, Axis, Legend, Title and Source; used when graphing

satellite image an image of Earth taken from space

scarcity the idea that resources are limited and that there is competition for them

seasonal migration moving from a place of origin to another place according to crop cycles and weather changes

secondary method a data-gathering activity undertaken outside of field studies, such as research; data collected by others outside of your research group

SHEEPT the acronym for Social, Historical, Environmental, Economic, Political and Technological factors; used when explaining the reasons why a spatial pattern occurs

sketch a simple drawing made to record data when in the field

social distancing a term developed to describe space required between individuals to limit the spread of disease; during the 2020 COVID-19 pandemic a social distance of 1.5 m was recommended

socially responsible acting morally and ethically towards others and towards the planet

spatial technology a computer system that interacts with real-world locations in some way

SPICESS an acronym that assists in using geographic language in responses; SPICESS stands for space, place, interconnection, change, environment, scale and sustainability

species richness the number of different species within a community

sustainable able to be maintained at a certain level, such as energy or air quality

southern hemisphere the half of the Earth south of the equator

temporal distance the amount of time that separates a person's present time and the time of an anticipated event in the future

TOASTIE the acronym for Title, Orientation, Annotations, Scale, Time, Information and Edge; used when making a field sketch

tourism the movement of people to a location other than their place of residence or work for more than 24 hours, but for less than a year

transect straight lines created on a surface (often with measurements marked on them like rulers) that help us observe a sample of the surface environment

urban sprawl the continual expansion of suburbs at the edges of large cities onto natural or farming land

United Nations (UN) an international organisation made up of 193 member states, aiming to maintain peace and security between and within nations

viral caused by or relating to a virus

weathering a process where rock is worn away, broken down or dissolved into smaller and smaller pieces by water, heat and cold; chemical weathering is a chemical change in a rock, while mechanical weathering involves rocks being broken into pieces

Index

Aboriginal peoples *see* First Nations Peoples
absolute location 5
agricultural exports
  Australia 50, 51
  Victoria 61
agricultural productivity
  global warming influence on 42–3, 46
  spatial technology use 60, 62–3
agriculture 38, 39, 41
  Australia 50–1, 60
  effect of 2019–20 bushfires on 54, 55
  First Nations Peoples 53
  Victoria 60
algal blooms 38
Amazon Rainforest 26
animal production
  Australia 50
  impact of 2019–20 bushfires on 54
animal products, exported 51
animals 24
annotations (sketch) 140
anthropogenic biomes of the world 38–9
aquatic biome 15, 19
aquatic environments, human impact on 38
Aral Sea 40
Artic regions, biomes in 15
atmosphere 14
atmospheric carbon dioxide 42, 43
atmospheric circulation and biomes 16
Australia
  agricultural production 50, 60
  biomes 22–3
  bushfires 2019–20, impacts 31, 54–5
  connection with other places and cultures 104–5
  exports 50, 51, 90, 102–3, 104
  global warming 51
  imports 90
  migration program 104, 105
  tourism 104, 105, 106, 107, 115
  trade partners 90, 104
  vegetation zones 22
axis (graphs) 144, 145

bar graphs 144, 145
beef production versus cropping 57
biased websites 122
biodiversity 26, 149
biomes 12, 15
  anthropogenic 38–9
  Australia 22–3
  distribution 16–17
  exploring virtually (fieldwork task 1) 126–7
  freshwater 40–1
  how they vary 20–1
  interconnections in 26, 27, 28–9
  net primary productivity 24–8
  North America 28–9
  variation on a global scale 18–19
biosphere 14, 15
bird richness, North and Central America 28
Blackpool, UK, and the tourism life cycle 112, 113
block diagrams 142
BOLTS (NA) 132–3
border (maps) 132
bus network, Melbourne 76–7
bushfires 2019–20 54
  and the climate crisis 31
  damage to Australian land 54–5
  effects on agriculture 54
  impact on food production 55
  recovery efforts 54
Butler's six-stage model of the tourism life cycle 112–13
  Blackpool example 112, 113

capital resources 56
carbon dioxide, in the atmosphere 42, 43
carbon fertilisation 42
carbon footprint 115
cartograms 6
change (SPICESS) 5
changes and implications 12, 36, 68, 96
child labourers 100
China, exports 91
choices and changes 96–117
choropleth maps 7
city foodbowl plan, Melbourne 58–9
civics + citizenship 30–1, 86–7
climate 7
  and Australia's biomes 22–3
  and biomes 16–20
  difference from weather 145
climate change 31, 43, 145
  versus global warming 42
climate graphs 145
climate maps 7
climate zones
  Australia 22
  North America 28, 29
collecting data 123
commodification of culture 110–11
communicate data 13, 37, 69, 97
communicating your findings 124
communities 26
  and place 72–3
community transmission 84
compass 134
compass points 134
computer intrusion 86
coniferous forest soil 20
connecting to different places 72–3
  through trade 88–9
connection to country 70
consumerism
  definition 98
  impact on people 100–1
  impact on places 98–9
  smartphones and the environment 99
contour lines 150
coronavirus 84
cotton production 39
COVID-19
  identification and spread 84
  what is it? 84
COVID-19 pandemic 54, 78
  downloaded apps 85
  impact on education industry 102–3
  impact on tourism 106–7, 108, 109
  impact on trade 91
  international travel controls 109
  internet traffic during 85
  and our reliance on technology 84–5
  preventing transmission 84
  reducing community transmission 84
  and remote learning 83, 85
  responding to the pandemic 84
cropping versus beef production 57
crops, genetically modified (GM) 51

cross-sections 150
drawing 151
culture, commodification of 110–11
cyber crime 86
Cyber Security Operations Centre 86
cyberbullying 87
legal protection against 87

deforestation 26
desert biome 15, 16, 17, 18, 23
desert soil 20
desertification 48
factors 49
Sahel drylands 48
digital and spatial technologies 13, 37, 69, 97
direction (maps) 134
droughts 43, 46, 48, 54
drylands 48
and soil degradation 49

e-waste 99
Earth's spheres 14
ECO certification in Australia 115
economic (SHEEPT) 138
economics 57
economics + business 56–7
economy
impact of fires, drought and COVID-19 pandemic on 54
resources in the 56–7
ecosystems 15, 16
ecotourism 114, 115
edge (sketches) 141
education industry, COVID-19 impact on 102–3
education services, as an export 102, 103
egg consumption throughout the world 136, 137
energy flows 26
entrepreneurship 57
environment
classification 14–15
impact of smartphones on 99
influence on Australia's productivity 50–1
SPICESS 4
environmental (SHEEPT) 139
equator 16, 134, 135
erosion 20
evaluation section (reports) 125
exception (PQE) 136, 137
exports
Australia 50, 51, 90, 102–3, 104
China 91
global 88–9
tertiary education 102–3
Victoria 61

factors of production 56–7
Fairtrade 100–1
farming, spatial technology use 60, 62–3
fertilisers 38
fertility rates 46
field sketches 140–1
fieldwork 122–30
choosing a location 122
communicating your findings 124
creating and answering a research question? 122–5
evaluating the success of your methods 125
pre-fieldwork research 122–3
preparation for 124
task 1: exploring biomes virtually 126–7
task 2: exploring our interconnection with place 128–9
writing a fieldwork introduction 130
writing research reports 125
First Nations Peoples
as Australia's first agriculturalists 53
connection to country 70
fire use 53
food security 53
land management 52–3
seasonal migration 52
tribal boundaries 52
fish kills 38
floods 46
flow diagrams 142
food insecurity 47
causes 46
Sahel drylands 48
food production
Australia 50
and global warming 51
impact of 2019–20 bushfires on 55
increases arithmetically 45
and opportunity food loss 57
food security
First Nations Peoples 53
in the future 48–9
global 44–5
spatial technology to improve 60
what is it? 44
food waste 50
forest biome 15, 16, 19, 28
freshwater biome 40–1
degradation 40
world 41

Ganges River 40
genetically modified (GM) crops 51
geographer, thinking like a 2
geographic characteristics 4
geographic concepts 4–5
Geographic Information Systems (GIS) 148
geography
how to study? 4–5
of interconnections 68
why it matters 2–3
glaciers 17
global carbon emissions 42, 114
global citizens 30
environmental and social issues 31
schools strike for climate 30, 31
global citizenship 30
global citizenship education 30
global exports 88–9
global food security 44–5
global mobile phone usage 98
Global Positioning System (GPS) 60, 62–3, 77, 148
global tourism 106–7
global trade routes 88–9
global warming 31, 42
and atmospheric carbon dioxide levels 42, 43
Australia 51
and food production 51
influence on agricultural productivity 42–3, 46
influence on productivity 42
goods 56
Google Maps (GPS) 148
graphing 144–5
grassland biome 15, 16, 18, 23
Great Depression 91
gross domestic product (GDP) 50, 88, 90

Hadley cell, and biomes 16, 17, 20
historical (SHEEPT) 138
human characteristics of place 70, 71
human impact 26
on the environment 38–9
human population growth
Africa 48
by regions 47
and fertility rates 46
increases exponentially 45
human world 36–63
hydrosphere 14
hypothesis 122

ice ages 42
imports, Australia 90
Indigenous Australians *see* First Nations Peoples
industrialisation 39
information (sketches) 141
Instagram 79
interconnections
  between places 74–5
  in biomes 26, 27, 28–9
  geography of 68
  SPICESS 5
international arrivals and departures 104
internet access
  during COVID-19 pandemic 85
  not evenly distributed around the world 81–2
  world 81
internet censorship 82
introduction (fieldwork report) 130
irrigation 41

labour 56
labour resources 56
latitude and longitude 134–5
latitudinal effect, and biomes 16
law enforcement in a virtual world 86, 87
leaching 20
Learning Ladder 8, 12–13, 36–7, 68–9, 96–7
legend (maps and graphs) 133, 144
less economically developed nations (LEDCs)
  cotton production 39
  internet access 81
  manufacturing outsourced to 100
line graphs 144, 145
lithosphere 14
local transport networks 74–7
location 71
longitude 134, 135
lotus diagram 122

Malthus, Thomas, 'point of crisis' theory 45
mangrove forests 25
manufacturing, in LEDCs 100
mapping with BOLTS (NA) 132–3
mapping skills 134–5
maps 6–7, 148
Melbourne
  buses to solve transport problems 76–7
  city foodbowl plan 58–9
  train network 74–5
  transport options to work 74
migration 104, 105
mobile devices 78
mobile phone usage, global 98
monocultures 53
more economically developed nations (MEDCs), internet access 81

natural characteristics of place 70, 71
natural resources 56
natural world 12–29
neatness and accuracy (maps) 133
net primary productivity 24–8
non-renewable resources 56
North America
  climate zones 28, 29
  productivity and biome interconnection 28–9
North Pole 135
northern hemisphere 134

opportunity cost 57
opportunity food loss 57
orientation (maps and sketches) 132, 140
overlay maps 148

pandemics 84
  see also COVID-19 pandemic
pattern (PQE) 136, 137
patterns and interconnections 12, 36, 68, 96
people
  consumerism impact on 100–1
  exploring our interconnection with place (fieldwork task 2) 128–9
  influences on connection to place 72–3
perceptions and places 68–91
pesticides 39
phenomena 4
Phillip Island
  impact of tourism on 116–17
  population change 116
photo essays 143
photosynthesis 16, 24, 42
physical connections between places 74–7
physical maps 6
pie charts 144
place(s) 70
  connecting through trade 88–9
  connecting to 72–3
  consumerism impact on 98–9
  described by location 71
  describing 70–1
  exploring our interconnection with place (fieldwork task 2) 128–9
  human characteristics 70, 71
  influences on people's perception of 72
  natural characteristics 70, 71
  physical connections between 74–7
  SHEEPT analysis 70
  SPICESS 5
  virtual connection between 78–9
plants 24
polar ice biomes 16, 18
political (SHEEPT) 139
pollution 40
population change, Phillip Island 116
population pyramids 146
  interpreting 146
poverty 45, 100
PQE analysis 136, 145
  starting PQE sentences 137
pre-fieldwork research 122–3
precipitation
  climate graphs 145
  effect on biomes 15, 16, 21
preparation for fieldwork 124
primary collection methods 123
primary producers 24, 26
  humans and animals reliance on 24, 26, 27
  impact of deforestation 26–7
prime meridian 135
producers 24
productivity of biomes 24–5
  global warming influence on 42
  importance of studying 26–7
  in North America 28–9
profit 56, 100
public transport, Melbourne
  bus network 76–7
  train network 74–5

qualifying data 136
qualitative data 123
quantify (PQE) 136, 137
quantifying data 136
quantitative data 123
quantitative measures 24

rainfall, effect on biomes 15, 21, 22
rainforest biome 17, 18, 23, 26

relative location 5
religious connections to place and other people 73
remote learning 82, 83
  during the COVID-19 pandemic 83, 85
renewable resources 56
research, pre-fieldwork 122–3
research question 122
research reports
  writing 125
  writing the introduction 130
resources
  in the economy 56
  and factors of production 56–7
  scarcity of 57

Sahel drylands, food scarcity 48–9
saline water 14
SALTS (graphs) 144, 145
satellite images 147
scale (maps, sketches, graphs and changes over time) 133, 135, 141, 144
scale (SPICESS) 5
scarcity of resources 57
'School of the Air' 83
schools strike for climate 30, 31
secondary collection methods 123
self-quarantining 84
services 56
'severe acute respiratory syndrome coronavirus 2' (SARS-CoV-2) 84
sexual harassment 87
SHEEPT analysis 70, 138–9
sketches and annotating 140–1
smartphones, environmental impact 99
snapmap 79
social (SHEEPT) 138
social distancing 84
social media 78, 79, 82, 86
social responsibility 30
soil degradation, drylands 49
soil type, and biomes 20
solar radiation, and biomes 16
source (maps and graphs) 133, 144
South Pole 135
southern hemisphere 134
space (SPICESS) 4
spatial distribution and patterns 12, 36, 68, 96
spatial scale 80, 81
spatial technology 148
  to improve food security 60
  to monitor productivity 60–2
  used in farming 62–3
species interactions 26
species richness 26, 27, 28
  birds, North and Central America 28
  North and South America 27
  trees, and productivity 27
SPICESS (geographic concepts) 4–5
sporting teams connection to community and place 73
starvation 45, 46
subarctic regions, biomes in 15
succession 26
sustainability 31
  SPICESS 5
  of tourism 114–15
sustainable future 46
  planning for, Melbourne 58–9

taiga (snow forest) biome 15, 16
technological (SHEEPT) 139
technology
  reducing temporal scale 80–3
  reducing the urban–rural divide 81
  reliance on during the COVID-19 pandemic 84–5
  and remote learning 82–3
  and uneven distribution of internet access 81–2
  and virtual connections 78–9
temperate regions, biomes in 15
temperature
  climate graphs 145
  effect on biomes 15, 21, 22
  and global warming 42
temporal distance, reduction through technology 80, 81
temporal scale 80–1
tertiary education as an export 102–3
time (sketches) 141
title (maps, sketches and graphs) 133, 140, 144
TOASTIE (field sketches) 141
tourism
  Australia 104, 105, 106, 107, 115
  and commodification of culture 110–11
  definition 105
  determining destinations 108–9
  ecotourism 114, 115
  global 106–7
  impact of COVID-19 106–7, 108, 109
  impact on Phillip Island 116–17
  life cycle, Butler's six-stage model 112–13
  positive and negative effects of 110–11
  sustainability of 114–15
trade
  benefits of 88–9
  impact of COVID-19 on 91
trade partners, Australia 90, 104
train network, Melbourne 74–5
transects 149
transport options to work in Melbourne 74, 76
travel
  determining where people travel? 108–9
  and global COVID-19 pandemic 108, 109
  impact on the culture of a region 110–11
  safest countries in the world 108–9
tree species richness graph 27
tropical rainforest soil 20
tropical regions, biomes in 15
tundra biome 15, 16, 19

United Nations (UN) 30
  deaths from starvation 45
urban–rural divide, reducing 81
urban sprawl 71
urbanisation 38

vegetation zones, Australia 22
Victoria
  agriculture 60
  food and fibre exports 61
viral infections 84
virtual connection between places 78
  everyday virtual connections 78–9
  shortening the temporal distance between locations 80–3
virtual fieldwork to explore biomes (fieldwork task 1) 126–7
virtual reality 78
virtual world, dangers in 86–7
visual communication 142–3

water degradation 38
water use, cotton production 39
weather 7, 145
weathering 20
websites for research 122
world
  anthropogenic biomes 38–9
  freshwater biomes 41
  internet access 81–2
writing research reports 125
  introduction 130

Acknowledgements

The author and publisher are grateful to the following for permission to reproduce copyright material:

PHOTOGRAPHS: AAP/Stefan Postles, **86**, /Tourism Australia/PR Handout Image, **105**; Alamy Stock Photo/Andrey Armyagov, **13**, /Curt Teich Postcard Archives/Heritage Images, **135** (bottom left), /Dorling Kindersley ltd, **i** (bottom right), **131**, /Mike Goldwater, **49** (top right), /H. Mark Weidman Photography , **56** (bottom right), /Takatoshi Kurikawa, **76**, /ERIC LAFFORGUE, **101**, /Rainer Lesniewski, **48**, /Mruk, **111** (left), /National Geographic Image Collection, **129** (top), /Ian Nichols/National Geographic Image Collection , **149**, /ncamerastock, **152–4**, /Niebrugge Images, **135** (bottom right), /picturelibrary, **117** (top), /Science Photo Library, **25**, /inga spence, **41**, /Steve Allen Travel Photography, **40** (bottom), /Sydney Photographer, **103**, /Taras Vyshnya, **37**, /samuel wordley, **87;** Ben Murphy, Tambo Valley Honey, **55** (bottom); Getty Images/Lynsey Addario, **46–7**, /Yann Arthus-Bertrand, **39** (top), /Chang Hsiu Huang/Moment, **119**, /D-BASE, **129** (bottom), /DigitalGlobe, **3**, **150**, /DusanBartolovic, **38** (left), /Jeff Greenberg, **79** (left), /Kelvin Kam/EyeEm, **i** (centre left), **67**, /kokkai, **70** (top), /Barry Kusuma, **73**, /mantaphoto, **97** (left), /pigphoto, **97** (right), /Andrea Pistolesi, **i** (top right), **35**, /anucha sirivisansuwan, **i** (centre right), **95**, /Ben Stansall/AFP, **30** (bottom), /ViewStock, **155–9**, / Westend61, **162**, /Yousuf Tushar, **69**; iStockphoto/35007, **17** (centre), **15** (centre left), **18** (top), /AlexanderYershov, **10**, **34**, **66**, **94**, **120** (arch icon), /AndreyPopov, **80**, /aydinmutlu, **89**, /Boonyachoat, **59** (top), /brytta, **15** (top centre), /James Dale, **112**, /damircudic, **82** (bottom left), /danieldep, **71** (top), /davidf, **23** (bottom), /Wenjie Dong, **79**, /dszc, **38–9** (bottom), /eclipse_images, **57** (left), /ewg3D, **143** (bottom right), /FiledIMAGE, **140** (bottom), /fotofritz16, **70–1**, /FrankRamspott, **134** (bottom), /Gearstd, **36**, /Richard Griffin, **141** (top), /hypergurl, **27** (top), /ImagineGolf, **143** (top left), /Ingrid_Hendriksen, **93**, /IPGGutenbergUKLtd, **73** (inset), /izf, **79** (profile), /jack0m, **142** (bottom left), /Kalistratova, **17** (top), /Kinwun, **63**, /leonello, **134** (top), /Rainer Lesniewski, **135** (top), /longtaildog, **15** (centre), /lucky-photographer, **28** (right), /LUHUANFENG, **134** (centre), /luoman, **160–1**, /malerapaso, **144** (top), /malxes, **17** (bottom), /Marccophoto, **43**, /marimo, **96**, /martin-dm, **108**, /mazzzur, **i** (top centre),**11**, /merc67, **15** (top right), /mihtiander, **19** (bottom), /mikulas1, **55** (top), /muha04, **21**, /PamelaJoeMcFarlane, i (bottom left), **121**, /Andrew Peacock, **i** (background), /PietroPazzi, **18** (bottom), /pixdeluxe, **114**, /richcarey, **22–3** (bottom), /Elizabeth M. Ruggiero, **19** (top left), /sanyanwuji, **143** (bottom left), /Saturated, **123**, /simonbradfield, **23** (top), /SolStock, **124**, /sturti, **56** (top), /subjug, **57** (right), /Tammy616, **15** (top left), /TropicalShapes, **84**, /valentinrussanov, **77** (top right), /Vladimir Vladimirov, **110**, /Wavebreakmedia, **31** (right), /wmaster890, **19** (top right), /Xsandra, **143** (top right), /ymgerman, **23** (centre), /zetter, **138–9**; NASA images courtesy Jeff Schmaltz LANCE/EOSDIS MODIS Rapid Response Team, GSFC, **147** (bottom left and right); NASA Earth Observatory, **3** (top and centre right), **40** (top centre and right), /NOAA NGDC, **147** (top); Image courtesy of the National Archives of Australia NAA: A1200 L25261, **82–3**; Shutterstock.com/Kim Britten, **74** (bottom right), /Michael R Evans, **91**, /Everett Historical, **117** (inset, bottom), /Forewer, **138** (top left), /Global Warming Images, **72**, /irin-k, **56** (bottom left), /Thampitakkull Jakkree, **4–5**, /mustbeyou, **i** (top left), **1**, /stihii, **126–7**, /Billy Stock, **113**, /sunsetman, **115**; Rose Stereograph Co, State Library of Victoria [H32492/8811], **116–7**; United Nations Environment Programme (UNEP), **40** (top left); Unsplash/Matthew Waring, **79** (Santorini); Wellcome Collection/John Linnell. Licensed under CC BY 4.0 Licence, **45**.

OTHER MATERIAL: Graph, From Liang et al. 'Positive biodiversity-productivity relationship predominant in global forests', Science, 14 Oct 2016: Vol. 354, Issue 6309, aaf8957, DOI: 10.1126/science.aaf895. Reprinted with permission from AAAS, **27** (top); Article extract, 'Bus lobby pitches to solve Melbourne's transport problems a 'hell of a lot sooner than rail'' by James Oaten and Ben Knight, **25** October 2018. Reproduced by permission of the Australian Broadcasting Corporation – Library Sales © 2018 ABC, **76**; Graph, based on ABS data, **74** (bottom left); Figure **25–5** from Reisinger, A., R.L. Kitching, F. Chiew, L. Hughes, P.C.D. Newton, S.S. Schuster, A. Tait, and P. Whetton, 2014: Australasia. In: Climate Change 2014: Impacts, Adaptation, and Vulnerability. Part B: Regional Aspects. Contribution of Working Group II to the Fifth Assessment Report of the Intergovernmental Panel on Climate Change [Barros, V.R., C.B. Field, D.J. Dokken, M.D. Mastrandrea, K.J. Mach, T.E. Bilir, M. Chatterjee, K.L. Ebi, Y.O. Estrada, R.C. Genova, B. Girma, E.S. Kissel, A.N. Levy, S. MacCracken, P.R. Mastrandrea, and L.L. White (eds.)]. Cambridge University Press, Cambridge, United Kingdom and New York, NY, USA, **51** (left); Diagram, © Carolina Biological Supply Company. Reprinted by permission of Carolina Biological Supply Company only, **26**; Infographic and graph, © Commonwealth of Australia. Licensed under CC BY 4.0 Licence, **102**, **103**; Graph, Commonwealth of Australia, ABARES 2017, Agricultural commodities: March quarter 2017. CC BY 3.0, **51** (right); Map data, Victorian Land Use Information System 2016 © Department of Jobs, Precincts and Regions and licensed for re-use under the Creative Commons Attribution Version 4.0 (international) creativecommons.org/licenses/by/4.0/, **60**; Advertisement, Contiki, **106**; Cartoon, www.CartoonStock.com, **111** (right); Table, Digital Agriculture Services (DAS), **55** (top); Map, Ellis EC, Ramankutty N: Putting people in the map: anthropogenic biomes of the world. Frontiers in Ecology and the Environment 2008, 6:439-447, **38–9** (centre); Map, European Environment Agency (EEA). Licensed under CC BY 2.5 DK licence, creativecommons.org/licenses/by/2.5/dk/, **42–3**, **92** (top); Logo, Éric St-Pierre/Fairtrade International, **100**; Map, Diane Fassino, **100–1**; Map, based on Freedom of the Net, 2019, Freedom House, **82** (top); Screenshots, David Heath, **62**; Population and household forecasts, 2016 to 2036, based on ABS data, prepared by .id the population experts, November 2017; Sourced from .id, the population experts. www.id.com.au, **116**; Screenshot, Instagram, **79** (right); Map, © Institute for Economics and Peace: Global Peace Index Report 2019, **108–9**; Graph, based on data from Macro Business, **105**; Cartoon, Chris Madden, **114** (inset); Map, Courtesy of Mecardo, **54**; Graph, NASA, **43**; Graph, based on data from National Bureau of Statistics of China Customs Statistics, **91**; Painting, National Library of Australia, Lycett, Joseph, 'Aborigines using fire to hunt kangaroos', 1817; nla.pc-an2962715-s20, **53**; Graph, © 2020 nbn co ltd., **85** (top); Illustration, Danielle O'Leary, **133** (bottom); Map, data © OpenStreetMap contributors and available under the Open Database Licence, **132–3**; Map, Esteban Ortiz-Ospina and Diana Beltekian (2018) "Trade and Globalization". Published online at ourworldindata.org/trade-and-globalization Data sourced from World Bank–World Development Indicators. Licensed under CC BY 4.0 Licence, **92** (bottom); Map, Hannah Ritchie (2017) "Meat and Dairy Production". Published online at ourworldindata.org/meat-production'. Data source: United Nations Food and Agricultural Organization (FAO). Licensed under CC BY 4.0 licence, **137**; Graph, Max Roser (2013) "Future Population Growth". Published online at ourworldindata.org/future-population-growth. Data sources: HYDE, UN and UN Population Division [2019 revision]. Licensed under CC BY 4.0 Licence, **65**; Map, Max Roser (2013) "Economic Growth". Published online at ourworldindata.org/economic-growth. Data source: Maddison Project Database, version 2018. Bolt, Jutta, Robert Inklaar, Herman de Jong and Jan Luiten van Zanden (2018), "Rebasing 'Maddison': new income comparisons and the shape of long-run economic development", Maddison Project Working paper **10**. Licensed under CC BY 4.0 Licence, **90**; Map, Max Roser (2017) "Tourism". Published online at ourworldindata.org/tourism. Data sourced from World Bank–World Development Indicators. Licensed under CC BY 4.0 Licence, **118** (top left); Graph, Max Roser, Hannah Ritchie and Esteban Ortiz-Ospina (2013) "World Population Growth". Published online at ourworldindata.org/world-population-growth. Data source: Gapminder,

HYDE (2016) and United Nations Population Division (2019). Licenced under CC BY 4.0 licence, **47** (top); Map, Max Roser, Hannah Ritchie, Esteban Ortiz-Ospina and Joe Hasell (2020) "Coronavirus Pandemic (COVID-19)". Published online at ourworldindata.org/coronavirus. Data source: Thomas Hale, Sam Webster, Anna Petherick, Toby Phillips, and Beatriz Kira (2020). Oxford COVID-19 Government Response Tracker, Blavatnik School of Government. Licensed under CC BY 4.0 Licence, **109** (top); Map, Figure 4, from Laliberte, Andrea S.; Ripple, William J., Range Contractions of North American Carnivores and Ungulates, BioScience, Vol 54, Iss. 2, 2004, by permission of Oxford University Press, **32** (bottom); Map, Pew Research Center, **81**; Graph, https://www.populationpyramid.net, **146**; Screenshot, Snap Inc., **79** (centre); Map, Socioeconomic Data and Applications Center (SEDAC), Center for International Earth Science Information Network, Colombia University New York 2008. Licensed under CC BY 3.0 Licence, **7**, **27** (bottom), **28** (left); Map, South Australian Museum © Tony Tindale and Beryl George, 1974, **52**; Infographic, © Copyright State Government of Victoria. Licensed under CC BY 2.0 Australia licence, **61**; Graph, statista.com/chart/21323/top-apps-in-the-us-android-apple/ Licensed under CC BY ND licence, **85** (bottom); Graph, statista.com/chart/21288/flights-tracked-daily-worldwide/ Licensed under CC BY ND licence, **107** (left); Graph, statista.com/chart/21391/total-contribution-of-travel-and-tourism-to-gdp-of-selected-countries/ Licensed under CC BY ND licence, **107** (right); Map, © Head, Transport for Victoria, all rights reserved. The Public Transport Victoria logo, myki logo, train and bus modal icons, and SmartBus logo are trade marks of Head, TfV and used under licence by Meanwhile Education Pty Ltd., **74–5**, **77** (top left); Diagram, based on World Atlas of Desertification, 2nd Edition, UNEP, 1997, **49** (top left); Diagram, U.S. Department of Agriculture, **142** (bottom right); Infographic, U.S. Environmental Protection Agency, **99**; Illustration, Foodprint Melbourne research project, the University of Melbourne, **58–9**; Graph, Weatherzone, **33**; Graph, Wikimedia, Navarras (CC0), **21**; Table, World Bank, World Development Indicators. Licensed under CC BY 4.0 Licence, **93**; Map, World Economic Forum, **118** (bottom); Map, based on World Tourism Highlights 2018 Edition, Published by the World Tourism Organization (UNWTO) 2018, **6** (bottom), **118** (top right); Illustration, "Figure 33.10", from DISCOVER BIOLOGY, THIRD EDITION by Michael L. Cain, Hans Damman, Robert A. Lue & Carol Kaesuk Loon, Editors. Copyright © 2007 by W.W. Norton & Company, Inc. 2002, 2000 by Sinauer Associates, Inc. Used by permission of W. W. Norton & Company, Inc., **15** (bottom).

While every care has been taken to trace and acknowledge copyright, the publisher tenders their apologies for any accidental infringement where copyright has proved untraceable. They would be pleased to come to a suitable arrangement with the rightful owner in each case.

Clarifying details for NBN graph, page 85: This graph shows TC-4 usage (measured in terabits per second for both upstream and downstream) over a 24-hour period (using Australian Eastern Standard/Daylight time on the dates shown in the key). It compares the results from those two dates against a corresponding 24-hour period from nbn's pre-COVID-19 baseline on 28 February 2020 (the last week of February). Each marker on the x axis represents an hour period in the day. The y axis shows, for each of the 60-minute periods in that 24-hour period: The downstream throughput measure calculated by recording the highest downstream throughput for our TC-4 service from the busiest 15-minute increment in that 60 minute period. The upstream throughput measure, calculated by recording the highest upstream throughput for our TC-4 service from the busiest 30-minute increment in that 60-minute period. The terabits per second (Tbps) value is rounded to one decimal place.
The percentage increase is rounded to the nearest whole number.

FLIP BOOK

good humanities

Flip your book over and begin your journey through **History**